Lecture Notes in Control and Information Sciences 324

Editors: M. Thoma, M. Morari

Panagiotis D. Christofides Nael H. El-Farra

Control of Nonlinear and Hybrid Process Systems

Designs for Uncertainty, Constraints and Time-Delays

Springer

Authors

Professor Panagiotis D. Christofides
Department of Chemical Engineering
University of California, Los Angeles
Los Angeles, CA 90095-1592
U.S.A.
E-mail: pdc@seas.ucla.edu

Professor Nael H. El-Farra
Department of Chemical Engineering
and Materials Science
University of California, Davis
Davis, CA 95616-5294
U.S.A.
E-mail: nhelfarra@ucdavis.edu

Library of Congress Control Number: 2005931474

ISSN print edition: 0170-8643
ISSN electronic edition: 1610-7411
ISBN-10 3-540-28456-7 Springer Berlin Heidelberg New York
ISBN-13 978-3-540-28456-7 Springer Berlin Heidelberg New York

Springer is a part of Springer Science+Business Media
springeronline.com

Printed in The Netherlands

Typesetting: by the authors and TechBooks using a Springer LaTeX macro package

Cover design: *design & production* GmbH, Heidelberg

Printed on acid-free paper SPIN: 11376316 89/TechBooks 5 4 3 2 1 0

Preface

The increasing demand for environmental safety, energy efficiency and high product quality specifications dictate the development of advanced control algorithms in order to improve the operation and performance of chemical processes. Most chemical processes are characterized by nonlinear and hybrid nature and cannot be effectively controlled by using controllers which are designed on the basis of approximate linear or linearized process models. Typical sources of nonlinear behavior arising in chemical processes include complex reaction mechanisms and the Arrhenius temperature dependence of reaction rates, while sources of hybrid behavior include phase changes, the use of discrete actuators and the coupling of feedback with logic-based supervisory control, to name a few. The limitations of traditional linear control methods in dealing with nonlinear and hybrid processes have become increasingly apparent as chemical processes may be required to operate over a wide range of conditions due to large process upsets or set-point changes and the control systems required to be increasingly fault-tolerant. Moreover, the increased computing power and the low cost availability of computer software and hardware allow the implementation of advanced control algorithms. These factors, together with recent advances in analysis of nonlinear systems, have made control of nonlinear and hybrid process systems one of the most active research areas within the chemical engineering community over the last twenty years.

This book presents general, yet practical, methods for the synthesis of nonlinear feedback control systems for chemical processes described by nonlinear and hybrid systems. The key practical issues of model uncertainty, constraints in the capacity of the control actuator and dead-time in the measurement sensors/control actuators are explicitly accounted in the process model and dealt in the controller design. The controllers use process measurements to achieve stabilization of the closed-loop system, attenuation of the effect of the uncertain variables (unknown processes parameters and external disturbances) on process outputs, and force the controlled output to follow the reference input. The synthesis of the controllers is primarily based on Lyapunov techniques.

Explicit formulas for the controllers which can be readily used for controller synthesis in specific applications are provided. The methods are applied to numerous examples in detail and are shown to outperform existing approaches. The book includes discussions of practical implementation issues that can help engineers and researchers understand the application of the methods in greater depth.

The book assumes a basic knowledge about linear systems, linear control and nonlinear systems and is intended for researchers, graduate students and process control engineers interested in nonlinear control theory and process control applications.

In closing, we would like to thank all the people who contributed in some way to this project. In particular, we would like to thank the faculty and graduate students at UCLA for creating a pleasant working environment, the staff of Springer-Verlag for excellent cooperation, and the Petroleum Research Fund, the United States National Science Foundation and the United States Office of Naval Research for financial support. Last, but not least, we would like to express our deepest gratitude to our families for their dedication, encouragement and support over the course of this project. We dedicate this book to them.

Los Angeles, California
September 2004

Panagiotis D. Christofides
Nael H. El-Farra

Contents

1

Introduction

1.1 Motivation

The last few decades have witnessed a dramatic change in the chemical process industries. Modern industrial processes have become now highly integrated with respect to material and energy flows, constrained ever more tightly by high quality product specifications, and subject to increasingly strict safety and environmental regulations. These more stringent operating conditions have placed new constraints on the operating flexibility of chemical processes and made the performance requirements for process plants increasingly difficult to satisfy. The increased emphasis placed on safe and efficient plant operation dictates the need for continuous monitoring of the operation of a chemical plant and effective external intervention (control) to guarantee satisfaction of the operational objectives. In this light, it is only natural that the subject of process control has become increasingly important in both the academic and industrial communities. In fact, without process control it would not be possible to operate most modern processes safely and profitably, while satisfying plant quality standards.

The design of effective, advanced process control and monitoring systems that can meet these demands, however, can be quite a challenging undertaking given the multitude of fundamental and practical problems that arise in process control systems and transcend the boundaries of specific applications. Although they may vary from one application to another and have different levels of significance, these issues remain generic in their relationship to the control design objectives. Central to these issues is the requirement that the control system provide satisfactory performance in the presence of strong process nonlinearities, model uncertainty, control actuator constraints and combined discrete-continuous (hybrid) dynamics. Process nonlinearities, plant-model mismatch, actuator constraints and hybrid dynamics represent some of the more salient features whose frequently-encountered co-presence in many chemical processes can lead to severe performance deterioration and even closed-loop instability, if not appropriately accounted for in the controller design. In the remainder of this chapter, we highlight the origin and implications of these problems in the context of chemical process control applications and review some of the relevant results in the literature.

1.2 Background on Analysis and Control of Nonlinear Systems

The objective of this section is to provide a review of results on nonlinear process control with special emphasis on nonlinear control of chemical processes with uncertainty and constraints. This review is not intended to be exhaustive; its objective is to provide the necessary background for the results of this book. For smoothness of the flow of the chapter, review of results on the analysis and control of continuous-time process systems with discrete events is deferred to Sect. 1.4 below.

Many chemical processes, including high purity distillation columns, highly exothermic chemical reactors, and batch systems, are inherently nonlinear and cannot be effectively controlled and monitored with controllers and estimators designed on the basis of approximate linear or linearized process models. Typical sources of nonlinear behavior arising in chemical processes include complex reaction mechanisms, the Arrhenius temperature dependence of reaction rates, and radiative heat transfer phenomena, to name a few. The limitations of traditional linear control and estimation methods in dealing with nonlinear chemical processes have become increasingly apparent as chemical processes may be required to operate over a wide range of conditions due to large process upsets or set point changes. Motivated by this, the area of nonlinear process control has been one of the most active research areas within the chemical engineering community over the last two decades. The main bulk of the research has focused on nonlinear lumped parameter processes (i.e., processes described by systems of nonlinear ordinary differential equations (ODEs)), where important contributions have been made including the synthesis of state feedback controllers [118, 121, 141, 158, 160]; the design of state estimators [142, 144, 248, 254] and output feedback controllers [50, 64, 65, 165]; the synthesis of well-conditioned controllers for nonlinear multiple-time-scale systems [49,50,54,56]; the synthesis of controllers for time-delay systems [9–11, 117, 162, 260]; the analysis and control of nonlinear systems using functional expansions [30, 111, 112]; control design using concepts from thermodynamics [92, 109, 298]; and the design of nonlinear model predictive controllers [94,102,122,192,213,225,228,241,242,244,280,300,302]. Successful experimental applications of the above nonlinear control methods to lumped parameter chemical processes have been also reported including control of distillation columns [174], polymerization reactors [247,255,256], pH neutralization processes [212,289,291] and anaerobic digestion processes [230]. Excellent reviews of results in the area of nonlinear process control can be found in [3, 35, 118, 157, 159, 160, 167].

In addition to nonlinear behavior, many industrial process models are characterized by the presence of time-varying uncertainties such as unknown process parameters and external disturbances which, if not accounted for in the controller design, may cause performance deterioration and even closed-loop instability. Motivated by the problems caused by model uncertainty on the closed-loop behavior, the problem of designing controllers for nonlinear systems with uncertain variables, that enforce stability and output tracking in the closed-loop system, has received significant attention in the past. For feedback linearizable nonlinear systems with constant uncertain variables, adaptive control techniques have been employed to design controllers that enforce asymptotic stability and output tracking (see, for example, [29, 72, 134, 164, 188, 238, 266, 296]). On the other hand, for feedback linearizable nonlinear systems with time-varying uncertain variables that satisfy the so-called matching condition, Lyapunov's direct method has been used to design robust state feedback controllers that enforce boundedness and arbitrary degree of asymptotic attenuation of the effect of uncertainty on the output (e.g., [4,15,52,58,161,172,224]; a review of results on controller design via Lyapunov's direct method can be found in [60,172]). More recently, robust output feedback controllers have been also designed through combination of robust state feedback controllers with high-gain observers [50, 146, 183].

While the above works provide systematic methods for adaptive and robust control design, they do not lead in general to controllers that are optimal with respect to a meaningful cost (i.e., these controllers do not guarantee achievement of the control objectives with the smallest possible control action). This is an important limitation especially in light of the fact that the capacity of control actuators used to regulate chemical processes is almost always limited. Such limitations arise typically due to the finite capacity of control actuators (e.g., bounds on the magnitude of the opening of valves). Input constraints restrict our ability to freely modify the dynamic behavior of a chemical process and compensate for the effect of model uncertainty through high-gain feedback control. The ill-effects due to actuator constraints manifest themselves, for example, in the form of sluggishness of response and loss of stability. Additional problems that arise in the case of dynamic controllers include undesired oscillations and overshoots, a phenomenon usually referred to as "windup". The problems caused by input constraints have consequently motivated many studies on the dynamics and control of chemical processes subject to input constraints. Notable contributions in this regard include controller design and stability analysis within the model predictive control framework [61,116,225,239,244,276], constrained linear [48] and nonlinear [186] quadratic-optimal control, the design of anti-windup schemes in order to prevent excessive performance deterioration of an already designed controller when the input saturates [42, 136, 139, 154, 274], the study of the nonlinear bounded control problem for a class of two and three state chemical reactors [7, 8], the characterization of regions of closed-loop stability under static state feedback linearizing controllers [135], and some general results on the dynamics of constrained nonlinear systems [137]. However, these control methods do not explicitly account for robust uncertainty attenuation.

An approach to address the design of robust optimal controllers is within the nonlinear H_∞ control framework (e.g., [214,277]). However, the practical applicability of this approach is still questionable because the explicit construction of the controllers requires the analytic solution of the steady-state Hamilton-Jacobi-Isaacs (HJI) equation which is not a feasible task except for simple problems. An alternative approach to robust optimal controller design which does not require solving the HJI equation is the inverse optimal approach proposed originally by Kalman [133] and introduced recently in the context of robust stabilization in [97]. Inverse problems have a long history in control theory (e.g., see [133,205,270]). The central idea of the inverse optimal approach is to compute a robust stabilizing control law together with the appropriate penalties that render the cost functional well-defined and meaningful in some sense. This approach provides a convenient route for robust optimal controller design and is well-motivated by the fact that the closed-loop robustness achieved as a result of controller optimality is largely independent of the specific choice of the cost functional [245] so long as this cost functional is a meaningful one. The appealing features of the inverse optimal approach have motivated its use for the design of robust optimal controllers in [96,97,163]. These controllers, however, do not lead to an arbitrary degree of attenuation of the effect of uncertainty on the closed-loop output. This is a particularly desirable feature in the case of non-vanishing uncertainty that changes the nominal equilibrium point of the system. Results on robust optimal control of chemical processes have been also derived within a model predictive control framework [22,23,239,302].

Summarizing, a close look at the available literature reveals that the existing process control methods lead to the synthesis of controllers that can deal with either model uncertainty or input constraints, but not simultaneously and effectively with both. This inability clearly limits the achievable control quality and closed-loop performance, especially in light of the frequently-encountered simultaneous presence of uncertainty and constraints in chemical processes. Therefore, the development of a unified framework for control of nonlinear systems that explicitly accounts for the presence of model uncertainty and input constraints is expected to have a significant impact on process control.

1.3 Examples of Nonlinear Chemical Processes

1.3.1 Non-Isothermal Continuous Stirred Tank Reactors

Consider two well–mixed, non–isothermal continuous stirred tank reactors (CSTRs) in series, where three parallel irreversible elementary exothermic reactions of the form $A \xrightarrow{k_1} B$, $A \xrightarrow{k_2} U$ and $A \xrightarrow{k_3} R$ take place, where A is the reactant species, B is the desired product and U, R are undesired byproducts.

The feed to CSTR 1 consists of pure A at flow rate F_0, molar concentration C_{A0} and temperature T_0, and the feed to CSTR 2 consists of the output of CSTR 1 and an additional fresh stream feeding pure A at flow rate F_3, molar concentration C_{A03} and temperature T_{03}. Due to the non–isothermal nature of the reactions, a jacket is used to remove/provide heat to both reactors. Under standard modeling assumptions, a mathematical model of the plant can be derived from material and energy balances and takes the following form:

$$\begin{aligned}
\frac{dT_1}{dt} &= \frac{F_0}{V_1}(T_0 - T_1) + \sum_{i=1}^{3}\frac{(-\Delta H_i)}{\rho c_p}R_i(C_{A1}, T_1) + \frac{Q_1}{\rho c_p V_1} \\
\frac{dC_{A1}}{dt} &= \frac{F_0}{V_1}(C_{A0} - C_{A1}) - \sum_{i=1}^{3} R_i(C_{A1}, T_1) \\
\frac{dT_2}{dt} &= \frac{F_1}{V_2}(T_1 - T_2) + \frac{F_3}{V_2}(T_{03} - T_2) + \sum_{i=1}^{3}\frac{(-\Delta H_i)}{\rho c_p}R_i(C_{A2}, T_2) + \frac{Q_2}{\rho c_p V_2} \\
\frac{dC_{A2}}{dt} &= \frac{F_1}{V_2}(C_{A1} - C_{A2}) + \frac{F_3}{V_2}(C_{A03} - C_{A2}) - \sum_{i=1}^{3} R_i(C_{A2}, T_2)
\end{aligned} \tag{1.1}$$

where $R_i(C_{Aj}, T_j) = k_{i0} \exp\left(\frac{-E_i}{RT_j}\right) C_{Aj}$, for $j = 1, 2$. T, C_A, Q, and V denote the temperature of the reactor, the concentration of species A, the rate of heat input/removal from the reactor, and the volume of reactor, respectively, with subscript 1 denoting CSTR 1 and subscript 2 denoting CSTR 2. ΔH_i, k_i, E_i, $i = 1, 2, 3$, denote the enthalpies, pre–exponential constants and activation energies of the three reactions, respectively, c_p and ρ denote the heat capacity and density of the fluid in the reactor. The above model consists of a set of continuous-time nonlinear differential equations; note in particular the nonlinear dependence of the reaction rates, $R_i(C_{Aj}, T_j)$, on reactor temperature. The above system typically exhibits steady-state multiplicity. For example, using typical values for the process parameters (see Chap. 8, Table 8.1), the system describing the first CSTR, with $Q_1 = 0$, has three steady–states: two locally asymptotically stable and one unstable. The unstable steady–state of the first CSTR also corresponds to three steady–states for the second CSTR (with $Q_2 = 0$), one of which is unstable.

A typical control problem for this process is to regulate the reactor temperature and/or reactant concentration by manipulating the rate of heat input subject to constraints. Typical sources of uncertainty include external time-varying disturbance in the feed temperature and/or inlet reactant concentration as well as parametric uncertainty in the model parameters.

1.3.2 Isothermal Continuous Crystallizer

Consider the following fifth-order nonlinear model describing the evolution of the first four moments of the crystal size distribution, and the solute concentration, in an isothermal continuous crystallizer:

$$\begin{aligned}
\dot{x}_0 &= -x_0 + (1 - x_3)Da \exp\left(\frac{-F}{y^2}\right) \\
\dot{x}_1 &= -x_1 + yx_0 \\
\dot{x}_2 &= -x_2 + yx_1 \\
\dot{x}_3 &= -x_3 + yx_2 \\
\dot{y} &= \frac{1 - y - (\alpha - y)yx_2}{1 - x_3} + \frac{u}{1 - x_3}
\end{aligned} \tag{1.2}$$

where x_i, $i = 0, 1, 2, 3$, are dimensionless moments of the crystal size distribution, y is a dimensionless concentration of the solute in the crystallizer, and u is a dimensionless concentration of the solute in the feed (the reader may refer to [47, 76] for a detailed process description, population balance modeling of the crystal size distribution and derivation of the moments model, and to [51] for further results and references in this area). In the above process model, a primary source of nonlinearity comes from the Arrhenius-like type dependence of the nucleation rate on the solute concentration. Using typical values of the dimensionless process parameters: $F = 3.0$, $\alpha = 40.0$ and $Da = 200.0$, it can be shown that, at the nominal operating condition of $u^{nom} = 0$, the above system has an unstable equilibrium point surrounded by a stable limit cycle. The open-loop behavior of the crystallizer is therefore characterized by sustained oscillations. A typical control objective is to suppress the oscillations and stabilize the system at the unstable equilibrium point (corresponding to a desired crystal size distribution) by manipulating the solute feed concentration, u, subject to constraints.

1.4 Background on Analysis and Control of Hybrid Systems

Traditionally, most of the research work in process control has been concerned predominantly with the control of continuous dynamic processes described by ordinary differential equations. Yet, there are many examples of chemical and biological processes where the dynamical properties of the system depend on a rather intricate interaction between discrete and continuous variables. These are referred to as hybrid processes because they involve coupled continuous dynamics and discrete events that together drive the overall system response in time and space. In many of these applications, the continuous behavior often arises as a manifestation of the underlying physical laws governing the process, such as momentum, mass, and energy conservation, and is modeled by continuous-time differential equations. Discrete behavior, on the other hand, is ubiquitously multi-faceted and can originate from a variety of sources, including: (1) inherent physico-chemical discontinuities in the continuous process dynamics, such as phase changes, flow reversals, shocks and transitions, (2) the use of measurement sensors and control actuators with discrete settings/positions (e.g., binary sensors, on/off valves, pumps, heaters with constant current, motors with speed control), and (3) the use of logic-based switching for supervisory and safety control tasks.

Another common source of hybrid behavior in chemical processes comes from the interaction of the process with its operating environment. Changes in raw materials, energy sources, and product specifications, together with fluctuations in market demands, forecasts and the concomitant adjustments in management decisions, lead invariably to the superposition of discrete events on the basically continuous process dynamics, in the form of controlled transitions between different operational regimes. Regardless of whether the hybrid (combined discrete-continuous) behavior arises as an inherent feature of the process itself, its operation, or its control system, the overall process behavior in all of these instances is characterized by structurally different dynamics in different situations (regimes), and is, therefore, more appropriately viewed as intervals of piecewise continuous behavior (corresponding to material and energy flows) interspersed by discrete transitions governed by a higher-level decision-making entity.

The traditional approach of dealing with these systems in many areas of industrial control has been to separate the continuous control from the discrete control. In recent years, however, it has become increasingly evident that the interaction of discrete events with even simple continuous dynamics can lead to complex unpredictable dynamics and, consequently, to very undesirable outcomes (particularly in safety-critical applications) if not explicitly accounted for in the control system design. As efficient and profitable process operation becomes more dependent on the control system, the need to design flexible, reliable and effective control systems that can explicitly handle the intermixture of continuous and discrete dynamics, is increasingly apparent. Such flexibility and responsiveness play a critical role in achieving optimum production rates, high-quality products, minimizing waste to the environment and operating as efficiently as possible. These considerations, together with the abundance of hybrid phenomena in chemical processes, provide a strong motivation for the development of analytical tools and systematic methods for the analysis and control of these systems in a way that explicitly captures the combined discrete-continuous interactions and their effects.

Even though tools for the analysis and control of purely continuous-time processes exist and, to a large extent, are well-developed, similar techniques for combined discrete-continuous systems are limited at present primarily due to the difficulty of extending the available concepts and tools to account for the hybrid nature of these systems and their changing dynamics, which makes them more difficult to describe, analyze, or control. These challenges, coupled with the abundance of hybrid phenomena in many engineering systems in general, have fostered a large and growing body of research work on a diverse array of problems, including the modeling (e.g., [28, 294]), simulation (e.g., [28]), optimization (e.g., [106]), stability analysis (e..g, [70, 120]), and control (e.g., [32, 91, 123, 156]) of several classes of hybrid systems. Continued progress notwithstanding, important theoretical and practical problems remain to be addressed in this area, including the development of a unified and practical approach for control that deals effectively with the co-presence of strong nonlinearities in the continuous dynamics, model uncertainty, actuator constraints, and combined discrete-continuous interactions.

1.5 Examples of Hybrid Process Systems

1.5.1 Fault-Tolerant Process Control Systems

The ability of the control system to deal with failure situations typically requires consideration of multiple control configurations and switching between them to preserve closed-loop stability in the event that the active control configuration fails. The occurrence of actuator faults, and the concomitant switching between different control configurations, give rise to hybrid closed-loop dynamics. As an example, consider a well–mixed, non–isothermal continuous stirred tank reactor where three parallel irreversible elementary exothermic reactions of the form $A \xrightarrow{k_1} B$, $A \xrightarrow{k_2} U$ and $A \xrightarrow{k_3} R$ take place, where A is the reactant species, B is the desired product and U, R are undesired byproducts. The feed to the reactor consists of pure A at flow rate F, molar concentration C_{A0} and temperature T_{A0}. Due to the non–isothermal nature of the reactions, a jacket is used to remove/provide heat to the reactor. Under standard modeling assumptions, a mathematical model of the process can be derived from material and energy balances and takes the following form:

$$
\begin{aligned}
\frac{dT}{dt} &= \frac{F}{V}(T_{A0} - T) + \sum_{i=1}^{3} \frac{(-\Delta H_i)}{\rho c_p} k_{i0} \exp\left(\frac{-E_i}{RT}\right) C_A + \frac{Q}{\rho c_p V} \\
\frac{dC_A}{dt} &= \frac{F}{V}(C_{A0} - C_A) - \sum_{i=1}^{3} k_{i0} \exp\left(\frac{-E_i}{RT}\right) C_A \qquad (1.3) \\
\frac{dC_B}{dt} &= -\frac{F}{V} C_B + k_{10} \exp\left(\frac{-E_1}{RT}\right) C_A
\end{aligned}
$$

where C_A and C_B denote the concentrations of the species A and B, respectively, T denotes the temperature of the reactor, Q denotes the rate of heat input to the reactor, V denotes the volume of the reactor, ΔH_i, k_i, E_i, $i = 1, 2, 3$, denote the enthalpies, pre–exponential constants and activation

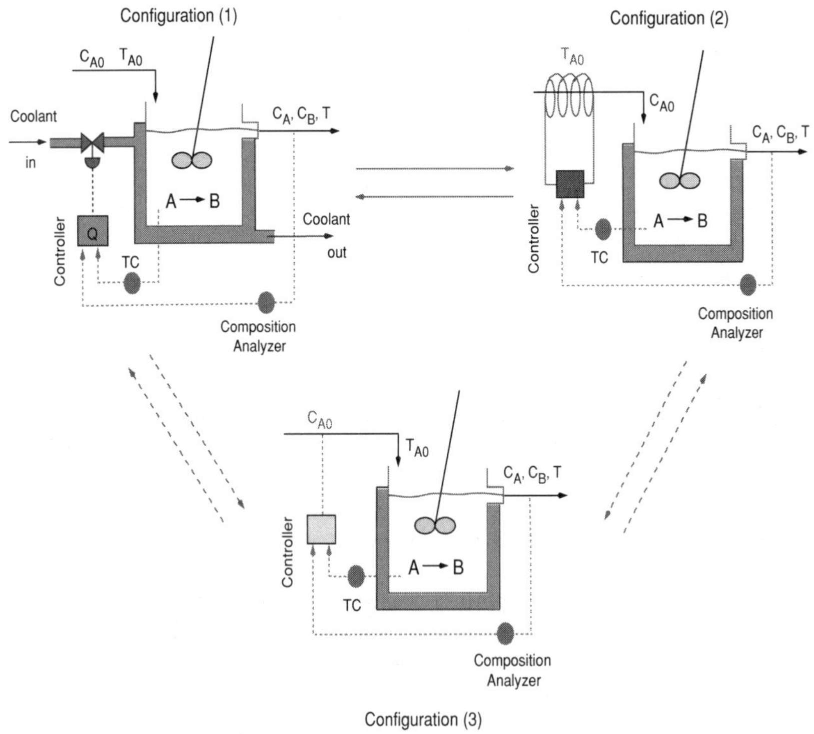

Fig. 1.1. Switching between multiple control configurations, each characterized by a different manipulated input, provides a mechanism for fault-tolerant control

energies of the three reactions, respectively, c_p and ρ denote the heat capacity and density of the fluid in the reactor. For typical values of the process parameters (see Chap. 8, Table 8.2), the process model has three steady–states: two locally asymptotically stable and one unstable.

The control objective is to stabilize the reactor at the open–loop unstable steady–state. Operation at this point is typically sought to avoid high temperatures while, simultaneously, achieving reasonable reactant conversion. To accomplish this objective in the presence of control system failures, we consider the following manipulated input candidates (see Fig. 1.1):

1. Rate of heat input, $u_1 = Q$, subject to the constraint $|Q| \leq u^1_{max} = 748\ KJ/s$.
2. Inlet stream temperature, $u_2 = T_{A0} - T^s_{A0}$, subject to the constraint $|u_2| \leq u^2_{max} = 100\ K$.
3. Inlet reactant concentration, $u_3 = C_{A0} - C^s_{A0}$, subject to the constraint $|u_3| \leq u^3_{max} = 4\ kmol/m^3$.

Each of the above manipulated inputs represents a continuous control configuration (or control–loop) that, by itself, can stabilize the reactor using available measurements of the reactor temperature, reactant and product concentrations. The first loop involving the heat input, Q, as the manipulated variable is considered as the primary control configuration. In the event of a total failure in this configuration, however, a higher-level supervisor will have to activate one of the other two fall–back configurations in order to maintain closed–loop stability. The main question in this case is typically how to design the supervisory switching logic in order to orchestrate safe transitions between the constituent configurations and guarantee stability of the overall switched closed-loop system.

1.5.2 Switched Biological Systems

The dynamics of many biological systems often involve switching between many qualitatively different modes of behavior. At the molecular level, for example, the fundamental process of inhibitor proteins turning off the transcription of genes by RNA polymerase reflects a switch between two continuous processes. An example of this is the classic genetic switch observed in the bacteriophage λ, where two distinct behaviors, lysis and lysogeny, each with different mathematical models, are seen. Also, at the cellular level, the cell growth and division in a eukaryotic cell is usually described as a sequence of four processes, each being a continuous process that is triggered by a set of conditions or events. As an example, consider a network of biochemical reactions, based on cyclin-dependent kinases and their associated proteins, which are involved in cell cycle control in frog egg development. A detailed description of this network is given in [210] where the authors used standard principles of biochemical kinetics and rate equations to construct a nonlinear dynamic model of the network that describes the time evolution of the key species including free cyclin, the M-phase promoting factor (MPF), and other regulatory enzymes. For illustration purposes, we will consider below the simplified network model derived by the authors (focusing only on the positive-feedback loops in the network) which captures the basic stages of frog egg development. The model is given by:

$$\begin{aligned}\frac{du}{dt} &= \frac{k_1'}{G} - \left(k_2' + k_2''u^2 + k_{wee}\right)u + \left(k_{25}' + k_{25}''u^2\right)\left(\frac{v}{G} - u\right) \\ \frac{dv}{dt} &= k_1' - (k_2' + k_2''u^2)v\end{aligned} \tag{1.4}$$

where $G = 1 + \frac{k_{INH}}{k_{CAK}}$, k_{INH} is the rate constant for inhibition of INH, a protein that negatively regulates MPF, k_{CAK} is the rate constant for activation of CAK, a cdc-2 activating kinase, u is a dimensionless concentration of active MPF and v is a dimensionless concentration of total cyclin, k_2' and k_2'' are rate constants for the low-activity and high activity forms, respectively, of cyclin degradation; k_{25}' and k_{25}'' are rate constants for the low-activity and high activity forms, respectively, of tyrosine dephosphorylation of MPF; k_1' is a rate constant for cyclin synthesis, k_{wee} is the rate constant for inhibition of Weel, an enzyme responsible for the tyrosine phosphorylation of MPF (which inhibits MPF activity). Bifurcation and phase-plane analysis of the above model [210] shows that, by changing the values of k_2', k_2'' and k_{wee}, the following four modes of behavior are predicted:

- A G2-arrested state (blocked before the G2-M transition) characterized by high cyclin concentration and little MPF activity. This corresponds to a unique, asymptotically stable steady-state ($k_2' = 0.01$, $k_2'' = 10$, $k_{wee} = 3.5$).
- An M-arrested state (blocked before the meta- to anaphase transition) state with lots of active MPF. This corresponds to a unique, asymptotically stable steady-state ($k_2' = 0.01$, $k_2'' = 0.5$, $k_{wee} = 2.0$).
- An oscillatory state (alternating phases of DNA synthesis and mitosis) exhibiting sustained, periodic fluctuation of MPF activity and total cyclin protein. This corresponds to a stable limit cycle surrounding an unstable equilibrium point ($k_2' = 0.01$, $k_2'' = 10$, $k_{wee} = 2.0$).
- Co-existing stable steady-states of G2 arrest and M arrest. This corresponds to three steady-states; one unstable and two locally asymptotically stable ($k_2' = 0.015$, $k_2'' = 0.1$, $k_{wee} = 3.5$).

The above analysis predicts that the network can be switched between different modes by changes in parameter values. For example, slight increases in k_2', k_{wee}, accompanied by a significant drop in k_2'' (which could be driven, for example, by down-regulation of cyclin degradation) can induce a transition from the oscillatory mode of MPF activity (early embryo stage) to the bi-stable mode. These parameter typically include rate constants and total enzyme concentrations that are under genetic control. Changing the expression of certain genes will change the parameter values of the model and move the network across bifurcation boundaries into regions of qualitatively different behavior. An important question that hybrid systems theory can be used to answer for the above system is how to determine when switching between two modes needs to take place in order to achieve a certain desired steady-state (e.g., whether the cell cycle will get arrested in the G2 or the M phase state upon switching from the oscillatory to the bistable mode).

1.6 Objectives and Organization of the Book

Motivated by the fact that many industrially-important processes are characterized by strong nonlinearities, model uncertainty, constraints and combined discrete-continuous dynamics, and the lack of general nonlinear and hybrid control methods for such systems, the broad objectives of this book are:

- To develop a rigorous, yet practical, unified framework for the design of nonlinear feedback control laws for processes modeled by nonlinear systems with uncertain variables and manipulated input constraints, that integrates robustness and explicit constraint-handling capabilities in the controller designs and provide an explicit characterization of the stability and performance properties of the designed controllers.
- To develop a general hybrid nonlinear control methodology, for hybrid nonlinear processes with switched, constrained and uncertain dynamics, that integrates the synthesis of lower-level nonlinear feedback controllers with the design of the upper-level supervisor switching logic.
- To provide fundamental understanding and insight into the nature of the control problem for nonlinear and hybrid processes as well as the limitations imposed by nonlinearities, uncertainty, constraints and the coupling of continuous dynamics with discrete events on our ability to steer the dynamics of such systems.
- To illustrate the application of the proposed controller design and analysis methods to chemical and biological systems of practical interest and document their effectiveness and advantages with respect to traditional control methods.

The rest of the book is organized as follows. Chapter 2 reviews some basic results on the analysis and control of nonlinear systems. Chapter 3 presents robust inverse optimal controller designs for input/output linearizable nonlinear systems with time-varying bounded uncertain variables. Under full state feedback, the controller designs are obtained by re-shaping the scalar nonlinear gain of Sontag's formula in a way that meets different robustness and optimality objectives under vanishing and non-vanishing uncertainty. Combination of the state feedback controllers with high-gain observers and appropriate saturation filters that eliminate observer peaking is then employed to design dynamic robust output feedback controllers that enforce semi-global closed-loop stability and achieve near-optimal performance, provided that the observer gain is sufficiently large. The implementation of the controllers is illustrated using models of chemical reactors with uncertainty.

Chapter 4 focuses on multivariable nonlinear systems, that include both time-varying uncertain variables and manipulated input constraints, and presents a unified framework that integrates robustness and explicit constraint-handling capabilities in the controller synthesis. Using a general state-space Lyapunov-based approach, the proposed framework leads to the derivation of explicit formulas for bounded robust nonlinear state feedback controllers with well-characterized stability and performance properties. The proposed controllers are shown to guarantee closed-loop stability and robust asymptotic set-point tracking with an arbitrary degree of attenuation of the effect of uncertainty on the output of the closed-loop system, and provide an explicit characterization of the region of closed-loop stability in terms of the size of the uncertainty and constraints. The state feedback control designs are subsequently combined with appropriate high-gain observers to yield output feedback controllers. The proposed control method is illustrated through chemical reactor examples and compared with more traditional process control strategies. The chapter concludes with a presentation of a technique for tuning classical controllers using nonlinear control methods.

Having laid the foundation for bounded Lyapunov-based control techniques in Chap. 4, we turn in Chaps. 5 and 6 to take advantage of the well-characterized stability properties of these designs, through hybrid control techniques, to aid the implementation of model predictive controllers. We start in Chap. 5 with linear time-invariant systems with input constraints and develop a hybrid predictive control structure that unites model predictive control (MPC) and bounded control in a way that reconciles the tradeoffs between their respective stability and performance properties. The basic idea is to embed the implementation of MPC within the stability region of the bounded controller and use this controller as a fall–back mechanism. In the event that the predictive controller is unable to stabilize the closed–loop system, supervisory switching from MPC to the bounded controller guarantees closed–loop stability. Extensions of the hybrid predictive control strategy to address the output feedback stabilization problem are also presented. The hybrid predictive control strategy is shown to provide, irrespective of the MPC formulation, a safety net for the practical implementation of MPC. Chapter 6 generalizes the hybrid predictive control structure to address the stabilization problem for nonlinear and uncertain systems, and demonstrates the efficacy of the proposed strategies through applications to chemical process examples.

Chapter 7 focuses on developing hybrid nonlinear control methodologies for various classes of hybrid nonlinear processes. Initially, switched processes whose dynamics are both constrained and uncertain are considered. These are systems that consist of a finite family of continuous uncertain nonlinear dynamical subsystems, subject to hard constraints on their manipulated inputs, together with a higher-level supervisor that governs the transitions between the constituent modes. The key feature of the proposed control methodology is the integrated synthesis, via multiple Lyapunov functions (MLFs), of: (1) a family of lower-level robust bounded nonlinear feedback controllers that enforce robust stability in the constituent uncertain modes, and provide an explicit characterization of the stability region for each mode under uncertainty and constraints, and (2) upper-level robust switching laws that orchestrate safe transitions between the modes in a way that guarantees robust stability in the overall switched closed-loop system. Next, switched nonlinear systems with scheduled mode transitions are considered. These are systems that transit between their constituent modes at some predetermined switching times, following a prescribed switching sequence. For such systems, we develop a Lyapunov-based predictive control strategy that enforces both the required switching schedule and closed-loop stability. The main idea is to design a Lyapunov–based predictive controller for each mode, and incorporate transition constraints in the predictive controller design to ensure that the prescribed transitions between the modes occur in a way that guarantees stability of the switched closed–loop system. The proposed control methods are demonstrated through applications to chemical process examples. Finally, the chapter concludes with a demonstration of how hybrid systems techniques – in particular, the idea of coupling the switching logic to stability regions – can be applied for the analysis of mode transitions in biological networks.

Chapter 8 presents an application of the hybrid control methodology developed in Chap. 7 to the problem of designing fault–tolerant control systems for plants with multiple, distributed interconnected processing units. The approach brings together tools from Lyapunov–based control and hybrid systems theory and is based on a hierarchical distributed architecture that integrates lower–level feedback control of the individual units with upper–level logic–based supervisory control over communication networks. The proposed approach provides explicit guidelines for managing the interplays between the coupled tasks of feedback control, fault–tolerance and communication. The efficacy of the proposed approach is demonstrated through chemical process examples. The Chapter concludes with an application of the developed concepts to the dynamic analysis of mode transitions in switched biological networks.

Chapter 9 presents a methodology for the synthesis of nonlinear output feedback controllers for nonlinear Differential Difference Equation (DDE) systems which include time delays in the states, the control actuator and the measurement sensor. Initially, DDE systems which only include state delays are considered and a novel combination of geometric and Lyapunov-based techniques is employed for the synthesis of nonlinear state feedback controllers that guarantee stability and enforce output tracking in the closed-loop system, independently of the size of the state delays. Then, the problem of designing nonlinear distributed state observers, which reconstruct the state of the DDE system while guaranteeing that the discrepancy between the actual and the estimated state tends exponentially to zero, is addressed and solved by using spectral decomposition techniques for DDE systems. The state feedback controllers and the distributed state observers are combined to yield distributed output feedback controllers that enforce stability and output tracking in the closed-loop system, independently of the size of the state delays. For DDE systems with state, control actuator and measurement delays, distributed output feedback controllers are synthesized on the basis of an auxiliary output constructed within a Smith-predictor framework.

Finally, the proofs of all the results are given in the appendix.

2

Background on Analysis and Control of Nonlinear Systems

In this chapter, we review some basic results on the analysis and control of nonlinear systems. The review is not intended to be exhaustive; its only purpose is to provide the reader with the necessary background for the results presented throughout subsequent chapters. Since most of the results given in this chapter are standard in the nonlinear systems and control literature, they are given here without proofs. For detailed proofs, the reader is referred to the classic books [126, 148].

2.1 Stability of Nonlinear Systems

For all control systems, stability is the primary requirement. One of the most widely used stability concepts in control theory is that of *Lyapunov stability*, which we employ throughout the book. In this section we briefly review basic facts from Lyapunov's stability theory. To begin with, we note that Lyapunov stability and asymptotic stability are properties not of a dynamical system as a whole, but rather of its individual solutions. We restrict our attention to the class of time-invariant nonlinear systems:

$$\dot{x} = f(x) \tag{2.1}$$

where $x \in \mathbb{R}^n$, $f : \mathbb{R}^n \to \mathbb{R}^n$ is a locally Lipshcitz function. The solution of (2.1), starting from x_0 at time $t_0 \in \mathbb{R}$, is denoted as $x(t; x_0, t_0)$, so that $x(t_0; x_0, t_0) = x_0$. Because the solutions of (2.1) are invariant under a translation of t_0, that is, $x(t + T; x_0, t_0 + T) = x(t; x_0, t_0)$, the stability properties of $x(t; x_0, t_0)$ are *uniform*, i.e., they do not depend on t_0. Therefore, without loss of generality, we assume $t_0 = 0$ and write $x(t; x_0)$ instead of $x(t; x_0, 0)$.

Lyapunov stability concepts describe continuity properties of $x(t; x_0, t_0)$ with respect to x_0. If the initial state x_0 is perturbed to $\tilde{x}_0$, then, for stability, the perturbed solution $\tilde{x}(t; x_0)$ is required to stay close to $x(t; x_0)$ for all $t \geq 0$. In addition, for asymptotic stability, the error $\tilde{x}(t; x_0) - x(t; x_0)$ is required to vanish as $t \longrightarrow \infty$. Some solutions of (2.1) may be stable and some unstable. We are particularly interested in studying and characterizing the stability properties of *equilibria*, that is, constant solutions $x(t; x_e) \equiv x_e$ satisfying $f(x_e) = 0$.

For convenience, we state all definitions and theorems for the case when the equilibrium point is at the origin of $\mathbb{R}^n$; that is, $x_e = 0$. There is no loss of generality in doing so since any equilibrium point under investigation can be translated to the origin via a change of variables. Suppose $x_e \neq 0$, and consider the change of variables, $z = x - x_e$. The derivative of z is given by:

$$\dot{z} = \dot{x} \; = \; f(x) \; = \; f(z + x_e) \; := \; g(z)$$

where $g(0) \; = \; 0$. In the new variable z, the system has equilibrium at the origin. Therefore, for simplicity and without loss of generality, we will always assume that $f(x)$ satisfies $f(0) = 0$ and confine our attention to the stability properties of the origin $x_e = 0$.

2.1.1 Stability Definitions

The origin is said to be a *stable* equilibrium point of the system of (2.1), in the sense of Lyapunov, if for every $\varepsilon > 0$ there exists a $\delta > 0$ such that we have:

$$\|x(0)\| \leq \delta \;\; \Longrightarrow \;\; \|x(t)\| \leq \varepsilon \quad \forall \, t \geq 0 \tag{2.2}$$

In this case we will also simply say that the system of (2.1) is stable. A similar convention will apply to other stability concepts introduced below. The origin is said to be *unstable* if it is not stable. The ε-δ requirement for stability takes a challenge-answer form. To demonstrate that the origin is stable, then, for every value of ε that a challenger may care to design, we must produce a value of δ, possibly dependent on ε, such that a trajectory starting in a δ neighborhood of the origin will never leave the ε neighborhood.

The origin of the system of (2.1) is said to be *asymptotically stable* if it is stable and δ in (2.2) can be chosen so that:

$$\|x(0)\| \leq \delta \;\; \Longrightarrow \;\; x(t) \longrightarrow 0 \;\; as \;\; t \longrightarrow \infty \tag{2.3}$$

When the origin is asymptotically stable, we are often interested in determining how far from the origin the trajectory can be and still converge to the origin as t approaches ∞. This gives rise to the definition of the *region of attraction* (also called *region of asymptotic stability*, *domain of attraction*, and *basin*). Let $\phi(t; x)$ be the solution of (2.1) that starts at initial state x at time $t = 0$. Then the region of attraction is defined as the set of all points x such that $\lim\limits_{t \to \infty} \phi(t; x) = 0$. Throughout the book, stability properties for which an estimate of the domain of attraction is given are referred to as *regional*. If the condition of (2.3) holds for all δ, i.e., if the origin is a stable equilibrium and its domain of attraction is the entire state-space, then the origin is called *globally asymptotically stable.*

If the system is not necessarily stable but has the property that all solutions with initial conditions in some neighborhood of the origin converge to the origin, then it is called (locally) attractive. We say that the system is *globally attractive* if its solutions converge to the origin from all initial conditions.

The system of (2.1) is called *exponentially stable* if there exist positive real constants δ, c, and λ such that all solutions of (2.1) with $\|x(0)\| \leq \delta$ satisfy the inequality:

$$\|x(t)\| \leq c\|x(0)\|e^{-\lambda t} \quad \forall\, t \geq 0 \tag{2.4}$$

If this exponential decay estimate holds for all δ, the system is said to be *globally exponentially stable.*

2.1.2 Stability Characterizations Using Function Classes $\mathcal{K}$, $\mathcal{K}_\infty$, and $\mathcal{KL}$

Scalar comparison functions, known as class $\mathcal{K}$, $\mathcal{K}_\infty$, and $\mathcal{KL}$, are important stability analysis tools that are frequently used to characterize the stability properties of a nonlinear system.

Definition 2.1. *A function $\alpha : [0, a) \longrightarrow [0, \infty)$ is said to be of class $\mathcal{K}$ if it is continuous, strictly increasing, and $\alpha(0) = 0$. It is said to belong to class $\mathcal{K}_\infty$ if $a = \infty$ and $\alpha(r) \longrightarrow \infty$ as $r \longrightarrow \infty$.*

Definition 2.2. *A function $\beta : [0, a) \times [0, \infty) \longrightarrow [0, \infty)$ is said to be of class $\mathcal{KL}$ if, for each fixed $t \geq 0$, the mapping $\beta(r, t)$ is of class $\mathcal{K}$ with respect to r and, for each fixed r, the mapping $\beta(r, t)$ is decreasing with respect to t and $\beta(r, t) \longrightarrow 0$ as $t \longrightarrow \infty$.*

We will write $\alpha \in \mathcal{K}$ and $\beta \in \mathcal{KL}$ to indicate that α is a class $\mathcal{K}$ function and β is a class $\mathcal{KL}$ function, respectively. As an immediate application of these function classes, we can rewrite the stability definitions of the previous section in a more compact way. For example, stability of the system of (2.1) is equivalent to the property that there exist a $\delta > 0$ and a class $\mathcal{K}$ function, α, such that all solutions with $\|x(0)\| \leq \delta$ satisfy:

$$\|x(t)\| \leq \alpha(\|x(0)\|) \quad \forall\, t \geq 0 \tag{2.5}$$

Asymptotic stability is equivalent to the existence of a $\delta > 0$ and a class $\mathcal{KL}$ function, β, such that all solutions with $\|x(0)\| \leq \delta$ satisfy:

$$\|x(t)\| \leq \beta(\|x(0)\|, t) \quad \forall\, t \geq 0 \tag{2.6}$$

Global asymptotic stability amounts to the existence of a class $\mathcal{KL}$ function, β, such that the inequality of (2.6) holds for all initial conditions. Exponential stability means that the function β takes the form $\beta(r, s) = cre^{-\lambda s}$ for some $c, \lambda > 0$.

2.1.3 Lyapunov's Direct (Second) Method

Having defined stability and asymptotic stability of equilibrium points, the next task is to find ways to determine stability. To be of practical interest, stability conditions must not require that we explicitly solve (2.1). The direct method of Lyapunov aims at determining the stability properties of an equilibrium point from the properties of $f(x)$ and its relationship with a positive-definite function $V(x)$.

Definition 2.3. *Consider a C^1 (i.e., continuously differentiable) function $V : \mathbb{R}^n \longrightarrow \mathbb{R}$. It is called positive-definite if $V(0) = 0$ and $V(x) > 0$ for all $x \neq 0$. If $V(x) \longrightarrow \infty$ as $\|x\| \longrightarrow \infty$, then V is said to be radially unbounded.*

If V is both positive-definite and radially unbounded, then there exist two class $\mathcal{K}_\infty$ functions α_1, α_2 such that V satisfies:

$$\alpha_1(\|x\|) \leq V(x) \leq \alpha_2(\|x\|) \tag{2.7}$$

for all x. We write $\dot{V}$ for the derivative of V along the solutions of the system of (2.1), i.e.:

$$\dot{V}(x) = \frac{\partial V}{\partial x} f(x) \tag{2.8}$$

The main result of Lyapunov's stability theory is expressed by the following statement.

Theorem 2.4. *(Lyapunov) Let $x = 0$ be an equilibrium point for the system of (2.1) and $D \subset \mathbb{R}^n$ be a domain containing $x = 0$ in its interior. Suppose that there exists a positive-definite C^1 function $V : \mathbb{R}^n \longrightarrow \mathbb{R}$ whose derivative along the solutions of the system of (2.1) satisfies:*

$$\dot{V}(x) \leq 0 \quad \forall\, x \in D \tag{2.9}$$

then the system of (2.1) is stable. If the derivative of V satisfies:

$$\dot{V}(x) < 0 \quad \forall\, x \in D - \{0\} \tag{2.10}$$

then the system of (2.1) is asymptotically stable. If in the latter case, V is also radially unbounded, then the system of (2.1) is globally asymptotically stable.

A continuously differentiable positive-definite function $V(x)$ satisfying (2.9) is called a *Lyapunov function*. The surface $V(x) = c$, for some $c > 0$, is called a *Lyapunov surface* or a level surface. The condition $\dot{v} \leq 0$ implies that when a trajectory crosses a Lyapunov surface $V(x) = c$, it moves inside the set $\Omega_c = \{x \in \mathbb{R}^n : v(x) \leq c\}$ and can never come out again. When $\dot{V} < 0$, the trajectory moves from one Lyapunov surface to an inner Lyapunov surface with smaller c. As c decreases, the Lyapunov surface $V(x) = c$ shrinks to the origin, showing that the trajectory approaches the origin as time progresses.

If we only know that $\dot{V}(x) \leq 0$, we cannot be sure that the trajectory will approach the origin (see, however, Sect. 2.1.4), but we can conclude that the origin is stable since the trajectory can be contained inside any ball, B_ϵ, by requiring that the initial state x_0 to lie inside a Lyapunov surface contained in that ball.

Various converse Lyapunov theorems show that the conditions of Theorem 2.4 are also necessary. For example, if the system is asymptotically stable, then there exists a positive-definite $\mathcal{C}^1$ function V that satisfies the inequality of (2.10).

Remark 2.5. It is well-known that for the linear-time invariant system:

$$\dot{x} = Ax \tag{2.11}$$

asymptotic stability, exponential stability, and their global versions are all equivalent and amount to the property that A is a Hurwitz matrix, i.e., all eigenvalues of A have negative real parts. Fixing an arbitrary positive-definite symmetric matrix Q and finding the unique positive-definite symmetric matrix P that satisfies the Lyapunov equation:

$$A^T P + PA = -Q$$

one obtains a quadratic Lyapunov function $V(x) = x^T P x$ whose derivative along the solutions of the system of (2.11) is $\dot{V} = -x^T Q x$. The explicit formula for P is

$$P = \int_0^\infty e^{A^T t} Q e^{At} dt$$

Indeed we have

$$A^T P + PA = \int_0^\infty \frac{d}{dt}\left(e^{A^T t} Q e^{At}\right) dt \;=\; -Q$$

because A is Hurwitz.

2.1.4 LaSalle's Invariance Principle

With some additional knowledge about the behavior of solutions, it is possible to prove asymptotic stability using a Lyapunov function which satisfies the nonstrict inequality of (2.9). This is facilitated by *LaSalle's invariance principle.* To state this principle, we first recall the definition of an invariant set.

Definition 2.6. *A set M is called (positively) invariant with respect to the given system if all solutions starting in M remain in M for all future times.*

We now state a version of LaSalle's theorem.

Theorem 2.7. *(LaSalle) Suppose that there exists a positive-definite C^1 function $V : \mathbb{R}^n \longrightarrow \mathbb{R}$ whose derivative along the solutions of the system of (2.1) satisfies the inequality of (2.9). Let M be the largest invariant set contained in the set $\{x : \dot{V}(x) = 0\}$. Then the system of (2.1) is stable and every solution that remains bounded for $t \geq 0$ approaches M as $t \longrightarrow 0$. In particular, if all solutions remain bounded and $M = \{0\}$, then the system of (2.1) is globally asymptotically stable.*

To deduce global asymptotic stability with the help of this result, one needs to check two conditions,. First, all solutions of the system must be bounded. This property follows automatically from the inequality of (2.9) if V is chosen to be radially unbounded; however, radial boundedness of V is not necessary when boundedness of solutions can be established by other means. The second condition is that V is not identically zero along any nonzero solution. We also remark that if one only wants to prove asymptotic convergence of bounded solutions to zero and is not concerned with Lyapunov stability of the origin, then positive-definiteness of V is not needed (this is in contrast to Theorem 2.4).

While Lyapunov's stability theorem readily generalizes to time-varying systems, for LaSalle's invariance principle this is not the case. Instead, one usually works with the weaker property that all solutions approach the set $\{x : \dot{V}(x) = 0\}$.

2.1.5 Lyapunov's Indirect (First) Method

Lyapunov's indirect method allows one to deduce stability properties of the nonlinear system of (2.1), where f is C^1, from stability properties of its *linearization*, which is the linear system of (2.11) with:

$$A := \frac{\partial f}{\partial x}(0) \tag{2.12}$$

By the mean value theorem, we can write:

$$f(x) = Ax + g(x)x$$

where g is given componentwise by $g_i(x) := \frac{\partial f_i}{\partial x}(z_i) - \frac{\partial f_i}{\partial x}(0)$ for some point, z_i, on the line segment connecting x to the origin, $i = 1, \ldots, n$. Since $\frac{\partial f}{\partial x}$ is continuous, we have $\|g(x)\| \longrightarrow 0$ as $x \longrightarrow 0$. From this it follows that if the matrix A is Hurwitz (i.e., all its eigenvalues in the open left half of the complex plane), then a quadratic Lyapunov function for the linearization serves – locally – as a Lyapunov function for the original nonlinear system. Moreover, its rate of decay in a neighborhood of the origin can be bounded below by a quadratic function, which implies that stability is in fact exponential. This is summarized by the following result.

Theorem 2.8. *If f is $\mathcal{C}^1$ and the matrix of (2.12) is Hurwitz, then the system of (2.1) is locally exponentially stable.*

It is also known that if the matrix A has at least one eigenvalue with a positive real part, the origin of the nonlinear system of (2.1) is not stable. If A has eigenvalues on the imaginary axis but no eigenvalues in the open right half-plane, the linearization test is inconclusive. However, in this critical case, the system of (2.1) cannot be exponentially stable, since exponential stability of the linearization is not only a sufficient but also a necessary condition for (local) exponential stability of the nonlinear system.

2.1.6 Input-to-State Stability

It is of interest to extend stability concepts to systems with disturbance inputs. In the linear case represented by the system:

$$\dot{x} = Ax + B\theta$$

It is well known that if the matrix A is Hurwitz, i.e., if the unforced system, $\dot{x} = Ax$, is asymptotically stable, then bounded inputs θ lead to bounded states while inputs converging to zero produce states converging to zero. Now, consider a nonlinear system of the form:

$$\dot{x} = f(x, \theta) \tag{2.13}$$

where θ is a measurable locally essentially bounded disturbance input. In general, global asymptotic stability of the unforced system $\dot{x} = f(x, 0)$ does not guarantee input-to-state properties of the kind mentioned above. For example, the scalar system:

$$\dot{x} = -x + x\theta \tag{2.14}$$

has unbounded trajectories under the bounded input $\theta \equiv 2$. This motivates the following important concept, introduced by Sontag.

Definition 2.9. *The system of (2.13) is called input-tot-state stable (ISS) with respect to θ if for some functions $\gamma \in \mathcal{K}_\infty$ and $\beta \in \mathcal{KL}$, for every initial state $x(0)$, and every input θ, the corresponding solution of the system of (2.13) satisfies the inequality:*

$$\|x(t)\| \leq \beta(\|x(0)\|, t) + \gamma(\|\theta\|^s_{[0,t]}) \tag{2.15}$$

where $\|\theta\|^s_{[0,t]} := ess.sup.\{\|\theta(s)\| : s \in [0,t]\}$ (supremum norm on $[0,t]$ except for a set of measure zero).

Since the system of (2.13) is time-invariant, the same property results if we write

$$\|x(t)\| \leq \beta(\|x(t_0)\|, t - t_0) + \gamma(\|\theta\|^s_{[t_0,t]}) \quad \forall\, t \geq t_0 \geq 0 \tag{2.16}$$

The ISS property admits the following Lyapunov-like equivalent characterization: the system of (2.13) is ISS if and only if there exists a positive-definite radially unbounded $\mathcal{C}^1$ function $V : \mathbb{R}^n \longrightarrow \mathbb{R}$ such that for some class $\mathcal{K}_\infty$ functions α and χ we have

$$\frac{\partial V}{\partial x} f(x,\theta) \leq -\alpha(\|x\|) + \chi(\|\theta\|) \quad \forall\, x, \theta \tag{2.17}$$

This is in turn equivalent to the following "gain margin" condition:

$$\|x\| \;\geq\; \rho(\|\theta\|) \Longrightarrow \frac{\partial V}{\partial x} f(x,\theta) \;\leq\; -\alpha(\|x\|) \tag{2.18}$$

where $\alpha, \rho \in \mathcal{K}_\infty$. Such functions V are called *ISS-Lyapunov functions.* If the system of (2.13) is ISS, then $\theta(t) \longrightarrow 0$ implies $x(t) \longrightarrow 0$.

The system of (2.13) is said to be *locally input-tot-state stable* (locally ISS) if the bound of (2.15) is valid for solutions with sufficiently small initial conditions and inputs, i.e. if there exists a $\delta > 0$ such that (2.15) is satisfied whenever $\|x(0)\| \leq \delta$ and $\|\theta\|^s_{[0,t]} \leq \delta$. It turns out that (local) asymptotic stability of the unforced system $\dot{x} = f(x, 0)$ implies local ISS.

2.2 Stabilization of Nonlinear Systems

This book is about control *design.* Our objective is to create closed-loop systems with desirable stability properties, rather than analyze the properties of a given system. For this reason, we are interested in an extension of the Lyapunov function concept, called a *control Lyapunov function* (CLF).

Suppose that our problem for the time-invariant system:

$$\dot{x} = f(x, u) \tag{2.19}$$

where $x \in \mathbb{R}^n$, $u \in \mathbb{R}$, $f(0,0) = 0$, is to design a feedback control law $\alpha(x)$ for the control variable u such that the equilibrium $x = 0$ of the closed-loop system:

$$\dot{x} = f(x, \alpha(x)) \tag{2.20}$$

is globally asymptotically stable. We can pick a function $V(x)$ as a Lyapunov candidate, and require that its derivative along the solutions of the system of (2.20) satisfy $\dot{V} \leq -W(x)$, where $W(x)$ is a positive-definite function. We therefore need to find $\alpha(x)$ to guarantee that for all $x \in \mathbb{R}^n$

$$\frac{\partial V}{\partial x}(x) f(x, \alpha(x)) \leq -W(x) \tag{2.21}$$

This is a difficult task. A stabilizing control law for the system of (2.19) may exist but we may fail to satisfy (2.21) because of a poor choice of $V(x)$ and $W(x)$. A system for which a good choice of $V(x)$ and $W(x)$ exists is said to possess a CLF. This notion is made more precise below.

Definition 2.10. *A smooth positive-definite radially unbounded function $V : \mathbb{R}^n \longrightarrow \mathbb{R}$ is called a control Lyapunov function (CLF) for the system of (2.19) if:*

$$\inf_{u \in \mathbb{R}} \left\{ \frac{\partial V}{\partial x}(x) f(x, u) \right\} < 0, \quad \forall\, x \neq 0 \tag{2.22}$$

The CLF concept of Artstein [17] and Sontag is a generalization of Lyapunov design results by Jacobson and Judjevic and Quinn. Artstein showed that (2.22) is not only necessary, but also sufficient for the existence of a control law satisfying (2.21), that is the existence of a CLF is equivalent to global asymptotic stabilizability.

For systems affine in the control:

$$\dot{x} = f(x) + g(x)u, \quad f(0) = 0 \tag{2.23}$$

and using the Lie derivative notation, $L_f V(x) = \frac{\partial V}{\partial x}(x) f(x)$ and $L_g V(x) = \frac{\partial V}{\partial x}(x) g(x)$, the CLF inequality of (2.21) becomes:

$$L_f V(x) + L_g V(x) u \leq -W(x) \tag{2.24}$$

If V is a CLF for the system of (2.23), then a particular stabilizing control law $\alpha(x)$, smooth for all $x \neq 0$, is given by Sontag's formula [252]:

$$u = \alpha_s(x) = \left\{ \begin{array}{ll} -\dfrac{L_f V(x) + \sqrt{(L_f V)^2(x) + (L_g V)^4(x)}}{(L_g V)^2(x)} L_g V(x), & L_g V(x) \neq 0 \\ 0, & L_g V(x) = 0 \end{array} \right\} \tag{2.25}$$

It should be noted that (2.24) can be satisfied only if:

$$L_g V(x) = 0 \Longrightarrow L_f V(x) < 0, \quad \forall\, x \neq 0 \tag{2.26}$$

and that in this case (2.25) results in:

$$W(x) = \sqrt{(L_f V)^2(x) + (L_g V)^4(x)} > 0, \quad \forall\, x \neq 0 \tag{2.27}$$

A further characterization of a stabilizing control law $\alpha(x)$ for the system of (2.23) with a given CLF V is that $\alpha(x)$ is continuous at $x = 0$ if and only if the CLF satisfies the *small control property*: For each $\varepsilon > 0$ there is a $\delta(\varepsilon) > 0$ such that, if $x \neq 0$ satisfies $|x| \leq \delta$, then there is some u with $|u| < \varepsilon$ such that:

$$L_f V(x) + L_g V(x) u < 0 \tag{2.28}$$

The main deficiency of the CLF concept as a design tool is that for most nonlinear systems a CLF is not known. The task of finding an appropriate CLF maybe as complex as that of designing a stabilizing feedback law. For several important classes of nonlinear systems, however, it is possible to systematically construct CLFs, and this issue will be elaborated on throughout the book.

2.3 Feedback Linearization and Zero Dynamics

One of the popular methods for nonlinear control design is *feedback linearization*, which employs a change of coordinates and feedback control to transform a nonlinear system into a system whose dynamics are linear (at least partially). A great deal of research has been devoted to this subject over the last three decades, as evidenced by the comprehensive books [126, 208] and the references therein. In this section, we briefly review some of the basic geometric concepts that will be used in subsequent chapters. While this book does not require the formalism of differential geometry, we will employ Lie derivatives only for notational convenience. If $f : \mathbb{R}^n \longrightarrow \mathbb{R}^n$ is a vector field and $h : \mathbb{R}^n \longrightarrow \mathbb{R}$ is a scalar function, the notation $L_f h$ is used for $\frac{\partial h}{\partial x} f(x)$. It is recursively extended to:

$$L_f^k h(x) = L_f(L_f^{k-1} h(x)) = \frac{\partial}{\partial x}(L_f^{k-1} h(x)) f(x)$$

Let us consider the nonlinear system:

$$\begin{aligned} \dot{x} &= f(x) + g(x)u \\ y &= h(x) \end{aligned} \tag{2.29}$$

where $x \in \mathbb{R}^n$, $u \in \mathbb{R}$, $y \in \mathbb{R}$, f, g, h are smooth (i.e., infinitely differentiable) vector functions. The derivative of the output $y = h(x)$ is given by:

$$\begin{aligned} \dot{y} &= \frac{\partial h}{\partial x}(x) f(x) + \frac{\partial h}{\partial x}(x) g(x) u \\ &= L_f h(x) + L_g h(x) u \end{aligned} \tag{2.30}$$

If $L_g h(x_0) \neq 0$, then the system of (2.29) is said to have *relative degree one at* x_0 (note that since the functions are smooth $L_g h(x_0) \neq 0$ implies that there exists a neighborhood of x_0 on which $L_g h(x) \neq 0$). In our terminology, this implies that the output y is separated form the input u by one integration only. If $L_g h(x_0) = 0$, there are two cases:

(i) If there exist points arbitrarily close to x_0 such that $L_g h(x) \neq 0$, then the system of (2.29) does not have a well-defined relative degree at x_0.

(ii) If there exists a neighborhood B_0 of x_0 such that $L_g h(x) = 0$ for all $x \in B_0$, then the relative degree of the system of (2.29) may be well-defined.

In case (ii), we define:

$$\psi_1(x) = h(x), \quad \psi_2(x) = L_f h(x) \tag{2.31}$$

and compute the second derivative of y:

$$\begin{aligned} \ddot{y} &= \frac{\partial \psi_2}{\partial x}(x) f(x) + \frac{\partial \psi_2}{\partial x}(x) g(x) u \\ &= L_f^2 h(x) + L_g L_f h(x) u \end{aligned} \tag{2.32}$$

If $L_gL_fh(x_0) \neq 0$, then the system of (2.29) is said to have *relative degree two at* x_0. If $L_gL_fh(x) = 0$ in a neighborhood of x_0, then we continue the differentiation procedure.

Definition 2.11. *The system of (2.29) is said to have relative degree r at the point x_0 if there exists a neighborhood B_0 of x_0 on which:*

$$L_gh(x) = L_gL_fh(x) = \cdots = L_gL_f^{r-1}h(x) = 0 \tag{2.33}$$

$$L_gL_f^rh(x) = 0 \tag{2.34}$$

If (2.33–2.34) are valid for all $x \in \mathbb{R}^n$, then the relative degree of the system of (2.29) is said to be globally defined.

Suppose now that the system of (2.29) has relative degree r at x_0. Then we can use a change of coordinates and feedback control to locally transform this system into the *cascade interconnection* of an r-dimensional linear system and an $(n-r)$-dimensional nonlinear system. In particular, after differentiating r times the output $y = h(x)$, the control appears:

$$y^{(r)} = L_f^rh(x) + L_gL_f^{r-1}h(x)u \tag{2.35}$$

Since $L_gL_f^{r-1}h(x) \neq 0$ in a neighborhood of x_0, we can linearize the input-output description of the system of (2.29) using feedback to cancel the nonlinearities in (2.35):

$$u = \frac{1}{L_gL_f^{r-1}h(x)}\left[-L_f^rh(x) + v\right] \tag{2.36}$$

Then the dynamics of y and its derivatives are governed by a chain of r integrators: $y^{(r)} = v$. Since our original system of (2.29) has dimension n, we need to account for the remaining $n-r$ states. Using differential geometric tools, it can be shown that it is always possible to find $n-r$ functions $\psi_{r+1}, \ldots, \psi_n(x)$ with $\frac{\partial \psi_i}{\partial x}(x)g(x) = L_gL_f^{i-1}h(x) = 0$, for $i = r+1, \ldots, n$ such that the change of coordinates:

$$\begin{gathered}\zeta_1 = y = h(x),\ \zeta_2 = \dot{y} = L_fh(x),\ \cdots,\ \zeta_r = y^{(r-1)} = L_f^{r-1}h(x)\\ \eta_1 = \psi_{r+1},\ \cdots,\ \eta_{n-r} = \psi_n(x)\end{gathered} \tag{2.37}$$

is locally invertible and transforms, along with the feedback law of (2.36), the system of (2.29) into:

$$
\begin{aligned}
\dot{\zeta}_1 &= \zeta_2 \\
&\vdots \\
\dot{\zeta}_r &= v \\
\dot{\eta}_1 &= \Psi_1(\zeta, \eta) \\
&\vdots \\
\dot{\eta}_{n-r} &= \Psi_{n-r}(\zeta, \eta) \\
y &= \zeta_1
\end{aligned}
\tag{2.38}
$$

where $\Psi_1(\zeta, \eta) = L_f^{r+1}h(x)$, $\Psi_{n-r}(\zeta, \eta) = L_f^n h(x)$.

The states $\eta_1, \ldots, \eta_{n-r}$ have been rendered *unobservable* from the output y by the control of (2.36). Hence, feedback linearization in this case is the nonlinear equivalent of placing $n-r$ poles of a linear system at the origin and cancelling the r zeros with the remaining poles. Of course, to guarantee stability, the cancelled zeros must be stable. In the nonlinear case, using the new control input v to stabilize the linear subsystem of (2.38) does not guarantee stability of the whole system, unless the stability of the nonlinear part of the system of (2.38) has been established separately.

When v is used to keep the output y equal to zero for all $t > 0$, that is, when $\zeta_1 \equiv \cdots \equiv \zeta_r \equiv 0$, the dynamics of $\eta_1, \ldots, \eta_{n-r}$ are described by:

$$
\begin{aligned}
\dot{\eta}_1 &= \Psi_1(0, \eta) \\
&\vdots \\
\dot{\eta}_{n-r} &= \Psi_{n-r}(0, \eta)
\end{aligned}
\tag{2.39}
$$

They are called the zero dynamics of the system of (2.29), because they evolve on the subset of the state-space on which the output of the system is identically zero. If the equilibrium at $\eta_1 = \cdots = \eta_{n-r} = 0$ of the zero dynamics of (2.39) is asymptotically stable, the system of (2.29) is said to be *minimum phase*.

Remark 2.12. Most nonlinear analytical controllers emanating from the area of geometric process control are input-output linearizing and induce a linear input-output response in the absence of constraints [126, 160]. For the class of processes modeled by equations of the form of (2.29) with relative order r and under the minimum phase assumption, the appropriate linearizing state feedback controller is given by:

$$
u = \frac{1}{L_g L_f^{r-1} h(x)} \left(v - L_f^r h(x) - \beta_1 L_f^{r-1} h(x) - \cdots - \beta_{r-1} L_f h(x) - \beta_r h(x) \right)
\tag{2.40}
$$

and induces the linear r-th order response:

$$\frac{d^r y}{dt^r} + \beta_1 \frac{d^{r-1}y}{dt^{r-1}} + \cdots + \beta_{r-1}\frac{dy}{dt} + \beta_r y = v \tag{2.41}$$

where the tunable parameters, $\beta_1, \cdots, \beta_r$, are essentially closed-loop time constants that influence and shape the output response. The nominal stability of the process is guaranteed by placing the roots of the polynomial $s^r + \beta_1 s^{r-1} + \cdots + \beta_{r-1}s + \beta_r$ in the open left-half of the complex plane.

2.4 Singularly Perturbed Systems

In this section, we review two results on the analysis of singularly perturbed nonlinear systems having a reduced system that is input-to-state stable with respect to disturbances. Specifically, these results establish robustness of this ISS property to uniformly globally asymptotically stable singular perturbations. To this end, we focus on singularly perturbed nonlinear systems with the following state-space description:

$$\begin{aligned} \dot{x} &= f(x, z, \theta(t), \epsilon) \\ \epsilon \dot{z} &= g(x, z, \theta(t), \epsilon) \end{aligned} \tag{2.42}$$

where $x \in \mathbb{R}^n$ and $z \in \mathbb{R}^p$ denote vectors of state variables, $\theta \in \mathbb{R}^q$ denotes the vector of the disturbances and ϵ is a small positive parameter. The functions f and g are locally Lipschitz on $\mathbb{R}^n \times \mathbb{R}^p \times \mathbb{R}^q \times [0, \bar{\epsilon})$, for some $\bar{\epsilon} > 0$. The input vector $\theta(t)$ in the system of (2.42) may represent constant or time-varying (not necessarily slowly) parameters, tracking signals and/or exogenous disturbances. In what follows, for simplicity, we will suppress the time-dependence in the notation of the vector of input variables $\theta(t)$.

A standard procedure that is followed for the analysis of systems in the form of (2.42) is the decomposition of the original system into separate reduced-order systems, each one associated with a different time-scale. This procedure is called two-time-scale decomposition [153]. Formally, it can be done by setting $\epsilon = 0$, in which case the dynamics of the state vector z becomes instantaneous and the system of (2.42) takes the form:

$$\begin{aligned} \dot{x} &= f(x, z_s, \theta, 0) \\ g(x, z_s, \theta, 0) &= 0 \end{aligned} \tag{2.43}$$

where z_s denotes a quasi-steady-state for the fast state vector z. Assumption 2.1 states that the singularly perturbed system in (2.42) is in *standard form.*

Assumption 2.1 *The algebraic equation $g(x, z_s, \theta, 0) = 0$ possesses a unique root:*

$$z_s = h(x, \theta) \tag{2.44}$$

with the properties that $h : \mathbb{R}^n \times \mathbb{R}^q \to \mathbb{R}^p$ and its partial derivatives $\left(\frac{\partial h}{\partial x}, \frac{\partial h}{\partial \theta}\right)$ are locally Lipschitz.

Substituting (2.44) into (2.43), the following locally Lipschitz system is obtained:

$$\dot{x} = f(x, h(x,\theta), \theta, 0) \tag{2.45}$$

The dynamical system in (2.45) is called the *reduced system* or *slow subsystem.* The inherent two-time-scale behavior of the system of (2.42) can be analyzed by defining a fast time-scale:

$$\tau = \frac{t}{\epsilon} \tag{2.46}$$

and the new coordinate $y := z - h(x,\theta)$. In the (x, y) coordinates, and with respect to the τ time scale, the singularly perturbed system in (2.42) takes the form:

$$\begin{aligned}\frac{dx}{d\tau} &= \epsilon f(x, h(x,\theta) + y, \theta, \epsilon) \\ \frac{dy}{d\tau} &= g(x, h(x,\theta) + y, \theta, \epsilon) - \epsilon \left[\frac{\partial h}{\partial x} f(x, h(x,\theta) + y, \theta, \epsilon) + \frac{\partial h}{\partial \theta}\dot{\theta}\right]\end{aligned} \tag{2.47}$$

Setting ϵ equal to zero, the following locally Lipschitz system is obtained:

$$\frac{dy}{d\tau} = g(x, h(x,\theta) + y, \theta, 0) \tag{2.48}$$

Here, x and θ are to be thought of as constant vectors. In what follows, we will refer to the dynamical system in (2.48) as the *fast subsystem* or the *boundary layer* system. The assumptions that follow state our stability requirements on the slow and fast subsystems.

Assumption 2.2 *The reduced system in (2.45) is input-to-state stable (ISS) with Lyapunov gain* γ.

Assumption 2.3 *The equilibrium* $y = 0$ *of the boundary layer system in (2.48) is globally asymptotically stable, uniformly in* $x \in \mathbb{R}^n$, $\theta \in \mathbb{R}^q$.

The main result of this section is given in the following theorem (the proof is given in [57]).

Theorem 2.13. *Consider the singularly perturbed system in (2.42) and suppose Assumptions 2.1–2.3 hold and that* $\theta(t)$ *is absolutely continuous. Define* $y = z - h(x,\theta)$ *and let* γ *be the function given by Assumption 2.2. Then there exist functions* β_x, β_y *of class* $\mathcal{KL}$, *and for each pair of positive real numbers* (δ, d), *there is an* $\epsilon^* > 0$ *such that if* $\max\{\|x(0)\|, \|y(0)\|, \|\theta\|^s, \|\dot{\theta}\|^s\} \leq \delta$ *and* $\epsilon \in (0, \epsilon^*]$, *then, for all* $t \geq 0$,

$$\begin{aligned}\|x(t)\| &\leq \beta_x(\|x(0)\|, t) + \gamma(\|\theta\|^s) + d \\ \|y(t)\| &\leq \beta_y\left(\|y(0)\|, \frac{t}{\epsilon}\right) + d\end{aligned} \tag{2.49}$$

Remark 2.14. When g does not depend on θ, $\theta(t)$ can simply be measurable and there is no requirement on $\|\dot{\theta}\|^s$ (if $\dot{\theta}$ exists).

Remark 2.15. The result of Theorem 2.13 can be applied to arbitrarily large initial conditions $(x(0), y(0))$, uncertainty $\theta(t)$, and rate of change of uncertainty $\dot{\theta}(t)$. Furthermore, this result (even in the undisturbed case, i.e., $\theta(t) \equiv 0$), imposes no growth or interconnection conditions to establish boundedness of the trajectories (compare with the discussion in Remark 2.17 below).

Remark 2.16. In principle, a value for ϵ^* can be extracted from the proof of the theorem. However, as is the case for most general singular perturbation results for nonlinear systems (e.g., [63, 235]), this value will typically be quite conservative.

Remark 2.17. Note that, even with $\theta(t) \equiv 0$, Theorem 2.13 does not provide any result concerning stability of or convergence to the origin. To guarantee such properties further assumptions are required. consider, for example, the singularly perturbed system:

$$\begin{aligned} \dot{x} &= z - x^3 \\ \epsilon \dot{z} &= -z + \epsilon x \end{aligned} \tag{2.50}$$

where $x \in \mathbb{R}$, $z \in \mathbb{R}$. One can easily see that the above system is in standard form, $h(x, \theta) = 0$ and its fast dynamics:

$$\frac{dy}{d\tau} = -y \tag{2.51}$$

possess a globally exponentially stable equilibrium ($y = 0$). Moreover, the reduced system takes the form:

$$\dot{x} = -x^3 \tag{2.52}$$

which clearly possesses a globally asymptotically stable equilibrium ($x = 0$). Thus, the assumptions of the Theorem 2.13 are satisfied and its result can be applied. However, the origin is not an asymptotically stable equilibrium for the system of (2.50). This can be easily seen considering the linearization of (2.50) around the origin:

$$\begin{aligned} \dot{x} &= z \\ \epsilon \dot{z} &= -z + \epsilon x \end{aligned} \tag{2.53}$$

which possesses an eigenvalue in the right half of the complex plane for all positive values of ϵ. It is clear from the above example that we cannot draw any conclusions about the stability properties of the equilibrium point of the full-order system from knowledge of the stability properties of the reduced-order systems without some type of additional interconnection condition. There are several possible interconnection conditions that can be imposed to guarantee stability of and convergence to the origin. Most efficient conditions are essentially related to the small gain theorem (see [187,301]) in one way or another. Saberi and Khalil provide Lyapunov conditions in [235] which are closely related to small gain conditions in an L_2 setting. An L_∞ nonlinear small gain condition is given in [268] and [132]. Perhaps the most straightforward, although conservative, interconnection condition is the assumption that, with $\theta(t) \equiv 0$, the origin of the slow and fast subsystems are locally exponentially stable. In this case, it can be shown [235] that there exists ϵ sufficiently small such that the origin of the system of (2.42) is locally exponentially stable and that the basin of attraction does not shrink as ϵ becomes small. In general, when a local interconnection condition is given that has this "nonshrinking" property, the result of Theorem 2.13 can be used to show stability and convergence from an arbitrary large compact set under Assumptions 2.1–2.3. This is obtained by choosing d_x, d_y sufficiently small to guarantee convergence in finite time to the domain of attraction of the origin.

Theorem 2.18 below is a natural extension of the result of Theorem 2.13, to nonlinear singularly perturbed systems for which the initial conditions of the fast states depend singularly on ϵ (the proof is given in [50].

Theorem 2.18. *Consider the following singularly perturbed system:*

$$\begin{aligned} \dot{x} &= \bar{f}(x, z_1, sat(z_2), \theta) \\ \epsilon \dot{z}_1 &= \bar{g}_1(x, z_1, sat(z_2), \theta) \\ \epsilon \dot{z}_2 &= A z_2 + \epsilon \bar{g}_2(x, z_1, sat(z_2), \theta) \end{aligned} \tag{2.54}$$

where $x \in \mathbb{R}^n$, $z_1 \in \mathbb{R}^{p_1}$, $z_2 \in \mathbb{R}^{p_2}$, $\theta \in \mathbb{R}^q$, $\bar{f}, \bar{g}_1, \bar{g}_2$ are locally Lipschitz on $\mathbb{R}^n \times \mathbb{R}^{p_1} \times \mathbb{R}^{p_2} \times \mathbb{R}^q$, and A is a constant matrix. Suppose that the algebraic equation $\bar{g}(x, z_1, 0, \theta) = 0$ possesses a unique root $z_1 = h_1(x, \theta)$ and that the slow subsystem:

$$\dot{x} = \bar{f}(x, h_1(x, \theta), 0, \theta) \tag{2.55}$$

is ISS with Lyapunov gain γ. Also, define $y_1 = z_1 - h_1(x, \theta)$ and suppose that the fast subsystem:

$$\frac{dy_1}{d\tau} = \bar{g}_1(x, h_1(x, \theta) + y_1, 0, \theta) \tag{2.56}$$

is globally asymptotically stable, uniformly in $x \in \mathbb{R}^n$, $\theta \in \mathbb{R}^q$. Finally, suppose that A is Hurwitz and that $\theta(t)$ is absolutely continuous, and define $z_2(0) = \dfrac{\xi(0)}{\epsilon^{p_2}}$, where $\xi \in \mathbb{R}^{p_2}$. Whenever the above assumptions are satisfied, there exist functions β_x, β_{y_1} of class $\mathcal{KL}$ and strictly positive constants K_1, a_1, and for each pair of positive real numbers (δ, d), there exists a positive real number ϵ^* such that if $\max\{\|x(0)\|, \|y_1(0)\|, \|\xi(0)\|, \|\theta\|^s, \|\dot{\theta}\|^s\} \leq \delta$, $\epsilon \in (0, \epsilon^*]$ then, for all $t \geq 0$,

$$\|x(t)\| \leq \beta_x(\|x(0)\|, t) + \gamma(\|\theta\|^s) + d$$

$$\|y_1(t)\| \leq \beta_{y_1}\left(\|y_1(0)\|, \frac{t}{\epsilon}\right) + d \tag{2.57}$$

$$\|z_2(t)\| \leq K_1\|z_2(0)\|e^{-a_1\frac{t}{\epsilon}} + d$$

Remark 2.19. Theorem 2.18 is particularly useful for the analysis of closed-loop systems under output feedback controllers that utilize nonlinear high-gain observers.

3

Control of Nonlinear Systems with Uncertainty

3.1 Introduction

In this chapter, we consider nonlinear systems with time-varying uncertain variables and present robust inverse optimal controller designs that enforce an arbitrary degree of attenuation of the effect of uncertainty on the output of the closed-loop system. The controller designs are obtained by re-shaping the scalar nonlinear gain of Sontag's formula in a way that guarantees the desired uncertainty attenuation properties in the closed-loop system. The proposed re-shaping is made different for vanishing and non-vanishing uncertainties so as to meet different robustness and optimality objectives.

The rest of the chapter is organized as follows. In Sect. 3.2 we introduce the class of uncertain nonlinear systems considered and review some preliminaries on inverse optimal controller design. Then in Sect. 3.3 we formulate the robust inverse optimal state feedback control problem, for both vanishing and non-vanishing uncertainties, and present the proposed gain re-shaping procedure, and the resulting control laws, for each case. We discuss some of the advantages of the proposed gains relative to others that have been introduced in the literature. In contrast to previous works, the uncertain variables considered do not obey state-dependent bounds and, in the non-vanishing case, are not only persistent over time but can also change the nominal equilibrium point of the system. A simulation example is presented to illustrate the performance of the state feedback controllers and compare it with other possible controller designs. In Sect. 3.4, we address the output feedback control problem by combining the state feedback controllers with appropriate high gain observers. We show that the output feedback controllers enforce closed-loop stability and robust asymptotic output tracking for initial conditions and uncertainty in arbitrarily large compact sets, as long as the observer gain is sufficiently large. Utilizing the inverse optimal control approach and singular perturbation techniques, this approach is shown to yield a near-optimal output feedback design in the sense that the performance of the resulting output feedback controllers can be made arbitrarily close to that of the robust inverse optimal state feedback controllers, when the observer gain is sufficiently large. The developed controllers are successfully applied to a chemical reactor example. Finally, in Sect. 3.5, we use a singular perturbation formulation to establish robustness of the output feedback controllers with respect to unmodeled dynamics. The state feedback control results in this chapter were first presented in [83], while the output feedback control results were first presented in [79].

3.2 Preliminaries

We consider single-input single-output (SISO), uncertain continuous-time nonlinear systems with the following state-space description:

$$\begin{aligned} \dot{x} &= f(x) + g(x)u + \sum_{k=1}^{q} w_k(x)\theta_k(t) \\ y &= h(x) \end{aligned} \tag{3.1}$$

where $x \in \mathbb{R}^n$ denotes the vector of state variables, $u \in \mathbb{R}$ denotes the manipulated input, $\theta_k(t) \in \mathcal{W}_k \subset \mathbb{R}$ denotes the k-th uncertain (possibly time-varying) but bounded variable taking values in a nonempty compact convex subset $\mathcal{W}_k$ of $\mathbb{R}$, and $y \in \mathbb{R}$ denotes the output to be controlled. Without loss of generality, we assume that the origin is an equilibrium point of the nominal system ($u(t) = \theta_k(t) \equiv 0$) of (3.1). This can always be achieved by working with deviation variables. The uncertain variable $\theta_k(t)$ may describe time-varying parametric uncertainty and/or exogenous disturbances. The vector functions $f(x)$, $w_k(x)$ and $g(x)$, and the scalar function $h(x)$ are assumed to be sufficiently smooth. Throughout the chapter, the notation $\|\theta\|^s$ is used to denote ess.sup. $\|\theta(t)\|$, $t \geq 0$ where the function θ is measurable (with respect to the Lebesgue measure). Given a measurable function $f : \mathcal{T} \to \mathbb{R}$, where $\mathcal{T}$ is a measure space with measure μ, the essential supremum is defined as ess.sup.$f(t) = \inf\{M : \mu\{t : f(t) > M\} = 0\}$, i.e. it is the smallest positive integer M such that $\|f\|$ is bounded by M almost everywhere. For simplicity, we will suppress the time-dependence in the notation of the uncertain variable $\theta_k(t)$.

Preparatory for its use as a tool for robust optimal controller design, we begin by reviewing the concept of inverse optimality introduced in the context of robust stabilization in [97]. To this end, consider the system of (3.1) with $q = 1$ and $w(0) = 0$. Also, let $l(x)$ and $R(x)$ be two continuous scalar functions such that $l(x) \geq 0$ and $R(x) > 0$ $\forall\, x \in \mathbb{R}^n$ and consider the problem of finding a feedback control law $u(x)$ for the system of (3.1) that achieves asymptotic stability of the origin and minimizes the infinite time cost functional:

$$J = \int_0^\infty (l(x) + uR(x)u)\, dt \tag{3.2}$$

The steady-state Hamilton-Jacobi-Isaacs (HJI) equation associated with the system of (3.1) and the cost of (3.2) is:

$$0 \equiv \inf_{u \in \mathbb{R}} \sup_{\theta \in \mathcal{W}} \left(l(x) + uR(x)u + L_f V + L_g V u + L_w V \theta\right) \tag{3.3}$$

where the value function V is the unknown. A smooth positive-definite solution V to this equation will lead to a continuous state feedback $u(x)$ of the form:

$$u = -p(x) \;=\; -\frac{1}{2} R^{-1}(x) L_g V \tag{3.4}$$

which provides stability, optimality, and robustness with respect to the disturbance θ. However, such a smooth solution may not exist or may be extremely difficult to compute. Suppose, instead, that we were able to find a positive-definite radially unbounded C^1 scalar function V such that:

$$\inf_{u \in \mathbb{R}} \sup_{\theta \in \mathcal{W}} \left(L_f V + L_g V u + L_w V \theta\right) < 0 \quad \forall\, x \neq 0 \tag{3.5}$$

Now, if we can find a meaningful cost functional (i.e., $l(x) \geq 0$, $R(x) > 0$) such that the given V is the corresponding value function, then we will have indirectly obtained a solution to the HJI equation and can therefore compute the optimal control law of (3.4). Such reasoning motivates following the inverse path. In the inverse approach, a stabilizing feedback control law is designed first and then shown to be optimal with respect to a well-defined and meaningful cost functional of the form of (3.2). The problem is inverse because the weights $l(x)$ and $R(x)$ in the cost functional are a posteriori computed from the chosen stabilizing feedback control law, rather than a priori specified by the designer.

A stabilizing control law $u(x)$ is said to be inverse optimal for the system of (3.1) if it can be expressed in the form of (3.4) where the negative-definiteness of $\dot{V}$ is achieved with the control $u^* = -\frac{1}{2}p(x)$, that is:

$$\begin{aligned} \sup_{\theta \in \mathcal{W}} \dot{V} &= \sup_{\theta \in \mathcal{W}} \left(L_f V + L_g V u^* + L_w V \theta\right) \\ &= L_f V - \frac{1}{2} L_g V p(x) + \|L_w V\| \theta_b \; < \; 0 \quad \forall\, x \neq 0 \end{aligned} \tag{3.6}$$

where the worst-case uncertainty (i.e., the one that maximizes $\dot{V}$) is given by $\theta = sgn[L_w V(x)]\theta_b$ where $\theta_b = \|\theta\|^s$ (which is an admissible uncertainty since it is both measurable and bounded). When the function $l(x)$ is set equal to $l(x) = -\sup_{\theta \in \mathcal{W}} \dot{V}$, then V is a solution to the following steady-state HJI equation:

$$0 \equiv l(x) + L_f V - \frac{1}{4} L_g V R^{-1}(x) L_g V + \|L_w V\| \theta_b \tag{3.7}$$

and the optimal (minimal) value of the cost J is $V(x(0))$. At this point, we need to recall the definition of a robust control Lyapunov function which will be used in the development of the main results of this chapter.

Definition 3.1. *[97] A smooth, proper, and positive-definite function* $V : \mathbb{R}^n \rightarrow \mathbb{R}_+$ *is called a robust control Lyapunov function for a system of the form* $\dot{x} = f(x) + g(x)u + w(x)\theta$ *when there exist a function,* $\alpha_v(\cdot)$*, of class* $\mathcal{K}$*, and* $c_v > 0$ *such that:*

$$\inf_{u \in \mathcal{U}} \sup_{\theta \in \mathcal{W}} [L_f V(x) + L_g V(x) u + L_w V(x)\theta + \alpha_v(x)] < 0 \tag{3.8}$$

whenever $V(x) > c_v$*, where* $L_f V(x) = \frac{\partial V}{\partial x} f(x)$*,* $L_g V(x) = \frac{\partial V}{\partial x} g(x)$ *and* $L_w V(x) = \frac{\partial V}{\partial x} w(x)$*.*

Finally, we recall the definition of input-to-state stability (ISS) for a system of the form of (3.1).

Definition 3.2. *[252] The system in (3.1) (with* $u \equiv 0$*) is said to be ISS with respect to* θ *if there exist a function* β *of class KL and a function* γ *of class K such that for each* $x_\circ \in \mathbb{R}^n$ *and for each measurable, essentially bounded input* $\theta(\cdot)$ *on* $[0, \infty)$ *the solution of (3.1) with* $x(0) = x_\circ$ *exists for each* $t \geq 0$ *and satisfies:*

$$\|x(t)\| \leq \beta(\|x(0)\|, t) + \gamma(\|\theta\|^s), \ \forall\, t \geq 0 \tag{3.9}$$

3.3 Robust Inverse Optimal State Feedback Controller Design

3.3.1 Control Problem Formulation

Referring to the uncertain nonlinear system of (3.1), we consider the following two control problems. In the first problem, we assume that the uncertain variables are vanishing (i.e., $w_k(0)\theta_k = 0$) which means that the origin is an equilibrium point of the uncertain system of (3.1). We focus on the synthesis of robust nonlinear state feedback controllers of the general form:

$$u = \mathcal{P}(x, \bar{v}) \tag{3.10}$$

where $\mathcal{P}(x, \bar{v})$ is a scalar function and $\bar{v} = [v \ v^{(1)} \ \cdots \ v^{(r)}]^T$ is a generalized reference input ($v^{(k)}$ denotes the k-th time-derivative of the reference input v which is assumed to be a sufficiently smooth function of time), that enforce global asymptotic stability and asymptotic output tracking with attenuation of the effect of the uncertainty on the output in the closed-loop system, and are optimal with respect to a meaningful, infinite time cost functional that imposes penalty on the control action. In the second control problem, we assume that the uncertain variables are non-vanishing (i.e., $w_k(0)\theta_k \neq 0$). With this assumption, the origin is no longer an equilibrium point for the uncertain system. The objective here is to synthesize robust nonlinear state feedback controllers of the form of (3.10) that guarantee global boundedness of the trajectories, enforce the discrepancy between the output and the reference input to be asymptotically arbitrarily small, and minimize a meaningful cost functional defined over a finite time interval. In both control problems, the controller is designed using combination of geometric and Lyapunov techniques. In particular, the design is carried out by combining a Sontag-like control law with a robust control design proposed in [58]. The analysis of the closed-loop system is performed by utilizing the concept of input-to-state stability and nonlinear small gain theorem-type arguments, and optimality is established using the inverse optimal control approach.

3.3.2 Controller Design Under Vanishing Uncertainties

In order to proceed with the synthesis of the controllers, we will impose the following assumptions on the system of (3.1). The first assumption allows transforming the system of (3.1) into a partially linear form and is motivated by the requirement of output tracking.

Assumption 3.1 *There exists an integer r and a set of coordinates:*

$$\begin{bmatrix} \zeta \\ \eta \end{bmatrix} = \begin{bmatrix} \zeta_1 \\ \zeta_2 \\ \vdots \\ \zeta_r \\ \eta_1 \\ \vdots \\ \eta_{n-r} \end{bmatrix} = T(x) = \begin{bmatrix} h(x) \\ L_f h(x) \\ \vdots \\ L_f^{r-1} h(x) \\ T_1(x) \\ \vdots \\ T_{n-r}(x) \end{bmatrix} \tag{3.11}$$

where $T_1(x), \ldots, T_{n-r}(x)$ are scalar functions such that the system of (3.1) takes the form:

$$\begin{aligned} \dot{\zeta} &= A\zeta + bL_f^r h(T^{-1}(\zeta,\eta)) + bL_g L_f^{r-1} h(T^{-1}(\zeta,\eta))u \\ &\quad + b\sum_{k=1}^{q} L_{wk} L_f^{r-1} h(T^{-1}(\zeta,\eta))\theta_k \end{aligned} \tag{3.12}$$

$$\dot{\eta} = \Psi(\zeta,\eta,\theta)$$

where Ψ is an $(n-r) \times 1$ nonlinear vector function,

$$A = \begin{bmatrix} 0 & 1 & 0 & \cdots & 0 \\ 0 & 0 & 1 & \cdots & 0 \\ \vdots & & & & \vdots \\ 0 & 0 & 0 & \cdots & 1 \\ 0 & 0 & 0 & \cdots & 0 \end{bmatrix}, \quad b = \begin{bmatrix} 0 \\ 0 \\ \vdots \\ 1 \end{bmatrix} \tag{3.13}$$

A is an $r \times r$ matrix, b is an $r \times 1$ vector, and $L_g L_f^{r-1} h(x) \neq 0$ for all $x \in \mathrm{IR}^n$. Moreover, for each $\theta \in \mathrm{IR}^q$, the states ζ, η are bounded if and only if the state x is bounded; and $(\zeta, \eta) \to (0, 0)$ if and only if $x \to 0$.

Assumption 3.1 includes the matching condition of the proposed robust control method; loosely speaking, it states that the uncertain variables cannot have a stronger effect on the controlled output than the manipulated input. We note that the change of variables of (3.11) is independent of θ (this is because the vector field, $g(x)$, that multiplies the input u is independent of θ) , and is invertible, since, for every x, the variables ζ, η are uniquely determined by (3.11). Introducing the notation $e = [e_1 \; e_2 \; \cdots \; e_r]^T$ where $e_i = \zeta_i - v^{(i-1)}$, $i = 1, \ldots, r$, the system of (3.12) can be re-written as:

$$\begin{aligned} \dot{e} &= \bar{f}(e, \eta, \bar{v}) + \bar{g}(e, \eta, \bar{v})u + \sum_{k=1}^{q} \bar{w}_k(e, \eta, \bar{v})\theta_k \\ \dot{\eta} &= \Psi(e, \eta, \theta, \bar{v}) \end{aligned} \tag{3.14}$$

where $\bar{f}(\cdot) = Ae + b\left(L_f^r h(T^{-1}(e, \eta, \bar{v})) - v^{(r)}\right)$, $\bar{g}(\cdot) = bL_g L_f^{r-1} h(T^{-1}(e, \eta, \bar{v}))$, $\bar{w}_k(\cdot) = bL_{wk} L_f^{r-1} h(T^{-1}(e, \eta, \bar{v}))$.

Following [58], the requirement of input-to-state stability of the η subsystem of (3.14) is imposed to allow the synthesis of a robust state feedback controller that enforces the requested properties in the closed-loop system for any initial condition. Assumption 3.2 that follows states this requirement.

Assumption 3.2 *The dynamical system:*

$$\dot{\eta} = \Psi(e, \eta, \theta, \bar{v}) \tag{3.15}$$

is ISS with respect to e uniformly in θ, $\bar{v}$.

In most practical applications, there exists partial information about the uncertain terms of the process model. Information of this kind may result from physical considerations, preliminary simulations, experimental data, etc. In order to achieve attenuation of the effect of the uncertain variables on the output, we will quantify the possible knowledge about the uncertain variables by assuming the existence of known bounds that capture the size of the uncertain variables for all times.

Assumption 3.3 *There exist known positive constants θ_{bk} such that $\|\theta_k(t)\|^s \leq \theta_{bk}$.*

In order to address the robust controller synthesis problem on the basis of the e-subsystem of (3.14), we need to construct an appropriate robust control Lyapunov function. This can be done in many different ways. One way, for example, is to use a quadratic function, $V = e^T Pe$, where P is a positive-definite matrix chosen to satisfy the following Ricatti equation:

$$A^T P + PA - PBB^T P = -Q \tag{3.16}$$

for some positive-definite matrix, Q. Computing the time-derivative of V along the trajectories of the e-subsystem of (4.6), and using the relation of (3.16), it can be shown that:

$$\begin{aligned}
\inf_{u\in\mathcal{U}} \sup_{\theta\in\mathcal{W}^q} \dot{V} &= \inf_{u\in\mathcal{U}} \sup_{\theta\in\mathcal{W}^q} \left[L_{\bar{f}}V + L_{\bar{g}}Vu + \sum_{k=1}^{q} L_{\bar{w}_k} V\theta_k \right] \\
&= \inf_{u\in\mathcal{U}} \sup_{\theta\in\mathcal{W}^q} \left[-e^T Qe + e^T Pb \left(b^T Pe + 2r(e,\eta,\bar{v}) + 2L_g L_f^{r-1} h(x)u \right.\right. \\
&\quad \left.\left. +2\sum_{k=1}^{q} L_{w_k} L_f^{r-1} h(x)\theta_k \right) \right]
\end{aligned} \tag{3.17}$$

where $r = L_f^r h(x) - v^{(r)}$, $L_{\bar{f}}V = -e^T Qe + e^T Pbb^T Pe + 2e^T Pbr$, $L_{\bar{g}}V = 2e^T PbL_g L_f^{r-1} h(x)$, and $L_{w_k}V = 2e^T PbL_{w_k} L_f^{r-1} h(x)$. Therefore, choosing P to satisfy (3.16) guarantees that when $L_{\bar{g}}V = 0$, we have $\inf_{u\in\mathcal{U}} \sup_{\theta\in\mathcal{W}^q} \dot{V} < 0$ for all $e \neq 0$.

Theorem 3.3 below provides a formula of the robust state feedback controller and states precise conditions under which the proposed controller enforces the desired properties in the closed-loop system. The proof of this theorem is given in Appendix A.

Theorem 3.3. *Consider the uncertain nonlinear system of (3.1), for which Assumptions 3.1-(3.3) hold, under the feedback control law:*

$$\begin{aligned}
u &= p(x, c_0, \rho, \chi, \phi, \theta_{bk}, \bar{v}) \\
&= -k(x, c_0, \rho, \chi, \phi, \theta_{bk}, \bar{v}) L_{\bar{g}}V
\end{aligned} \tag{3.18}$$

where

$$k(\cdot) = \begin{cases} c_0 + r_s(x) + r_d(x,\rho,\chi,\phi,\theta_b) & , L_{\bar{g}}V \neq 0 \\ c_0 + r_c(x,\rho,\chi,\phi,\theta_{bk}) & , L_{\bar{g}}V = 0 \end{cases} \tag{3.19}$$

and

$$r_s(x) = \left(\frac{L_{\bar{f}}V + \sqrt{(L_{\bar{f}}V)^2 + (L_{\bar{g}}V)^4}}{(L_{\bar{g}}V)^2} \right) \tag{3.20}$$

$$r_d(x,\rho,\chi,\phi,\theta_b) = \left(\frac{\rho + \chi \sum_{k=1}^{q} \theta_{bk} \| L_{wk} L_f^{r-1} h(x) \|}{(L_g L_f^{r-1} h(x))^2 (\frac{\|L_{\bar{g}}V\|}{\|L_g L_f^{r-1} h(x)\|} + \phi)} \right) \tag{3.21}$$

$$r_c(x, \rho, \chi, \phi, \theta_{bk}) = \left(\frac{\rho + \chi \sum_{k=1}^{q} \theta_{bk} \| L_{w_k} L_f^{r-1} h(x) \|}{(L_g L_f^{r-1} h(x))^2 \phi} \right) \tag{3.22}$$

where $V = e^T P e$, P is a positive-definite matrix that satisfies $A^T P + PA - Pbb^T P < 0$, and c_0, ρ, χ, and ϕ are adjustable parameters that satisfy $c_0 > 0$, $\rho > 0$, $\chi > 2$, and $\phi > 0$. Furthermore, assume that the uncertain variables in (3.1) are vanishing in the sense that there exists positive real numbers δ_k such that $\|\bar{w}_k(e, \eta, \bar{v})\| \le \delta_k \|2b^T P e\| \; \forall \; e \in D$, where $D = \{e \in \mathbb{R}^r : \|2b^T P e\| \le \phi(\frac{1}{2}\chi - 1)^{-1}\}$. Then for any initial condition, there exists $\phi^* > 0$ such that if $\phi \le \phi^*$, the following holds:
(1) The origin of the closed-loop system is asymptotically stable.
(2) The output of the closed-loop system satisfies a relation of the form:

$$\limsup_{t \to \infty} \|y(t) - v(t)\| = 0 \tag{3.23}$$

(3) The control law of (3.18–3.21) minimizes the cost functional:

$$J_v = \int_0^\infty (l(e) + uR(x)u)dt \tag{3.24}$$

where $R(x) = \frac{1}{2}[c_0 + r_s(x) + r_d(x, \rho, \chi, \phi, \theta_b)]^{-1} > 0 \;\forall\; x \in \mathbb{R}^n$, $l(e) = -L_{\bar{f}} V + \frac{1}{4} L_{\bar{g}} V R^{-1}(x) L_{\bar{g}} V - \sum_{k=1}^{q} \|L_{w_k} V\| \theta_{bk} \ge k \|2b^T P e\|^2$ for some real number, $k > 0$, and the minimum cost is $J_v^* = V(e(0))$.

Remark 3.4. Referring to the controller of (3.18–3.21) (when $L_{\bar{g}} V \neq 0$), one can observe that it is comprised of two components. The first component:

$$u_s = -[c_0 + r_s(x)] L_{\bar{g}} V \tag{3.25}$$

is responsible for achieving stabilization and reference-input tracking in the nominal closed-loop system (i.e., with $\theta(t) \equiv 0$). This component, with $c_0 = 0$, is the so-called Sontag's formula proposed in [253]. The second component of the controller:

$$u_r = -r_d(x, \rho, \chi, \phi, \theta_b) L_{\bar{g}} V \tag{3.26}$$

enforces, on the other hand, output tracking with an arbitrary degree of asymptotic attenuation of the effect of θ on y. This component is also continuous everywhere since $L_g L_f^{r-1} h(x) \neq 0$ for all x (from Assumption 3.1) and $\phi > 0$. A key feature of the uncertainty compensator of (3.26) is the presence of a scaling term of the form $\dfrac{\|x\|}{\|x\| + \phi}$ which allows us, not only to tune the degree of uncertainty attenuation (by adjusting ϕ), but also to use smaller and smaller control effort to cancel the uncertainties as the state gets closer and closer to the equilibrium point. The full weight (gain) of the compensator is used only when the state is far from the equilibrium point (since ϕ is small).

Remark 3.5. Since the term $r(x)$ of (3.20) is undefined when $L_{\bar{g}}V = 0$, the $L_{\bar{g}}V$ controller gain, $k(\cdot)$, must be set to some value at those points where $L_{\bar{g}}V = 0$. Choosing $k(\cdot) = c_0 + r_c(x)$ when $L_{\bar{g}}V = 0$ ensures that the controller gain is continuous when $L_{\bar{g}}V = 0$ (note that $r_c(x)$ is the limit of $r_d(x)$ as $L_{\bar{g}}V$ approaches zero). Continuity of the controller gain ensures that the weight on the control penalty, $R(x)$, in the cost functional of (3.24) is bounded on compact sets, which is necessary for the optimality properties of the controller to be meaningful (see Remark 3.8 below).

Remark 3.6. The re-shaping of the scalar gain of the nominal L_gV controller of (3.25), via the addition of the uncertainty compensator of (3.26), allows us to achieve attenuation of the effect of θ on y without using unreasonably large (high-gain) control action. The use of a high-gain controller of the form $u = \dfrac{1}{\epsilon}u_s$ where ϵ is a small positive parameter could also lead to uncertainty attenuation at the expense of employing unnecessarily large control action (see the example in Sect. 3.3.4 for a numerical verification of this fact).

Remark 3.7. The control law of (3.18–3.21) possesses four adjustable parameters that directly influence the performance of the output of the closed-loop system and the achievable level of uncertainty attenuation. The positive parameter c_0 allows a certain degree of flexibility in shaping the dynamic behavior of the closed-loop output response as desired. To see this, we note that by computing the time-derivative of V along the trajectories of the closed-loop system, the following estimate can be obtained (see Part 1 of the Proof of Theorem 3.3):

$$\dot{V} \leq -c_0(L_{\bar{g}}V)^2 - \sqrt{(L_{\bar{f}}V)^2 + (L_{\bar{g}}V)^4} \tag{3.27}$$

From the above equation, it is transparent that the choice of c_0 affects how negative the time-derivative of V is along the closed-loop trajectories and can therefore be made so as to make this derivative more or less negative as requested. For example, a large value for c_0 will make $\dot{V}$ more negative and therefore generate a faster transient response. Note that a positive value of c_0 is not necessary for stabilization (since $\dot{V}$ is negative-definite even when $c_0 = 0$ as can be seen from (3.27)); however, a positive value of c_0 helps both in shaping $\dot{V}$ and in ensuring the positivity of the penalty weights in the cost functional of (3.24) (see Part 2 of the Proof of Theorem 3.3 for the detailed calculations). The positive parameter ρ also helps in ensuring the positivity of the penalty weights; however, its central role is to provide an additional robustness margin that increases the negativity of $\dot{V}$ and, thus, provides a sufficient counterbalance to the undesirable effect of the uncertainty on $\dot{V}$ to guarantee global asymptotic stability. As can be seen from the calculations in Part 1 of the Proof of Theorem 3.3, a zero value of ρ guarantees only boundedness of the closed-loop state (but not asymptotic stability). The remaining two parameters, χ and ϕ, are primarily responsible for achieving the desired degree of attenuation of the effect of uncertainty on the closed-loop output.

A significant degree of attenuation can be achieved by selecting ϕ to be sufficiently small and/or χ to be sufficiently large. While robust stability can be achieved using any value of χ satisfying $\chi > 1$, a value satisfying $\chi > 2$ is necessary to ensure the positive-definiteness of the weight function, $l(e)$, in the cost functional of (3.24), and thus to ensure meaningful optimality.

Remark 3.8. Regarding the properties of the cost functional of (3.24), we observe that J_v is well-defined according to the formulation of the optimal control problem for nonlinear systems [245] because of the following properties of $l(e)$ and $R(x)$. The function $l(e)$ is continuous, positive-definite, and bounded from below by a class $\mathcal{K}$ function of the norm of e (see Part 2 of the Proof of Theorem 3.3). The function $R(x)$ is continuous, strictly positive, and the term $uR(x)u$ has a unique global minimum at $u = 0$ implying a larger control penalty for u farther away from zero. Moreover, owing to the vanishing nature of the uncertain variables and the global asymptotic stability of the origin of the closed-loop system, J_v is defined on the infinite-time interval and is finite.

Remark 3.9. Note that since $l(e) \geq k\|2b^T Pe\|^2$, where k is a positive constant (details of this calculation are given in the Part 2 of the Proof of Theorem 3.3 in the Appendix), the cost functional of (3.24) includes penalty on the tracking error $e = v - y$ and its time-derivatives up to order r, and not on the full state of the closed-loop system. This is consistent with the requirement of output tracking. The term $-\sum_{k=1}^{q}\|L_{wk}V\|\theta_{bk}$ in (3.24) is a direct consequence of the worst-case uncertainty formulation of the HJI equation (see (3.3)).

Remark 3.10. The linear growth bound on the functions $\bar{w}_k(e, \eta, \bar{v})$ is imposed to ensure that the results of Theorem 3.3 hold globally. Note, however, that this condition is required to hold only over the set D (containing the origin) whose size can be made arbitrarily small by appropriate selection of the tuning parameters, ϕ and χ. This bound is consistent with the vanishing nature of the uncertain variables considered in this section, since it implies that the functions $\bar{w}_k(e, \eta, \bar{v})$ vanish when $e = 0$. We observe that locally (i.e., in a sufficiently small compact neighborhood of the origin), this growth condition is automatically satisfied due to the local Lipschitz properties of the functions $\bar{w}_k$. Finally, we note that the imposed growth assumption can be readily relaxed via a slight modification of the controller synthesis formula. One such modification, for example, is to replace the term $\|L_{wk}L_f^{r-1}h(x)\|$ in (3.18) by the term $\|L_{wk}L_f^{r-1}h(x)\|^2$. Using a standard Lyapunov argument, one can show that the resulting controller globally asymptotically stabilizes the system without imposing the linear growth condition on the functions $\bar{w}_k$ over D. However, we choose not to modify the original controller design because such modification will possibly increase the gain of the controller, prompting the expenditure of unnecessarily large control action.

Remark 3.11. In contrast to feedback linearizing controllers, the controller of (3.18–3.21) has two desirable properties not present in the feedback linearizing design. The first property is the fact that the controller of (3.18–3.21) recognizes the beneficial effect of the term $L_{\bar{f}}V$ when $L_{\bar{f}}V < 0$ and prevents its unnecessary cancellation. In this case, the term $L_{\bar{f}}V$ is a stabilizing term whose cancellation may generate positive feedback and destabilize the process. Such a situation can arise whenever the process under consideration is open-loop stable or contains beneficial nonlinearities that help drive the process towards the desired steady-state. Under such circumstances, the term $(L_{\bar{f}}V)^2$ in (3.18–3.21) guards against the wasteful cancellation of such nonlinearities (by essentially deactivating the cancelling term $L_{\bar{f}}V$) and helps the controller avoid the expenditure of large control effort. This is particularly important when the process operating conditions start relatively far away from the equilibrium point in which case the feedback linearizing designs may generate a huge control effort to cancel process nonlinearities. The second property that the controller of (3.18–3.21) possesses is the fact that it dominates the term $L_{\bar{f}}V$, instead of cancelling it, when $L_{\bar{f}}V > 0$. This situation arises, for example, when the process is open-loop unstable. In this case, the term $L_{\bar{f}}V$ is a destabilizing one that must be eliminated. The controller of (3.18–3.21), however, eliminates the term by domination rather than by cancellation. This property guards against the non-robustness of cancellation designs which increases the risk of instability due to the presence of other uncertainty not taken into account in the controller design. This means that input perturbations (or, equivalently, an error in implementing the control law) will be tolerated by the design presented in Theorem 3.3 in the sense that the trajectories will remain bounded.

Remark 3.12. Referring to the practical implementation of the controller of (3.18–3.21), we note that even though the controller is continuous at the origin, it is not smooth at that point and, therefore, when implemented in simulations, the numerical integration could result in a chattering-like behavior for the control input near the origin (this problem arises because of the piecewise, or "non-smooth," form of the Sontag's formula). In practice, this problem can be circumvented via a slight modification of the control law, whereby a sufficiently small positive real number, ϵ, is added to the $(L_{\bar{g}}V)^2$ term in the denominator of Sontag's formula in order to smoothen out the control action. The addition of this parameter obviously destroys the asymptotic stability capability of the controller and leads to some offset in the closed-loop response. However, this offset can be made arbitrarily small by choosing ϵ sufficiently small (practical stability). Clearly, a tradeoff exists between the smoothness of the control action (which favors a relatively large ϵ) and the size of the offset (a small offset favors a relatively small ϵ). Therefore, in tuning the controller, ϵ should be chosen in a way that balances this tradeoff.

3.3.3 Controller Design Under Non-Vanishing Uncertainties

To proceed with the main results for the case of non-vanishing uncertain variables, we let Assumptions 3.1 and 3.3 hold and modify Assumption 3.2 to the following one.

Assumption 3.4 *The dynamical system of (3.15) is ISS with respect to* e, θ, *uniformly in* $\bar{v}$.

Theorem 3.13. *Let Assumptions 3.1, 3.3, and 3.4 hold and assume that the uncertain variables in (3.1) are non-vanishing in the sense that there exist positive real numbers* $\bar{\delta}_k$, μ_k *such that* $\|\bar{w}_k(e,\eta)\| \leq \bar{\delta}_k \|2b^T Pe\| + \mu_k \ \forall\, e \in D$. *Consider the uncertain nonlinear system of (3.1) under the control law:*

$$u = -p(x, \rho, \chi, \phi, \theta_{bk}, \bar{v}) \tag{3.28}$$

where $p(\cdot)$ *was defined in Theorem 3.3. Then for any initial condition and for every positive real number* d, *there exists* $\phi^*(d) > 0$ *such that if* $\phi \in (0, \phi^*(d)]$, *the following holds:*
(1) The trajectories of the closed-loop system are bounded.
(2) The output of the closed-loop system satisfies a relation of the form:

$$\limsup_{t\to\infty} \|y(t) - v(t)\| \leq d \tag{3.29}$$

(3) The control law of (3.18) minimizes the cost functional:

$$J_n = \lim_{t \to T_f} V(e(t)) + \int_0^{T_f} (\bar{l}(e) + u\bar{R}(x)u)dt \tag{3.30}$$

with a minimum cost, $J_n^* = V(e(0))$, *where* $\bar{l}(e(t)) = -L_{\bar{f}}V + \dfrac{1}{4}L_{\bar{g}}V\bar{R}^{-1}(x)$ $L_{\bar{g}}V - \sum_{k=1}^{q} \|L_{wk}V\|\theta_{bk} > 0 \ \ \forall\, t \in [0, T_f]$, $T_f = \inf\{T \geq 0 : e(t) \in \Gamma \ \forall\, t \geq T\}$, $\Gamma = \{e \in \mathbb{R}^r : e^T Pe \leq \lambda_{max}(P)\epsilon^2\}$, $\epsilon = \max\{\alpha_1^{-1}(2\bar{\phi}^*), \alpha_2^{-1}(2\bar{\phi}^*)\}$, $\bar{\phi}^* = \sum_{k=1}^{q} \mu_k \theta_{bk}\phi^*$, $\alpha_1(\cdot)$, $\alpha_2(\cdot)$ *are class* $\mathcal{K}$ *functions that satisfy, respectively,* $\alpha_1(\|e\|) \leq \sqrt{(L_{\bar{f}}V)^2 + (L_{\bar{g}}V)^4}$ *and* $\alpha_2(\|e\|) \leq \frac{1}{2}\left[-L_{\bar{f}}V + \sqrt{(L_{\bar{f}}V)^2 + (L_{\bar{g}}V)^4}\right]$, *and* $\bar{R}(x) = -\frac{1}{2}\left[c_0 + r_s(x) + r_d(x, \rho, \chi, \phi, \theta_b)\right]^{-1} > 0$.

Remark 3.14. Note that the cost functional of (3.30) differs from that of (3.24) in two ways. First, owing to the persistent nature of the uncertainty in this case ($\bar{w}_k(0, \eta, \bar{v}) \neq 0$) and the fact that asymptotic convergence to the equilibrium point of the nominal system is no longer possible, J_n cannot achieve a finite value over the infinite time interval. Therefore, the cost functional of (3.24) is defined only over a finite time interval $[0, T_f]$. The size of this interval is given by the time required for the tracking error trajectory to reach and enter a compact ball Γ, centered around the origin, without ever leaving again. In the Proof of Theorem 3.13, we show that $\dot{V} < 0$ outside this ball, which implies that – once inside – the closed-loop trajectory cannot escape. The size of Γ scales with $\phi(\chi - 1)^{-1}$ and, depending on the desired degree of attenuation of the effect of the uncertainty on the output, can be made arbitrarily small by adjusting the controller tuning parameters ϕ and χ. Therefore one can practically get as close as desired to the nominal equilibrium point. In addition to the terminal time, T_f, the cost functional of (3.30) imposes a terminal penalty given by the term $\lim_{t \to T_f} V(e)$ to penalize the tracking error trajectory for its inability to converge to the origin for $t > T_f$. Note that since the error trajectory of the closed-loop system converges to the ball Γ as time tends to T_f, the cost functional of (3.30) is meaningful in the sense that $\bar{l}(e) \geq \bar{k}\|2b^T Pe\|^2$ and $\bar{R}(x) > 0 \;\; \forall \, t \in [0, T_f]$.

Remark 3.15. It is important to compare the controller design proposed in Theorem 3.13 with the following robust nonlinear controller based on the design proposed in [58]:

$$u = \frac{1}{L_g L_f^{r-1} h(x)} \sum_{i=0}^{r} \frac{\beta_i}{\beta_r} (v^{(i)} - L_f^i h(x)) + u_r \tag{3.31}$$

where u_r was defined in (3.26), which is continuous (since $L_g L_f^{r-1} h(x) \neq 0$ and $\phi > 0$) and guarantees boundedness of the trajectories of the closed-loop system and output tracking with an arbitrary degree of asymptotic attenuation of the effect of θ on y. The controller of (3.31) consists of a term that cancels the nonlinearities of the system of (3.12) with $\theta(t) \equiv 0$ that can be cancelled by feedback, and a term that compensates for the effect of θ on y. While the control law of (3.31) can be shown to be optimal with respect to some performance index, it remains to be seen whether the index is meaningful or not. In particular, we note that the feedback linearizing component of the controller of (3.31) may generate unnecessarily large control action to cancel beneficial nonlinearities and, therefore, may not in general be optimal with respect to a meaningful performance index. Consequently, we use the normal form of (3.42) not for the design of the controller but only for the construction of V. The choice of $V = e^T Pe$ is made to simplify the explicit formula of the controller in the context of a specific application. Of course, this choice for V is not unique and many other positive-definite functions whose time-derivative along the trajectories of the closed-loop system can be rendered negative-definite via feedback could be used.

Remark 3.16. An alternative to the gain-reshaping procedure underlying the controller designs in Theorems 3.3 and 3.13 is the one proposed in [163] which involves incorporating the uncertainty compensator within (rather than adding it to) the "nominal" gain of Sontag's formula. The resulting L_gV controller in this case has a scalar gain that is structurally-similar to that of

Sontag's formula, except that the term L_fV appears with the uncertainty compensator added to it. In particular, when the compensator of (3.26) is used, this gain-reshaping approach yields the following control law:

$$u = -\left(\frac{L_{\bar{f}}^{*}V + \sqrt{(L_{\bar{f}}^{*}V)^2 + (L_{\bar{g}}V)^4}}{(L_{\bar{g}}V)^2}\right) L_{\bar{g}}V \tag{3.32}$$

where

$$L_{\bar{f}}^{*}V = L_{\bar{f}}V + \chi\sum_{k=1}^{q} \theta_{bk}\|L_{\bar{w}k}V\|\left(\frac{\|2Pe\|}{\|2Pe\| + \phi}\right) \tag{3.33}$$

Using calculations similar to those in the appendix, one can show that the above control law is also globally robustly stabilizing and inverse optimal with respect to a meaningful cost. However, a potential drawback of this design is the fact that it can limit the ability of the controller to fully "cancel out" the uncertainty, especially when $L_{\bar{f}}^{*}V$ is negative. This in turn can lead either to a level of asymptotic uncertainty attenuation smaller than that achieved by the controllers of (3.18) and (3.28), or to larger control action in order to enforce the same attenuation level. To illustrate this point, consider the following system:

$$\dot{x} = u + \theta \tag{3.34}$$

Since the system is scalar, we take $V = \frac{1}{2}x^2$ and get $L_gV = L_wV = x$ and . For this choice, the control law based on (3.28) is:

$$u = -\left(1 + \frac{\chi\theta_b}{\|x\| + \phi}\right) x \tag{3.35}$$

while the control law based on (3.32–3.33) is:

$$u = -\left(\frac{\chi\theta_b}{\|x\| + \phi} + \sqrt{1 + \left(\frac{\chi\theta_b}{\|x\| + \phi}\right)^2}\right) x \tag{3.36}$$

It is straightforward to verify that, for the same choice of tuning parameters ϕ and χ, both controllers enforce the same level of asymptotic uncertainty attenuation, with an ultimate bound on the state of $\|x\| \leq d$ where $d = \phi(\chi - 1)^{-1}$. However, it is also clear that the controller gain in (3.36) is higher, point-wise, than that of the control law in (3.35). The discrepancy between the two gains reflects the cost of incorporating the uncertainty compensator within a Sontag-type gain in (3.36) as opposed to adding it to the nominal gain of Sontag's formula as done in (3.35). Note that to reduce the controller gain in (3.36) requires an increase in ϕ and/or reduction in χ which, in either case, implies a lesser level of uncertainty attenuation (larger residual set).

Remark 3.17. Referring to the practical applications of the result of Theorem 3.13, one has to initially verify whether Assumptions 3.1 and 3.4 hold for the process under consideration. Then, given the bounds θ_{bk}, (3.28) can be used to compute the explicit formula for the controller. Finally, given the asymptotic tracking error desired, d, which can be chosen arbitrarily close to zero, the value of ϕ^* should be computed (usually through simulations) to achieve $\limsup_{t\to\infty}\|y(t) - v(t)\| \leq d$.

3.3.4 Illustrative Example

In order to illustrate the performance of the developed control law, we consider a continuous stirred tank reactor where an irreversible first-order exothermic reaction of the form $A \xrightarrow{k} B$ takes place. The inlet stream consists of pure A at flow rate $F = 100\ L/min$, concentration $C_{A0} = 1.0\ mol/L$ and temperature $T_{A0} = 300\ K$. Under standard modeling assumptions, the mathematical model for the process takes the form:

$$\begin{aligned}\frac{dC_A}{dt} &= \frac{F}{V}(C_{A0} - C_A) - k_0 \exp\left(\frac{-E}{RT}\right) C_A \\ \frac{dT}{dt} &= \frac{F}{V}(T_{A0} - T) + \frac{(-\Delta H_{nom})}{\rho_m c_p} k_0 \exp\left(\frac{-E}{RT}\right) C_A + \frac{Q}{\rho_m c_p V}\end{aligned} \tag{3.37}$$

where C_A denotes the reactant concentration, T denotes the reactor temperature, Q denotes the rate of heat input to the reactor, $V = 100\ L$ is the volume of the reactor, $k_0 = 7.0\ min^{-1}$, $E = 27.4\ KJ/mol$, $\Delta H_{nom} = 718\ KJ/mol$ are the pre-exponential constant, the activation energy, and the enthalpy of the reaction, $c_p = 0.239\ J/g \cdot K$ and $\rho_m = 1000\ g/L$, are the heat capacity and fluid density in the reactor. For these values of process parameters, the process (with $Q = 0$) has an unstable equilibrium point at $(C_{As}, T_s) = (0.731\ mol/L, 1098.5\ K)$. The control objective is the regulation of the reactant concentration at the (open-loop) unstable equilibrium point, in the presence of time-varying, persistent disturbances in the feed temperature and uncertainty in the heat of reaction, by manipulating the rate of heat input Q.

Defining the variables $x_1 = (C_A - C_{As})/C_{As}$, $x_2 = (T - T_s)/T_s$, $u = Q/(\rho_m c_p V T_s)$, $\theta_1(t) = T_{A0} - T_{A0s}$, $\theta_2(t) = \Delta H - \Delta H_{nom}$, $y = x_1$, where the subscript s denotes the steady-state values, it is straightforward to verify that the process model of (3.37) can be cast in the form of (3.1) with:

$$\begin{aligned}f(x) &= \begin{bmatrix} 0.368 - x_1 - \delta(x_1, x_2) \\ -0.736 - x2 + 2\delta(x_1, x_2) \end{bmatrix}, \quad g(x) = \begin{bmatrix} 0 \\ 1 \end{bmatrix}, \\ w_1(x) &= \begin{bmatrix} 0 \\ \\ \dfrac{1}{T_s} \end{bmatrix}, \quad w_2(x) = \begin{bmatrix} 0 \\ \dfrac{C_{As}}{\rho_m c_p T_s} \delta(x_1, x_2) \end{bmatrix}\end{aligned} \tag{3.38}$$

where $\delta(x_1, x_2) = (1 + x_1)\exp\left(2 - \dfrac{3}{x_2 + 1}\right)$, and has a relative degree $r = 2$. To simulate the effect of the uncertainty, we consider the time-varying functions $\theta_k(t) = \theta_{bk} sin(t),\ \ k = 1, 2$ with upper bounds $\theta_{b1} = 30\ K$ and $\theta_{b2} = 28.7\ KJ/mol$, respectively. To proceed with the design of the controller, we initially use the coordinate transformation $e_1 = x_1$, $e_2 = 0.368 - x_1 - \delta(x_1, x_2)$ to cast the input/output dynamics of the system of (3.37) in the following form:

$$\dot{e} = \bar{f}(e) + \bar{g}(e)u + \bar{w}_1(e)\theta_1 + \bar{w}_2(e)\theta_2 \tag{3.39}$$

where the functions $\bar{f}$, $\bar{g}$, $\bar{w}_1$, $\bar{w}_2$ can be computed directly from (3.14). Note that the uncertainties considered are non-vanishing and, therefore, we use the result of Theorem 3.13 to design the necessary controller which takes the form of (3.18) with $\rho = 0$, $q = 2$, $r = 2$, and $P = \begin{bmatrix} 1 & 0.45 \\ 0.45 & 1 \end{bmatrix}$. Moreover, the following values are used for the tuning parameters of the controller: $\phi = 0.0001$, $\chi = 1.6$, and $c_0 = 0.0001$ to guarantee that the output of the closed-loop system satisfies a relation of the form $\limsup_{t\to\infty} \|y - v\| \leq 0.005$.

Closed-loop simulations were performed to evaluate the robustness and optimality properties of the controller. The results are shown by the solid lines in Fig. 3.1, which depict the controlled output (top) and manipulated input (bottom) profiles. One can immediately see that the robust optimal controller drives the closed-loop output close to the desired steady-state, while attenuating the effect of the uncertainty on the output. Included also in Fig. 3.1 is the closed-loop output profile (top plot, dashed-dotted line) and corresponding input profile (bottom plot, dashed-dotted line) when the controller of (3.18) is implemented without the uncertainty compensation component ($\chi = 0$). It is clear from the result that the effect of uncertainty is significant and leads to poor transient performance and offset.

For the sake of comparison, we also considered the robust nonlinear controller of (3.31) (see Remark 3.6 above) as well as the nonlinear high-gain controller of the form $u = \dfrac{1}{\epsilon}[c_0 + r_s(x)]L_{\bar{g}}V$ where no uncertainty compensation is included and ϵ is a small positive number. The results for these two cases (starting from the same initial condition) are shown in Fig. 3.1 by the dashed and dotted lines, respectively, for a choice of the tuning parameters: $\beta_0 = 4$, $\beta_1 = 9$, $\beta_2 = 1$, $\epsilon = 0.05$. It is clear from the input profiles that, compared with these two controllers, the controller of (3.18) starts by requesting significantly smaller control action to achieve a relatively faster closed-loop response and enforce, asymptotically, the same level of attenuation of the effect of uncertainty on x_1. Finally, in order to establish the fact that the robust optimal controller of (3.18) does not use unnecessary control action compared with the controller of (3.31), we compared the control actions generated by the two controllers for a broad range of x_1 and x_2. The results are shown in Fig. 3.2 and clearly indicate that the controller of (3.18) (dashed lines) uses smaller control action, point-wise, than the controller of (3.31) (solid lines).

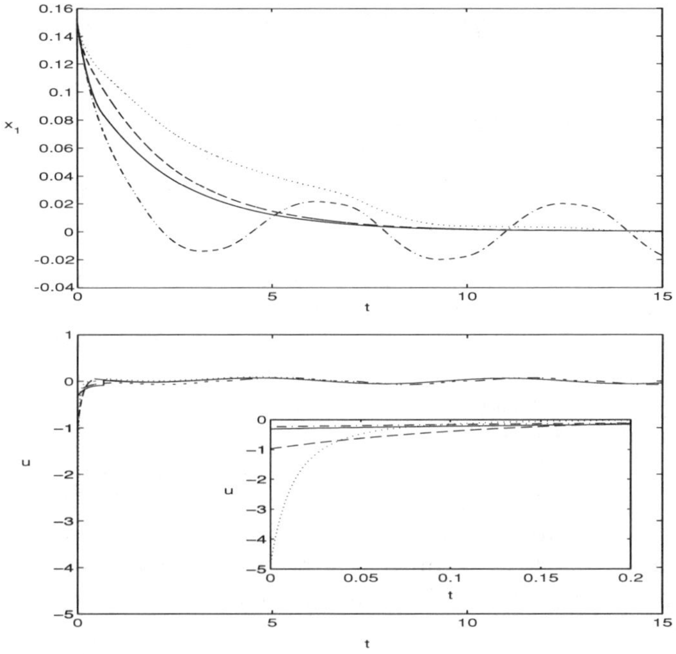

Fig. 3.1. Closed-loop output and manipulated input profiles under the controller of (3.18–3.21) (*solid lines*), the controller of (3.18–3.21) with $\chi = 0$ (*dashed-dotted lines*), the controller of (3.31) (*dashed lines*), and the high-gain controller (*dotted lines*)

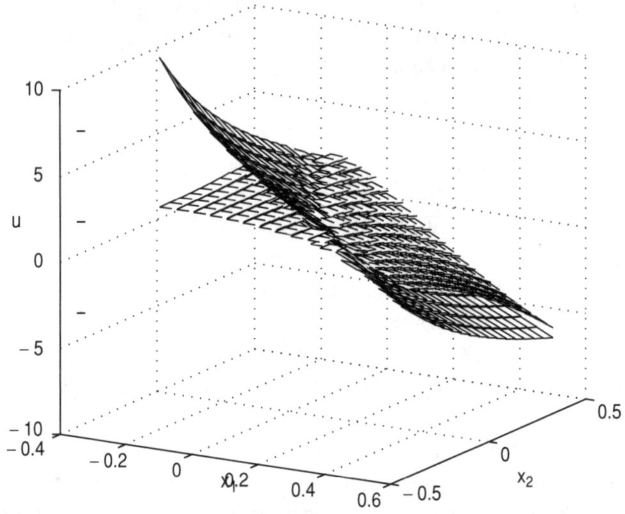

Fig. 3.2. Control action computed by the controller of (3.18–3.21) (*dashed lines*) and the controller of (3.31) (*solid lines*)

3.4 Robust Near-Optimal Output Feedback Controller Design

In Sect. 3.3, we addressed the problem of synthesizing robust optimal *state* feedback control laws for a large class of nonlinear systems. The practical applicability of these control laws, however, rests on the assumption of accessibility of all process states for measurement. In chemical process control – more so than in most other control areas – however, the complete state cannot be measured in general (e.g., concentrations of certain species are not accessible on-line) and therefore the state feedback controllers proposed in Chap. 3.3 may not be directly suited for practical applications. In the past few years, significant advances have been made in the direction of output feedback controller design for the purpose of robustly stabilizing nonlinear systems. Important contributions in this area include general results on robust output feedback stabilization using the controller-observer combination approach (e.g., [267,268]), adaptive output feedback control (e.g., [188,189]), and robust output feedback controller design for various classes of nonlinear systems (e.g., [50,146,183]). In addition to these approaches, other methods for constructing robust stabilizers via output feedback, which do not necessarily appeal to the principle of "composing a state feedback with estimates of the states" have been explored, including output feedback control within the H_∞ control framework (e.g., [127,129,130]) and the recursive stabilization scheme proposed in [128].

In this section, we address the problem of synthesizing robust near-optimal *output* feedback controllers for a broad class of nonlinear systems with time-varying bounded uncertain variables. Under the assumptions of input-to-state stable (ISS) inverse dynamics and vanishing uncertainty, a dynamic controller is synthesized through combination of a high-gain observer with a robust optimal state feedback controller designed via Lyapunov's direct method and shown through the inverse optimal approach to be optimal with respect to a meaningful cost defined on the infinite time-interval. The dynamic output feedback controller enforces exponential stability and asymptotic output tracking with attenuation of the effect of the uncertain variables on the output of the closed-loop system for initial conditions and uncertainty in arbitrarily large compact sets, as long as the observer gain is sufficiently large. Utilizing the inverse optimal control approach and standard singular perturbation techniques, this approach is shown to yield a near-optimal output feedback design in the sense that the performance of the resulting output feedback controller can be made arbitrarily close to that of the robust optimal state feedback controller on the infinite time-interval, when the observer gain is sufficiently large. For systems with non-vanishing uncertainty, the same controllers are shown to ensure boundedness of the states, robust asymptotic output tracking, and near-optimality over a finite time-interval. The developed controllers are successfully applied to a chemical reactor example.

3.4.1 Control Problem Formulation

Referring to the system of (3.1), we consider two control problems with different control objectives. In the first problem, we consider the class of systems described by (3.1) where the uncertain variable terms are assumed to be vanishing (i.e., $w_k(0)\theta_k = 0$ for any $\theta_k \in \mathcal{W}_k$ where the origin is the equilibrium point of the system $\dot{x} = f(x)$). In this case, the origin is also an equilibrium point for the uncertain system of (3.1). The objective in this problem is to synthesize robust nonlinear dynamic output feedback controllers of the form:

$$\begin{aligned} \dot{\omega} &= \mathcal{F}(\omega, y, \bar{v}) \\ u &= \mathcal{P}(\omega, y, \bar{v}, t) \end{aligned} \tag{3.40}$$

where $\omega \in \mathrm{I\!R}^s$ is a state, $\mathcal{F}(\omega, y, \bar{v})$ is a vector function, $\mathcal{P}(\omega, y, \bar{v}, t)$ is a scalar function, $\bar{v} = [v \ v^{(1)} \ \cdots \ v^{(r)}]^T$ is a generalized reference input ($v^{(k)}$ denotes the k-th time derivative of the reference input v, which is assumed to be a sufficiently smooth function of time), that: (a) enforce exponential stability in the closed-loop system, (b) guarantee asymptotic robust output tracking with an arbitrary degree of attenuation of the effect of the uncertainty on the output, and (c) are near-optimal with respect to a meaningful cost functional, defined over the infinite time-interval, that imposes penalty on the control action and the worst-case uncertainty. The near-optimality property, to be made more precise mathematically in Theorem 3.20 below, is sought in the sense that the performance of the dynamic output feedback controller can be made arbitrarily close to the optimal performance of the corresponding robust optimal state feedback controller over the infinite time-interval.

In the second control problem, we again consider the system of (3.1). In this case, however, we assume that the uncertain variable terms are non-vanishing (i.e., $w_k(0)\theta_k \neq 0$). Under this assumption, the origin is no longer an equilibrium point of the system of (3.1). The objective in this problem is to synthesize robust nonlinear dynamic output feedback controllers of the form of (3.40) that: (a) guarantee boundedness of the closed-loop trajectories, (b) enforce the discrepancy between the output and reference input to be asymptotically arbitrarily small, and (c) are near-optimal with respect to a meaningful cost defined over a finite time-interval, in the sense that their performance can be made arbitrarily close to the optimal performance of the corresponding robust optimal state feedback controllers over the same time-interval.

In both control problems, the design of the dynamic controllers is carried out using combination of a high-gain observer and the robust optimal state feedback controllers proposed in Sect. 3.3. In particular, referring to (3.40), the system $\dot{\omega} = \mathcal{F}(\omega, y, \bar{v})$ is synthesized to provide estimates of the system state variables, while the static component, $\mathcal{P}(\omega, y, \bar{v}, t)$, is synthesized to enforce the requested properties in the closed-loop system. The analysis of the closed-loop system employs standard singular perturbation techniques (due to the high-gain nature of the observer) and utilizes the concept of input-to-state stability (see Definition 3.2) and nonlinear small gain theorem-type arguments. Near-optimality is established through the inverse optimal approach (see Sect. 3.2) and using standard singular perturbation results. The requested closed-loop properties in both control problems are enforced for arbitrarily large initial conditions and uncertainty provided that the gain of the observer is sufficiently large (semi-global type result).

In order to proceed with the design of the desired output feedback controllers, we need to impose the following three assumptions on the system of (3.1). The first assumption is motivated by the requirement of output tracking and allows transforming the system of (3.1) into a partially linear form.

Assumption 3.5 *There exists an integer r and a set of coordinates:*

$$\begin{bmatrix} \zeta \\ \eta \end{bmatrix} = \begin{bmatrix} \zeta_1 \\ \zeta_2 \\ \vdots \\ \zeta_r \\ \eta_1 \\ \vdots \\ \eta_{n-r} \end{bmatrix} = \mathcal{X}(x) = \begin{bmatrix} h(x) \\ L_f h(x) \\ \vdots \\ L_f^{r-1} h(x) \\ \chi_1(x) \\ \vdots \\ \chi_{n-r}(x) \end{bmatrix} \tag{3.41}$$

where $\chi_1(x), \ldots, \chi_{n-r}(x)$ are nonlinear scalar functions of x, such that the system of (3.1) takes the form:

$$\begin{aligned} \dot{\zeta}_1 &= \zeta_2 \\ &\vdots \\ \dot{\zeta}_{r-1} &= \zeta_r \\ \dot{\zeta}_r &= L_f^r h(\mathcal{X}^{-1}(\zeta,\eta)) + L_g L_f^{r-1} h(\mathcal{X}^{-1}(\zeta,\eta))u + \sum_{k=1}^{q} L_{w_k} L_f^{r-1} h(\mathcal{X}^{-1}(\zeta,\eta))\theta_k \\ \dot{\eta}_1 &= \Psi_1(\zeta,\eta) \\ &\vdots \\ \dot{\eta}_{n-r} &= \Psi_{n-r}(\zeta,\eta) \\ y &= \zeta_1 \end{aligned} \tag{3.42}$$

where $L_g L_f^{r-1} h(x) \neq 0$ for all $x \in \mathbb{R}^n$, $\theta \in \mathbb{R}^q$. Moreover, for each $\theta \in \mathbb{R}^q$, the states ζ, η are bounded if and only if the state x is bounded; and $(\zeta, \eta) \to (0,0)$ if and only if $x \to 0$.

Remark 3.18. We note that the change of variables of (3.41) is independent of θ and invertible, since, for every x, the variables ζ, η are uniquely determined by (3.41). This implies that if we can estimate the values of ζ, η for all times, using appropriate state observers, then we automatically obtain estimates of x for all times. This property will be exploited later to synthesize a state estimator for the system of (3.1) on the basis of the system of (3.42). We also note that Assumption 3.5 includes the matching condition of our robust control method. In particular, we consider systems of the form (3.1) for which the uncertain variables enter the system in the same equation with the manipulated input. This assumption is more restrictive than Assumption 3.1 used in Sect. 3.3.2 to solve the same robust control problem via state feedback and is motivated by our requirement to eliminate the presence of θ in the η-subsystem of the system of (3.42). This requirement and the stability requirement of Assumption 3.6 below will allow including in the controller a replica of the η-subsystem of (3.42) which provides estimates of the η states (see Theorem 3.20 below).

Introducing the notation $e = [e_1 \; e_2 \; \cdots \; e_r]^T$ where $e_i = \zeta_i - v^{(i-1)}$, $i = 1, \ldots, r$, the system of (3.42) can be re-written as:

$$\begin{aligned} \dot{e} &= \bar{f}(e, \eta, \bar{v}) + \bar{g}(e, \eta, \bar{v})u + \sum_{k=1}^{q} \bar{w}_k(e, \eta, \bar{v})\theta_k \\ \dot{\eta} &= \Psi(e, \eta, \bar{v}) \end{aligned} \tag{3.43}$$

where $\bar{f}(\cdot) = Ae + b\left(L_f^r h(\mathcal{X}^{-1}(e, \eta, \bar{v})) - v^{(r)}\right)$, $\bar{g}(\cdot) = bL_g L_f^{r-1} h(\mathcal{X}^{-1}(e, \eta, \bar{v}))$, $\bar{w}_k(\cdot) = bL_{wk} L_f^{r-1} h(\mathcal{X}^{-1}(e, \eta, \bar{v}))$, the matrix A and column vector b are defined in (3.13). Following the same development presented in Sect. 3.3, we use the above normal form of (3.43) to construct a quadratic Lyapunov function, $V = e^T Pe$, for our controller design, where the positive-definite matrix P is chosen to satisfy the following Riccati inequality:

$$A^T P + PA - Pbb^T P < 0 \tag{3.44}$$

Assumption 3.6 *The system:*

$$\begin{aligned} \dot{\eta}_1 &= \Psi_1(e, \eta, \bar{v}) \\ &\vdots \\ \dot{\eta}_{n-r} &= \Psi_{n-r}(e, \eta, \bar{v}) \end{aligned} \tag{3.45}$$

is ISS with respect to e, uniformly in $\bar{v}$, with $\beta_\eta(\|\eta(0)\|, t) = K_\eta \|\eta(0)\| e^{-at}$ where $K_\eta \geq 1$, $a > 0$ are real numbers, and exponentially stable when $e = 0$.

Remark 3.19. Following [58], the requirement of input-to-state stability of the system of (3.45) with respect to ζ is imposed to allow the synthesis of a robust state feedback controller that enforces the requested properties in the closed-loop system for arbitrarily large initial conditions and uncertain variables. On the other hand, the requirement that $\beta_\eta(\|\eta(0)\|, t) = K_\eta e^{-at}$ allows incorporating in the robust output feedback controller a dynamical system identical to the one of (3.45) that provides estimates of the variables η. Assumption 3.6 is satisfied by many chemical processes (see, for example, [54] and the chemical reactor example of Sect. 3.4.4).

We are now in a position to state the main results of this chapter. In what follows we present the results for the case of vanishing uncertainty first, and then provide the parallel treatment for case of non-vanishing uncertainty.

3.4.2 Near-Optimality Over the Infinite Horizon

Theorem 3.20 below provides an explicit formula for the robust near-optimal output feedback controller and states precise conditions under which the proposed controller enforces the desired properties in the closed-loop system. The proof of this theorem is given in Appendix A.

Theorem 3.20. *Consider the uncertain nonlinear system of (3.1), for which Assumptions 3.3, 3.5 and 3.6 hold, under the robust output feedback controller:*

$$\begin{aligned}
\dot{\tilde{y}} &= \begin{bmatrix} -La_1 & 1\,0\cdots 0 \\ -L^2a_2 & 0\,1\cdots 0 \\ \vdots & \vdots\;\vdots\;\ddots\;\vdots \\ -L^{r-1}a_{r-1} & 0\,0\cdots 1 \\ -L^r a_r & 0\,0\cdots 0 \end{bmatrix}\tilde{y} + \begin{bmatrix} La_1 \\ L^2a_2 \\ \vdots \\ L^{r-1}a_{r-1} \\ L^r a_r \end{bmatrix} y \\
\dot{\omega}_1 &= \Psi_1(sat(\tilde{y}),\omega) \\
&\;\;\vdots \\
\dot{\omega}_{n-r} &= \Psi_{n-r}(sat(\tilde{y}),\omega) \\
u &= p(\hat{x},c_0,\rho,\chi,\phi,\theta_b,\bar{v})
\end{aligned} \tag{3.46}$$

where the parameters, $a_1, \cdots, a_r$ are chosen such that the polynomial $s^r + a_1 s^{r-1} + a_2 s^{r-2} + \cdots + a_1 = 0$ is Hurwitz and $p(\cdot)$ was defined in Theorem 3.3, $\hat{x} = \mathcal{X}^{-1}(sat(\tilde{y}),\omega))$, $V = e^T Pe$, P is a positive-definite matrix that satisfies (3.44), and c_0, ρ, χ, and ϕ are adjustable parameters that satisfy $c_0 > 0$, $\rho > 0$, $\chi > 2$, and $\phi > 0$. Furthermore, assume that the uncertain variables are vanishing in the sense defined in Theorem 3.3 and let $\epsilon = \frac{1}{L}$. Then, for each set of positive real numbers $\delta_x, \delta_\theta, \delta_{\bar{v}}$, there exists $\phi^* > 0$ and for each $\phi \in (0,\phi^*]$, there exists an $\epsilon^*(\phi) > 0$, such that if $\phi \in (0,\phi^*]$, $\epsilon \in (0,\epsilon^*(\phi)]$, $sat(\cdot) = \min\{1, \frac{\zeta_{max}}{\|\cdot\|}\}(\cdot)$ with ζ_{max} being the maximum value of the vector $[\zeta_1\ \zeta_2\ \cdots\ \zeta_r]$ for $\|\zeta\| \le \beta_\zeta(\delta_\zeta, 0)$ where β_ζ is a class $\mathcal{KL}$ function and δ_ζ is the maximum value of the vector $[h(x)\ L_f h(x)\ \cdots L_f^{r-1}h(x)]$ for $\|x\| \le \delta_x$, $\|x(0)\| \le \delta_x$, $\|\theta\|^s \le \delta_\theta$, $\|\bar{v}\|^s \le \delta_{\bar{v}}$, $\|\tilde{y}(0)\| \le \delta_\zeta$, $\omega(0) = \eta(0) + O(\epsilon)$, the

following holds:
(1) The origin of the closed-loop system is exponentially stable.
(2) The output of the closed-loop system satisfies a relation of the form:

$$\limsup_{t \longrightarrow \infty} \|y(t) - v(t)\| = 0 \tag{3.47}$$

(3) The output feedback controller of (3.46) is near-optimal with respect to the cost functional of (3.24) in the sense that $J^*_{v,o} \longrightarrow V(e(0))$ *as* $\epsilon \longrightarrow 0$*, where* $J^*_{v,o} = J_v(u = p(\hat{x}))$.

Remark 3.21. The robust output feedback controller of (3.46) consists of a high-gain observer which provides estimates of the derivatives of the output y up to order $r-1$, denoted as $\tilde{y}_0, \tilde{y}_1, \ldots, \tilde{y}_{r-1}$, and thus estimates of the variables $\zeta_1, \ldots, \zeta_r$ (note that from Assumption 3.5 it follows directly that $\zeta_k = \dfrac{d^{k-1}y}{dt^{k-1}}, k = 1, \ldots, r$), an observer that simulates the inverse dynamics of the system of (3.42), and a static state feedback controller (see discussion in remark 3.2 below) that attenuates the effect of the uncertain variables on the output and enforces reference input tracking. To eliminate the peaking phenomenon associated with the high-gain observer, we use a standard saturation function, *sat*, to eliminate wrong estimates of the output derivatives for short times. We choose to saturate $\tilde{y}_0, \tilde{y}_1, \ldots, \tilde{y}_{r-1}$ instead of the control action as was proposed in [149], because in most practical applications, it is possible to use knowledge of process operating conditions to derive nonconservative bounds on the actual values of the output derivatives.

Remark 3.22. The static component of the controller of (3.46) is the same as that given in Theorem 3.3, except that it is now a function of the state estimate and not the actual state. The first component is a generalization of Sontag's formula proposed in [253] where the parameter c_0 was introduced to ensure the strict positivity of $R(x)$. Recall from Theorem 3.3 that the static component of the feedback controller of (3.18) (with $\hat{x} = x$) is optimal with respect to a meaningful cost of the form of (3.24) and that the minimum cost achieved under the state feedback controller is $V(e(0))$.

Remark 3.23. The near-optimality property established in Theorem 3.20 for the output feedback controller of (3.46) is defined in the sense that the infinite-horizon cost incurred by implementing this controller on the system of (3.1) tends to the minimum cost achieved by implementing the robust optimal state feedback controller (i.e., u of (3.46) with $\hat{x} = x$) when the gain of the observer is chosen to be sufficiently large. It is important therefore to realize that near-optimality of the controller of (3.46) is a consequence of two integral factors. The first is the optimality of the static component of the controller (state feedback problem), and the second is the high-gain nature of the observer which can be exploited to make the performance of the output feedback controller arbitrarily close to that of the state feedback controller over the infinite time-interval. Instrumental in this regard is the use of the saturation function (see

Remark 3.21) which allows the use of arbitrarily large values of the observer gain to achieve the desired degree of near-optimality without the detrimental effects of observer peaking. Combining other types of observers that do not possess these properties with the static component of (3.46), therefore, will not lead to a near-optimal feedback controller design in the sense of Theorem 3.20.

Remark 3.24. A stronger near-optimality result than the one established in Theorem 3.20 can be obtained in a sufficiently small neighborhood of the origin of the system of (3.1), where the functions $l(e)$ and $R(x)$ satisfy the Lipschitz property. Over such a neighborhood, it can be shown, using the local Lipschitz properties, that the cost computed using the output feedback controller of (3.46) is, in fact, $O(\epsilon)$ close to the optimal cost attained under the optimal state feedback controller (i.e. $J_v^* = V(e(0)) + O(\epsilon)$).

Remark 3.25. Owing to the presence of the fast (high-gain) observer in the dynamical system of (3.46), the closed-loop system can be cast as a two time-scale system and, therefore, represented in the following singularly perturbed form, where $\epsilon = \dfrac{1}{L}$ is the singular perturbation parameter:

$$\begin{aligned}
\epsilon \dot{e}_o &= Ae_o + \epsilon b\Omega(x, \hat{x}, \theta, \phi, \bar{v}) \\
\dot{\omega}_1 &= \Psi_1(sat(\tilde{y}), \omega) \\
&\vdots \\
\dot{\omega}_{n-r} &= \Psi_{n-r}(sat(\tilde{y}), \omega) \\
\dot{x} &= f(x) + g(x)\mathcal{A}(\hat{x}, \bar{v}, \phi) + \sum_{k=1}^{q} w_k(x)\theta_k(t)
\end{aligned} \tag{3.48}$$

where e_o is a vector of the auxiliary error variables $\hat{e}_i = L^{r-i}(y^{(i-1)} - \tilde{y}_i)$ (see part 1 of the proof of Theorem 3.20 in the appendix for the details of the notation used). It is clear from the above representation that, within the singular perturbation formulation, the observer states, e_o, which are directly related to the estimates of the output and its derivatives up to order $r-1$, constitute the fast states of the singularly perturbed system of (3.48), while the ω states of the observer and the states of the original system of (3.1) represent the slow states. Owing to the dependence of the controller of (3.46) on both the slow and fast states, the control action computed by the static component in (3.46) is not $O(\epsilon)$ close to that computed by the state feedback controller for all times. After the decay of the boundary layer term e_o, however, the static component in (3.46) approximates the state feedback controller to within $O(\epsilon)$.

It is important to point out that the result of Theorem 3.20 is novel even in the case of no plant-model mismatch, where $\theta(t) \equiv 0$. The following corollary states the result for this case:

Corollary 3.26. *Suppose the assumptions of Theorem 3.20 are satisfied for the nonlinear system of (3.1) with $\theta_k(t) \equiv 0$, then the closed-loop system properties (1)-(3) given in Theorem 3.20 hold under the output feedback controller of (3.46) with $\rho = \chi = 0$.*

3.4.3 Near-Optimality Over the Finite Horizon

We now turn to the second control problem posed earlier where the uncertain variable terms in (3.1) are non-vanishing. To this end, we modify Assumption 3.6 to the following one.

Assumption 3.7 *The system of (3.45) is ISS with respect to e, uniformly in $\bar{v}$, with $\beta_\eta(\|\eta(0)\|, t) = K_\eta \|\eta(0)\| e^{-at}$ where $K_\eta \geq 1$, $a > 0$ are real numbers.*

Theorem 3.27 below provides an explicit formula for the construction of the necessary robust near-optimal output feedback controller and states precise conditions under which the proposed controller enforces the desired properties in the closed-loop system. The proof of this theorem is given in Appendix A.

Theorem 3.27. *Consider the uncertain nonlinear system of (3.1), for which Assumptions 3.3, 3.5 and 3.7 hold, under the robust output feedback controller:*

$$
\begin{aligned}
\dot{\tilde{y}} &= \begin{bmatrix} -La_1 & 1\,0 \cdots 0 \\ -L^2 a_2 & 0\,1 \cdots 0 \\ \vdots & \vdots\;\vdots\;\ddots\;\vdots \\ -L^{r-1} a_{r-1} & 0\,0 \cdots 1 \\ -L^r a_r & 0\,0 \cdots 0 \end{bmatrix} \tilde{y} + \begin{bmatrix} La_1 \\ L^2 a_2 \\ \vdots \\ L^{r-1} a_{r-1} \\ L^r a_r \end{bmatrix} y \\
\dot{\omega}_1 &= \Psi_1(sat(\tilde{y}), \omega) \\
&\;\;\vdots \\
\dot{\omega}_{n-r} &= \Psi_{n-r}(sat(\tilde{y}), \omega) \\
u &= p(\hat{x}, c_0, \rho, \chi, \phi, \theta_b)
\end{aligned}
\tag{3.49}
$$

where the parameters, a_1, $\cdots$, a_r are chosen such that the polynomial $s^r + a_1 s^{r-1} + a_2 s^{r-2} + \cdots + a_1 = 0$ is Hurwitz and $p(\cdot)$ was defined in Theorem 3.3. Furthermore, assume that the functions $\bar{w}_k(\zeta, \eta)$ are non-vanishing in the sense defined in Theorem 3.13 and let $\epsilon = \frac{1}{L}$. Then, for each set of positive real numbers δ_x, δ_θ, $\delta_{\bar{v}}$, d, there exists $\phi^* > 0$ and for each $\phi \in (0, \phi^*]$, there exists an $\epsilon^*(\phi) > 0$, such that if $\phi \in (0, \phi^*]$, $\epsilon \in (0, \epsilon^*(\phi)]$, $sat(\cdot) = \min\{1, \frac{\zeta_{max}}{\|\cdot\|}\}(\cdot)$ with ζ_{max} being the maximum value of the vector $[\zeta_1\ \zeta_2\ \cdots\ \zeta_r]$ for $\|\zeta\| \leq \beta_\zeta(\delta_\zeta, 0) + d$ where β_ζ is a class $\mathcal{KL}$ function and δ_ζ is the maximum value of the vector $[h(x)\ L_f h(x)\ \cdots L_f^{r-1} h(x)]$ for $\|x\| \leq \delta_x$, $\|x(0)\| \leq \delta_x$, $\|\theta\|^s \leq \delta_\theta$, $\|\bar{v}\|^s \leq \delta_{\bar{v}}$, $\|\tilde{y}(0)\| \leq \delta_\zeta$, $\omega(0) = \eta(0) + O(\epsilon)$, the following holds:
(1) The trajectories of the closed-loop system are bounded.
(2) The output of the closed-loop system satisfies a relation of the form:

$$\limsup_{t\to\infty}\|y(t) - v(t)\| \le d \tag{3.50}$$

(3) The output feedback controller of (3.49) is near-optimal with respect to the cost functional of (3.30) in the sense that $J^*_{n,o} \longrightarrow V(e(0))$ *as* $\epsilon \longrightarrow 0$*, where* $J^*_{n,o} = J_n(u = \tilde{p}(\hat{x}))$*.*

Remark 3.28. Note that owing to the persistent nature of the uncertainty in this case, asymptotic convergence of the closed-loop system to the equilibrium point of the nominal system (i.e. $\theta(t) \equiv 0$) is no longer possible. Therefore, the cost functional J cannot achieve a finite value over the infinite time-interval. Instead, J is defined over a finite time-interval $[0, T_f]$ whose size is determined by the time required for the tracking error trajectory to reach and enter the ball Γ without ever leaving again. Depending on the desired degree of uncertainty attenuation, the size of this ball can be tuned by adjusting the parameters ϕ and χ. An additional consequence of the non-vanishing uncertainty is the appearance of the terminal penalty term $\lim_{t\to T_f} V(e(t))$ in the cost functional of (3.30) which accounts for the fact that the closed-loop trajectories do not converge asymptotically to the origin for $t > T_f$.

Remark 3.29. In contrast to the output feedback controller design of Theorem 3.20, the controller design of Theorem 3.27 is near-optimal in the sense that the cost incurred by implementing this controller to the system of (3.1) can be made arbitrarily close to the optimal cost achieved by the corresponding robust optimal state feedback controller, for times in the finite interval $[0, T_f]$, by choosing the gain of the observer to be sufficiently large. As a consequence of the non-vanishing uncertainty, near-optimality can be established on the finite time-interval only and not for all times.

Remark 3.30. Referring to the practical applications of the result of Theorem 3.27, one has to initially verify whether Assumptions 3.5 and 3.7 hold for the process under consideration and compute the bounds θ_{bk}. Then, (3.46) can be used to compute the explicit formula of the controller. Finally, given the ultimate uncertainty attenuation level desired, d, the values of ϕ^* and the observer gain L should be computed (usually through simulations) to achieve $\limsup_{t\to\infty}\|y(t) - v(t)\| \le d$ and guarantee the near-optimal performance of the controller.

3.4.4 Application to a Non-Isothermal Chemical Reactor

Process Description and Control Problem Formulation

To illustrate the implementation and performance of the output feedback controller design presented in the previous subsection, we consider a well-mixed continuous stirred tank reactor where three parallel irreversible elementary endothermic reactions of the form $A \xrightarrow{k_1} D$, $A \xrightarrow{k_2} U$ and $A \xrightarrow{k_3} R$ take place, where A is the reactant species, D is the desired product and U, R are undesired byproducts. The feed to the reactor consists of pure A at flow rate F, molar concentration C_{A0} and temperature T_{A0}. Due to the endothermic nature of the reactions, a heating jacket is used to provide heat to the reactor. Under standard modeling assumptions, a mathematical model of the process can be derived from material and energy balances and takes the following form:

$$
\begin{aligned}
V\frac{dC_A}{dt} &= F(C_{A0} - C_A) - \sum_{i=1}^{3} k_{i0} \exp\left(\frac{-E_i}{RT}\right) C_A V \\
V\frac{dC_D}{dt} &= -FC_D + k_{10} \exp\left(\frac{-E_1}{RT}\right) C_A V \\
V\frac{dT}{dt} &= F(T_{A0} - T) + \sum_{i=1}^{3} \frac{(-\Delta H_i)}{\rho c_p} k_{i0} \exp\left(\frac{-E_i}{RT}\right) C_A V + \frac{UA}{\rho c_p}(T_c - T)
\end{aligned}
\tag{3.51}
$$

where C_A and C_D denote the concentrations of the species A and D, T denotes the temperature of the reactor, T_c denotes the temperature of the heating jacket, V denotes the volume of the reactor, ΔH_i, k_{i0}, E_i, $i = 1, 2, 3$, denote the enthalpies, pre-exponential constants and activation energies of the three reactions, respectively, c_p and ρ denote the heat capacity and density of the reactor, and U denotes the heat transfer coefficient between the reactor and the heating jacket. The values of the process parameters and the corresponding steady-state values are given in Table 3.1. It was verified that these conditions correspond to a stable equilibrium point of the system of (3.51). The control problem is formulated as the one of regulating the concentration of the desired product C_D by manipulating the temperature of the fluid in the heating jacket T_c. Within the posed control problem we distinguish and address the following two scenarios. In the first scenario, the control objective is achieved in the absence of any model uncertainty or exogenous disturbances in the process. By contrast, the control objective in the second scenario is accomplished in the presence of uncertainty. In the latter scenario, the enthalpies of the three reactions $\Delta H_i, i = 1, 2, 3$ and the feed temperature T_{A0} are assumed to be the main uncertain variables present. Defining $x_1 = C_A - C_{As}$, $x_2 = C_D - C_{Ds}$, $x_3 = T - T_s$, $u = T_c - T_{cs}$, $\theta_1 = \Delta H_1 - \Delta H_{10}$, $\theta_2 = \Delta H_2 - \Delta H_{20}$, $\theta_3 = \Delta H_3 - \Delta H_{30}$, $\theta_4 = T_{A0} - T_{A0s}$, $y = C_D - C_{Ds}$, where the subscript s denotes the steady-state values and ΔH_{D0}, ΔH_{U0}, ΔH_{R0} are the nominal values for the enthalpies, the process model of (3.51) can be written in the form of (3.1) where:

Table 3.1. Process parameters and steady-state values for the reactor of (3.51).

V	$= 1.00$	m^3
A	$= 6.0$	m^2
U	$= 1000.0$	$kcal\ hr^{-1}\ m^{-2}\ K^{-1}$
R	$= 1.987$	$kcal\ kmol^{-1}\ K^{-1}$
C_{A0s}	$= 3.75$	$kmol\ m^{-3}$
T_{A0s}	$= 310.0$	K
T_{cs}	$= 320.0$	K
ΔH_{10}	$= 5.4 \times 10^3$	$kcal\ kmol^{-1}$
ΔH_{20}	$= 5.1 \times 10^3$	$kcal\ kmol^{-1}$
ΔH_{30}	$= 5.0 \times 10^4$	$kcal\ kmol^{-1}$
k_{10}	$= 3.36 \times 10^6$	hr^{-1}
k_{20}	$= 7.21 \times 10^6$	hr^{-1}
k_{30}	$= 1.25 \times 10^7$	hr^{-1}
E_1	$= 8.0 \times 10^3$	$kcal\ kmol^{-1}$
E_2	$= 9.0 \times 10^3$	$kcal\ kmol^{-1}$
E_3	$= 9.5 \times 10^3$	$kcal\ kmol^{-1}$
c_p	$= 0.231$	$kcal\ kg^{-1}\ K^{-1}$
ρ	$= 900.0$	$kg\ m^{-3}$
F	$= 3.0$	$m^3\ hr^{-1}$
C_{AS}	$= 0.913$	$kmol\ m^{-3}$
C_{DS}	$= 1.66$	$kmol\ m^{-3}$
T_s	$= 302.0$	K

$$f(x) = \begin{bmatrix} \dfrac{F}{V}(C_{A0} - C_{As} - x_1) - \displaystyle\sum_{i=1}^{3} k_{i0} \exp\left(\dfrac{-E_i}{R(x_3 + T_s)}\right)(x_1 + C_{As}) \\[2em] -\dfrac{F}{V}(C_{Ds} + x_2) + k_{10} \exp\left(\dfrac{-E_1}{R(x_3 + T_s)}\right)(x_1 + C_{As}) \\[2em] \dfrac{F}{V}(T_{A0s} - T_s - x_3) + \displaystyle\sum_{i=1}^{3} \dfrac{(-\Delta H_i)}{\rho c_p} k_{i0} \exp\left(\dfrac{-E_i}{R(x_3 + T_s)}\right)(x_1 + C_{As}) \\ + \dfrac{UA}{\rho c_p}(T_{cs} - T_s - x_3) \end{bmatrix}$$

$$g(x) = \begin{bmatrix} 0 \\[1em] 0 \\[1em] \dfrac{UA}{\rho c_p} \end{bmatrix}, \quad w_i(x) = \begin{bmatrix} 0 \\[1em] 0 \\[1em] k_{i0} \exp\left(\dfrac{-E_i}{R(x_3 + T_s)}\right)(x_1 + C_{As}) \end{bmatrix}, \ i = 1, 2, 3$$

$$w_4(x) = \begin{bmatrix} 0 \\ 0 \\ \frac{F}{V} \end{bmatrix}, \quad h(x) = \begin{bmatrix} x_2 \end{bmatrix}$$

Using the following coordinate transformation:

$$\begin{bmatrix} \zeta_1 \\ \zeta_2 \\ \eta_1 \end{bmatrix} = \begin{bmatrix} h(x) \\ L_f h(x) \\ t(x) \end{bmatrix} = \begin{bmatrix} x_2 \\ -\frac{F}{V}(C_{Ds} + x_2) + k_{10} \exp\left(\frac{-E_1}{R(x_3 + T_s)}\right)(x_1 + C_{As}) \\ x_1 \end{bmatrix} \tag{3.52}$$

and the notation $\bar{f}(e, \eta, \bar{v}) = [e_2 \;\; L_f^2 h(x) - v^{(2)}]^T$, $\bar{g}(e, \eta, \bar{v}) = [0 \;\; L_g L_f h(x)]^T$, $\bar{w}_k(e, \eta, \bar{v}) = [0 \;\; L_{wk} L_f h(x)]^T$, $k = 1, 2, 3, 4$, the process model can be cast in the following form:

$$\begin{aligned} \dot{e} &= \bar{f}(e, \eta, \bar{v}) + \bar{g}(e, \eta, \bar{v})u + \sum_{k=1}^{4} \bar{w}_k(e, \eta, \bar{v})\theta_k \\ \dot{\eta} &= \Psi(e, \eta, \bar{v}) \\ y &= e_1 + v \end{aligned} \tag{3.53}$$

Controller Synthesis and Simulation Results

We now proceed with the design of the controller and begin with the first scenario where no uncertainty is assumed to be present in the process model (i.e. $\theta_k \equiv 0$). Owing to the absence of uncertainty in this case, we use the result of Corollary 3.26 to design the controller. It can be easily verified that the system of (3.53) satisfies the Assumptions of the corollary and therefore the necessary output feedback controller (whose practical implementation requires measurements of C_D only) takes the form:

$$\dot{\tilde{y}}_1 = \tilde{y}_2 + La_1(y - \tilde{y}_1)$$

$$\dot{\tilde{y}}_2 = L^2 a_2(y - \tilde{y}_1)$$

$$\dot{\omega}_1 = \frac{F}{V}(C_{A0} - C_{As} - \hat{x}_1) - \sum_{i=1}^{3} k_{i0} \exp\left(\frac{-E_i}{R(\hat{x}_3 + T_s)}\right)(\hat{x}_1 + C_{As}) \quad (3.54)$$

$$u = -\left(c_0 + \frac{L_{\hat{f}}V + \sqrt{(L_{\hat{f}}V)^2 + (L_{\hat{g}}V)^4}}{(L_{\hat{g}}V)^2}\right) L_{\hat{g}}V$$

where

$$V = e^T P e, \quad P = \begin{bmatrix} 1 & c \\ c & 1 \end{bmatrix}, \quad c \in (0, 1)$$

$$\begin{aligned} L_{\hat{f}}V &= 2\left((\hat{x}_2 - v) + c\left(L_f h(\hat{x}) - v^{(1)}\right)\right)\left(L_f h(\hat{x}) - v^{(1)}\right) \\ &\quad + 2\left(c(\hat{x}_2 - v) + \left(L_f h(\hat{x}) - v^{(1)}\right)\right)\left(L_f^2 h(\hat{x}) - v^{(2)}\right) \end{aligned} \quad (3.55)$$

$$L_{\hat{g}}V = 2\left(c(\hat{x}_2 - v) + \left(L_f h(\hat{x}) - v^{(1)}\right)\right) L_g L_f h(\hat{x})$$

For the sake of comparison, we consider also the nonlinear output feedback controller synthesized based on the concepts of feedback linearization and proposed in [50] with the static component:

$$u = -\frac{1}{L_g L_f h(\hat{x})}\left\{\frac{\beta_0}{\beta_2}(\hat{x}_2 - v) + \frac{\beta_1}{\beta_2}(L_f h(\hat{x}) - v^{(1)}) + (L_f^2 h(\hat{x}) - v^{(2)})\right\} \quad (3.56)$$

Before we present the simulation results for the first scenario, we first establish the fact that the static *state* feedback component of the controller of (3.54) does not expend unnecessary control effort in comparison with the static controller of (3.56). To this end, we compared the control actions generated by the two controllers for a broad range of T and C_D. The results are shown in Figure 3.3 for several values of C_A. Clearly, the controller of (3.54) utilizes smaller control action than that of (3.56).

Two sets of simulation runs were performed to evaluate the performance and near-optimality properties of the dynamic output feedback controller of (3.54). The following values were used for the controller and observer parameters: $c_0 = 0.1$, $\alpha = 1.0$, $c = 0.99$, $L = 3000$, $a_1 = 10$, $a_2 = 20$, $a_m = 1.0$.

In the first set of simulation runs, we evaluated the capability of the controller to steer the process to the steady-state given in Table 3.1 starting from arbitrary initial conditions. Figure 3.4 shows the controlled output profile and

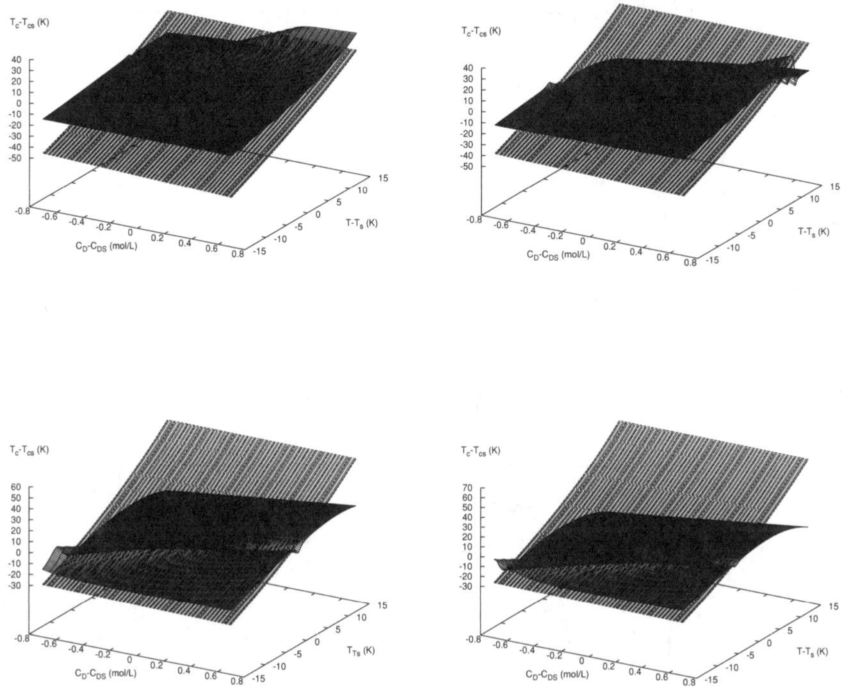

Fig. 3.3. Control actions computed by the static component of the controller of (3.54) (*solid lines*) and the controller of (3.56) (*dashed lines*) for $C_A - C_{AS} = -0.4$ (*top left*), -0.2 (*top right*), 0.2 (*bottom left*), 0.4 (*bottom right*)

the corresponding manipulated input profile. Clearly, the controller successfully steers the process to the desired steady-state. Also shown in the figure is a comparison between the output feedback controller of (3.54) and the corresponding state feedback controller (i.e., the static component of (3.54) with $\hat{x} = x$) synthesized under the assumption that all process states are available for measurement. It was shown in Sect. 3.3.2 that the state feedback controller is optimal with respect to a meaningful cost of the form of (3.24). From Fig. 3.4, one can immediately observe that the controlled output and manipulated input profiles obtained under the output feedback controller are very close to the profiles obtained under the state feedback controller. This comparison then makes clear the fact that the controller of (3.54) is near-optimal in the sense that its performance approaches the optimal performance of the state feedback controller when the observer gain is sufficiently high. To further illustrate the near-optimality result, we compared the costs associated with both the state feedback and output feedback controllers. The costs were computed and found to be 0.250 and 0.257, respectively, further confirming the near-optimality of the controller of (3.54).

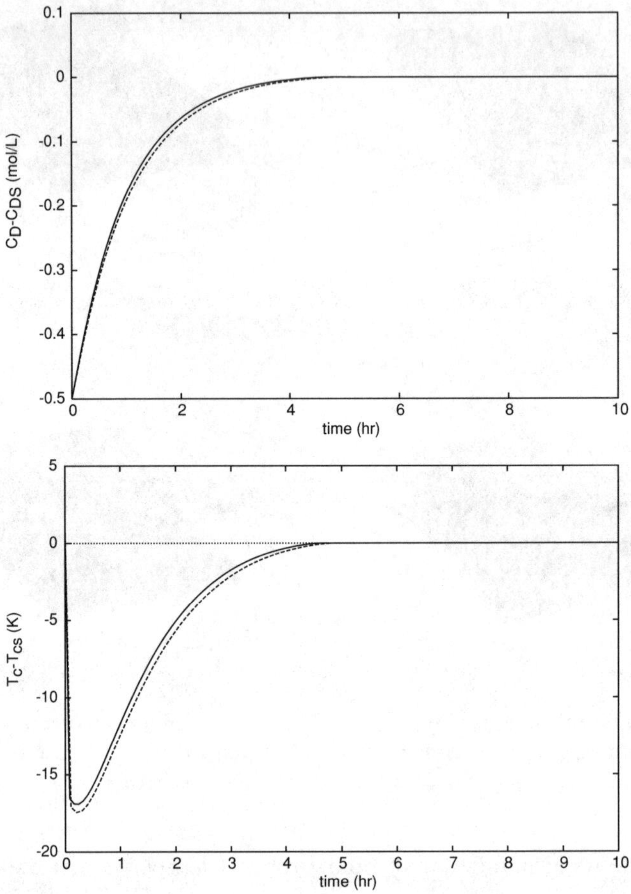

Fig. 3.4. Closed-loop output and manipulated input profiles for stabilization under the output feedback controller of (3.54) (*dashed lines*) and the optimal state feedback controller (*solid lines*) when no uncertainty is present

In the second set of simulation runs, we tested the output tracking capabilities of the output feedback controller of (3.54). The process was initially assumed to be at the steady-state given in Table 3.1 and then a 0.5 mol/L decrease in the value of the reference input was imposed ($v = -0.5$). Figure 3.5 shows the profiles of the controlled output of the process and of the corresponding manipulated input. One can immediately observe that the controller successfully drives the output to the desired new reference input value. The figure also shows the closeness of the controlled output and manipulated input profiles obtained under the output feedback and state feedback controllers. It is clear that the tracking capabilities of the output feedback controller approach those of the optimal state feedback controller and therefore the controller of (3.54) is near-optimal.

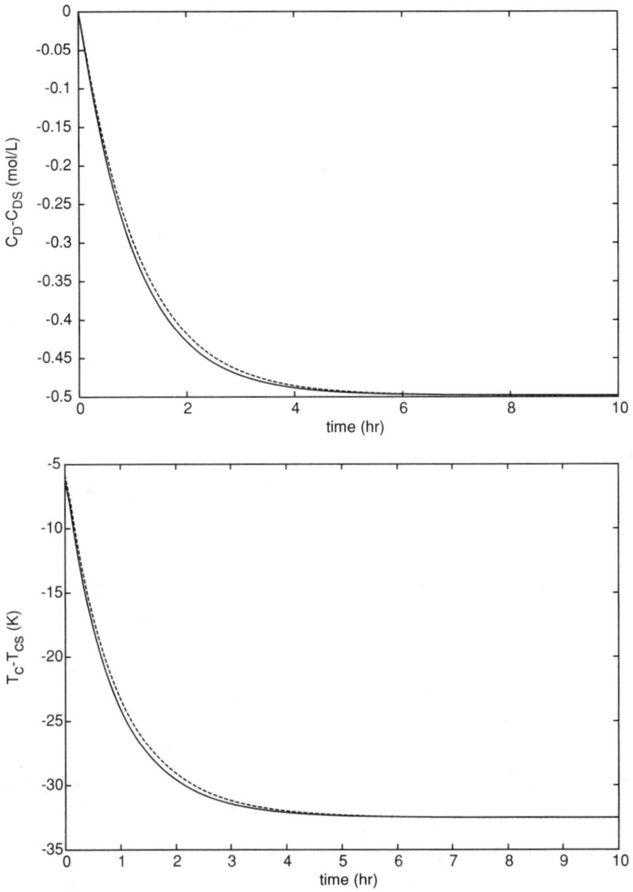

Fig. 3.5. Closed-loop output and manipulated input profiles for reference input tracking under the output feedback controller of (3.54) (*dashed lines*) and the optimal state feedback controller (*solid lines*) when no uncertainty is present

We now consider the second scenario where we account for the presence of uncertainty in the process. The following time-varying uncertain variables were considered in the simulation runs:

$$\begin{aligned} \theta_i(t) &= 0.4\,(-\Delta H_{i0})\,[1 + sin(2t)]\,, \quad i = 1, 2, 3 \\ \theta_4(t) &= 0.07 T_{A0s}\,[1 + sin(2t)] \end{aligned} \tag{3.57}$$

The bounds on the uncertain variables were taken to be $\theta_{bi} = 0.4\,|(-\Delta H_{i0})|$, for $i = 1, 2, 3$, $\theta_{b4} = 0.07 T_{A0s}$. Moreover, the following values were used for the tuning parameters of the controller and the observer: $c_0 = 0.1$, $c = 0.99$, $\phi = 0.005$, $\chi = 1.1$, $L = 30000$, $a_1 = 10$, $a_2 = 20$, $a_m = 1.0$ to guarantee that the output of the closed-loop system satisfies a relation of the following form:

$$\limsup_{t \to \infty} \|y - v\| \leq 0.005 \tag{3.58}$$

To design the necessary controller in this case, we note that since the $\bar{w}_k$ functions defined in (3.53) do not vanish at the origin, the uncertain variables considered here are non-vanishing. Therefore, we use the result of Theorem 3.27 to construct the necessary robust output feedback controller which takes the following form:

$$\begin{aligned}
\dot{\tilde{y}}_1 &= \tilde{y}_2 + La_1(y - \tilde{y}_1) \\
\dot{\tilde{y}}_2 &= L^2 a_2(y - \tilde{y}_1) \\
\dot{\omega}_1 &= \frac{F}{V}(C_{A0} - C_{As} - \hat{x}_1) - \sum_{i=1}^{3} k_{i0} \exp\left(\frac{-E_i}{R(\hat{x}_3 + T_s)}\right)(\hat{x}_1 + C_{As}) \\
u &= -\left(c_0 + \frac{L_{\hat{f}}V + \sqrt{(L_{\hat{f}}V)^2 + (L_{\hat{g}}V)^4}}{(L_{\hat{g}}V)^2}\right) L_{\hat{g}}V \\
&\quad - \left(\frac{\chi \sum_{k=1}^{4} \theta_{bk} \|L_{wk} L_f h(\hat{x})\|}{(L_g L_f h(\hat{x}))^2 (\frac{\|L_{\hat{g}}V\|}{\|L_g L_f h(\hat{x})\|} + \phi)}\right) L_{\hat{g}}V
\end{aligned} \tag{3.59}$$

where V, $L_{\hat{f}}V$, $L_{\hat{g}}V$ were defined in (3.55).

Several sets of simulation studies were performed to assess the performance, robustness, and near-optimality properties of the dynamic robust output feedback controller of (3.59). In the first set of simulations, we tested the ability of the controller to drive the output of the process close to the desired steady-state starting from arbitrary initial conditions despite the presence of uncertainty. Figure 3.6 shows the controlled output profile and the manipulated input profile. One can immediately see that the effect of the uncertainty has been significantly reduced (compare with the output of the open-loop system) and the output of the process remains very close to the desired steady-state satisfying the requirement of (3.58). Included in the figure also are the controlled output and manipulated input profiles for the process under the optimal state feedback controller (i.e., the static component of (3.59) with $\hat{x} = x$) which was shown in Theorem 3.13 to be optimal with respect to the cost functional of (3.30) with $\hat{x} = x$. It is clear from the figure that the profiles obtained under the output feedback controller follow closely those obtained

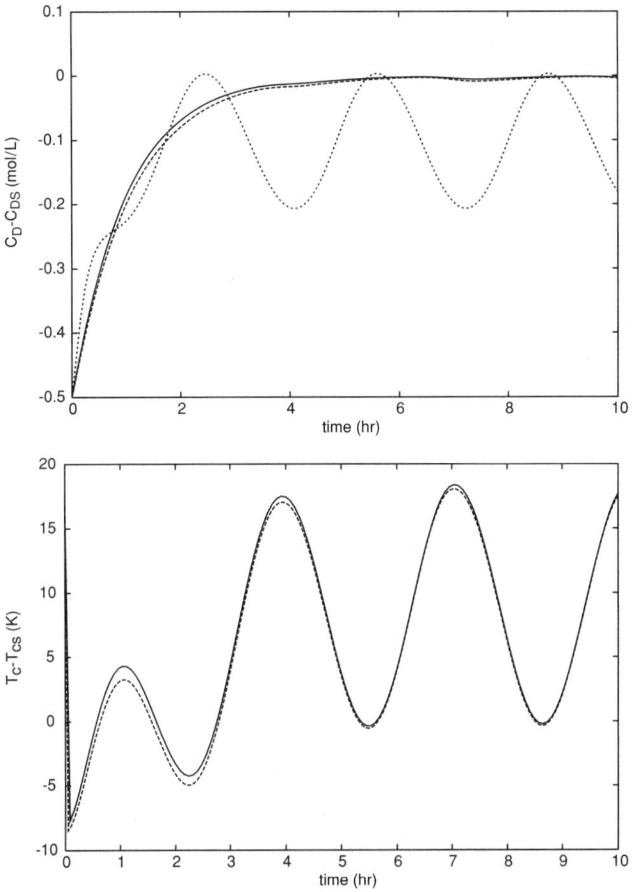

Fig. 3.6. Closed-loop output and manipulated input profiles for stabilization under the robust output feedback controller of (3.59) (*dashed lines*), the robust optimal state feedback controller (*solid lines*) and open-loop profile (*dotted line*)

under the optimal state feedback controller and, therefore, the robust output feedback controller of (3.59) is near-optimal.

In the second set of simulations, we investigated the significance of the term responsible for uncertainty compensation in the controller of (3.59). To this end, we implemented the controller of (3.59) on the process without the uncertainty compensation component. The closed-loop output and manipulated input profiles for this simulation run are given in Fig. 3.7. It is clear that the effect of the uncertain variables is appreciable, leading to poor transient performance and offset and that uncertainty compensation is required to achieve the control objective.

In the last set of simulations performed, we tested the output tracking capabilities of the controller of (3.59) for a 0.5 mol/L decrease in the value of the

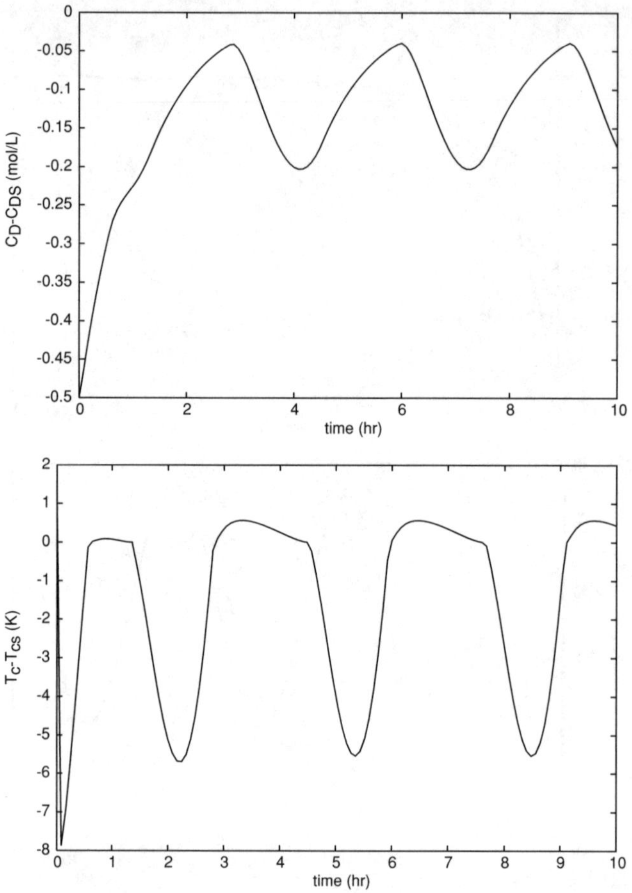

Fig. 3.7. Closed-loop output and manipulated input profiles for stabilization under the output feedback controller of (3.59) with no uncertainty compensation ($\chi = 0$)

reference input. The resulting closed-loop output and manipulated input profiles, shown in Fig. 3.8, clearly establish the capability of the output feedback controller of (3.59) to enforce the requirement of (3.58), despite the presence of uncertainty. Furthermore, Fig. 3.8 demonsttrates the near-optimality of the output feedback controller within the context of robust output tracking. This is evident from the closeness of the performance of the robust output feedback controller of (3.59) to that of the corresponding robust optimal state feedback controller.

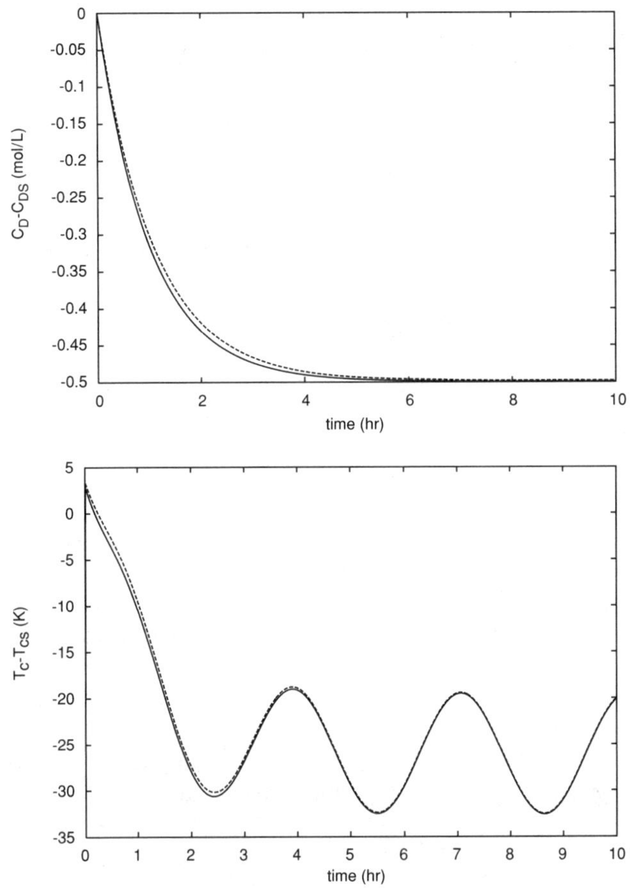

Fig. 3.8. Closed-loop output and manipulated input profiles for reference input tracking under the robust output feedback controller of (3.59) (*dashed lines*) and the robust optimal state feedback controller (*solid lines*)

3.5 Robust Control of Nonlinear Singulary-Perturbed Systems

In addition to parametric uncertainty and external disturbances, unmodeled dynamics constitute another common source of uncertainty. In the modeling of physical and chemical systems, it is common practice to neglect stable, high-frequency dynamics (e.g., actuators/sensors with small time constants) that increase the order of the model in order to reduce model complexity for the analysis and controller design tasks. An important consideration in controller design is to ensure robustness of the controller designed on the basis of the simplified model with respect to the unmodeld dynamics. Singular perturbation theory provides a natural framework for addressing this problem given that the presence of unmodeld dynamics induces time-scale multiplicity. Singular perturbation methods allow decomposing a multiple-time-scale system into separate reduced-order systems that evolve in different time-scales, and inferring its asymptotic properties from the knowledge of the behavior of the reduced-order systems [152]. In this Sect., we study, using singular perturbations techniques, the robustness of the output feedback controller design presented in Sect. 3.4 with respect to unmodeled dynamics. To this end, we consider single-input single-output nonlinear singularly perturbed systems with uncertain variables of the form:

$$\begin{aligned} \dot{x} &= f_1(x) + Q_1(x)z + g_1(x)u + w_1(x)\theta(t) \\ \epsilon\dot{z} &= f_2(x) + Q_2(x)z + g_2(x)u + w_2(x)\theta(t) \\ y &= h(x) \end{aligned} \tag{3.60}$$

where $x \in \mathbb{R}^n$ and $z \in \mathbb{R}^p$ denote vectors of state variables, $u \in \mathbb{R}$ denotes the manipulated input, $\theta \in \mathbb{R}$ denotes the uncertain time-varying variable, which may include disturbances and unknown process parameters, and $y \in \mathbb{R}$ denotes the controlled output, and ϵ is a small positive parameter that quantifies the speed ratio of the slow versus the fast dynamics of the system. The functions, $f_1(x)$, $f_2(x)$, $g_1(x)$ and $g_2(x)$, $w_1(x)$ and $w_2(x)$, are sufficiently smooth vector functions, $Q_1(x)$ and $Q_2(x)$ are sufficiently smooth matrices, and $h(x)$ is a sufficiently smooth scalar function.

Setting $\epsilon = 0$ in the system of (3.60) and assuming that $Q_2(x)$ is invertible, uniformly in $x \in \mathbb{R}^n$, the following slow subsystem is obtained:

$$\begin{aligned} \dot{x} &= f(x) + g(x)u + w(x)\theta(t) \\ y &= h(x) \end{aligned} \tag{3.61}$$

where

$$\begin{aligned} f(x) &= f_1(x) - Q_1(x)Q_2^{-1}(x)f_2(x) \\ g(x) &= g_1(x) - Q_1(x)Q_2^{-1}(x)g_2(x) \\ w(x) &= w_1(x) - Q_1(x)Q_2^{-1}(x)w_2(x) \end{aligned} \tag{3.62}$$

Defining the fast time-scale, $\tau = \dfrac{t}{\epsilon}$, deriving the representation of the system of (3.60) in the τ time scale, and setting $\epsilon = 0$, the following fast subsystem is obtained:

$$\frac{dz}{d\tau} = f_2(x) + Q_2(x)z + g_2(x)u + w_2(x)\theta(t) \tag{3.63}$$

where x can be considered equal to its initial value, $x(0)$, and θ can be viewed as constant. In this section, we consider systems of the form of (3.60), for which the corresponding fast subsystem is globally asymptotically stable (the reader may refer to [49] for a general method for output feedback controller design for nonlinear singularly perturbed systems with unstable fast dynamics).

Assumption 3.8 *The matrix $Q_2(x)$ is Hurwitz uniformly in $x \in \mathbb{R}^n$.*

3.5.1 Control Problem Formulation

Referring to the system of (3.60), the objective of this section is to synthesize robust dynamic output feedback controllers of the form:

$$\begin{aligned} \dot{\omega} &= \mathcal{F}(\omega, y, \bar{v}) \\ u &= \mathcal{P}(\omega, y, \bar{v}, t) \end{aligned} \tag{3.64}$$

where $\omega \in \mathbb{R}^s$ is a state, $\mathcal{F}(\omega, y, \bar{v})$ is a vector function, $\mathcal{P}(\omega, y, \bar{v}, t)$ is a scalar function, $\bar{v} = [v \; v^{(1)} \; \cdots \; v^{(r)}]^T$ is a generalized reference input ($v^{(k)}$ denotes the $k-th$ time derivative of the reference input v, which is assumed to be a sufficiently smooth function of time), that enforce the following properties in the closed-loop system: (a) boundedness of the trajectories, (b) arbitrary degree of asymptotic attenuation of the effect of the uncertainty on the output, and robust output tracking for changes in the reference input. The above properties are enforced in the closed-loop system for arbitrarily large initial conditions, uncertainty and rate of change of uncertainty, provided that ϵ is sufficiently small (semi-global type result).

The controller of (3.64) consists of the dynamical system, $\dot{\omega} = \mathcal{F}(\omega, y, \bar{v})$, which will be synthesized to provide estimates of the state variables, and the static component, $\mathcal{P}(\omega, y, \bar{v}, t)$, which will be synthesized to enforce the requested properties in the closed-loop system.

Motivated by the assumption of global asymptotic stability of the fast dynamics of the system of (3.60), the synthesis of a controller of the form of (3.64) will be carried out on the basis of the slow subsystem of (3.61). Note that the slow subsystem is in a form similar to that of (3.1) with $q = 1$. To this end, we will impose on the slow subsystem of (3.61) assumptions which are similar to the ones used in addressing the robust output feedback problem for the system of (3.1).

3.5.2 Robust Output Feedback Controller Synthesis

Theorem 3.31 below provides a formula of the robust output feedback controller and states precise conditions under which the proposed controller enforces the desired properties in the closed-loop system. The proof is given in Appendix A.

Theorem 3.31. *Consider the uncertain singularly perturbed nonlinear system of (3.60), for which Assumptions 3.3, 3.5, 3.7, 3.8 hold, under the robust output feedback controller:*

$$\begin{aligned}
\dot{\tilde{y}} &= \begin{bmatrix} -La_1 & 1\,0 \cdots 0\,0 \\ -L^2a_2 & 0\,1 \cdots 0\,0 \\ \vdots & \vdots\,\vdots\,\ddots\,\vdots\,\vdots \\ -L^{r-1}a_{r-1} & 0\,0 \cdots 0\,1 \\ -L^ra_r & 0\,0 \cdots 0\,0 \end{bmatrix} \tilde{y} + \begin{bmatrix} La_1 \\ L^2a_2 \\ \vdots \\ L^{r-1}a_{r-1} \\ L^ra_r \end{bmatrix} y \\
\dot{\omega}_1 &= \Psi_1(sat(\tilde{y}), \omega) \\
&\vdots \\
\dot{\omega}_{n-r} &= \Psi_{n-r}(sat(\tilde{y}), \omega) \\
u &= p(\hat{x}, c_0, \rho, \chi, \phi, \theta_b)
\end{aligned} \tag{3.65}$$

where the parameters, $a_1, \cdots, a_r$ are chosen such that the polynomial $s^r + a_1 s^{r-1} + a_2 s^{r-2} + \cdots + a_1 = 0$ is Hurwitz, $p(\cdot)$ was defined in Theorem 3.3, $\hat{x} = \mathcal{X}^{-1}(sat(\tilde{y}), \omega))$. Then, for each set of positive real numbers δ_x, δ_z, δ_θ, $\delta_{\dot{\theta}}$, $\delta_{\bar{v}}$, d, there exists $\phi^*(\chi) > 0$ and for each $\phi \in (0, \phi^*(\chi)]$, there exists $\epsilon^*(\phi) > 0$, such that if $\phi \in (0, \phi^*]$, $\bar{\epsilon} = \max\{\epsilon, \frac{1}{L}\} \in (0, \epsilon^*(\phi)]$, $sat(\cdot) = \min\{1, \frac{\zeta_{max}}{\|\cdot\|}\}(\cdot)$ with ζ_{max} being the maximum value of the vector $[\zeta_1\ \zeta_2\ \cdots\ \zeta_r]$ for $\|\zeta\| \leq \beta_\zeta(\delta_\zeta, 0) + d$ where β_ζ is a class $\mathcal{KL}$ function and δ_ζ is the maximum value of the vector $[h(x)\ L_f h(x)\ \cdots\ L_f^{r-1}h(x)]$ for $\|x\| \leq \delta_x$, $\|x(0)\| \leq \delta_x$, $\|z(0)\| \leq \delta_z$, $\|\theta\|^s \leq \delta_\theta$, $\|\dot{\theta}\|^s \leq \delta_{\dot{\theta}}$, $\|\bar{v}\|^s \leq \delta_{\bar{v}}$, $\|\tilde{y}(0)\| \leq \delta_\zeta$, $\omega(0) = \eta(0) + O(\bar{\epsilon})$, the output of the closed-loop system satisfies a relation of the form:

$$\limsup_{t\to\infty} \|y(t) - v(t)\| \leq d \tag{3.66}$$

Remark 3.32. Note that the output feedback controller of (3.65) is the same as the one proposed in Theorem 3.27 where it has been established that this output feedback controller enforces stability and robust output tracking for arbitrarily large initial conditions and uncertainty, provided that the observer gain is sufficiently large. The result of Theorem 3.31 establishes that the same output feedback controller, which is designed on the basis of the reduced system of (3.61), continues to enforce stability and robust output tracking for the full singularly perturbed closed-loop system for arbitrarily large initial conditions, uncertainty and rate of change of uncertainty, provided that $\bar{\epsilon} := \max\{\epsilon, \frac{1}{L}\}$ is sufficiently small. The result therefore establishes robustness of the output feedback controller design of (3.65) with respect to stable and sufficiently fast unmodeled dynamics.

3.6 Conclusions

In this chapter, we presented robust inverse optimal controller designs for input/output linearizable nonlinear systems with time-varying bounded uncertain variables. Under full state feedback, the controller designs were obtained by re-shaping the scalar nonlinear gain of Sontag's formula in a way that guarantees the desired uncertainty attenuation properties in the closed-loop system. The proposed re-shaping was made different for vanishing and non-vanishing uncertainties so as to meet different robustness and optimality objectives. For systems with vanishing uncertainty, the proposed controller design was shown to enforce global asymptotic stability, robust asymptotic output tracking and inverse optimality with respect to an infinite-time cost functional that includes penalty on the control effort. For systems with non-vanishing uncertainty, the designed controllers were shown to ensure boundedness of the states, robust asymptotic output tracking with an arbitrary degree of attenuation of the effect of uncertainty, and optimality with respect to a meaningful cost defined on a finite time-interval. In the absence of full state measurements, combination of the state feedback controllers with high-gain observers and appropriate saturation filters (to eliminate observer peaking) was employed to design dynamic robust output feedback controllers. Under vanishing uncertainty, the output feedback controllers were shown to enforce asymptotic stability stability, robust asymptotic output tracking with uncertainty attenuation, and near-optimal performance over the infinite time-interval for arbitrarily large initial conditions and uncertainty, provided that the observer gain is sufficiently large. Under persistent uncertainty, the same controllers were shown to guarantee boundedness of the states, robust output tracking with uncertainty attenuation, and near-optimality over a finite time-interval. The developed controllers were successfully tested on non-isothermal stirred tank reactors with uncertainty.

4

Control of Nonlinear Systems with Uncertainty and Constraints

4.1 Introduction

In addition to nonlinear behavior and model uncertainty, one of the ubiquitous characteristics of process control systems is the presence of hard constraints on the manipulated inputs. At this stage, most existing process control methods lead to the synthesis of controllers that can deal with either model uncertainty or input constraints, but not simultaneously or effectively with both. This clearly limits the achievable control quality and closed-loop performance, especially in view of the commonly-encountered co-presence of uncertainty and constraints in chemical processes. Therefore, the development of a unified framework for control of nonlinear systems that explicitly accounts for the presence of model uncertainty and input constraints is expected to have a significant impact on process control.

While the individual presence of either model uncertainty or input constraints poses its own unique set of problems that must be dealt with at the stage of controller design, the combined presence of both uncertainty and constraints is far more problematic for process stability and performance. The difficulty here emanates not only from the cumulative effect of the co-presence of the two phenomena but also from the additional issues that arise from the interaction of the two. At the core of these issues are the following two problems:

- The co-presence of model uncertainty and input constraints creates an inherent conflict in the controller design objectives of the control policy to be implemented. On the one hand, suppressing the influence of significant external disturbances typically requires large (high-gain) control action. On the other hand, the availability of such action is often limited by the presence of input constraints. Failure to resolve this conflict will render any potential control strategy essentially ineffective.
- The set of feasible process operating conditions under which the process can be operated safely and reliably is significantly restricted by the

co-presence of uncertainty and constraints. While input constraints by themselves place fundamental limitations on the size of this set (and consequently on our ability to achieve certain control objectives), these limitations are even stronger when uncertainty is present. Intuitively, many of the feasible operating conditions permitted by constraints under nominal conditions (i.e., predicted using a nominal model of the process) cannot be expected to continue to be feasible in the presence of significant plant-model mismatch. The success of a control strategy, therefore, hinges not only on the design of effective controllers, but also on the ability to predict a priori the feasible conditions under which the designed controllers are guaranteed to work in the presence of both uncertainty and constraints.

A natural approach to resolve the apparent conflict between the need to compensate for model uncertainty through high-gain control action and the presence of input constraints that limit the availability of such action is the design of robust controllers which expend reasonable control effort to achieve stabilization and uncertainty attenuation. This problem was addressed in Chap. 3 where we synthesized, through Lyapunov's direct method, robust inverse optimal nonlinear controllers that use reasonable control effort to enforce stability and asymptotic output tracking with attenuation of the effect of the uncertain variables on the output of the closed-loop system. The resulting controllers, although better equipped to handle input constraints than feedback linearizing controllers, do not explicitly account for input constraints and, therefore, offer no a priori guarantees regarding the desired closed-loop stability and performance in the presence of arbitrary input constraints.

Motivated by these considerations, we focus in this chapter on state and output feedback control of multi-input multi-output (MIMO) nonlinear processes with model uncertainty and input constraints. A Lyapunov-based nonlinear controller design approach that accounts explicitly and simultaneously for process nonlinearities, model uncertainty, input constraints, multivariable interactions, and the lack of full state measurements is proposed. When measurements of the full state are available, the approach leads to the explicit synthesis of bounded robust nonlinear state feedback controllers that enforce stability and robust asymptotic set-point tracking in the constrained uncertain closed-loop system and provide, at the same time, an explicit characterization of the region of guaranteed closed-loop stability. When complete state measurements are not available, a combination of the state feedback controllers with high-gain state observes and appropriate saturation filters, is employed to synthesize bounded robust output feedback controllers that require only measurements of the outputs for practical implementation. The resulting output feedback design enforces the same closed-loop stability and performance properties of the state feedback controllers and, in addition, practically preserves the region of guaranteed closed-loop stability obtained under state feedback, provided that the observer gain is sufficiently large. The developed state and output feedback controllers are applied successfully to non-isothermal multivariable chemical reactor examples with uncertainty, input constraints, and incomplete state measurements. Finally, we conclude the chapter with a discussion on how nonlinear control tools can be used to provide improved tuning guidelines for classical controllers. The results of this chapter were first presented in [81].

4.2 Preliminaries

We consider multi-input multi-output (MIMO) nonlinear processes with uncertain variables and input constraints, with the following state-space description:

$$\begin{aligned} \dot{x} &= f(x) + \sum_{i=1}^{m} g_i(x)u_i + \sum_{k=1}^{q} w_k(x)\theta_k(t) \\ u(t) &\in \mathcal{U} \\ y_i &= h_i(x), \quad i = 1, \ldots, m \end{aligned} \tag{4.1}$$

where $x \in \mathbb{R}^n$ denotes the vector of process state variables, $u=[u_1 \; u_2 \; \cdots \; u_m]^T$ denotes the vector of constrained manipulated inputs taking values in a nonempty convex subset, $\mathcal{U} = \{u \in \mathbb{R}^m : \|u\| \leq u_{max}\}$, $\|\cdot\|$ is the Euclidean norm of a vector, u_{max} is a positive real number that captures the maximum magnitude of the Euclidean norm of the vector of manipulated inputs allowed by the constraints, $\theta_k(t) \in \mathcal{W} \subset \mathbb{R}$ denotes the k-th uncertain (possibly time-varying) but bounded variable taking values in a nonempty compact convex subset of $\mathbb{R}$, $y_i \in \mathbb{R}$ denotes the i-th output to be controlled. The uncertain variable, $\theta_k(t)$, may describe time-varying parametric uncertainty and/or exogenous disturbances. It is assumed that the origin is the equilibrium point of the nominal (i.e., with $u(t) = \theta_k(t) \equiv 0$) system of (4.1). The vector functions, $f(x)$, $g_i(x)$, $w_k(x)$, and the scalar functions $h_i(x)$ are assumed to be sufficiently smooth on their domains of definition. In the remainder of this chapter, for simplicity, we will suppress the time-dependence in the notation of the uncertain variable, $\theta_k(t)$. The class of systems described by (4.1) is general enough to be of practical interest (see the application studies throughout the chapter), yet specific enough to allow the meaningful synthesis of controllers.

Throughout the chapter, the Lie derivative notation will used. In particular, $L_f\bar{h}$ denotes the standard Lie derivative of a scalar function $\bar{h}(x)$ with respect to the vector function $f(x)$, $L_f^k\bar{h}$ denotes the k-th order Lie derivative and $L_{g_i}L_f^{k-1}\bar{h}$ denotes the mixed Lie derivative where $g_i(x)$ is a vector function. We will also need the definition of a class $\mathcal{KL}$ function. A function $\beta(s,t)$ is said to be of class $\mathcal{KL}$ if, for each fixed t, the function $\beta(s,\cdot)$ is continuous, increasing, and zero at zero and, for each fixed s, the function $\beta(\cdot,t)$ is nonincreasing and tends to zero at infinity.

Referring to the system of (4.1), we define the relative order of the output, y_i, with respect to the vector of manipulated inputs, u, as the smallest integer r_i for which:

$$\left[L_{g_1} L_f^{r_i-1} h_i(x) \cdots L_{g_m} L_f^{r_i-1} h_i(x)\right] \not\equiv [0 \cdots 0] \tag{4.2}$$

or $r_i = \infty$ if such an integer does not exist. We also define the characteristic matrix [126]:

$$C(x) = \begin{bmatrix} L_{g_1} L_f^{r_1-1} h_1(x) & \cdots & L_{g_m} L_f^{r_1-1} h_1(x) \\ \vdots & \cdots & \vdots \\ L_{g_1} L_f^{r_m-1} h_m(x) & \cdots & L_{g_m} L_f^{r_m-1} h_m(x) \end{bmatrix} \tag{4.3}$$

4.3 Bounded Robust State Feedback Control

In this section, we focus on the state feedback control problem where the full process state is assumed to be available for measurement. The output feedback control problem will be discussed in Sect. 4.5. We begin in Sect. 4.3.1 with the control problem formulation and then present the controller synthesis results in Sect. 4.3.2.

4.3.1 Control Problem Formulation

In order to formulate our control problem, we consider the MIMO nonlinear process of (4.1) and assume initially that the uncertain variable terms are vanishing in the sense that $w_k(0) = 0$ (note that this does not require the variable θ_k itself to vanish over time). Under this assumption, the origin – which is an a nominal equilibrium point – continues to be an equilibrium point for the uncertain process. The results for the case when the uncertain variables are non-vanishing (i.e., perturb the nominal equilibrium point) are discussed later on (see Remark 4.14 below). For the control problem at hand, our objective is two fold. The first is to synthesize, via Lyapunov-based control methods, a bounded robust multivariable nonlinear state feedback control law of the general form:

$$u = \mathcal{P}(x, u_{max}, \bar{v}) \tag{4.4}$$

that enforces the following closed-loop properties in the presence of uncertainty and input constraints, including: (a) asymptotic stability, and (b) robust asymptotic reference-input tracking with an arbitrary degree of attenuation of the effect of the uncertainty on the outputs of the closed-loop system. In (4.4), $\mathcal{P}(\cdot)$ is an m-dimensional bounded vector function (i.e., $\|u\| \leq u_{max}$), $\bar{v}$ is a generalized reference input which is assumed to be a sufficiently smooth function of time. Our second objective is to explicitly characterize the region of guaranteed closed-loop stability associated with the controller, namely the set of admissible initial states starting from where, stability and robust output tracking can be guaranteed in the constrained uncertain closed-loop system.

4.3.2 Controller Synthesis

To proceed with the controller synthesis task, we will impose the following three assumptions on the process of (4.1). We initially assume that there exists a coordinate transformation that renders the system of (4.1) partially linear. This assumption is motivated by the requirement of robust output tracking and is formulated precisely below.

Assumption 4.1 *There exist a set of integers,* $\{r_1, r_2, \ldots, r_m\}$*, and a coordinate transformation* $(\zeta, \eta) = T(x)$ *such that the representation of the system of (4.1), in the* (ζ, η) *coordinates, takes the form:*

$$\begin{aligned}
\dot{\zeta}_1^{(i)} &= \zeta_2^{(i)} \\
&\vdots \\
\dot{\zeta}_{r_i-1}^{(i)} &= \zeta_{r_i}^{(i)} \\
\dot{\zeta}_{r_i}^{(i)} &= L_f^{r_i} h_i(x) + \sum_{j=1}^{m} L_{g_j} L_f^{r_i-1} h_i(x) u_j + \sum_{k=1}^{q} L_{w_k} L_f^{r_i-1} h_i(x) \theta_k \\
\dot{\eta}_1 &= \Psi_1(\zeta, \eta, \theta) \\
&\vdots \\
\dot{\eta}_{n-\sum_i r_i} &= \Psi_{n-\sum_i r_i}(\zeta, \eta, \theta) \\
y_i &= \zeta_1^{(i)}, \quad i = 1, \ldots, m
\end{aligned} \tag{4.5}$$

where $x = T^{-1}(\zeta, \eta)$*,* $\zeta = [\zeta^{(1)^T} \cdots \zeta^{(m)^T}]^T$*,* $\eta = [\eta_1 \cdots \eta_{n-\sum_i r_i}]^T$*,* $\theta = [\theta_1 \cdots \theta_q]^T$*.*

Assumption 4.1 includes the matching condition of our robust control methodology. In particular, we consider systems of the form (4.1) for which the time-derivatives of the outputs, y_i, up to order $r_i - 1$, are independent of the uncertain variables, θ_k. Notice that this condition is different from the standard one which restricts the uncertainty to enter the system of (4.1) in the same equation with the manipulated inputs, u_i. Defining the tracking error variables, $e_k^{(i)} = \zeta_k^{(i)} - v_i^{(k-1)}$, and introducing the vector notation, $e^{(i)} = [e_1^{(i)} \; e_2^{(i)} \cdots e_{r_i}^{(i)}]^T$, $e = [e^{(1)^T} \; e^{(2)^T} \cdots e^{(m)^T}]^T$, where $i = 1, \ldots, m$, $k = 1, \ldots, r_i$, the ζ-subsystem of (4.5) can be re-written in the following more compact form:

$$\dot{e} = Ae + B\left[r(e,\eta,\bar{v}) + C_1(T^{-1}(e,\eta,\bar{v}))u + C_2(T^{-1}(e,\eta,\bar{v}))\theta\right] \tag{4.6}$$

where A is a constant, $(\sum_{i=1}^{m} r_i) \times (\sum_{i=1}^{m} r_i)$ block-diagonal matrix, whose i-th block is an $r_i \times r_i$ matrix of the form:

$$A_i = \begin{bmatrix} 0\,1\,0 \cdots 0 \\ 0\,0\,1 \cdots 0 \\ \vdots \qquad \vdots \\ 0\,0\,0 \cdots 1 \\ 0\,0\,0 \cdots 0 \end{bmatrix} \tag{4.7}$$

B is an a constant matrix of dimension $(\sum_{i=1}^{m} r_i) \times m$. The i-th column of B has the general form: $b_i = [b_i^{(0)^T} \; b_i^{(1)^T} \; b_i^{(2)^T}]^T$, where $b_i^{(0)}$ and $b_i^{(2)}$ are zero row vectors of dimensions $1 \times \left(\sum_{j=1}^{i-1} r_j\right)$ and $1 \times \left(\sum_{j=i+1}^{m} r_j\right)$, respectively, and $b_i^{(1)} = [0 \; 0 \; \cdots \; 1]$ is a $1 \times r_i$ row vector. The function $r(\cdot)$ is a $(\sum_{i=1}^{m} r_i) \times 1$ continuous nonlinear vector function of the form $[L_f^{r_1} h_1(x) - v_1^{r_1} \; L_f^{r_2} h_2(x) - v_2^{r_2} \; \cdots \; L_f^{r_m} h_m(x) - v_m^{r_m}]^T$, and $\bar{v} = [\bar{v}_1^T \; \bar{v}_2^T \; \cdots \; \bar{v}_m^T]^T$ where $\bar{v}_i = [v_i \; v_i^{(1)} \; \cdots \; v_i^{(r_i)}]^T$ is a smooth vector function, and $v_i^{(k)}$ is the k-th time derivative of the external reference input v_i (which is assumed to be a smooth function of time). The $m \times m$ matrix, $C_1(\cdot)$, is the characteristic matrix of the system of (4.1) defined in (4.3) with $x = T^{-1}(e,\eta,\bar{v})$, while $C_2(\cdot)$ is an $m \times q$ matrix of the form:

$$C_2(x) = \begin{bmatrix} L_{w_1} L_f^{r_1-1} h_1(x) & \cdots & L_{w_q} L_f^{r_1-1} h_1(x) \\ \vdots & \cdots & \vdots \\ L_{w_1} L_f^{r_m-1} h_m(x) & \cdots & L_{w_k} L_f^{r_m-1} h_m(x) \end{bmatrix} \tag{4.8}$$

In order to simplify the presentation of the main results, the matrix $C_1(x)$ is assumed to be nonsingular, uniformly in x. This assumption can be relaxed if robust dynamic state feedback, instead of robust static state feedback, is used to solve the control problem (see [126] for details). Finally, we define the function $\bar{f}(e,\eta,\bar{v}) = Ae + Br(e,\eta,\bar{v})$, denote by $\bar{g}_i$ the i-th column of the matrix function $G := BC_1$, $i = 1, \cdots m$, and denote by $\bar{w}_k$ the k-th column of the matrix function $W := BC_2$, $k = 1, \cdots q$.

Following [78], the requirement of input-to-state stability is imposed on the η-subsystem of (4.5) to ensure bounded stability of the internal dynamics and allow the synthesis of the desired controller on the basis of the e-subsystem. This assumption, stated below, is standard in the process control literature concerned with enforcing output tracking and is satisfied by many practical systems, including the chemical reactor models studied in this chapter which do not exhibit non-minimum phase behavior (see also [58, 78] for additional examples).

Assumption 4.2 *The dynamical system:*

$$\begin{aligned} \dot{\eta}_1 &= \Psi_1(e,\eta,\theta,\bar{v}) \\ &\vdots \\ \dot{\eta}_{n-\sum_i r_i} &= \Psi_{n-r}(e,\eta,\theta,\bar{v}) \end{aligned} \tag{4.9}$$

is ISS with respect to e *uniformly in* θ, $\bar{v}$.

In order to achieve attenuation of the effect of the uncertain variables on the outputs, we assume the existence of known (but not necessarily small) bounds that capture the size of the uncertain variables for all times.

Assumption 4.3 *There exist known constants* θ_{bk} *such that* $\|\theta_k\|^s \leq \theta_{bk}$ *where* $\|\theta_k\|^s$ *denotes ess.sup.* $\|\theta(t)\|$, $t \geq 0$.

In order to address the robust controller synthesis problem on the basis of the e-subsystem of (4.6), we need to construct an appropriate robust control Lyapunov function. For systems of the form of (4.6), this can be done in many different ways. One way, for example, is to use a quadratic function, $V = e^T Pe$, where P is a positive-definite matrix chosen to satisfy the following Ricatti equation:

$$A^T P + PA - PBB^T P = -Q \tag{4.10}$$

for some positive-definite matrix, Q. Computing the time-derivative of V along the trajectories of the system of (4.6), and using the relation of (4.10), it can be shown that:

$$\begin{aligned} \inf_{u\in\mathcal{U}} \sup_{\theta\in\mathcal{W}^q} \dot{V} &= \inf_{u\in\mathcal{U}} \sup_{\theta\in\mathcal{W}^q} \left[L_{\bar{f}}V + L_G V u + L_W V\theta\right] \\ &= \inf_{u\in\mathcal{U}} \sup_{\theta\in\mathcal{W}^q} \left[-e^T Qe + e^T PB\big(B^T Pe + 2r(e,\eta,\bar{v}) + 2C_1 u + 2C_2\theta\big)\right] \end{aligned} \tag{4.11}$$

where $L_{\bar{f}}V = -e^T Qe + e^T PBB^T Pe + 2e^T PBr$, $L_G V = 2e^T PBC_1$, and $L_W V = 2e^T PBC_2$. Therefore, choosing P to satisfy (4.10) guarantees that when $L_G V = 0$ (i.e., when $e^T PB = 0$, since C_1 is nonsingular), we have $\inf_{u\in\mathcal{U}} \sup_{\theta\in\mathcal{W}^q} \dot{V} < 0$ for all $e \neq 0$.

Theorem 4.1 that follows provides the explicit synthesis formula for the desired bounded robust multivariable state feedback control law and states precise conditions that guarantee closed-loop stability and robust asymptotic output tracking in the presence of uncertainty and input constraints. The key idea in the proposed controller design is that of bounding a robust Lyapunov-based controller, and is inspired by the results on bounded control in [177] for systems without uncertainty. Analysis of the closed-loop system utilizes combination of standard Lyapunov techniques together with the concept of input-to-state stability and nonlinear small-gain type arguments. The proof of the theorem can be found in Appendix B.

Theorem 4.1. *Consider the constrained uncertain nonlinear process of (4.1), for which Assumptions 4.1–4.3 hold, under the static state feedback law:*

$$u = -k(x, u_{max}, \theta_b, \rho, \chi, \phi)\,(L_G V)^T \tag{4.12}$$

where

$$k(x, u_{max}, \theta_b, \rho, \chi, \phi) = \frac{L_{\bar{f}}^{*} V + \sqrt{\left(L_{\bar{f}}^{**} V\right)^2 + (u_{max}\|(L_G V)^T\|)^4}}{\|(L_G V)^T\|^2 \left[1 + \sqrt{1 + (u_{max}\|(L_G V)^T\|)^2}\right]} \tag{4.13}$$

when $L_G V \neq 0$ and $k(\cdot) = 0$ when $L_G V = 0$; and

$$\begin{aligned} L_{\bar{f}}^{*} V &= L_{\bar{f}} V + \left(\rho\|2Pe\| + \chi\sum_{k=1}^{q}\theta_{bk}\|L_{\bar{w}k}V\|\right)\left(\frac{\|2Pe\|}{\|2Pe\| + \phi}\right) \\ L_{\bar{f}}^{**} V &= L_{\bar{f}} V + \rho\|2Pe\| + \chi\sum_{k=1}^{q}\theta_{bk}\|L_{\bar{w}k}V\| \end{aligned} \tag{4.14}$$

$V = e^T Pe$, P is a symmetric, positive-definite matrix that satisfies (4.10), $L_G V = [L_{\bar{g}_1} V \ \cdots \ L_{\bar{g}_m} V]$ is a row vector, and ρ, χ and ϕ are adjustable parameters that satisfy $\rho > 0$, $\chi > 1$ and $\phi > 0$. Let Π be the set defined by:

$$\begin{aligned} &\Pi(\rho, \chi, \theta_b, u_{max}) \\ &= \{x \in \mathrm{IR}^n : L_{\bar{f}} V + \rho\|2Pe\| + \chi\sum_{k=1}^{q}\theta_{bk}\|L_{\bar{w}k}V\| \ \leq \ u_{max}\|(L_G V)^T\|\} \end{aligned} \tag{4.15}$$

and, assuming the origin is contained in the interior of Π, let $\delta_s > 0$ be chosen such that the set $\Omega = \{x \in \mathrm{IR}^n : \|x\| \leq \delta_s\}$ is an invariant subset of Π. Then given any initial condition, $x_0 \in \Omega$, there exists $\phi^* > 0$ such that if $\phi \in (0, \phi^*]$:

(1) The origin of the constrained closed-loop system is asymptotically stable (i.e., there exists a function β of class $\mathcal{KL}$ such that $\|x(t)\| \leq \beta(\|x_0\|, t), \forall\, t \geq 0$).
(2) The outputs of the closed-loop system satisfy a relation of the form:

$$\limsup_{t \to \infty} \|y_i(t) - v_i(t)\| = 0, \ i = 1, \ldots, m \tag{4.16}$$

Remark 4.2. Theorem 4.1 proposes a direct robust nonlinear controller design method that accounts simultaneously for closed-loop performance and stability in the presence of uncertainty and input constraints. Note that the bounded control law of (4.12–4.14) uses explicitly the available knowledge of both the size of the uncertainty (i.e., θ_{bk}) and the manipulated input constraints (i.e., u_{max}) to generate the necessary control action. This is in contrast to the two-step approach typically employed in process control strategies which first involves the design of a controller for the unconstrained process and then accounts for the input constraints through a suitable anti-windup modification to attenuate the adverse effects of improperly handled input constraints.

Remark 4.3. In addition to the explicit controller synthesis formula, Theorem 4.1 provides an explicit characterization of the region in the state-space where the desired closed-loop stability and set-point tracking properties are guaranteed under the proposed control law. This characterization can be obtained from the inequality in (4.15) which describes a closed state-space region where: (1) the control action satisfies the input constraints, and (2) the time-derivative of the Lyapunov function is guaranteed to be negative-definite, along the trajectories of the constrained uncertain closed-loop system. Any closed-loop trajectory that evolves (i.e., starts and remains) within this region is guaranteed to converge asymptotically to the origin. In fact, it is not difficult to see how this inequality is closely related to the classical Lyapunov stability condition:

$$\dot{V} = L_{\bar{f}}V + L_G V u + \sum_{k=1}^{q} L_{\bar{w}_k} V \theta_k \leq 0 \tag{4.17}$$

when the additional requirements that $\|u\| \leq u_{max}$ and $\|\theta_k\|^s \leq \theta_{bk}$ are imposed. The key idea here is that by taking the constraints directly into account (i.e., bounding the controller), we automatically obtain information about the region where both stability is guaranteed and the input constraints are respected. The inequality of (4.15) can therefore be used to identify a priori (before implementing the controller) the set of admissible initial conditions starting from where closed-loop stability is guaranteed (region of closed-loop stability). This aspect of the proposed design has important practical implications for efficient process operation since it provides plant operators with a systematic and easy-to-implement guide to identify feasible initial conditions for process operation. Considering the fact that the presence of disturbances and constraints limits the conditions under which the process can be operated safely and reliably, the task of identifying these conditions becomes a central one. This is particularly significant in the case of unstable plants where lack of such a priori knowledge can lead to undesirable consequences.

Remark 4.4. Referring to the region described by the set Π in (4.15), it is important to note that even though a trajectory starting in Π will move from one Lyapunov surface to an inner Lyapunov surface with lower energy (because $\dot{V} < 0$), there is no guarantee that the trajectory will remain forever in Π, since it is not necessarily a region of invariance. Once the trajectory leaves Π, however, there is no guarantee that $\dot{V} < 0$. Therefore, in order to use the inequality in (4.15) to identify the admissible initial conditions starting from where closed-loop stability is guaranteed, we need to find (or estimate) an invariant subset – preferably the largest – within Π to guarantee that a trajectory starting in Π remains in the region for all future times. For notational convenience, this invariant subset is represented in Theorem 4.1 by a Euclidean ball, Ω, of size δ_s. The stability results of the theorem, however, hold for any invariant subset (which can be of arbitrary shape) that can be constructed within Π and contains the origin in its interior. For example, a more general estimate of the stability region can be obtained using the level set:

$$\Omega_c = \{x \in \mathbb{R}^n : \bar{V}(x) \leq c\}$$

when Ω_c is bounded and contained in Π and $\dot{\bar{V}}(x) < 0$ over Π. This estimate can be made less conservative by selecting the value of c sufficiently large. Referring to the above definition of the invariant set, Ω_c, it is important to note that the Lyapunov function, $\bar{V}$, is in general not the same as the Lyapunov function, V, given in the statement of Theorem 4.1. The function $\bar{V}$ is based on the full system (i.e., the ζ-η interconnected system of (4.5), while V is based only on the ζ-subsystem. The reason for the difference is the fact that owing to the ISS property of the η-subsystem (see Assumption 4.2), only a Lyapunov function based on the ζ-subsystem, namely $V = e^T P e$, is needed and used to design a control law that stabilizes the full closed-loop system. However, in constructing Ω_c, we need to guarantee that the evolution of x (and, hence, both the ζ and η subsystems) is confined within Π. Therefore, the Lyapunov function in this case must account for the evolution of the η states as well. One possible choice for $\bar{V}$ is a composite Lyapunov function, $\bar{V} = V_e + V_\eta$ whose time-derivative is negative-definite at all points satisfying (4.15), where V_e is taken to be the same as V given in Theorem 4.1 and V_η is another Lyapunov function for the η-subsystem. The existence of V_η is guaranteed from converse Lyapunov theorems owing to the ISS and asymptotic stability properties of the η-subsystem, and can be computed given the particular structure for this system. For nonlinear systems with relative degrees $\sum_i r_i = n$, the choice $\bar{V} = V = e^T P e$ is sufficient.

Remark 4.5. Note that, since the set Π is closed, the assumption that the origin lies in the interior of Π assures the existence of an invariant subset around the origin. This assumption can be checked prior to the practical implementation of the results of the theorem. However, it can be shown by means of local Lipschitz arguments that this assumption is automatically satisfied if the vector fields are sufficiently smooth on their domains of definition and the adjustable parameters in (4.15) are chosen properly.

Remark 4.6. By inspection of the inequality in (4.15), it is clear that the size of the set, Π, depends on the size of the constraints and the size of the uncertainty. The tighter the input constraints (i.e., smaller u_{max}) and/or the larger the plant-model mismatch (i.e., larger θ_{bk}), the smaller the closed-loop stability region. This is quite intuitive since under such conditions, fewer initial conditions will satisfy the inequality of (4.15). Under extremely tight constraints (u_{max} close to zero), this set may contain only the origin (particularly in the case of open-loop unstable processes where $L_{\bar{f}}V > 0$). In this case, the origin could be the only admissible initial condition for which closed-loop stability can be guaranteed. In most applications of practical interest, however, the stability region contains the origin in its interior and is not empty (see the chemical reactor examples studied in Sects. 4.4 and 4.6 for a demonstration of this fact). Another important feature of the inequality of (4.15) is that it suggests a general way for estimating the stability region, which can be used in conjunction with other control approaches (such as optimization-based approaches; see, for example, Chaps. 5 and 6 of this book) to provide the necessary a priori closed-loop stability guarantees. Although the inequality of (4.15) pertains specifically to the bounded robust controller of (4.12–4.14), similar ideas can be used to identify the feasible initial conditions for other control approaches.

Remark 4.7. The inequality in (4.15) captures the classical tradeoff between stability and performance. In particular, analysis of this inequality reveals an interesting interplay between the controller design parameters and the uncertainty in influencing the size of the closed-loop stability region. To this end, note the multiplicative appearance of the parameter, χ, and the uncertainty bound, θ_{bk}, in (4.15). In the presence of significant disturbances, one typically selects a large value for χ to achieve an acceptable level of robust performance of the controller. According to the inequality of (4.15), this comes at the expense of obtaining a smaller region of closed-loop stability. Alternatively, if one desires to expand the region of closed-loop stability by selecting a small value for χ, this may be achieved at the expense of obtaining an unsatisfactory degree of uncertainty attenuation or will be limited to cases where the uncertainty is not too large (i.e., small θ_{bk}) where a large value for χ is not needed. Therefore, while the presence of the design parameter χ in (4.15) offers the possibility of enlarging the region of guaranteed closed-loop stability, a balance must always be maintained between the desired level of uncertainty attenuation and the desired size of the region of closed-loop stability.

Remark 4.8. Theorem 4.1 explains some of the advantages of using a Lyapunov framework for our development and why it is a natural framework to address the problem within an analytical setting. Owing to the versatility of the Lyapunov approach in serving both as a useful robust control design tool (see Chap. 3) and an efficient analysis tool, the proposed controller design method integrates explicitly the two seemingly separate tasks of controller design and closed-loop stability analysis into one task. This is in contrast to other controller design approaches where a control law is designed first to meet certain design specifications and then a Lyapunov function is found to analyze the closed-loop stability characteristics. Using a Lyapunov function to examine the closed-loop stability under a predetermined non-Lyapunov based control law (e.g., a feedback linearizing controller) usually results in a rather conservative stability analysis and yields conservative estimates of the region of closed-loop stability (see Sect. 4.4 for an example). In the approach proposed by Theorem 4.1, however, the Lyapunov function used to characterize the region of guaranteed closed-loop stability under the control law of (4.12–4.14) is the same one used to design the controller and is therefore a natural choice to analyze the closed-loop system.

Remark 4.9. In the special case when the term $L_{\bar{f}}V$ is negative and no uncertainty is present (i.e., $\theta_{bk} = 0$), the closed-loop stability and performance properties outlined in Theorem 4.1 can be made to hold globally by setting $\rho = 0$. This follows from the fact that when $L_{\bar{f}}V < 0$, $\theta_{bk} = 0$ and $\rho = 0$, the inequality in (4.15) is automatically satisfied everywhere in the state-space. Consequently, the closed-loop stability and set-point tracking properties are satisfied for any initial condition; hence the global nature of the result. The significance of this result is that it establishes the fact that, for open-loop stable nonlinear systems of the form of (4.1) (where $L_{\bar{f}}V < 0$), asymptotic stability and output tracking can always be achieved globally in the presence of arbitrary input constraints using the class of bounded control laws of (4.12–4.14), both in the case when no uncertainty is present and in the case when the uncertainty is mild enough that (4.15) continues to be satisfied.

Remark 4.10. Despite some conceptual similarities between the controller of (4.12–4.14) and that of (3.18), the two control laws differ in several fundamental respects. First, the control law of (4.12–4.14) is inherently bounded by u_{max} and generates control action that satisfies the input constraints within the region described by (4.15). By contrast, the control law of (3.18) is unbounded and may compute larger control action that violates the constraints within the same region. Second, the control law of (4.12–4.14) possesses a well-defined region of guaranteed stability and performance properties in the state-space. One can, therefore, determine a priori whether a particular initial condition is feasible. No explicit characterization of this region is given for the control law of (3.18). Finally, while the tasks of nominal stabilization and uncertainty attenuation could be conceptually distinguished and assigned separately to different components of the control law of (3.18), this is no longer possible for the controller of (4.12–4.14) due to the presence of the term $\chi\theta_{bk}\|L_{\bar{w}k}V\|$ in the radicand, which is a direct consequence of the bounding procedure of the controller.

Remark 4.11. Similar to the control law of (3.18), the control law of (4.12–4.14) recognizes the beneficial effects of process nonlinearities and does not expend unnecessary control effort to cancel them. However, unlike the controller of (3.18), the one of (4.12–4.14) has the additional ability to recognize the extent of the effect of uncertainty on the process and prevent its cancellation if this effect is not significant. To see this, recall that the controller of (3.18) prevents the unnecessary cancellation of the term $L_{\bar{f}}V$ – which does not include the uncertainty – when it is negative. This controller, therefore, compensates for the presence of any model uncertainty it detects and does not discriminate between small or large uncertainty. In contrast, the controller of (4.12–4.14) recognizes the beneficial effect of the entire term $L_{\bar{f}}V + \chi\theta_{bk}\|L_{\bar{w}k}V\|$ (which includes the uncertainty) when it is negative and prevents its unnecessary and wasteful cancellation. Therefore, if the plant-model mismatch is not too significant such that the term, $L_{\bar{f}}V + \chi\theta_{bk}\|L_{\bar{w}k}V\|$, remains negative in the presence of such uncertainty, the controller will prevent the expenditure of unnecessary control effort to cancel such uncertainty. Essentially the controller realizes, through the negative sign of this term, that the uncertainty present in the process does not have a strong enough adverse effect to warrant its cancellation. Therefore the control law of (4.12–4.14) has the ability to discriminate between small and large model uncertainty and assess the need for reasonable control effort expenditure accordingly.

Remark 4.12. The bounded nonlinear controller of (4.12–4.14) possesses certain optimality properties characteristic of its ability to use reasonable control action to accomplish the desired closed-loop objectives. Using techniques from the inverse optimal control approach (see, for example, [97,245] and Chap. 3), one can prove that, within a well-defined subset of the set Π, this controller is optimal with respect to a meaningful, infinite-time cost functional of the form:

$$J = \int_0^{\infty} (l(e) + u^T R(x) u) dt \tag{4.18}$$

where $l(e)$ is a positive-definite penalty on the tracking error and its time-derivatives, up to order r_{i-1}, that is bounded below by a quadratic function of the norm of the tracking error, and $R(x) = \frac{1}{2}k(x) > 0$ is a positive-definite penalty weight on the control action. It can also be shown that the minimum cost achieved by the state feedback controller is $V(e(0))$. Note that, unlike the stability and set-point tracking properties which are guaranteed provided that the closed-loop trajectory remains within Π, inverse optimality is guaranteed provided that the evolution of the closed-loop trajectory is confined within a smaller subset of Π. The fact that the optimality region is, in general, a subset of the stability region can be understood in light of the bounded nature of the controller. In contrast to their unbounded counterparts (e.g., the controller of (3.18)), bounded controllers tend to experience a reduction in their stability margins near the boundary of the stability region (note in particular that $\dot{V}$ becomes increasingly less negative as we approach the boundary of the region described by Π). These margins, however, are needed in the inverse optimal approach to establish meaningful optimality (i.e., the positive-definiteness of the penalty functions). Therefore, to guarantee optimality, one must step back from the boundary of the region described by the set, Π, and restrict the evolution of the closed-loop trajectories in a smaller region where stability margins are sufficient to render the cost functional meaningful. The detailed mathematical proof of this point can be found in [78] for the case of SISO nonlinear systems and will not be repeated here.

Remark 4.13. An important problem, frequently encountered in the control of multivariable processes with input constraints, is the problem of directionality. This is a performance deterioration problem that arises when a multivariable controller, designed on the basis of the unconstrained MIMO process, is "clipped" to achieve a feasible plant input. Unlike the SISO case, clipping an unconstrained controller output to achieve a feasible plant input may not lead in MIMO systems to a plant output that is closest to the unconstrained plant output, thus steering the plant in the wrong directions and leading to performance deterioration. Controllers that do not account explicitly for input constraints suffer from this problem and often require the design of directionality compensators to prevent the resulting excessive performance deterioration (see, for example, [257]). It is important to note that the Lyapunov-based control approach proposed in Theorem 4.1 avoids the directionality problem by taking input constraints directly into account in the controller design. In particular, note that the controller of (4.12–4.14) inherently respects the constraints within the region described by (4.15) and, therefore, starting from any initial condition in Ω, there is never a mismatch between the controller output and actual plant input at any time.

Remark 4.14. The result of Theorem 4.1 can be extended to the case when the uncertain variables in (4.1) are non-vanishing. In this case, starting from any initial state within the invariant set Ω, the bounded robust controller of (4.12–4.14) enforces boundedness of the process states and robust asymptotic output tracking with an arbitrary degree of attenuation of the effect of uncertainty on the outputs of the closed-loop system. Owing to the persistent nature of the uncertainty, asymptotic convergence to the origin in this case is not possible. Instead, the controller sends the closed-loop trajectory, in finite-time, into a small neighborhood of the origin. The size of this neighborhood (i.e., the asymptotic tracking error) can be made arbitrarily small by selecting the tuning parameters ϕ to be sufficiently small and/or selecting the tuning parameter χ to be sufficiently large. The problem of non-vanishing uncertainty was addressed in detail in [78] for SISO nonlinear processes. It is worth noting that the use of a Lyapunov-based control approach allows the synthesis of a robust controller that can effectively attenuate the effect of both constant and time-varying persistent uncertainty on the closed-loop outputs, which cannot be achieved using classical uncertainty compensation techniques, including integral action and parameter adaptation in the controller. For nonlinear controller design, Lyapunov methods provide useful and systematic tools (see, for example, [141]).

Remark 4.15. Regarding the continuity properties of the controller of (4.12–4.14), it should be noted that the controller is continuous, away from $L_GV = 0$, because the functions $L_{\bar{f}}^*V$, $L_{\bar{f}}^{**}V$, L_GV are continuous everywhere on their domains of definition. However, similar to other L_GV-controller designs, since the gain function in (4.13) is undefined when $L_GV = 0$, the controller gain must be set to some value that satisfies the constraints whenever $L_GV = 0$.

A simple choice is to set the gain to zero at those points. While this choice can make the controller discontinuous at $L_G V = 0$, continuity at those points can be ensured by setting the gain to the limit of the function in (4.13) as $L_G V$ approaches zero. Note, however, that regardless of the value chosen, stability is unaffected owing to the property that the time-derivative of V is negative-definite whenever $L_G V = 0$ (see (4.11)). Furthermore, to circumvent any potential numerical difficulties that may arise when implementing the controller (for example, chattering of the control input near the points where $L_G V = 0$), a continuous approximation of the controller can be implemented, whereby a sufficiently small positive real number, ϵ, is added to the $\|(L_G V)^T\|^2$ term in the denominator of (4.13) in order to smoothen out the control action near $L_G V = 0$ (see the simulation example in Sect. 4.4). As noted in Remark 3.12, this modification reduces the ability of the controller to cancel out, or dominate, the undesirable terms, $L_{\bar{f}} V$ and $\|L_{\bar{w}_k} V\| \theta_{bk}$, and must therefore be compensated for by increasing the controller gain in the numerator (for example, through proper selection of χ, ρ) to ensure practical stability (see Remark 3.12 for other possible modifications).

Remark 4.16. Referring to the practical implementation of the proposed control approach, we initially verify whether Assumptions (4.1–4.2) hold for the nonlinear process under consideration, and determine the available bounds on the uncertain variables. Next, given the constraints on the manipulated inputs, the inequality of (4.15) is used to compute the estimate of the region of guaranteed closed-loop stability, Ω, and check whether a given initial condition is admissible. Then, the synthesis formula of (4.12–4.14) (with the appropriate modifications discussed in Remark 4.15) is used to design the controller which is then implemented on the process. Note that the actual controller synthesis requires only off-line computations, including differentiation and algebraic calculations to compute the various terms in the analytical controller formula. This task can be easily coded into a computer using available software for symbolic manipulations (e.g., MATHEMATICA). Furthermore, the approach provides explicit guidelines for the selection of the tuning parameters that guarantee closed-loop stability and the desired robust performance. For example, the tuning parameters, χ and ϕ, are responsible for achieving the desired degree of uncertainty attenuation. A large value for χ (greater than one) and a small value for ϕ (close to zero) lead to significant attenuation of the effect of uncertainty on the closed-loop outputs.

4.4 Application to an Exothermic Chemical Reactor

Consider a well-mixed continuous stirred tank reactor where an irreversible elementary exothermic reaction of the form $A \xrightarrow{k} B$ takes place. The feed to the reactor consists of pure A at flow rate F, molar concentration C_{A0} and temperature T_{A0}. Under standard modeling assumptions, the process model takes the following form:

$$
\begin{aligned}
V\frac{dC_A}{dt} &= F(C_{A0}-C_A) - k_0 \exp\left(\frac{-E}{RT}\right) C_A V \\
V\frac{dT}{dt} &= F(T_{A0}-T) + \frac{(-\Delta H)}{\rho c_p} k_0 \exp\left(\frac{-E}{RT}\right) C_A V + \frac{Q}{\rho c_p}
\end{aligned}
\tag{4.19}
$$

where C_A denotes the concentration of species A, T denotes the temperature of the reactor, Q denotes the rate of heat input to the reactor, V denotes the volume of the reactor, k_0, E, ΔH denote the pre-exponential constant, the activation energy, and the enthalpy of the reaction, c_p and ρ, denote the heat capacity and density of the fluid in the reactor. The steady-state values and process parameters are given in Table 4.1. For these parameters, it was verified that the given equilibrium point is an unstable one (the system also possesses two locally asymptotically stable equilibrium points).

Table 4.1. Process parameters and steady-state values for the reactor of (4.19).

V	$= 100.0$	L
R	$= 8.314$	$J/mol.K$
C_{A0}	$= 1.0$	mol/L
T_{A0s}	$= 310.0$	K
ΔH	$= -4.78 \times 10^4$	J/mol
k_0	$= 7.20 \times 10^{10}$	min^{-1}
E	$= 8.31 \times 10^4$	J/mol
c_p	$= 0.239$	$J/g.K$
ρ	$= 1000.0$	g/L
F	$= 100.0$	L/min
C_{As}	$= 0.577$	mol/L
T_s	$= 395.3$	K

The control objective is to regulate both the reactor temperature and reactant concentration at the (open-loop) unstable equilibrium point by manipulating both the rate of heat input/removal and the inlet reactant concentration. The control objective is to be accomplished in the presence of: (1) time-varying persistent disturbances in the feed stream temperature, (2) parametric uncertainty in the value of the heat of reaction, and (3) hard constraints on the manipulated inputs. Defining $x_1 = C_A$, $x_2 = T$, $u_1 = C_{A0} - C_{A0s}$, $u_2 = Q$, $\theta_1(t) = T_{A0} - T_{A0s}$, $\theta_2(t) = \Delta H - \Delta H_{nom}$, $y_1 = C_A$, $y_2 = T$, where the subscript s denotes the steady-state value and ΔH_{nom} denotes the nominal value of the heat of reaction, the process model of (4.19) can be cast in the form of (4.1) with:

$$f(x) = \begin{bmatrix} \dfrac{F}{V}(C_{A0s} - C_A) - k_0 \exp\left(\dfrac{-E}{RT}\right) C_A \\ \\ \dfrac{F}{V}(T_{A0s} - T) + \dfrac{(-\Delta H_{nom})}{\rho c_p} k_0 \exp\left(\dfrac{-E}{RT}\right) C_A \end{bmatrix}, \; g_1(x) = \begin{bmatrix} \dfrac{F}{V} \\ 0 \end{bmatrix}$$

$$g_2(x) = \begin{bmatrix} 0 \\ \\ \dfrac{1}{\rho c_p V} \end{bmatrix}, \; w_1(x) = \begin{bmatrix} 0 \\ \\ \dfrac{F}{V} \end{bmatrix}, w_2(x) = \begin{bmatrix} 0 \\ \\ k_0 \exp\left(\dfrac{-E}{RT}\right) C_A \end{bmatrix} \tag{4.20}$$

$h_1(x) = x_1$, $h_2(x) = x_2$. In all simulation runs, the following time-varying function was considered to simulate the effect of exogenous disturbances in the feed temperature:

$$\theta_1(t) = \theta_0 sin(3t) \tag{4.21}$$

where $\theta_0 = 0.08T_{A0s}$. In addition, a parametric uncertainty of 50% in the heat of reaction was considered, i.e., $\theta_2(t) = 0.5\,(-\Delta H_{nom})$. The upper bounds on the uncertain variables were therefore taken to be $\theta_{b1} = 0.08T_{A0s}$, $\theta_{b2} = 0.5\,|(-\Delta H_{nom})|$. Also, the following constraints were imposed on the manipulated inputs: $\|u_1\| \leq 1.0\; mol/L$ and $\|u_2\| \leq 92\; KJ/s$.

A quadratic Lyapunov function of the form $V = (C_A - C_{As})^2 + (T - T_s)^2$ was used to design the multivariable bounded robust controller of (4.12–4.14) and to compute the associated region of guaranteed closed-loop stability with the aid of (4.15). Since the uncertainty considered is non-vanishing (see Remark 4.14), the following values were used for the tuning parameters: $\chi = 8.0$, $\phi = 0.0001$, $\rho = 0.01$ to guarantee that the outputs of the closed-loop system satisfy a relation of the form $\limsup_{t \to \infty} \|y_i - v_i\| \leq 0.0005$, $i = 1, 2$. Also, the term $\|(L_G V)^T\|^2$ in the denominator of the control law of (4.12–4.13) was replaced by the number, $\nu = 0.001$, when the system trajectory reached close to the desired steady-state to alleviate chattering of the control action (note that, for this example, $\|(L_G V)^T\| = 0$ if and only if $(C_A, T) = (C_{As}, T_s)$).

Several closed-loop simulation runs were performed to evaluate the robustness and constraint-handling capabilities of the multi-variable controller. Figure 4.1 depicts the startup reactor temperature and reactant concentration profiles for an initial condition that lies within the stability region computed from (4.15). The solid lines represent the process response when the bounded robust controller is tuned properly and implemented on the process. The dashed lines correspond to the response when the controller is designed and implemented without the robust uncertainty compensation component (i.e., $\chi = 0$). Finally, the dotted lines represent the open-loop response. When compared with the open-loop profiles, it is clear that, starting from the given admissible initial condition, the properly tuned bounded robust controller successfully drives both outputs to the desired steady-state, in the presence of input constraints, while simultaneously attenuating the effect of the uncertain

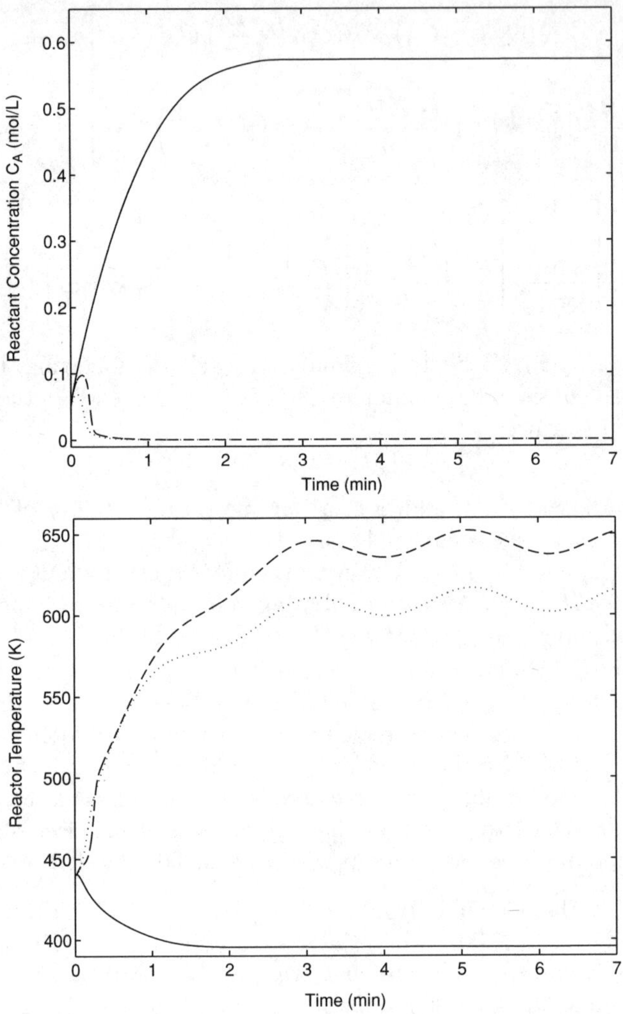

Fig. 4.1. Controlled outputs: reactant concentration (*top*) and reactor temperature (*bottom*) profiles under the bounded multivariable nonlinear state feedback controller of (4.12–4.14) with uncertainty compensation and initial condition inside the stability region (*solid*), without uncertainty compensation (*dashed*) and under open-loop conditions (*dotted*)

variables on the outputs. It should be noted that this conclusion was reached a priori (before controller implementation) based on the characterization of the stability region given in Theorem 4.1. From the dashed profiles we see that the uncertain variables have a significant effect on the process outputs and that failure to explicitly account for them in the controller design leads to instability and poor transient performance. Figure 4.2 shows the correspond-

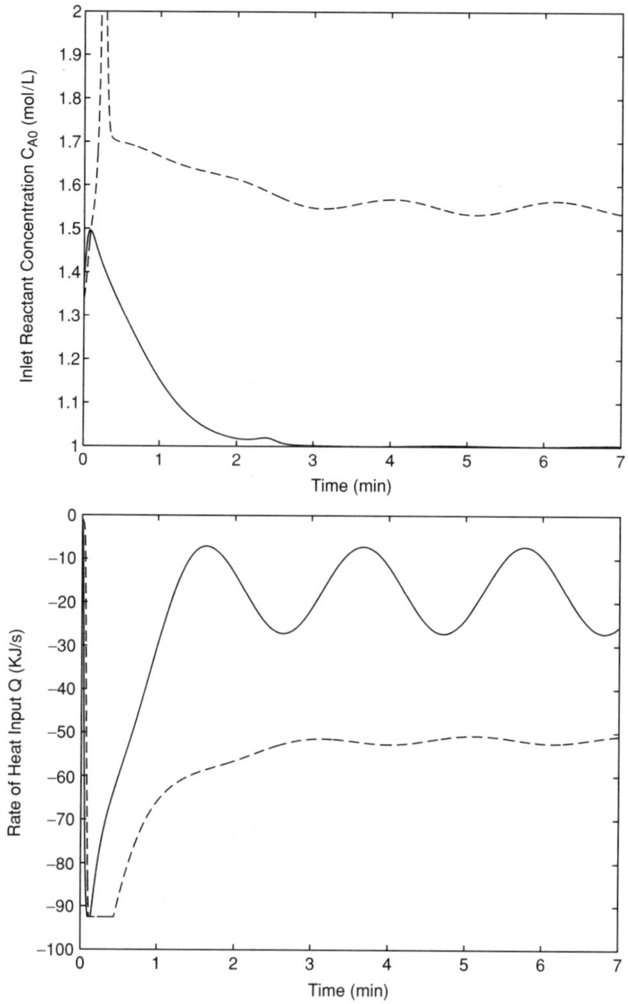

Fig. 4.2. Manipulated inputs: inlet reactant concentration (*top*) and rate of heat input (*bottom*) profiles under the bounded multivariable nonlinear state feedback controller of (4.12–4.14) with uncertainty compensation and initial condition inside the stability region (*solid*), without uncertainty compensation (*dashed*) and under open-loop conditions (*dotted*)

ing profiles for the manipulated inputs. Note that since the initial condition is chosen within the region of guaranteed stability, the well-tuned multivariable controller generates, as expected, control action (solid lines) that respects the constraints imposed.

In Fig. 4.3, we tested the ability of the controller to robustly stabilize the process at the desired steady-state, starting from initial conditions that

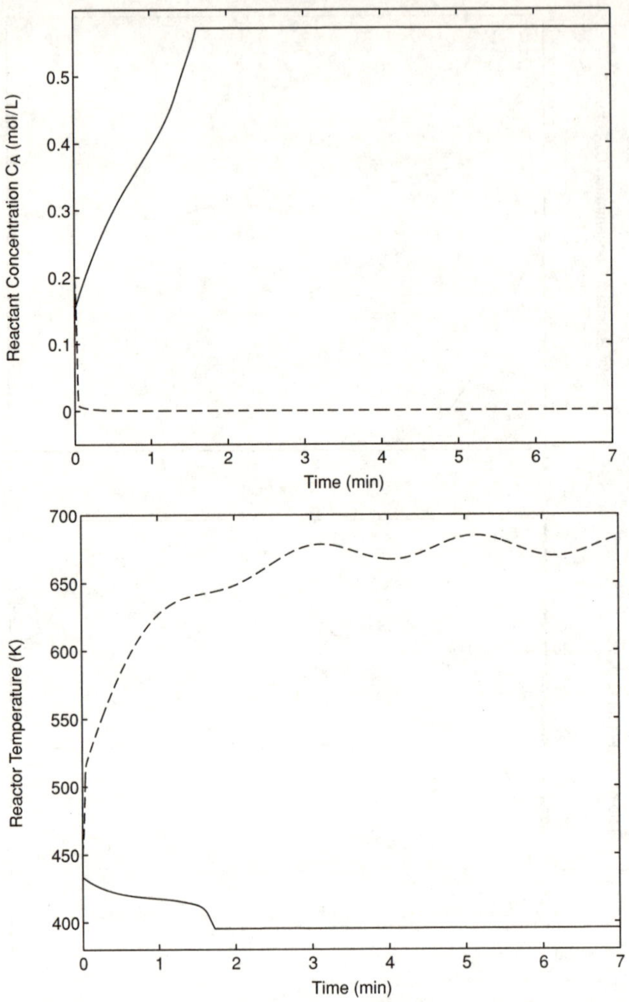

Fig. 4.3. Controlled outputs: reactant concentration (*top*) and reactor temperature (*bottom*) profiles under the bounded robust multivariable nonlinear state feedback controller of (4.12–4.14) for initial conditions outside the region of guaranteed closed-loop stability

lie outside the region of guaranteed closed-loop stability. In this case, no a priori guarantees can be made regarding closed-loop stability. In fact, from the process response denoted by the dashed lines in Fig. 4.3, we see that, starting from the given initial condition, the controller is unable to successfully stabilize the process or attenuate the effect of uncertainty on the process outputs. The reason for this can be seen from the corresponding manipulated inputs' profiles in Fig. 4.4 (dashed lines) which show that both the inlet reactant concentration and rate of heat input stay saturated for all times, indicating that stabilizing

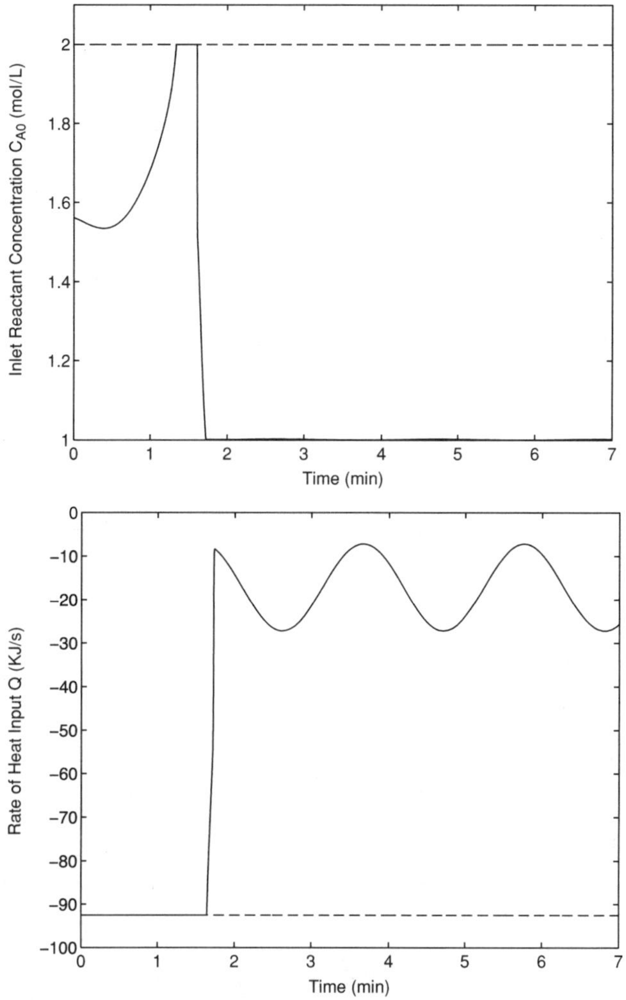

Fig. 4.4. Manipulated inputs: inlet reactant concentration (*top*) and rate of heat input (*bottom*) profiles under the bounded multivariable nonlinear state feedback controller of (4.12–4.14) for initial conditions outside the region of guaranteed closed-loop stability

the process from the given initial condition requires significantly larger control action, which the controller is unable to provide due to the constraints. The combination of insufficient reactant material in the incoming feed stream and insufficient cooling of the reactor prompt an increase in the reaction rate and, subsequently, an increase in the reactor temperature and depletion of the reactant material in the reactor. Note that since the region of guaranteed closed-loop stability given in Theorem 4.1 does not necessarily capture all the admissible initial conditions (this remains an unresolved problem in control),

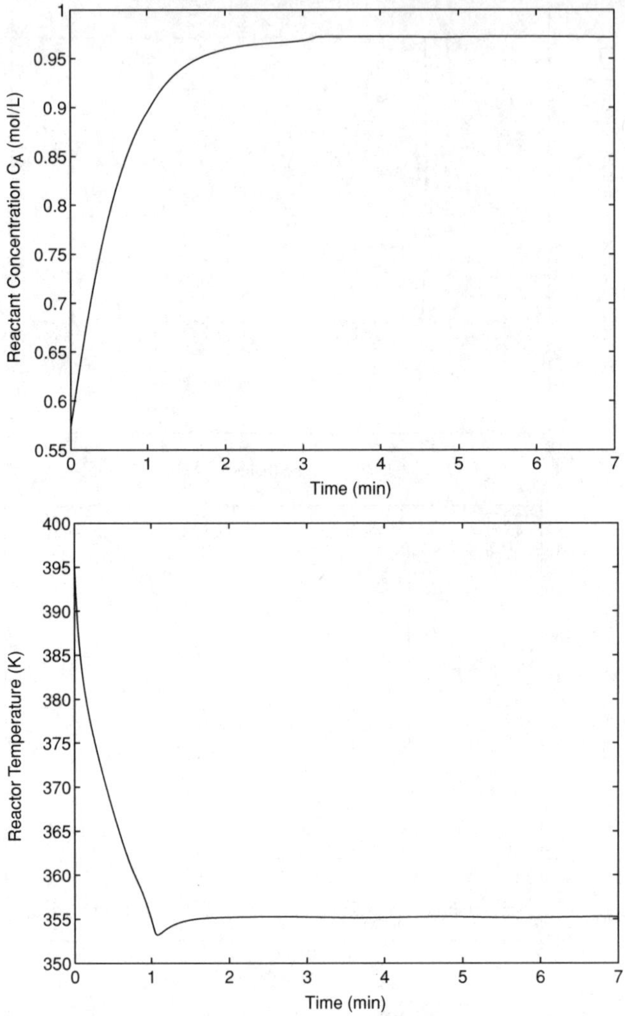

Fig. 4.5. Controlled outputs: reactant concentration (*top*) and reactor temperature (*bottom*) profiles under the bounded robust multivariable nonlinear state feedback controller of (4.12–4.14) for a set-point increase of 0.4 mol/L in the reactant concentration and a set-point decrease of 40 K in the reactor temperature

it is possible for the controller to robustly stabilize the process starting from some initial conditions outside this region. An example of this is shown by the controlled outputs and manipulated inputs' profiles denoted by solid lines in Figs. 4.3–4.4, respectively.

In addition to robust stabilization, we also tested the robust set-point tracking capabilities of the controller under uncertainty and constraints. The results for this case are shown in Figs. 4.5–4.6 which depict the profiles of

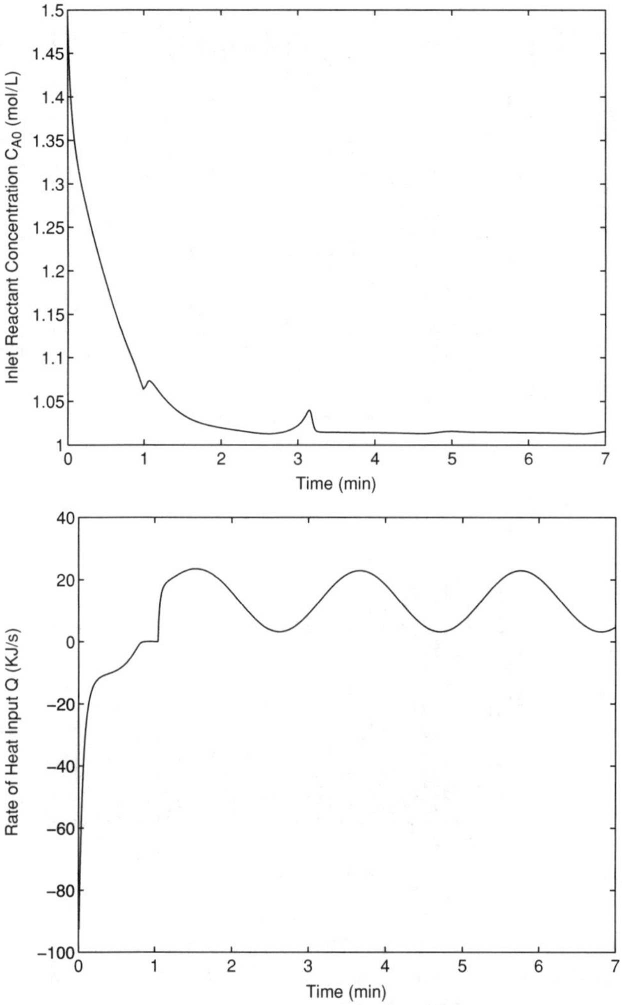

Fig. 4.6. Manipulated inputs: inlet reactant concentration (*top*) and rate of heat input (*bottom*) profiles under the bounded multivariable nonlinear state feedback controller of (4.12–4.14) for a set-point increase of 0.4 mol/L in the reactant concentration and a set-point decrease of 40 K in the reactor temperature

the controlled outputs and manipulated inputs, respectively, in response to a 0.4 mol/L increase in the reactant concentration set-point and 40 K decrease in the reactor temperature set-point. It is clear from the figures that the controller successfully achieves the requested tracking while simultaneously attenuating the effect of uncertainty and generating control action that respects the constraints imposed. Finally, the state-space region where the bounded robust multivariable controller satisfies the input constraints was

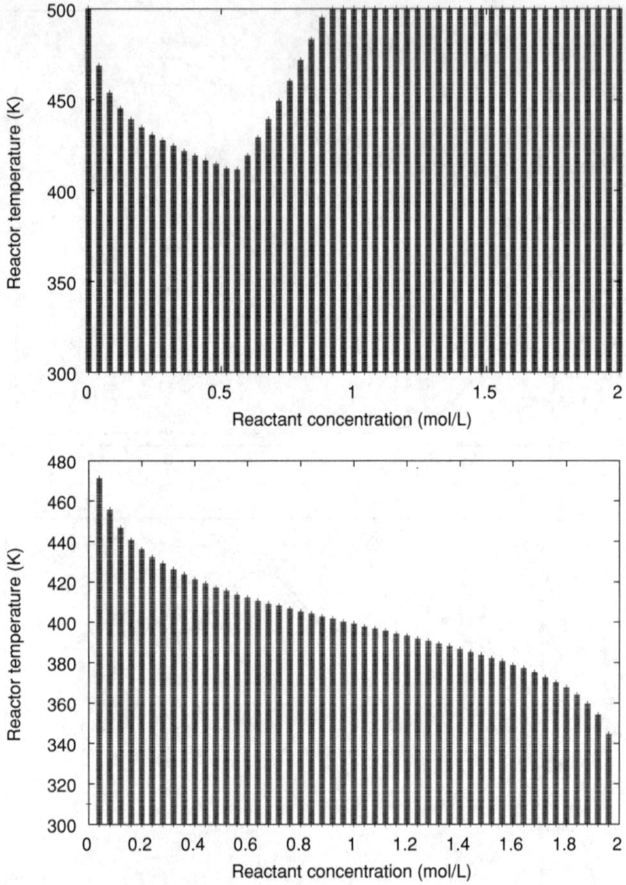

Fig. 4.7. Comparison between the state-space regions where the bounded robust multivariable controller of (4.12–4.14) (*top*) and a multivariable input-output linearizing controller (*bottom*) satisfy the input constraints. The regions are used to provide estimates of the corresponding stability regions

computed using (4.15) and is depicted in Fig. 4.7 (top plot). For the sake of comparison, we included in Fig. 4.7 (bottom plot) the corresponding region for a multivariable input-output linearizing controller that cancels process nonlinearities (the stabilizing linear terms were not included since the associated stabilization cost of these terms would yield an even smaller region). Evolution of the closed-loop trajectories within either region (which is insured by constructing the largest invariant subregion within) guarantees closed-loop stability. It is clear from the comparison that the bounded robust multivariable controller satisfies the constraints for a wider range of process operating conditions. This is a consequence of the fact that, unlike the input-output linearizing controller, the bounded multivariable controller accounts explicitly for constraints. Furthermore, the cost of cancelling process nonlinearities by the linearizing controller renders many initial conditions that are far from the equilibrium point infeasible. In contrast, the bounded controller avoids the unnecessary cancellation of nonlinearities and employs reasonably smaller control action to stabilize the process. This, in turn, results in a larger set of operating conditions that satisfy the constraints.

4.5 State Estimation and Output Feedback Control

The feedback controller of (4.12–4.14) was designed under the assumption of accessibility of all process states for measurement. In chemical process control, however, more so than in other control areas, the complete state often cannot be measured. For example, the concentration of certain intermediates in chemical reactors may not be accessible for online measurements and therefore cannot be used directly for feedback purposes. In the past few years, significant advances have been made in the direction of output feedback controller design for the purpose of robustly stabilizing nonlinear systems. In this section, we address the problem of synthesizing bounded robust *output* feedback controllers for the class of constrained uncertain multivariable nonlinear processes in (4.1). Owing to the explicit presence of time-varying uncertain variables in the process of (4.1), the approach that is followed for output feedback controller design is based on combination of high-gain observers and state feedback controllers (see also [50,146,183,267] for results on output feedback control for unconstrained nonlinear systems). To this end, we initially formulate the control problem in Sect. 4.5.1 and then present it solution in Sect. 4.5.2.

4.5.1 Control Problem Formulation

Referring to the system of (4.1), our objective is to synthesize a bounded robust nonlinear dynamic output feedback controller of the form:

$$\begin{aligned} \dot{\omega} &= \mathcal{F}(\omega, y, \bar{v}) \\ u &= \mathcal{P}(\omega, y, \bar{v}) \end{aligned} \tag{4.22}$$

where $\omega \in \mathbb{R}^s$ is an observer state, $\mathcal{F}(\omega, y, \bar{v})$ is a vector function, $\mathcal{P}(\omega, y, \bar{v})$ is a *bounded* vector function, $\bar{v}$ is a generalized reference input, that: (a) enforces, in the presence of actuator constraints, exponential stability and asymptotic robust output tracking with arbitrary degree of attenuation of the effect of the uncertainty on the output, and (b) provides an explicit characterization of the region where the aforementioned properties are guaranteed.

The design of the dynamic controller is carried out using combination of a high-gain state observer and the bounded robust state feedback controller proposed in Sect. 4.3. In particular, the system $\dot{\omega} = \mathcal{F}(\omega, y, \bar{v})$ in (4.22) is constructed to provide estimates of the process state variables from the measured outputs, while the bounded static component, $\mathcal{P}(\omega, y, \bar{v})$, is synthesized to enforce the requested properties in the closed-loop system and at the same time provide the necessary explicit characterization of the region of guaranteed closed-loop stability. The stability analysis of the closed-loop system employs singular perturbation techniques (due to the high-gain nature of the observer) and utilizes the concept of input-to-state stability and nonlinear small gain theorem-type arguments.

4.5.2 Controller Synthesis

In order to proceed with the controller synthesis task, we need to slightly modify Assumption 4.1 to the following one.

Assumption 4.4 *There exist a set of integers, $\{r_1, r_2, \ldots, r_m\}$, and a coordinate transformation, $(\zeta, \eta) = T(x)$, such that the representation of the system of (4.1), in the (ζ, η) coordinates, takes the form of (4.5) with:*

$$\begin{aligned} \dot{\eta}_1 &= \Psi_1(\zeta, \eta) \\ &\vdots \\ \dot{\eta}_{n-\sum_i r_i} &= \Psi_{n-\sum_i r_i}(\zeta, \eta) \end{aligned} \tag{4.23}$$

We note that the change of variables of (4.5) and (4.23) is independent of θ and invertible since, for every x, the variables, ζ and η, are uniquely determined by (4.5) and (4.23). This implies that if we can estimate the values of ζ, η for all times, using appropriate state observers, then we automatically obtain estimates of x for all times. This property will be exploited later to synthesize a state estimator for the system of (4.1) on the basis of the system of (4.5). We also note that Assumption 4.4 includes the matching condition of our robust control method. In particular, we consider systems of the form (4.1) for which the uncertain variables enter the system in the same equation with the manipulated inputs. This assumption is motivated by our requirement to eliminate the presence of θ in the η-subsystem. This requirement and the stability requirement of Assumption 4.5 below will allow including in the controller a replica of the η-subsystem of (4.23) which provides estimates of the η states.

Assumption 4.5 *The system of (4.23) is ISS with respect to ζ, with $\beta_\eta(\|\eta(0)\|, t) = K_\eta \|\eta(0)\| e^{-at}$ where K_η, a are positive real numbers and $K_\eta \geq 1$.*

We are now ready to state the main result of this section. Theorem 4.17 that follows provides the explicit synthesis formula for the desired bounded robust output feedback controller and states precise conditions that guarantee closed-loop stability and robust asymptotic output tracking in the presence of uncertainty and input constraints. The proof of the theorem can be found in the Appendix B.

Theorem 4.17. *Consider the constrained uncertain nonlinear system of (4.1), for which Assumptions 4.3–4.5 hold, under the output feedback controller:*

$$\begin{aligned}\dot{\tilde{y}}^{(i)} &= A\tilde{y}^{(i)} + F_i(y_i - y_0^{(i)}) \\ \dot{\omega}_1 &= \Psi_1(sat(\tilde{y}), \omega) \\ &\vdots \\ \dot{\omega}_{n-\sum r_i} &= \Psi_{n-\sum r_i}(sat(\tilde{y}), \omega) \\ u &= -k(\hat{x}, u_{max}, \theta_b, \rho, \chi, \phi)\left(\widehat{L_G V}\right)^T\end{aligned} \qquad (4.24)$$

where $\hat{x} = T^{-1}(sat(\tilde{y}), \omega)$, $\tilde{y} = [\tilde{y}^{(1)^T} \cdots \tilde{y}^{(m)^T}]^T$, $\omega = [\omega_1 \cdots \omega_{n-\sum_i r_i}]^T$, $F_i = [L_i a_1^{(i)}\ L_i^2 a_2^{(i)} \cdots L_i^{r_i} a_{r_i}^{(i)}]^T, i = 1, \ldots, m$. *Let* $\bar{\epsilon} = max\{1/L_i\}$. *Then for each pair of positive real numbers* (δ_b, d) *such that* $\beta(\delta_b, 0) + d \leq \delta_s$, *where* $\beta(\cdot,\cdot)$ *and* δ_s *were defined in Theorem 4.1, and for each pair of positive real numbers* $(\delta_\theta, \delta_{\bar{v}})$, *there exists* $\phi^* > 0$, *and for each* $\phi \in (0, \phi^*]$, *there exists* $\bar{\epsilon}^*(\phi) > 0$, *such that if* $\phi \in (0, \phi^*]$, $\bar{\epsilon} \in (0, \bar{\epsilon}^*(\phi)]$, $sat(\cdot) = min\{1, \zeta_{max}/\|\cdot\|\}(\cdot)$ *with* ζ_{max} *being the maximum value of the vector* ζ *for* $\|\zeta\| \leq \beta_\zeta(\delta_\zeta, 0)$ *where* β_ζ *is a class* $\mathcal{KL}$ *function and* δ_ζ *is the maximum value of the vector* $[l_1^T(x)\ l_2^T(x) \cdots l_m^T(x)]^T$ *for* $\|x\| \leq \delta_b$, *where* $l_i(x) = [h_i(x)\ L_f h_i(x) \cdots L_f^{r_i-1} h_i(x)]^T$, $\|x(0)\| \leq \delta_b$, $\|\theta\|^s \leq \delta_\theta$, $\|\bar{v}\|^s \leq \delta_{\bar{v}}$, $\|\tilde{y}(0)\| \leq \delta_\zeta$, $\omega(0) = \eta(0) + O(\epsilon)$, *the following holds in the presence of constraints:*

(1) The origin of the closed-loop system is asymptotically (and locally exponentially) stable.

(2) The outputs of the closed-loop system satisfy a relation of the form:

$$\limsup_{t\to\infty} \|y_i(t) - v_i(t)\| = 0, \ i = 1, \ldots, m \qquad (4.25)$$

Remark 4.18. The robust output feedback controller of (4.24) consists of: (1) m high-gain observers, each of which provides estimates of the derivatives of one of the m controlled outputs, y_i, up to order $r_i - 1$, and thus estimates of the variables $\zeta_1^{(i)}, \ldots, \zeta_{r_i}^{(i)}$, (2) an observer that simulates the inverse dynamics of the system of (4.23), and (3) a bounded robust static feedback controller (see Theorem 4.1) that uses measurements of the outputs and estimates of the states to attenuate the effect of the uncertain variables on the process outputs and enforce reference input tracking. The use of high-gain observers allows us to achieve a certain degree of separation in the output feedback controller design, where the observer design is carried out independently of the state feedback design. This is possible because of the disturbance rejection properties of high-gain observers that allow asymptotic recovery of the performance achieved under state feedback, where "asymptotic" here refers to the behavior of the system as the poles of the observer approach infinity. It is important to note here that while such "separation principle" holds for the class of feedback linearizable nonlinear systems considered in this chapter, one should not expect the task of observer design to be independent from the state feedback design for more general nonlinear systems. In fact, even in linear control design, when model uncertainties are taken into consideration, the design of the observer cannot be separated from the design of the state feedback control.

Remark 4.19. In designing the output feedback controller of (4.24), we use a standard saturation function, *sat*, to eliminate the peaking phenomenon typically exhibited by high-gain observers in their transient behavior. The origin of this phenomenon owes to the fact that as the observer poles approach infinity, its exponential modes will decay to zero arbitrarily fast, but the amplitude of these modes will approach infinity, thus producing impulsive-like behavior. To eliminate observer peaking, we use the saturation filter in conjunction with the high-gain observers to eliminate (or filter out) wrong estimates of the process output derivatives provided by the observer for short times. The idea here is to exploit our knowledge of an estimate of the stability region obtained under state feedback (Ω) – where the process states evolve – to derive bounds on the actual values of the outputs' derivatives, and then use these bounds to design the saturation filter. Therefore, during the short transient period when the estimates of the high-gain observers exhibit peaking, the saturation filter eliminates those estimates which exceed the state feedback bounds, thus preventing peaking from being transmitted to the plant. Over the same period, the estimation error decays to small values, while the state of the plant remains close to its initial value. The validity of this idea is justified via asymptotic analysis from singular perturbation theory (see Proof of Theorem 4.17 in Appendix B).

Remark 4.20. An important consequence of the combined use of high-gain observers and saturation filters is that the region of closed-loop stability obtained under state feedback remains practically preserved under output feedback. Specifically, starting from any compact subset of initial conditions (whose size is fixed by δ_b) within the state feedback region, there always exist observer gains such that the dynamic *output* feedback controller of (4.12) continues to enforce asymptotic stability and reference-input tracking in the constrained uncertain closed-loop system, when the observer gains are selected to be sufficiently large. Instrumental in deriving this result is the use of the saturation filter, which allows us to use arbitrarily large observer gains, in order to recover the state feedback region, without suffering the detrimental effects of observer peaking. Note that the size of the output feedback region (δ_b) can be made close to that of the state feedback region (δ_s) by selecting d to be sufficiently small which, in turn, can be done by making $\bar{\epsilon}$ sufficiently small. Therefore, although combination of the bounded state feedback controller with the observer results in some loss (given by d) in the size of the region of guaranteed closed-loop stability, this loss can be made small by selecting $\bar{\epsilon}$ to be sufficiently small. As expected, the nature of this semi-regional result is consistent with the semi-global result obtained in Chap. 3 for the unconstrained case.

Remark 4.21. In addition to preserving the stability region, the controller-observer combination of (4.24) practically preserves the optimality properties of the state feedback controller explained in Remark 4.12. The output feedback controller design is near-optimal in the sense that the cost incurred by implementing this controller tends to the optimal cost achieved by implementing the bounded state feedback controller when the observer gains are taken to be sufficiently large. Using standard singular perturbation arguments, one can show that cost associated with the output feedback controller is $O(\bar{\epsilon})$ close to the optimal cost associated with the state feedback controller (i.e., $J_{min} = V(e(0)) + O(\bar{\epsilon})$). The basic reason for near-optimality is the fact that by choosing $\bar{\epsilon}$ to be sufficiently small, the observer states can be made to converge quickly to the process states. This fact can be exploited to make the performance of the output feedback controller arbitrarily close to that of the optimal state feedback controller (see Chap. 3 for an analogous result for the unconstrained case).

Remark 4.22. Owing to the presence of the fast (high-gain) observer in the dynamical system of (4.24), the closed-loop system can be cast as a singularly perturbed system, where $\bar{\epsilon} = \max\{1/L_i\}$ is the singular perturbation parameter. Within this system, the states of the high-gain observers, which provide estimates of the outputs and their derivatives, constitute the fast states, while the ω states of the observer and the states of the original system of (4.1) under state feedback represent the slow states. Owing to the dependence of the controller of (4.24) on both the slow and fast states, the control action computed by the static component in (4.24) is not $O(\bar{\epsilon})$ close to that computed by the state feedback controller for all times. After the decay of the boundary layer term (fast transients of the high-gain observers), however, the static component in (4.24) approximates the state feedback controller to within $O(\bar{\epsilon})$.

Remark 4.23. It is important to note that the asymptotic stability results of Theorem 4.1 (Theorem 4.17) are regional (semi-regional) and nonlocal. By definition, a local result is one that is valid provided that the initial conditions are sufficiently small. In this regard, neither the result of Theorem 4.1 nor that of Theorem 4.17 requires the initial conditions to be sufficiently small in order for stability to be guaranteed. The apparent limitations imposed on the size of the initial conditions that guarantee closed-loop stability follow from the fundamental limitations imposed by the input constraints and uncertainty, as can be seen from (4.15). In the absence of constraints, for example, one can establish asymptotic stability globally for the state feedback problem and semi-globally for the output feedback case. In fact, one of the valuable features of the results in Theorems 4.1 and 4.17 is that they provide an explicit procedure for constructing reasonably large estimates of the stability region that depend only on the magnitude of the constraints and the size of the uncertainty, which therefore allows one to start from initial conditions farther away (from the equilibrium point) than would be possible using approaches where no explicit characterization of the stability region is available and consequently one often has to restrict the initial conditions to be sufficiently close to the origin.

4.6 Robust Stabilization of a Chemical Reactor Via Output Feedback Control

To illustrate an application of the output feedback controller design presented in the previous section, we consider in this section a well-mixed continuous stirred tank reactor where three parallel irreversible elementary exothermic reactions of the form $A \overset{k_1}{\to} D$, $A \overset{k_2}{\to} U$ and $A \overset{k_3}{\to} R$ take place, where A is the reactant species, D is the desired product and U, R are undesired byproducts. The feed to the reactor consists of pure A at flow rate F, molar concentration C_{A0} and temperature T_{A0}. Due to the non-isothermal nature of the reactions, a jacket is used to remove/provide heat to the reactor. Under standard modeling assumptions, a mathematical model of the process can be derived from material and energy balances and takes the following form:

$$\begin{aligned}
V\frac{dT}{dt} &= F(T_{A0} - T) + \sum_{i=1}^{3} \frac{(-\Delta H_i)}{\rho c_p} k_{i0} \exp\left(\frac{-E_i}{RT}\right) C_A V + \frac{Q}{\rho c_p} \\
V\frac{dC_A}{dt} &= F(C_{A0} - C_A) - \sum_{i=1}^{3} k_{i0} \exp\left(\frac{-E_i}{RT}\right) C_A V \qquad (4.26) \\
V\frac{dC_D}{dt} &= -FC_D + k_{10} \exp\left(\frac{-E_1}{RT}\right) C_A V
\end{aligned}$$

where C_A and C_D denote the concentrations of the species A and D, T denotes the temperature of the reactor, Q denotes rate of heat input/removal from

Table 4.2. Process parameters and steady-state values for the reactor of (4.26).

V	$= 1000.0$	L
R	$= 8.314$	$J/mol.K$
C_{A0s}	$= 4.0$	mol/L
T_{A0s}	$= 300.0$	K
ΔH_1	$= -5.0 \times 10^4$	J/mol
ΔH_2	$= -5.2 \times 10^4$	J/mol
ΔH_3	$= -5.4 \times 10^4$	J/mol
k_{10}	$= 5.0 \times 10^4$	min^{-1}
k_{20}	$= 5.0 \times 10^3$	min^{-1}
k_{30}	$= 5.0 \times 10^3$	min^{-1}
E_1	$= 5.0 \times 10^4$	J/mol
E_2	$= 7.53 \times 10^4$	J/mol
E_2	$= 7.53 \times 10^4$	J/mol
c_p	$= 0.231$	$J/g.K$
ρ	$= 1000.0$	g/L
F	$= 83.3$	L/min
T_s	$= 390.97$	K
C_{As}	$= 3.58$	mol/L
C_{Ds}	$= 0.42$	mol/L

the reactor, V denotes the volume of the reactor, ΔH_i, k_i, E_i, $i = 1, 2, 3$, denote the enthalpies, pre-exponential constants and activation energies of the three reactions, respectively, c_p and ρ denote the heat capacity and density of the reactor. The values of the process parameters and the corresponding steady-state values are given in Table 4.2. It was verified that these conditions correspond to an unstable equilibrium point of the process of (4.26).

The control problem is formulated as the one of regulating both the concentration of the desired product, C_D, and the reactor temperature, T, at the unstable steady-state by manipulating the inlet reactant concentration, C_{A0}, and the rate of heat input, Q, provided by the jacket. The control objective is to be accomplished in the presence of: (1) exogenous time-varying disturbances in the feed stream temperature, (2) parametric uncertainty in the enthalpy of the three reaction, and (3) hard constraints on the manipulated inputs. Defining $x_1 = T$, $x_2 = C_A$, $x_3 = C_D$, $u_1 = Q$, $u_2 = C_{A0} - C_{A0s}$, $\theta_i = \Delta H_i - \Delta H_{i0}$, $i = 1, 2, 3$, $\theta_4 = T_{A0} - T_{A0s}$, $y_1 = x_1$, $y_2 = x_3$, where the subscript s denotes the steady-state values and ΔH_{i0} are the nominal values for the enthalpies, the process model of (4.26) can be written in the form of (4.1) with:

$$f(x) = \begin{bmatrix} \frac{F}{V}(T_{A0s} - T) + \sum_{i=1}^{3} \frac{(-\Delta H_{i0})}{\rho c_p} k_{i0} \exp\left(\frac{-E_i}{RT}\right) C_A \\ \frac{F}{V}(C_{A0s} - C_A) - \sum_{i=1}^{3} k_{i0} \exp\left(\frac{-E_i}{RT}\right) C_A \\ -\frac{F}{V} C_D + k_{10} \exp\left(\frac{-E_1}{RT}\right) C_A \end{bmatrix},$$

$$g_1(x) = \begin{bmatrix} \frac{1}{\rho c_p V} \\ 0 \\ 0 \end{bmatrix}, \quad g_2(x) = \begin{bmatrix} 0 \\ \frac{F}{V} \\ 0 \end{bmatrix},$$

$$w_i(x) = \begin{bmatrix} k_{i0} \exp\left(\frac{-E_i}{RT}\right) C_A \\ 0 \\ 0 \end{bmatrix}, \quad i = 1, 2, 3, \quad w_4(x) = \begin{bmatrix} \frac{F}{V} \\ 0 \\ 0 \end{bmatrix}$$

$h_1(x) = x_1$, $h_2(x) = x_3$. To simulate the effect of uncertainty on the process outputs, we consider a time-varying function of the form of (4.21), with $\theta_0 = 0.03T_{A0}$, to simulate the effect of external disturbances in the feed temperature. We also consider a parametric uncertainty of 50% in the values of the enthalpies. Therefore, the bounds on the uncertain variables are taken to be $\theta_{bk} = 0.5\,|(-\Delta H_{k0})|$, $k = 1, 2, 3$, $\theta_{b4} = 0.03T_{A0s}$. Also, the following constraints are imposed on the manipulated inputs: $\|u_1\| \leq 25\ KJ/s$ and $\|u_2\| \leq 4.0\ mol/L$.

For this process, the relative degrees of the process outputs, with respect to the vector of manipulated inputs, are $r_1 = 1$, $r_2 = 2$, respectively. Using (4.3), it can be verified that the decoupling matrix $C(x)$ is nonsingular and that the assumptions of Theorem 4.17 are satisfied. In order to proceed with controller synthesis, we initially use the following coordinate transformation (in error variables form):

$$\begin{bmatrix} e_1 \\ e_2 \\ e_3 \end{bmatrix} = \begin{bmatrix} \zeta_1^{(1)} - v_1 \\ \zeta_1^{(2)} - v_2 \\ \zeta_2^{(2)} - v_2^{(1)} \end{bmatrix} = \begin{bmatrix} x_1 - v_1 \\ x_3 - v_2 \\ -\frac{F}{V} C_D + k_{30} \exp\left(\frac{-E_3}{RT}\right) C_A \end{bmatrix}$$

to cast the process model in its input-output form of (4.6), which is used directly for output feedback controller design. The necessary controller, whose practical implementation requires measurements of T and C_D only, consists of a combination of two high-gain observers (equipped with appropriate saturation filters) that provide estimates of the process states and a bounded robust static feedback component. The static component is designed, with the aid of the explicit synthesis formula in (4.12–4.14), using a quadratic Lyapunov function of the form $V = e^T Pe$, where:

$$P = \begin{bmatrix} 1 & 0 & 0 \\ 0 & \sqrt{3} & 1 \\ 0 & 1 & \sqrt{3} \end{bmatrix} \tag{4.27}$$

which is also used to compute an estimate of the region of guaranteed closed-loop stability using (4.15). The high-gain observers are designed using (4.24) and consist of a replica of the linear part of the system of (4.6) plus a linear gain multiplying the discrepancy between the actual and the estimated values of the outputs. The output feedback controller takes the following form:

$$\begin{aligned} \dot{\tilde{y}}_1^{(1)} &= L_1 a_1^{(1)}(y_1 - \tilde{y}_1^{(1)}) \\ \dot{\tilde{y}}_1^{(2)} &= \tilde{y}_2^{(2)} + L_2 a_1^{(2)}(y_2 - \tilde{y}_1^{(2)}) \\ \dot{\tilde{y}}_2^{(2)} &= L_2^2 a_2^{(2)}(y_2 - \tilde{y}_1^{(2)}) \\ u &= -k(sat(\tilde{y}))\left(\widehat{L_{\bar{g}}V}\right)^T \end{aligned} \tag{4.28}$$

where $\tilde{y}_1^{(1)}$, $\tilde{y}_1^{(2)}$, $\tilde{y}_2^{(2)}$ are the estimates of the reactor temperature, reactant concentration, and desired product concentration, respectively, and the saturation filter is defined as:

$$sat(\tilde{y}_i^{(j)}) = \left\{ \begin{array}{ll} a_{m,i}^{(j)}, & \tilde{y}_i^{(j)} \geq a_m^{(j)} \\ \tilde{y}_i^{(j)}, & -a_{m,i}^{(j)} \leq \tilde{y}_i^{(j)} \leq a_{m,i}^{(j)} \\ -a_{m,i}^{(j)}, & \tilde{y}_i^{(j)} \leq -a_{m,i}^{(j)} \end{array} \right\} \tag{4.29}$$

The following values were used for the controller and observer parameters: $\chi = 3.5$, $\phi = 0.0001$, $\rho = 0.001$, $L_1 = 100$, $a_1^{(1)} = 10$, $L_2 = 400$, $a_1^{(2)} = 40$, $a_2^{(2)} = 400$, $a_{m,1}^{(1)} = 395$, $a_{m,1}^{(2)} = 4$, $a_{m,2}^{(2)} = 0.5$ to ensure that the process outputs satisfy a relation of the form $\limsup_{t\to\infty}\|y_i - v_i\| \leq 0.0005$, $i = 1, 2$.

Closed-loop simulations were performed to evaluate the robust stabilization capabilities of the output feedback controller, starting from an initial condition inside the stability region, and compare its performance with that

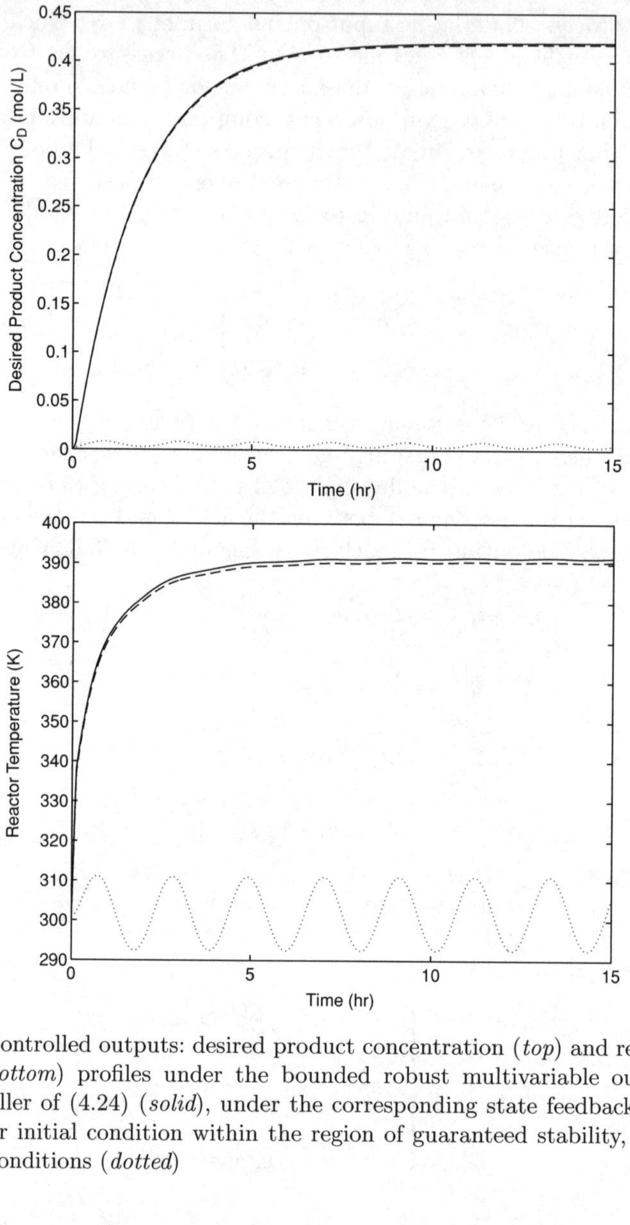

Fig. 4.8. Controlled outputs: desired product concentration (*top*) and reactor temperature (*bottom*) profiles under the bounded robust multivariable output feedback controller of (4.24) (*solid*), under the corresponding state feedback controller (*dashed*), for initial condition within the region of guaranteed stability, and under open-loop conditions (*dotted*)

of the state feedback controller and the open-loop response. Figures 4.8 and 4.9 depict the controlled outputs (desired product concentration and reactor temperature) and manipulated inputs (inlet reactant concentration and rate of heat input) profiles, respectively, under output feedback control (solid lines), under state feedback control (dashed lines) assuming all process states are available for measurement, and under no control (dotted lines). It is clear

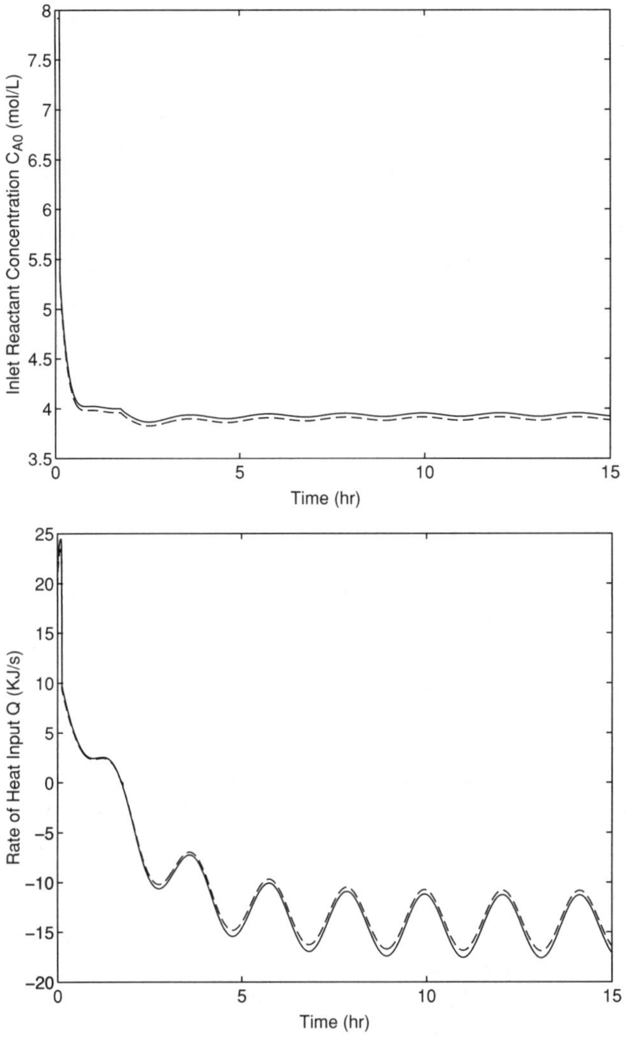

Fig. 4.9. Manipulated inputs: inlet reactant concentration (*top*) and rate of heat input (*bottom*) profiles under the bounded robust multivariable output feedback controller of (4.24) (*solid*) and under the corresponding state feedback controller (*dashed*), for an initial condition within the region of guaranteed stability

from the comparison with the open-loop profiles that the output feedback controller successfully drives both controlled outputs close to the desired steady-state while simultaneously attenuating the effect of disturbances and model uncertainty on the process outputs and generating control action that respects the constraints imposed. Note that the controlled outputs and manipulated inputs' profiles obtained under the output feedback controller (solid lines) are very close to the profiles obtained under the state feedback controller (dashed lines). This closeness, starting from the same initial condition, illustrates two important features of the output feedback design. The first is the fact that, by selecting the observer gains to be sufficiently large, the performance of the output feedback controller approaches (or recovers) that of the state feedback controller. Since the process response under state feedback control can be shown to be optimal (in the inverse sense) with respect to a meaningful cost (see Remark 4.12), the performance of the output feedback controller then is near-optimal with respect to the same cost. The second feature is that the set of admissible initial conditions, starting from where stability of the constrained closed-loop system is guaranteed under state feedback, remains practically preserved when the observer gain is chosen sufficiently large.

4.7 Connections with Classical Control

4.7.1 Motivation and Background

The majority (over 90%) of the regulatory loops in the process industries use conventional Proportional-Integral-Derivative (PID) controllers. Owing to the abundance of PID controllers in practice and the varied nature of processes that the PID controllers regulate, extensive research studies have been dedicated to the analysis of the closed–loop properties of PID controllers and to devising new and improved tuning guidelines for them, focusing on closed–loop stability, performance and robustness (see, for example, [19, 169, 215, 232, 249, 272, 303] and the survey papers [20, 59]). Most of the tuning rules are based on obtaining linear models of the system, either through running step tests or by linearizing a nonlinear model around the operating steady-state, and then computing values of the controller parameters that incorporate stability, performance and robustness objectives in the closed–loop system.

While the use of linear models for the PID controller tuning makes the tuning process easy, the underlying dynamics of many processes are often highly complex due, for example, to the inherent nonlinearity of the underlying chemical reaction or due to operating issues such as actuator constraints, time–delays and disturbances. Ignoring the inherent nonlinearity of the process when setting the values of the controller parameters may result in the controller's inability to stabilize the closed–loop system and may call for extensive re-tuning of the controller parameters.

The shortcomings of classical controllers in dealing with complex process dynamics, together with the abundance of such complexities in modern–day processes, have been an important driving force behind the significant and growing body of research work within the area of nonlinear process control over the past two decades, leading to the development of several practically–implementable nonlinear control strategies that can deal effectively with a wide range of process control problems such as nonlinearities, constraints, uncertainties, and time–delays (see, for example, [9, 78, 81, 177] and the books [126,148,245]). While process control practice has the potential to benefit from these advances through the direct implementation of the developed nonlinear controllers, an equally important direction in which process control practice stands to gain from these developments lies in investigating how nonlinear control techniques can be utilized for the improved tuning of classical PID controllers. This is an appealing goal because it allows control engineers to potentially take advantage of the improved stability and performance properties provided by nonlinear control without actually forsaking the ubiquitous conventional PID controllers or re-designing the control system hardware.

There has been some research effort towards incorporating nonlinear control tools in the design of PID controllers. For example, in [290] it is shown that controllers resulting from nonlinear model-based control theory can be put in a form that looks like the PI or PID controllers for first and second-order systems. Other examples include [34] where adaptive PID controllers are designed using a backstepping procedure, and [44] where a self-tuning PID controller is derived using Lyapunov techniques. In these works, however, even though the resulting controller has the same structure as that of a PID controller, the controller parameters (gain: K_c, integral time–constant: τ_I, and derivative time–constant: τ_D) are not constant but functions of the error or process states. While such analysis provides useful analogies between nonlinear controllers and PID controllers, implementation of these control designs would require changing the control hardware in a way that allows the tuning parameter values to be continuously changed while the process is in operation.

Motivated by the considerations above, we present in this section a two-level, optimization-based method for the derivation of tuning guidelines for PID controllers that take nonlinear process behavior explicitly into account. The central idea behind the proposed method is the selection of the tuning parameters in a way that has the PID controller emulate, as closely as possible, the control action and closed–loop response obtained under a given nonlinear controller, for a broad set of initial conditions and set-point changes. To this end, classical tuning guidelines (typically derived on the basis of linear approximations, running open or closed–loop tests) are initially used in the first level to obtain reasonable bounds on the range of stabilizing tuning parameters over which the search for the parameters best matching the PID and nonlinear controllers is to be conducted. In addition to stability, performance and robustness considerations for the linearized closed-loop system can be introduced in the first level to further narrow down the parameter search range. The bounds obtained from the first level are then incorporated as constraints on the optimization problem solved at the second level to yield a set of tuning parameter values that enforce closed-loop behavior under the PID controller that closely matches the closed-loop behavior under the nonlinear controller. Implications of the proposed method, as a transparent and meaningful link between the classical and nonlinear control domains, as well as possible extensions of the tuning guidelines and other implementation issues are discussed. Finally, the proposed tuning method is demonstrated through a chemical reactor example.

4.7.2 A PID Controller Tuning Method Using Nonlinear Control Tools

We consider continuous–time single–input single–output (SISO) nonlinear systems, with the following state-space description

$$\begin{aligned} \dot{x}(t) &= f(x(t)) + g(x(t))u(t) \\ y &= h(x) \end{aligned} \tag{4.30}$$

where $x = [x_1 \cdots x_n]^T \in \mathbb{R}^n$ denotes the vector of state variables and x^T denotes the transpose of x, $y \in \mathbb{R}$ is the process output, $u \in \mathbb{R}$ is the manipulated input, $f(\cdot)$ is a sufficiently smooth nonlinear vector function with $f(0) = 0$, $g(\cdot)$ is a sufficiently smooth nonlinear vector function and $h(\cdot)$ is a sufficiently smooth nonlinear function with $h(0) = 0$.

The basic idea behind the proposed approach is the design (but not implementation) of a nonlinear controller that achieves the desired process response, and then, the tuning of the PID controller parameters so as to best "emulate" the control action and closed–loop process response under the nonlinear controller, subject to constraints derived from classical PID tuning rules. These ideas are described algorithmically below (see also Fig. 4.10):

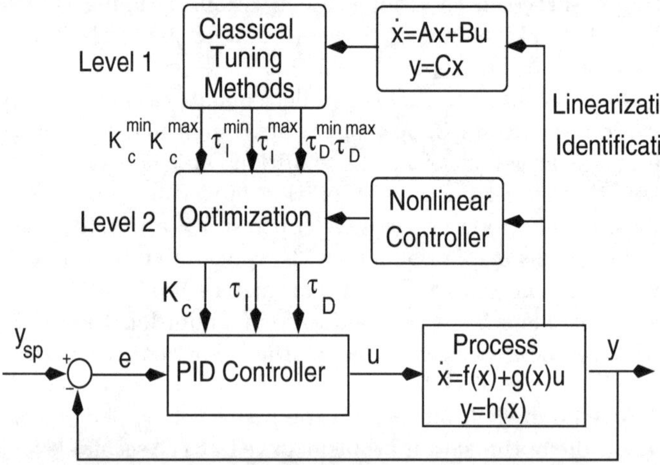

Fig. 4.10. Schematic representation of the implementation of the proposed two-level, optimization-based tuning method

1. Construct a nonlinear process model and derive a linear model around the operating steady-state (either through linearization or by running step tests).
2. On the basis of the linear model, use classical tuning guidelines to determine bounds on the values of K_c, τ_I and τ_D.
3. Using the nonlinear process model and desired process response, design a nonlinear controller.
4. For a set-point change, compute off-line, through simulations, the input trajectory ($u_{nl}(t)$) prescribed by the nonlinear controller over the time (t_{final}) that it takes to achieve the set-point change and the corresponding output profile under the nonlinear controller, $y_{nl}(t)$.
5. Compute PID tuning parameters (K_c, τ_I *and* τ_D) as the solution to the following optimization problem

$$J = \int_0^{t_{final}} \left[(y_{nl}(t) - y_{PID}(t))^2 + (u_{nl}(t) - u_{PID}(t))^2\right] dt \tag{4.31}$$

$$\begin{aligned} s.t.\ u_{PID}(t) &= K_c\left(e + \frac{\int_0^t e(t')dt'}{\tau_I} + \tau_D\frac{de}{dt}\right) \\ e(t) &= y_{sp} - y_{PID} \\ \dot{x}(t) &= f(x(t)) + g(x(t))u_{PID}(t) \\ y_{PID} &= h(x) \\ \alpha_1 K_c^c &\leq K_c \leq \alpha_4 K_c^c \\ \alpha_2 \tau_I^c &\leq \tau_I \leq \alpha_5 \tau_I^c \\ \alpha_3 \tau_D^c &\leq \tau_D \leq \alpha_6 \tau_D^c \\ (K_c,\ \tau_I,\ \tau_D) &= argmin(J) \end{aligned} \tag{4.32}$$

where y_{sp} is the desired set-point, y_{nl}, u_{nl} are the closed–loop process response and control action under the nonlinear controller respectively, K_c^c, τ_I^c and τ_D^c are the parameter values obtained by using tuning rules based on linear models and $0 \leq \alpha_i < 1,\ i = 1, 2, 3$ and $1 < \alpha_i < \infty,\ i = 4, 5, 6$ are design parameters.

Remark 4.24. The optimization problem of (4.31–4.32) computes values for K_c, τ_I, τ_D such that the closed–loop control action and process response under the PID controller are similar to those under the nonlinear controller, while being within acceptable ranges of the values obtained from the classical tuning methods. The method, therefore, allows fine-tuning the closed–loop performance under the PID controller to mimic that of the nonlinear controller.

Note that, in principle, the PID controller parameters could have been tuned to mimic any desired arbitrarily–chosen closed–loop behavior. Mimicking the behavior of the nonlinear controller, however, is a meaningful objective since the nonlinear controller utilizes the nonlinear system dynamics in generating the prescribed input and output response, and therefore provides a "target" closed–loop system behavior that is realizable. Note also that the performance index of (4.31) can be generalized to include a weight factor (in front of the input penalty term) in order to quantify the relative importance of the two main terms: the first one that aims at minimizing the closed-loop output mismatch associated with the nonlinear vs. the PID controller, and the second input penalty term. Furthermore, the inclusion of a weight factor would ensure consistency of units for the two terms in the performance functional.

Remark 4.25. It should be noted that the controller actually implemented in the closed–loop system is the PID controller with fixed parameter values, and that it may not always be possible for the PID controller to exactly match the closed–loop behavior under the nonlinear controller. The purpose behind introducing the first-level tuning is twofold: (1) to ensure that important objectives (such as closed-loop stability) are not sacrificed in the (possibly unsuccessful) quest for a nonlinear controller like behavior (which is accomplished through the optimization), and (2) to provide a rational way of constructing a range of the tuning parameter values over which the optimization is performed. In relation to the first objective, we note that the essence of the second-level optimization is to try to find the best tuning parameter values that make the PID controller emulate the behavior of the nonlinear controller. However, this objective should not come at the expense of more overriding objectives such as closed–loop stability. In particular, if the optimization problem were to be carried out without imposing any constraints on the parameter values, the solution may indeed lead to a closer match, but no longer guarantee that the PID controller enforces closed–loop stability when implemented. This is one reason why the proposed method includes the first level whose purpose, in part, is to make use of existing methods for PID controller tuning in order to first determine the range of tuning parameters for which closed-loop stability of the linearized process model under the PID controller is guaranteed. For such a range of values, the PID controller is expected to enforce (local) closed–loop stability when implemented on the nonlinear process.

Remark 4.26. Having obtained the stabilizing parameter range and incorporated it as a constraint on the optimization, the search for the "optimal" gain then can take place only over this range. However, the stabilizing range may be too large to search over and the designer may wish to limit this range further by incorporating additional performance and robustness considerations. The use of first-level methods (which are based on the use of linear models) provides a rational, though not necessarily unique, way of constructing an appropriate subrange to work with. For example, if certain robustness margins can be obtained (and quantified explicitly as ranges on the tuning parameters) through the use of existing methods based on the linearized model, these margins can be incorporated as constraints that further limit the search range. This of course does not guarantee that the PID controller will exhibit such robustness or performance when implemented on the nonlinear process since the margins are based on the linearized model. However, at a minimum, this is a meaningful way to go about constructing or, more precisely, narrowing down the range over which the optimization is done (by requesting that the performance of the linearized model be consistent with what existing methods based on linear models yield). Ultimately, it is the "closeness" of the PID controller to the nonlinear controller resulting from the second-level optimization (not the first-level methods) that is essentially responsible for the performance properties exhibited by the PID controller when implemented on the nonlinear process. Since different first-level tuning methods lead to different performance properties and yield different parameter values, the designer can examine the values obtained from different methods to form a reasonable idea about what an acceptable range might be around these nominal values and then construct such a range (through choosing appropriate α_i values) and implement it as constraints on the optimization (see the simulation studies in Sect. 4.7.3 for an example). If the parameter values obtained after performing the optimization do not yield satisfactory performance (tested through simulations), then the parameter range could be expanded further (but still within the stabilizing range determined initially) in an iterative procedure.

Remark 4.27. The α_i's in (4.32) are introduced into the optimization as design parameters that allow the designer flexibility in tightening or relaxing the range of parameter values over which the optimization is carried out. If a given method yields satisfactory performance, and it is desired that the tuning parameters not be changed appreciably, this can be enforced by using values of the design parameters, α_i, close to 1. The tuning parameters resulting from the solution to the optimization problem in this case, while changed to mimic the nonlinear control action, will be close to the ones obtained from the classical tuning method considered in the first level. If, on the other hand, it is decided that some further improvement is warranted, then, at a minimum, the α_i values should be chosen to reflect the range (or a subset of the range) within which the parameter values can be changed by the optimization without losing stability. If the designer wishes to constrain the search over a smaller range (using, for example, certain performance or robustness margins obtained from the first-level methods for the linearized closed-loop system), then the α_i values can be modified to reflect the new range. In general, the choice of α_i varies depending on the particular process under consideration and on the particular tuning methods that are being considered in the first level.

Remark 4.28. Regarding the performance properties of the proposed method in relation to those of the first-level methods, we first note that the first-level tuning guidelines are derived on the basis of the linearized process model, and therefore the robustness and performance properties obtained when using these methods to tune the PID controller are not guaranteed to carry over when the PID controller is implemented on the nonlinear system. So even if re-tuning of the first-level methods may bring about further performance improvement in the linear case (by possibly sacrificing stability and robustness margins), this does not imply that a similar improvement should be expected in the nonlinear setting. In general, there is no systematic way in which such methods can be re-tuned (if at all possible) to improve the performance in the nonlinear case. By contrast, the proposed method aims to improve the performance of the PID controller in the nonlinear setting by explicitly accounting for process nonlinearity through the optimization (an objective not shared by the first-level approaches). Whether this necessarily means that the resulting parameters will always yield performance that is "better" than what a given tuning method might yield is difficult to judge, and, more importantly, is not a point that the method is intended to address. The point is that the proposed approach is a meaningful way of tuning PID controllers that can yield good performance when implemented on the nonlinear system. From this perspective, the first-level methods serve as a rational starting point for the construction of the search ranges as discussed in Remark 4.25.

Remark 4.29. The optimization problem of (4.31–4.32) is solved off-line as part of the design procedure to compute the optimal values of the tuning parameters. Also, the above optimization problem can be carried out over a range of initial conditions and set-point changes that are locally representative of the process operation to obtain PID tuning parameters that allow the PID controller to approximate, in an average (with respect to initial conditions and set-point changes) sense, the closed–loop response under the nonlinear controller. If the process is required to operate at an equilibrium point that is very far from the operating point that the parameters are tuned for, then it is best to perform the optimization again around the new, desired operating point to yield new tuning parameter values (as is done in classical tuning also). Regarding the optimization complexity issue, we note that possible complexity of the proposed optimization is mainly a function of the model complexity (e.g., nonlinearity, model order, etc.). The increase in computational demand expected in the case of higher-order and highly nonlinear systems is primarily due to the need to solve a higher-order system of nonlinear differential equations. However, with current computational capabilities, this does not pose unduely significant limitations on the practical implementation prospects of the proposed method especially when compared with the computational complexity encountered in typical nonlinear optimization problems (e.g., nonlinear MPC). Furthermore, the approximations discussed in Remark 4.32 below provide possible means that can help manage potential complexities even further. Finally, we note that since the method involves a form of nonlinear optimization, it is expected that, in general, multiple optimal solutions may exist.

Remark 4.30. The basic idea behind the proposed PID controller tuning methodology, i.e., that of tuning the PID controller to emulate some other well–designed controller that handles complex dynamics effectively, can be used to develop conceptually similar tuning methods for PID control of processes with other sources of complexities (besides nonlinearity) such as uncertainty, time–delays, and manipulated input constraints. The logic behind such extensions is based on the following intuitive parallel: just as a nonlinear controller is a meaningful guide to be emulated by a PID controller being implemented on a nonlinear process, a controller that handles constraints, uncertainty and/or time–delays effectively can also be a meaningful guide to be emulated by a PID controller that is being implemented on a process with these characteristics. In principle, the extensions can be realized by adequately accounting for the complex characteristics of these processes within both levels of the tuning method. For example, for systems with uncertainty, classical tuning methods that provide sufficient robustness margins can be used to come up with the first–level parameter values. Then a robust nonlinear controller (for example, [78]) can be designed and the closed–loop profiles, obtained under the robust nonlinear controller for a sufficient number of realizations of the uncertainty (which may be simulated, for instance, using random number generators) may be computed. Finally, the parameter values obtained from the first level tuning method may be improved upon by solving an optimization problem that minimizes the error over the profiles in the representative set. In a conceptually similar fashion, for systems with constraints, an anti-windup scheme could be used initially to obtain the first level parameter values. A nonlinear controller design that handles input constraints can then be chosen for the second–level optimization. The PID controller tuning guidelines can then serve to "carry over" the constraint handling properties of this nonlinear controller and improve upon the first level tuning methods in two ways: (1) through the objective function, by requiring the control action and closed–loop process response under PID control to mimic that under the constrained nonlinear controller, and (2) through the incorporation of the input constraints directly into the optimization problem. It should be noted however that, while such extensions are intuitively appealing, a detailed assessment and characterization of their potential requires further investigation.

Remark 4.31. Note that the derivative part of the PID controller is often implemented using a filter. This feature can be easily incorporated in the optimization problem by explicitly accounting for the filter dynamics. Constraints on the filter time-constant, τ_f, obtained empirically through knowledge of the nature of noise in the process, can be imposed to ensure that the filtering action restricts the process noise from being transmitted to the control action.

Remark 4.32. To allow for simple computations, approximations can be introduced in solving the optimization problem of (4.31–4.32). For instance, in the computation of the control action, the error, $e(t)$, may be approximated by simply taking the difference between the set-point, y_{sp}, and the process output under the nonlinear controller, $y_{nl}(t)$, leading to a simpler optimization problem that can be solved easily using numerical solvers such as Microsoft Excel (for a given choice of the decision variables, the objective function can be computed algebraically and does not involve integrating the process dynamics). The justification behind this being that if the resulting value of $u_{PID}(t)$ is "close enough" to $u_{nl}(t)$, then this approximation holds (see the simulation examples in Sect. 4.7.3 for a demonstration). If the solution of the optimization problem does not yield a sufficiently small value for the objective function (indicating that u_{PID}, and hence y_{PID}, is significantly different from u_{nl}, and y_{nl}), this approximation may not be valid anymore. In this case, one could revert to using $e(t) = y_{sp} - y_{PID}(t)$ in the optimization problem, where y_{PID} is the closed–loop process response under the PID controller. Note also that in some cases, particularly for low–dimensional systems with real analytic vector fields, the value of the performance index may be calculated explicitly as an algebraic function of the controller parameters (leading to a static finite–dimensional optimization problem) by solving Zubov's partial differential equation using techniques similar to those presented in [143]. Those techniques can also be used in designing an optimally–tuned nonlinear controller that serves as a meaningful target to be emulated by the PID controller.

Remark 4.33. The characteristics of the closed–loop system, analyzed on the basis of the linearized model, can also be used as guidelines in setting up the optimization problem. For instance, the gain and phase margins can be computed for the parameter values prescribed by the proposed method and changed if not found to be adequate. Note that the value of the gain margin depends on the value of tuning parameters which in turn depends on the response of the nonlinear controller that the PID controller is required to mimic. A highly "aggressive" response prescribed by the nonlinear controller will likely lead to "aggressively" tuned parameter values and a possibly low gain margin. The value of the gain margin can be altered by changing the requested response of the nonlinear controller, making it less or more "aggressive" as the need may be.

Remark 4.34. Finally, we note that the proposed method *does not* turn the PID controller into a nonlinear controller. The tuning method can only serve to improve upon the process response of the PID controller for operating conditions for which PID control action can be used to stabilize the process. If the process is highly nonlinear, or a complex process response is desired, it may be possible that the PID controller structure is not adequate and, in this case, the appropriate nonlinear controller should be implemented in the closed–loop to achieve the desired closed–loop properties.

4.7.3 Application to a Chemical Reactor Example

We consider a continuous stirred tank reactor where an irreversible, first-order reaction of the form $A \xrightarrow{k} B$ takes place. The inlet stream consists of pure species A at flow rate F, concentration C_{A0} and temperature T_{A0}. Under standard modeling assumptions, the mathematical model for the process takes the form

$$\begin{aligned}\dot{C_A} &= \frac{F}{V}(C_{A0} - C_A) - k_0 \exp\left(\frac{-E}{RT_R}\right) C_A \\ \dot{T_R} &= \frac{F}{V}(T_{A0} - T_R) + \frac{(-\Delta H)}{\rho c_p} k_0 \exp\left(\frac{-E}{RT_R}\right) C_A + \frac{UA}{\rho c_p V}(T_j - T_R)\end{aligned} \tag{4.33}$$

where C_A denotes the concentration of the species A, T_R denotes the temperature of the reactor, T_j is the temperature of the fluid in the surrounding jacket, U is the heat–transfer coefficient, A is the jacket area, V is the volume of the reactor, k_0, E, ΔH are the pre-exponential constant, the activation energy, and the enthalpy of the reaction respectively, and c_p and ρ, are the heat capacity and fluid density in the reactor respectively. The values of all process parameters are given in Table 4.3. At the nominal operating condition of $T_j^{nom} = 493.87\ K$, the reactor is operating at the unique, stable steady-state $(C_A^s,\ T_R^s) = (0.52, 398.97)$. The control objective is to implement set-point changes in the reactor temperature using the jacket fluid temperature, T_j, as the manipulated input, using a P, PI or PID controller.

To proceed with our controller tuning method, we initially design an input/output linearizing nonlinear controller. Note that the linearizing controller design is used in the simulation example only for the purpose of illustration, and any other nonlinear controller design deemed fit for the problem at hand can be used as part of the proposed controller tuning method.

Table 4.3. Process parameters and steady–state values for the reactor of (4.33).

V	$= 100.0$	L
E/R	$= 8000$	K
C_{A0}	$= 1.0$	mol/L
T_{A0}	$= 400.0$	K
ΔH	$= 2.0 \times 10^5$	J/mol
k_0	$= 4.71 \times 10^8$	min^{-1}
c_p	$= 1.0$	$J/g.K$
ρ	$= 1000.0$	g/L
UA	$= 1.0 \times 10^5$	$J/min.K$
F	$= 100.0$	L/min
C_A^s	$= 0.52$	mol/L
T_R^s	$= 398.97$	K
T_j^{nom}	$= 493.87$	K

Defining $x = [C_A - C_A^s,\ T_R - T_R^s]'$ and $u = T_j - T_j^{nom}$, the process of (4.33) can be recast in the form of (4.30) where the explicit form of $f(\cdot)$ and $g(\cdot)$ are omitted for brevity. Consider the control law given by:

$$u = \frac{\nu - y(t) - \gamma L_f h(x)}{\gamma L_g h(x)} \tag{4.34}$$

where $L_f h(x)$ and $L_g h(x)$ are the Lie derivatives of the function $h(x)$ with respect to the vector functions $f(x)$ and $g(x)$, respectively, γ, a positive real number, is a design parameter and ν is the set-point. Taking the derivative of the output in (4.30) with respect to time, we get

$$\dot{y} = L_f h(x) + L_g h(x) u \tag{4.35}$$

Substituting the linearizing control law of (4.34), we get

$$\dot{y} = \frac{\nu - y}{\gamma} \tag{4.36}$$

Under the control law of (4.34), the controlled output y evolves linearly, to achieve the prescribed value of ν, with the design parameter γ being the time–constant of the closed–loop response.

It is well known that when a first–order closed–loop response, with a given time–constant, is requested for a linear first-order process, the method of direct synthesis yields a PI controller. Note that the relative order of the controller output, T_R, with respect to the manipulated input, T_j, in the example of (4.33) is also one. Even though the nonlinear linearizing controller is a static controller, and the PI controller is dynamic, both controllers are capable of generating closed–loop behaviors that are of the same kind (a linear first order response with a prescribed closed–loop time constant). This motivates using a PI controller and tuning the controller parameters to achieve the prescribed first-order response.

For the purpose of tuning the PI controller, the nonlinear process response generated under the nonlinear controller, using a value of $\gamma = 0.25$, was used in the optimization problem. An appropriate range for the tuning parameters was derived from the K_c and τ_I suggested by the IMC-based and Ziegler-Nichols tuning rules (where the parameters $K_{cu} = 6.36$ and $P_u = 5.0$ are obtained using the method of relay auto tuning [18]). In particular, the constraints on the values of the parameters were chosen as follows: for a given parameter, the largest and the smallest values prescribed by the available tuning methods (in this case the IMC-based and Ziegler Nichols) were chosen and the upper bound on the parameters was chosen as twice the maximum value, and the lower bound was chosen as half the minimum value; $0.9 \leq K_c \leq 5.7$ and $0.2 \leq \tau_I \leq 8.2$. The values of the parameters, computed using the IMC method, Ziegler-Nichols and the two-level PI tuning method are reported in Table 4.4.

Table 4.4. PI controller tuning parameters

Tuning Method	K_c	τ_I
IMC	1.81	0.403
Ziegler-Nichols	2.86	4.16
Two-level optimization method	5.56	0.297

The solid lines in Figs. 4.11(a)–(b) show the closed–loop response of the output and the manipulated input under the nonlinear controller of (4.34). Note that the value of γ was chosen as 0.25 to yield a smooth, fast transition to the desired set-point. The optimization problem was solved approximately, using the closed–loop process response under the nonlinear controller to compute $e(t)$, and the objective function only included penalties on the difference between the control actions under the PI controller and the nonlinear controller (see Remark 4.32). The dashed–line shows the response of the PI controller tuned using the proposed optimization-based method. The result shows that the response under the PI controller is close to that under the nonlinear controller and demonstrates the feasibility of using a PI controller to generate a closed–loop response that mimics the response of the nonlinear controller.

In Fig. 4.12(a), we present the closed–loop responses when the controller parameters computed using the IMC-based tuning rules and Ziegler-Nichols are implemented. As can be seen, the transition to the new set-point under the PID controller tuned using the proposed method (dashed lines) is fastest when compared to a classical PI controller tuned using IMC tuning rules (solid line) and Ziegler-Nichols tuning rules (dotted line). The corresponding manipulated input profiles are shown in Fig. 4.12b.

We now demonstrate the application of the proposed method to the same system, but with C_A as the controlled variable and T_j as the manipulated variable. As in the previous case, we initially design an input/output linearizing nonlinear controller to yield a second-order linear input-output response in the closed–loop system of the form:

$$\tau_{cl}^2 \ddot{y} + \frac{2\xi}{\tau_{cl}} \dot{y} + y = \nu \tag{4.37}$$

where τ_{cl} and ξ are design parameters and were chosen as $\tau_{cl} = 0.2$ and $\xi = 1.05$ (implying that the closed–loop system is a slightly over-damped second-order system).

The following tuning methods were used for the first level: (1) IMC-based tuning rule, where a step test is run to approximate the system by a first-order + time–delay process, hence referred to as IMC-I, (2) IMC-based tuning rule, where the process is linearized around the operating steady-state to obtain a second-order linear model, hence referred to as IMC-II, and (3) Ziegler-Nichols tuning rules, where the parameters $K_{cu} = -34543$ and $P_u = 0.223$

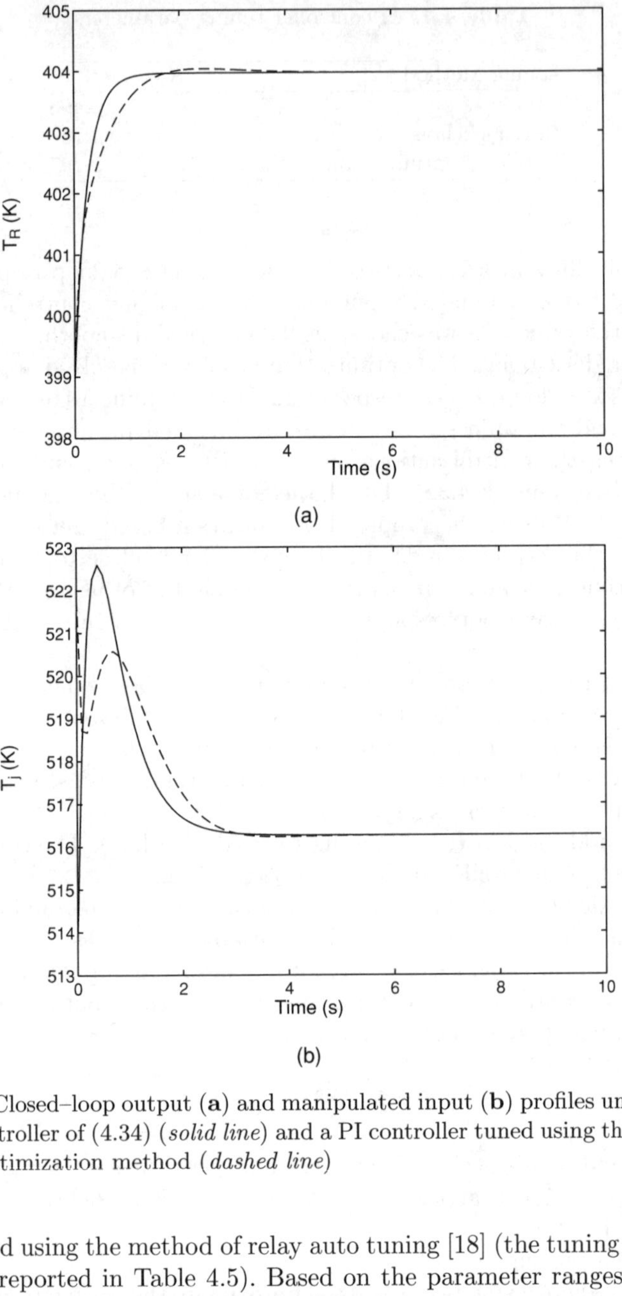

Fig. 4.11. Closed–loop output (**a**) and manipulated input (**b**) profiles under the linearizing controller of (4.34) (*solid line*) and a PI controller tuned using the proposed, two-level optimization method (*dashed line*)

are obtained using the method of relay auto tuning [18] (the tuning parameter values are reported in Table 4.5). Based on the parameter ranges suggested by the first level tuning methods, the following constraints were used in the optimization problem set up to compute K_c, τ_I and τ_D: $-2072.0 \leq K_c \leq -678$, $0.149 \leq \tau_I \leq 1$ and $0.0114 \leq \tau_D \leq 0.052$. The derivative part of

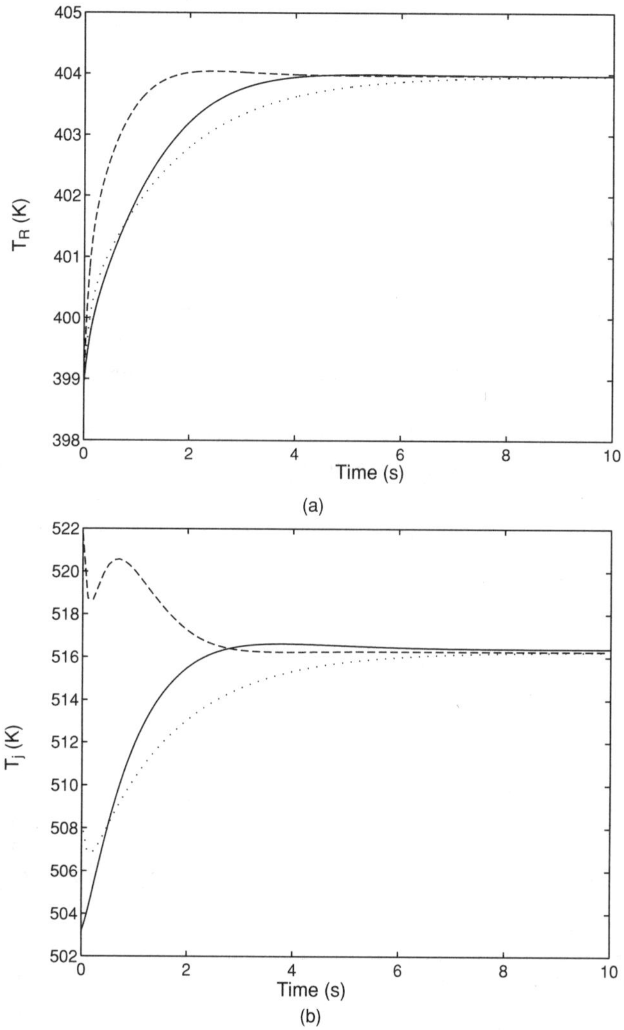

Fig. 4.12. Closed–loop output (**a**) and manipulated input (**b**) profiles using IMC tuning rules for the PI controller (*solid line*), using Ziegler-Nichols tuning rules (*dotted line*) and the proposed, two-level optimization method (*dashed line*)

the controller was implemented using a first order filter with time–constant $\tau_f = 0.1$.

The solid lines in Figs. 4.13(a)–(b) show the closed–loop response of the output and the manipulated input under the linearizing controller design. The dashed–line shows the response of the PID controller tuned using the proposed method, which is close to the response of the nonlinear controller. As is clear

Table 4.5. PID controller tuning parameters.

Tuning Method	K_c	τ_I	τ_D
IMC-I	−678.8	0.95	0.0524
IMC-II	−1208.9	1.00	0.114
Ziegler-Nichols	−2072.0	0.149	0.028
Two-level optimization method	−951.21	0.978	0.114

from Fig. 4.13, the resulting PID controller yields a response that is close enough to that of the nonlinear controller.

In Fig. 4.14(a), we present the closed–loop responses when the controller parameters computed using the classical tuning rules are implemented. The values suggested by Ziegler-Nichols tuning lead to an oscillatory closed–loop response. In the simulation, a smaller value for $K_c = -518.4$ and a larger

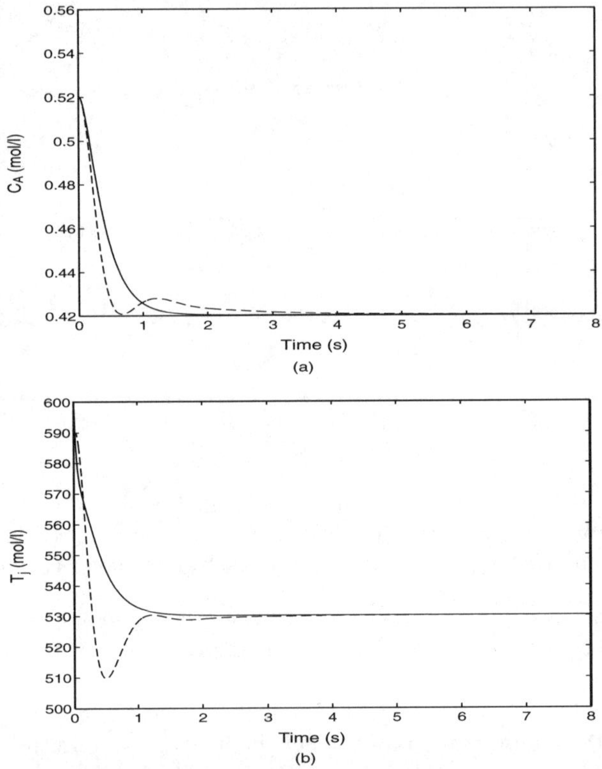

Fig. 4.13. Closed–loop output (**a**) and manipulated input (**b**) profiles under a linearizing controller (*solid line*) and a PID controller tuned using the proposed, two-level optimization method (*dashed line*)

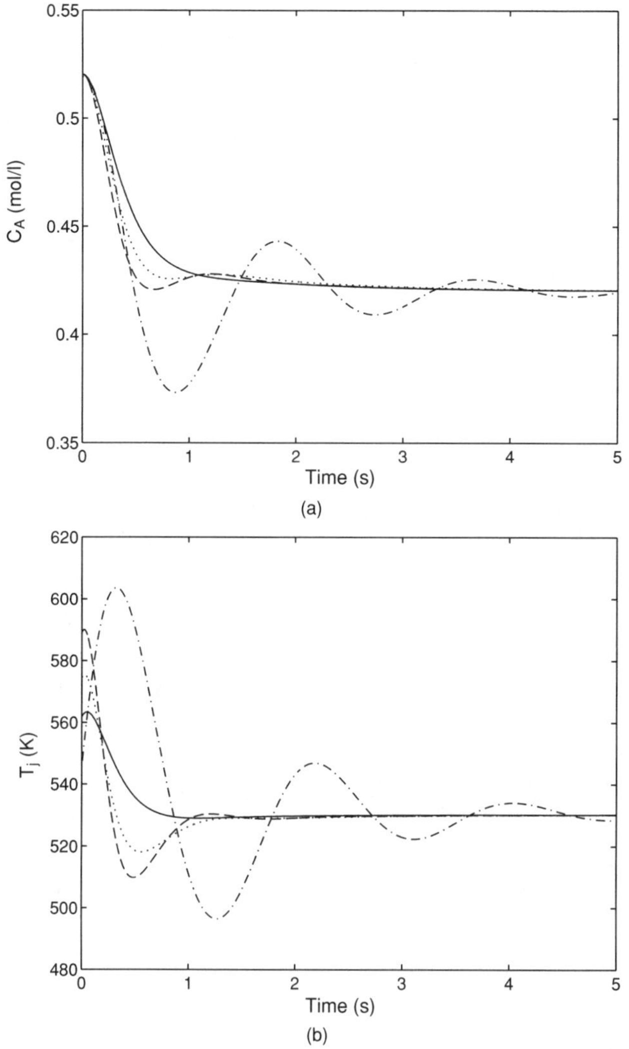

Fig. 4.14. Closed–loop output (**a**) and manipulated input (**b**) profiles using IMC tuning rules I (*dotted line*), IMC tuning rules II (*solid line*), Ziegler-Nichols tuning rules (*dash-dotted line*) and the proposed, two-level optimization method (*dashed line*)

$\tau_I = 0.14$ were used. As can be seen, the transition to the new set-point using the proposed tuning method (dashed lines in Fig. 4.14) compares favorably with that obtained when using the IMC-based tuning rules-I and II (dotted and solid lines, respectively) and the Ziegler-Nichols tuning rules (dash-dotted line). The corresponding manipulated input profiles are shown in Fig. 4.14(b).

4.8 Conclusions

A Lyapunov-based nonlinear controller design methodology, for MIMO nonlinear processes with uncertain dynamics, actuator constraints, and incomplete state measurements, was developed. Under the assumption that all process states are accessible for measurement, the approach led to the explicit synthesis of bounded robust multivariable nonlinear feedback controllers that enforce stability and robust asymptotic reference-input tracking in the constrained uncertain closed-loop system and provide, at the same time, an explicit
characterization of the region of guaranteed closed-loop stability. When complete state measurements are not available, a combination of the bounded robust state feedback controllers with high-gain state observes and appropriate saturation filters, was employed to synthesize bounded robust multivariable output feedback controllers that require only measurements of the outputs for practical implementation. The resulting output feedback design was shown to inherit the same closed-loop stability and performance properties of the state feedback controller and, in addition, practically preserve the region of guaranteed closed-loop stability obtained under state feedback. The developed state and output feedback controllers were applied successfully to non-isothermal chemical reactor examples with uncertainty, input constraints, and incomplete state measurements. Finally, the chapter concluded with an approach on how nonlinear control tools can be used to provide improved tuning guidelines for classical controllers.

5

Hybrid Predictive Control of Constrained Linear Systems

5.1 Introduction

Model Predictive Control (MPC), also known as receding horizon control (RHC), is a popular control method for handling constraints (both on manipulated inputs and state variables) within an optimal control setting [231]. In MPC, the control action is obtained by solving repeatedly, on–line, a finite–horizon constrained open–loop optimal control problem. The popularity of this approach stems largely from its ability to handle, among other issues, multi–variable interactions, constraints on controls and states, and optimization requirements, all in a consistent, systematic manner. Its success in many commercial applications is also well–documented in the literature (see, for example, [99, 223]). These considerations have motivated numerous research investigations into the stability properties of model predictive controllers and led to a plethora of MPC formulations that focus on closed-loop stability (see, for example, [68, 101, 145, 228] and the review paper [191]).

The significant progress in understanding and improving the stability properties of MPC notwithstanding, the issue of obtaining, *a priori* (i.e., before controller implementation), an explicit characterization of the region of constrained closed–loop stability for model predictive controllers remains to be adequately addressed. The difficulty in this direction owes in part to the fact that the stability of MPC feedback loops depends on a complex interplay between several factors such as the choice of the horizon length, the penalties in the performance index, the initial condition and the constraints on the state variables and manipulated inputs. *A priori* knowledge of the stability region requires an explicit characterization of these interplays which is a very difficult task. This difficulty can impact on the practical implementation of MPC by imposing the need for extensive closed–loop simulations over the whole set of possible initial conditions to check for closed–loop stability, or by potentially limiting operation within an unnecessarily small neighborhood of the nominal equilibrium point.

The desire to implement control approaches that allow for an explicit characterization of their stability properties has motivated significant work on the design of stabilizing control laws, using Lyapunov techniques, that provide explicitly–defined, large regions of attraction for the closed–loop system; the reader may refer to [151] for a survey of results in this area. In Chap. 4, a class of Lyapunov–based bounded robust nonlinear controllers, inspired by the results on bounded control originally presented in [177], was developed. The controllers enforce robust stability in the closed–loop system and provide, at the same time, an explicit characterization of the region of guaranteed closed–loop stability. Despite their well–characterized stability and constraint–handling properties, those Lyapunov–based controllers are not guaranteed to be optimal with respect to an arbitrary performance criterion (in Chap. 4 we show that the proposed controllers are inverse optimal with respect to meaningful cost functionals).

However, the fact that Lyapunov–based control methods provide an easily implementable controller with an explicitly characterized stability region for the closed–loop system motivates developing control techniques that use the a priori guarantees of stability provided by the bounded controller as a safety net for the implementation of the high performance model predictive controller. Reconciliation of different controllers (designed to satisfy different control objectives) calls for invoking the hybrid control paradigm, where the control structure consists of a blend of continuous (i.e., classical) controllers and discrete components (for example, a logic–based supervisor that orchestrates switching between the various continuous controllers; see Fig. 5.1).

Motivated by the above considerations, we develop in this chapter a hybrid predictive control structure that seamlessly unites MPC and bounded control, for the stabilization of linear systems with input constraints, in a way that allows both approaches to complement the stability and performance properties of each other. The guiding principle in developing the hybrid predictive control structure is the idea of decoupling performance requirements from the task of characterizing the region of constrained closed-loop stability. Specifically, by relaxing the optimality requirement, an explicit bounded feedback control law is designed and an explicit large estimate of the region of constrained closed-loop stability – which is not unnecessarily conservative – is

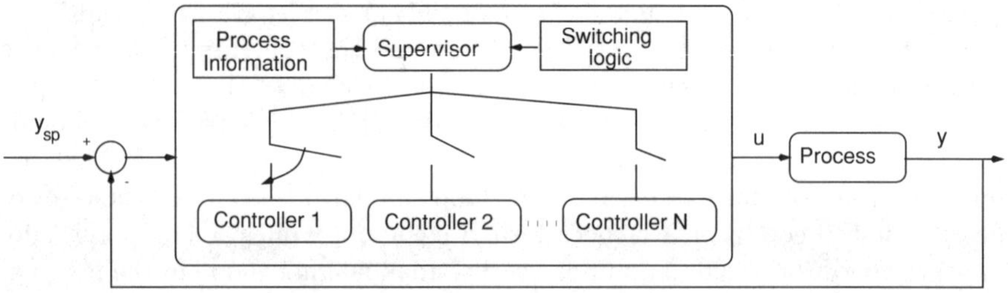

Fig. 5.1. Schematic representation of a hybrid control structure

computed. The predictive controller – which minimizes a given cost functional subject to the same constraints – is then designed and implemented within the stability region of the bounded controller, which is used a "fall–back" in the event that MPC is unable to achieve closed–loop stability (due, for example, to improper tuning of MPC parameters). Switching laws, that place appropriate restrictions on the evolution of the closed–loop trajectory under MPC, are then constructed to orchestrate the transition between the two controllers in a way that reconciles the tradeoffs between their respective stability and performance properties and guarantees closed–loop stability for all initial conditions within the stability region of the bounded controller. The switching scheme is shown to provide a safety net for the practical implementation of MPC by providing, through off–line computations, a priori knowledge of a large set of initial conditions for which closed-loop stability is guaranteed.

The general idea of switching, between different controllers or models, for the purpose of achieving some objective that either cannot be achieved or is more difficult to achieve using a single controller (or model) has been widely used in the literature, and in a variety of contexts. Examples include controller switching in gain scheduled control (e.g., [234]), logic-based switching in adaptive control (e.g., [119]), hybrid control of mixed logical dynamical (MLD) systems [32], the use of multiple linear models for transition control (e.g., [25, 259]) and scheduled predictive control (e.g., [21]) of nonlinear processes. The developed hybrid predictive control structure developed here differs from other hybrid control structures found in the literature, in the sense that it employs *structurally different* controllers as a tool for reconciling the objectives of optimal stabilization of the constrained closed-loop system (through MPC) and the a priori (off-line) determination of set of initial conditions for which closed-loop stability is guaranteed (through bounded control).

The rest of the chapter is organized as follows. In Sect. 5.2, we present some preliminaries that describe the class of systems considered and review briefly how the constrained control problem is addressed in both the bounded control and model predictive control approaches. We then proceed in Sect. 5.3 to formulate the hybrid control problem within the framework of switched systems and present the hybrid predictive control structure under the assumption of full state feedback. The theoretical underpinnings and practical implications of the proposed hybrid predictive control structure are highlighted, and possible extensions of the supervisory switching logic, that address a variety of practical implementation issues, are discussed. The implementation of the various switching schemes proposed is demonstrated through numerical simulations. In Sect. 5.4, we extend the hybrid predictive control structure to address the problem of output feedback stabilization. The state feedback control results of this chapter were first presented in [89] while the output feedback control results were presented in [195].

5.2 Preliminaries

In this chapter, we consider the problem of asymptotic stabilization of continuous-time linear time-invariant (LTI) systems with input constraints, with the following state-space description:

$$\dot{x}(t) = Ax(t) + Bu(t) \tag{5.1}$$

$$u(t) \in \mathcal{U} \subset \mathrm{I\!R}^m \tag{5.2}$$

where $x = [x_1 \cdots x_n]^T \in \mathrm{I\!R}^n$ denotes the vector of state variables, $u = [u_1 \cdots u_m]^T$ is the vector of manipulated inputs, taking values in a compact and convex subset of $\mathrm{I\!R}^m$, $\mathcal{U} := \{u \in \mathrm{I\!R}^m : \|u\| \leq u_{max}\}$, that contains the origin in its interior. The matrices A and B are constant $n \times n$ and $n \times m$ matrices, respectively. The pair (A, B) is assumed to be controllable. Throughout the chapter, the notation $\|\cdot\|$ is used to denote the standard Euclidean norm of a vector, while the notation $\|\cdot\|_Q$ refers to the weighted norm, defined by $\|x\|_Q^2 = x'Qx$ for all $x \in \mathrm{I\!R}^n$, where Q is a positive-definite symmetric matrix and x' denotes the transpose of x. Furthermore, the notation $x(T^-)$ is used to denote the limit of the trajectory, $x(t)$, as T is approached from the left, i.e., $x(T^-) = \lim_{t \to T^-} x(t)$.

In order to provide the necessary background for the main results of this chapter, we will briefly review in the remainder of this section the design procedure for, and the stability properties of, both the model predictive and bounded controllers which constitute the two components of the hybrid predictive control structure.

5.2.1 Model Predictive Control

We consider model predictive control of the system described by (5.1), subject to the control constraints of (5.2). The control action at time t is conventionally obtained by solving, on–line, a finite horizon optimal control problem [192] of the form:

$$P(x,t) : \min\{J(x,t,u(\cdot)) | u(\cdot) \in S\} \tag{5.3}$$

where $S = S(t,T)$ is the family of piecewise continuous functions (functions continuous from the right), with period Δ, mapping $[t, t+T]$ into $\mathcal{U}$ and T is the specified horizon. A control $u(\cdot)$ in S is characterized by the sequence $\{u[k]\}$ where $u[k] := u(k\Delta)$. A control $u(\cdot)$ in S satisfies $u(t) = u[k]$ for all $t \in [k\Delta, (k+1)\Delta)$. The performance index is given by:

$$J(x,t,u(\cdot)) = \int_t^{t+T} \left[\|x^u(s;x,t)\|_Q^2 + \|u(s)\|_R^2\right] ds + F(x(t+T)) \tag{5.4}$$

where R and Q are strictly positive–definite symmetric matrices, $x^u(s;x,t)$ denotes the solution of (5.1), due to control u, with initial state x at time t, and

$F(\cdot)$ denotes the terminal penalty. In addition to penalties on the state and control action, the objective function may also include penalties on the rate of change of the input, reflecting limitations on actuator speed (for example, a large valve requiring few seconds to change position). The minimizing control. $u^0(\cdot) \in S$, is then applied to the process over the interval $[k\Delta, (k+1)\Delta)$ and the procedure is repeated indefinitely. This defines an implicit model predictive control law:

$$M(x) := u^0(t; x, t) \tag{5.5}$$

It is well known that the control law defined by (5.3–5.5) is not necessarily stabilizing. To achieve closed-loop stability, early versions of MPC focused on tuning the horizon length, T, and/or increasing the terminal penalty (see [37] for a survey of these approaches), while more recent formulations focused on imposing stability constraints on the optimization (see [2, 32, 191] for surveys of different constraints proposed in the literature and the concomitant theoretical issues) or using off–line computations to come up with explicit model predictive control laws (see, for example, [219]). The additional stability constraints serve either to enforce convergence of the states of the closed–loop system to the equilibrium point, or to force the states to reach some invariant terminal set at the end of the horizon.

Remark 5.1. By incorporating stability conditions directly as part of the optimization problem, asymptotic stability under state feedback MPC is guaranteed provided that the initial condition is chosen so that the optimization yields a feasible solution. However, the implicit nature of the MPC control law, obtained through repeated on–line optimization, limits our ability to obtain, a priori, an explicit characterization of the admissible initial conditions starting from where the given model predictive controller (with fixed horizon length) is guaranteed to be feasible and enforce asymptotic stability. This set is a complex function of the constraints and the horizon length. Estimates of sufficiently large horizon lengths that ensure stability (see, for example, [48]) are typically conservative and, if used, may lead to a significant computational burden due to the increased size of the optimization problem. Therefore, in practice, the initial conditions and/or horizon lengths are usually chosen using ad hoc criteria and tested through closed–loop simulations which can add to the computational burden in implementing the model predictive controller. This motivates implementing MPC within a hybrid control structure that provides a "fall–back" controller for which a region of constrained closed–loop stability can be obtained off–line. Lyapunov–based controller design techniques provide a natural framework for the design of a stabilizing "fall–back" controller for which an explicit characterization of the region of closed–loop stability can be obtained.

5.2.2 Bounded Lyapunov-Based Control

Consider the Lyapunov function candidate, $V = x'Px$, where P is a positive–definite symmetric matrix that satisfies the Riccati equation:

$$A'P + PA - PBB'P = -\bar{Q} \tag{5.6}$$

for some positive–definite matrix $\bar{Q}$. Using this Lyapunov function, we can construct, using a modification of Sontag's formula for bounded controls proposed in [177] (see also Chap. 4), the following bounded nonlinear controller:

$$u(x) = -2k(x)B'Px \; := \; b(x) \tag{5.7}$$

where

$$k(x) = \left(\frac{L_f^*V + \sqrt{\left(L_f^*V\right)^2 + (u_{max}\|(L_gV)'\|)^4}}{\|(L_gV)'\|^2 \left[1 + \sqrt{1 + (u_{max}\|(L_gV)'\|)^2}\right]} \right) \tag{5.8}$$

when $L_gV \neq 0$, and $k(x) = 0$ when $L_gV = 0$, with $L_f^*V = x'(A'P + PA)x + \rho x'Px$, $(L_gV)' = 2B'Px$ and $\rho > 0$. This controller is continuous everywhere in the state–space and smooth away from the origin. Using a Lyapunov argument, one can show that whenever the closed–loop state trajectory evolves within the state–space region described by the set:

$$\Phi(u_{max}) = \{x \in \mathbb{R}^n : L_f^*V < u_{max}\|(L_gV)'\|\} \tag{5.9}$$

the resulting control action respects the constraints (i.e., $\|u\| \leq u_{max}$) and enforces, simultaneously, the negative–definiteness of the time–derivative of V along the trajectories of the closed–loop system (see Appendix C). Note that the size of the set, $\Phi(u_{max})$, depends on the magnitude of the constraints in a way such that the tighter the constraints, the smaller the region described by this set. Starting from any initial state within $\Phi(u_{max})$, asymptotic stability of the constrained closed–loop system can be guaranteed, provided that the closed–loop trajectory remains within the region described by $\Phi(u_{max})$. To ensure this, we consider initial conditions that belong to an invariant subset–preferably the largest – which we denote by $\Omega(u_{max})$. A similar idea was used in Chap. 4 in the context of bounded robust control of constrained nonlinear systems. One way of constructing such a subset is using the level sets of V (see Chap. 4 in [148] for details), i.e.:

$$\Omega(u_{max}) = \{x \in \mathbb{R}^n : x'Px \leq c_{max}\} \tag{5.10}$$

where $c_{max} > 0$ is the largest number for which all nonzero elements of $\Omega(u_{max})$ are contained within $\Phi(u_{max})$. The invariant region described by the set $\Omega(u_{max})$ provides an estimate of the stability region, starting from where the origin of the constrained closed–loop system under the control law of (5.7–5.8) is guaranteed to be asymptotically stable.

Remark 5.2. The bounded control law of (5.7–5.8) will be used in Sect. 5.3 to illustrate the basic idea of the proposed hybrid control scheme. Our choice of using this particular design is motivated by its explicit structure and well-defined region of stability. However, the hybrid predictive control structure is not restricted to this choice of bounded controllers; and any other analytical bounded control law, with an explicit structure and well-defined region of closed–loop stability, can also be used in implementing the proposed control strategy including, for example, the bounded controls developed in [265] for certain classes of constrained linear systems.

Remark 5.3. It is worth noting that the bounded nonlinear control law of (5.7–5.8) can be viewed as a "linear" controller with a variable (state-dependent) nonlinear gain. The fact that the nonlinear gain is shaped to account explicitly for the presence of input constraints allows one to obtain a larger region, in the state space, where the controller respects the constraints, than would be obtained, for example, by bounding (clipping) a constant-gain linear controller that has already been designed without taking the constraints into account. To illustrate this point, consider, for example, the scalar system $\dot{x} = x + u$ for which the LQR controller $u = -2x$, which minimizes the cost functional $J = \int_0^{\infty} u^2(t)dt$, has already been designed in the absence of constraints. When input constraints with magnitude $\|u\| \leq u_{max}$ are imposed, it is clear that this controller respects these constraints for all $\|x\| \leq \frac{u_{max}}{2}$. In contrast, we see from (5.9) that the bounded control law of (5.7–5.8) respects the same constraints for all $\|x\| \leq u_{max}$.

Remark 5.4. The stability region under a given stabilizing control law is typically only a subset of the null controllable region (the exact computation of which remains an open research problem). Furthermore, while the level sets of V provide only an estimate of the stability region under the bounded controller, less conservative estimates that capture larger portions of the null controllable region can be obtained using, for example, a combination of several Lyapunov functions (see, for example, [124] and Chap. 6 for some examples). In general, well–tuned bounded control laws provide larger estimates of the stability region than controllers designed without taking the constraints into account (see Remark 5.3 and Chap. 4 for further details on this issue).

Remark 5.5. Referring to the optimality properties of the bounded controller, it should be noted that, within a well–defined subset of Ω, this class of controllers can be shown to be inverse optimal with respect to some meaningful, yet unspecified *a priori*, cost functional that imposes meaningful penalties on the state and control action (see the Proof of Theorem 3.3 in Appendix A for an analogous proof and further details). However, unless this cost functional (which is determined only after controller design) coincides with the actual cost functional considered by the control system designer (this conclusion cannot be ascertained *a priori*), the performance of the bounded controller will not be optimal with respect to the actual cost functional. An explicit quantitative characterization of the cost incurred by implementing the bounded controller in this case is difficult to obtain. Furthermore, given an arbitrary cost functional, there is no transparent way of "tuning" the bounded controller to achieve optimality since, by design, the goal of the bounded controller is to achieve stabilization for an explicitly–defined set of initial conditions.

Remark 5.6. When comparing the optimality properties of the bounded and predictive controllers, it is important to realize that, in general, it is possible that the total cost of steering the system from the initial state to the origin under the bounded controller could be less than the total cost incurred under MPC. Keep in mind that MPC formulations typically consider minimizing a given cost over some finite horizon. Therefore, for a given objective function of the form of (5.4) for example, by solving the optimization problem once and implementing the resulting input trajectory in its entirety, the cost of taking the system from the initial state to the state at the end of the horizon under MPC is guaranteed to be smaller than the corresponding cost for the bounded controller. However, since MPC implementation involves repeated optimizations (i.e., only the first control move of the input trajectory is implemented each time), one could in some cases end up with a higher total cost for MPC. This possibility, however, cannot be ascertained a priori (i.e., before implementing both controllers and computing the total cost).

5.3 Hybrid Predictive State Feedback Control

By comparing the bounded and MPC controller designs reviewed in the previous section, some tradeoffs with respect to their stability and optimality properties are observed. For example, while the bounded controller possesses a well-defined region of admissible initial conditions that guarantee closed-loop stability in the presence of constraints, the performance of this controller is not guaranteed to be optimal with respect to an arbitrary performance criterion. On the other hand, MPC is well-suited for handling constraints within an optimal control setting; however, the analytical characterization of its set of admissible initial conditions is a more difficult task than it is through bounded control. In this section, we show how to reconcile the two approaches by means of a hybrid switching scheme that combines the desirable properties of both approaches.

To clearly present our approach, we will focus in this section only on the state feedback control problem where measurements of the entire state, $x(t)$, are assumed to be available for all t. The case when state measurements are available only at discrete sampling instants is discussed in Sect. 5.3.5, while the case of output feedback control is treated in Sect. 5.4.

5.3.1 Problem Formulation: Uniting MPC and Bounded Control

Consider the linear time-invariant system of (5.1), subject to input constraints, $\|u\| \leq u_{max}$, for which the bounded controller of (5.7–5.8) and the predictive controller of (5.3–5.5) have been designed. We formulate the controller switching problem as the one of designing a set of switching laws that orchestrate the transition between MPC and the bounded controller in a way that: (1) respects input constraints, (2) guarantees asymptotic stability of the origin of the closed-loop system starting from any initial condition in the set $\Omega(u_{max})$ defined in (5.10), and (3) guarantees recovery of the MPC performance whenever the pertinent stability criteria are met. For a precise statement of the problem, we first cast the system of (5.1) as a switched system of the form:

$$\begin{aligned} \dot{x} &= Ax + Bu_{i(t)} \\ \|u_i\| &\leq u_{max} \\ i(t) &\in \{1, 2\} \end{aligned} \tag{5.11}$$

where $i : [0, \infty) \rightarrow \{1, 2\}$ is the switching signal which is assumed to be a piecewise continuous (from the right) function of time, implying that only a finite number of switches between the two controllers is allowed on any finite interval of time. The index, $i(t)$, which takes values in the finite set, $\{1, 2\}$, represents a discrete state that indexes the control input, u, with the understanding that $i(t) = 1$ if and only if $u_i(x(t)) = M(x(t))$ (i.e., MPC is used) and $i(t) = 2$ if and only if $u_i(x(t)) = b(x(t))$ (i.e., bounded control is used). The value of $i(t)$ is determined by a higher-level supervisor responsible for executing the transition between the two controllers. Our goal is to construct a switching law:

$$i(t) = \psi(x(t), t) \tag{5.12}$$

that provides the supervisor with the switching times that ensure stabilizing transitions between the two controllers. This, in turn, determines the time-course of the discrete state, $i(t)$. A schematic representation of the proposed hybrid predictive control structure is depicted in Fig. 5.2.

In the remainder of this section, we present three switching schemes that address this problem. The first scheme is formalized in Theorem 5.7 (Sect. 5.3.2) and focuses primarily on closed-loop stability, while the second scheme, given in Theorem 5.14 (Sect. 5.3.3), extends the first one to accommodate closed-loop performance considerations as well. For illustration purposes, both the stability-based and performance-based schemes use a classical MPC formulation. Finally, in Sect. 5.3.4, we present the third switching scheme (feasibility-based switching) which shows how more advanced MPC formulations can also be accommodated by appropriate choice of the switching logic. The proofs of Theorems 5.7 and 5.14 are given in Appendix C.

5.3.2 Stability-Based Controller Switching

Theorem 5.7. *Consider the constrained LTI system of (5.11), with an initial condition $x(0) := x_0 \in \Omega(u_{max})$, where $\Omega(u_{max})$ was defined in (5.10), under*

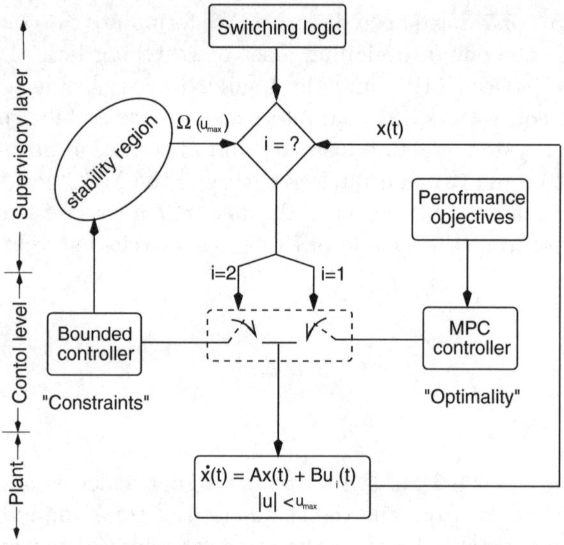

Fig. 5.2. A schematic representation of the hierarchical hybrid predictive control structure uniting MPC and bounded control for the stabilization of constrained LTI systems

the model predictive controller of (5.3–5.5). Also let $T_s > 0$ *be the earliest time for which the closed-loop state satisfies:*

$$- \| x(T_s^-) \|_Q^2 + \| B'Px(T_s^-) \|^2 + 2x'(T_s^-)PBu(T_s^-) \geq 0 \tag{5.13}$$

and suppose that:

$$i(t) = \left\{ \begin{array}{l} 1, \;\; 0 \leq t < T_s \\ \\ 2, \;\; t \geq T_s \end{array} \right\} \tag{5.14}$$

where $i(t) = 1 \Leftrightarrow u_i(x(t)) = M(x(t))$ *and* $i(t) = 2 \Leftrightarrow u_i(x(t)) = b(x(t))$*. Then the origin of the closed-loop system is asymptotically stable in the presence of input constraints.*

Remark 5.8. Theorem 5.7 describes a stability-based switching strategy for control of linear systems with input constraints. The three main components of this strategy include the bounded controller, the model predictive controller, and a higher-level supervisor that orchestrates the switching between the two controllers. The implementation of this hybrid control strategy is best understood through the following stepwise procedure (see also Fig. 5.3):

- Given the system model of (5.1) and the constraints on the input, design the bounded controller using (5.7–5.8). Given the performance objective and a choice of the horizon length, design the MPC controller.

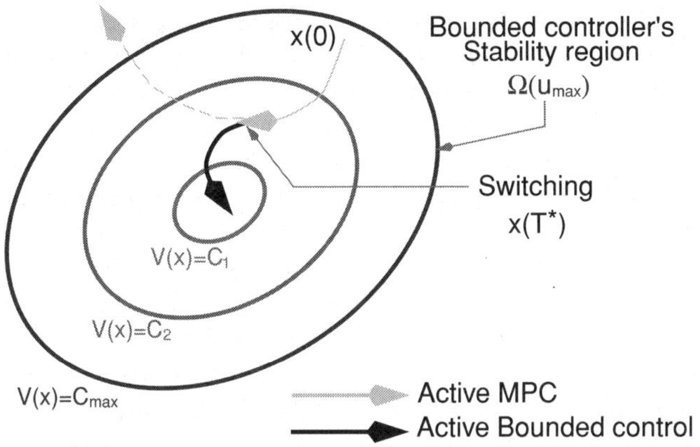

Fig. 5.3. A schematic representation of the implementation of the stability-based controller switching proposed in Theorem 5.7

- Compute the stability region estimate for the bounded controller, $\Omega(u_{max})$, using (5.9–5.10).
- Initialize the closed-loop system, using MPC, at an initial condition, x_0, within $\Omega(u_{max})$.
- Monitor the evolution of the closed-loop trajectory (by checking (5.13) at each time) until the earliest time, T_s, that (5.13) is met.
- At the earliest time that (5.13) is met, discontinue MPC implementation, switch to the bounded controller and implement it for all future times.

Remark 5.9. The relations of (5.13–5.14) represent the switching rule that the supervisor observes when deciding if a switch between the two controllers is needed. The implementation of this rule requires only the evaluation of the algebraic expression on the left hand-side in (5.13), which is the rate at which the Lyapunov function (used in designing the bounded controller and characterizing its stability region) grows or decays along the trajectories of the closed-loop system. By observing this rule, the supervisor is essentially tracking the temporal evolution of V under MPC, so that whenever an increase in V is detected after its initial implementation, MPC is disengaged from the closed-loop and the bounded controller is switched in, thus steering the closed-loop trajectory to the origin asymptotically. This logic guarantees that, under MPC, the closed-loop trajectory never escapes $\Omega(u_{max})$ before the bounded controller can be switched in.

Remark 5.10. Note that in the case when the condition in (5.13) is never fulfilled – i.e.. the Lyapunov function continues to decay monotonically along the closed-loop trajectories under MPC – the switching rule of (5.14) ensures that only MPC is implemented for all times (no switching occurs) since it is asymptotically stabilizing. In this case, the MPC performance is fully recovered.

Remark 5.11. Note that the proposed approach does not turn $\Omega(u_{max})$ into a stability region for MPC. What the approach does, however, is turn $\Omega(u_{max})$ into a stability region for the switched closed-loop system. The value of this can be understood in light of the difficulty in obtaining, a priori, an analytical characterization of the set of admissible initial conditions that the MPC controller can steer to the origin in the presence of input constraints. By using the bounded controller as a fall-back controller, the switching scheme allows us to safely initialize the closed-loop system anywhere within $\Omega(u_{max})$ using MPC, with the guarantee that the bounded controller can always intervene (through switching) to "rescue" closed-loop stability in case the closed-loop starts to become unstable under MPC (due, for example, to a poor choice of the initial condition or improper tuning of MPC). This safety feature distinguishes the bounded controller from other fall-back controllers that could be used, such as PID controllers, which do not provide a priori knowledge of the constrained stability region and, therefore, do not guarantee a safe transition in the case of unstable plants. For these controllers, a safe transition is critically dependent on issues such as whether or not the operator is able to properly tune the controller online, which, if not achieved, can possibly result in instability and/or performance degradation. The transition from the MPC to the bounded controller, on the other hand, is not fraught with such uncertainty and is always safe because of the a priori knowledge of the stability region. This aspect helps make plant operation safer and smoother.

Remark 5.12. When compared with Lyapunov-based MPC approaches (e.g., contractive MPC [220], CLF-based RHC [221]), we find that the proposed hybrid scheme also employs a Lyapunov stability condition to guarantee closed-loop asymptotic stability. However, the Lyapunov stability condition is enforced at the supervisory level, via continuous monitoring of the temporal evolution of V, rather than being incorporated as an inequality constraint on the optimization problem, as is customarily done in Lyapunov-based MPC approaches. This fact may help reduce the complexity of the optimization problem in some cases, without loss of stability, by reducing the number of constraints on the optimization. The underlying idea here is that of decoupling the stability requirement from optimality. In the proposed hybrid scheme, it is switching to the bounded controller, with its well-defined stability region, that safeguards against potential closed-loop instability arising as a consequence of implementing the MPC controller, for which the stability region is not known a priori. On the other hand, the MPC controller provides, by design, the desired high performance under constraints. The switching logic then ensures that any such "optimal" solution is implemented only when it is asymptotically stabilizing.

Remark 5.13. Note that no assumption is made regarding how fast the optimization needs to be solved for the MPC controller to be implemented because even if this time is relatively significant and the plant dynamics are unstable, implementation of the switching rule in Theorem 5.7 guarantees an "instantaneous" switch to the stabilizing bounded controller.

5.3.3 Performance-Driven Switching

While the switching scheme proposed in Theorem 5.7 guarantees closed-loop stability for all initial conditions within $\Omega(u_{max})$, the implementation of this scheme can, in some cases, place limitations on the achievable closed-loop performance by (unnecessarily) restricting the implementation of MPC and using the bounded controller in its place. In particular, note that the switching rule of (5.13–5.14) does not permit any transient increase in V under MPC and requires that, at the first instance such an increase is detected, the supervisor immediately terminate the implementation of MPC and switch to the bounded controller instead. Given, however, that a temporary, finite increase in V does not necessarily translate into closed-loop instability, this particular switching rule is restrictive in the sense that it does not give MPC any further chance to stay in the closed-loop despite the fact that the observed initial increase in V might actually be followed by monotonic decay (in which case MPC would be stabilizing, yet not implemented because of the particular switching rule used). Clearly in such a scenario, and since MPC provides the high performance controller, it would be desirable to keep MPC in the closed-loop instead of switching to the bounded controller as Theorem 5.7 would require. To allow for greater flexibility in recovering the performance of MPC, without concern for loss of stability, we propose in this section an extension of the switching strategy of Theorem 5.7 by relaxing the switching rule. This is described in the Theorem 5.14 below and shown schematically in Fig. 5.4. The proof of the theorem can be found in Appendix C

Theorem 5.14. *Consider the constrained LTI system of (5.11), under the model predictive controller of (5.3–5.4), with an initial condition* $x(0) := x_0 \in \Omega(u_{max})$*, as defined in Theorem 5.7, and let* $T_b > 0$ *be the earliest time for which the closed-loop state satisfies:*

$$x^T(T_b^-)Px(T_b^-) = c_{max} \tag{5.15}$$

Also let $\{T_i, i = 1, \dots, N,\ N < \infty\}$ *be a set of times such that, if exist:*

$$- \parallel x(T_i^-) \parallel_{\bar{Q}}^2 + \parallel B'Px(T_i^-) \parallel^2 + 2x'(T_i^-)PBu(T_i^-) = 0 \tag{5.16}$$

Then, the switching rule:

$$i(t) = \left\{ \begin{array}{l} 1,\ 0 \le t < \min\{T_b, T_N\} \\ \\ 2,\ t \ge \min\{T_b, T_N\} \end{array} \right\} \tag{5.17}$$

where $i(t) = 1 \Leftrightarrow u_i(x(t)) = M(x(t))$ *and* $i(t) = 2 \Leftrightarrow u_i(x(t)) = b(x(t))$*, guarantees that the origin of the closed-loop system is asymptotically stable in the presence of input constraints.*

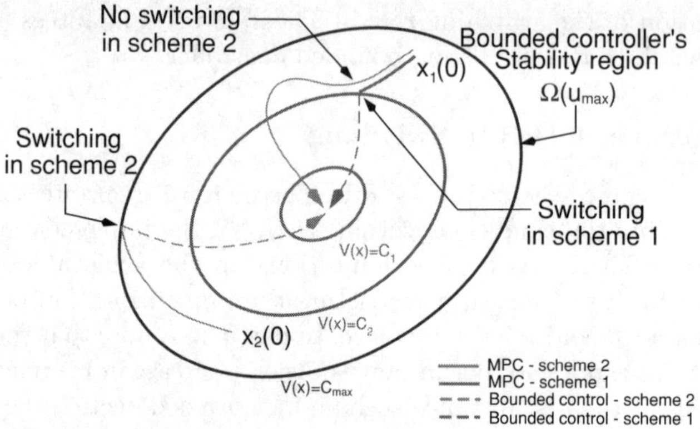

Fig. 5.4. A schematic representation of the implementation of the controller switching strategy proposed in Theorem 5.14 and how it compares with the scheme proposed in Theorem 5.7. Starting at $x_1(0)$, scheme 2 permits a temporary increase in V under MPC without switching thus allowing MPC to remain in the closed-loop for all times (*solid trajectory*). Starting at $x_1(0)$, scheme 1 does not allow the increase in V and switches immediately to the bounded controller instead (*solid followed by dashed trajectory*). Starting at $x_2(0)$, scheme 2 enforces a switch from MPC to the bounded controller when the closed-loop trajectory under MPC is about to leave Ω

Remark 5.15. The basic idea of the switching scheme of Theorem 5.14 is to allow MPC to remain in the closed-loop system even if V increases for some time, provided that the closed-loop trajectory does not leave $\Omega(u_{max})$ during this time. This requirement is necessary because this region is not invariant under MPC and, therefore, once the trajectory leaves, there are no guarantees that it can ever be steered back to the origin, whether using MPC or even the bounded controller. Note, however, that this requirement by itself guarantees only boundedness of the closed-loop trajectory. To ensure asymptotic convergence to the origin, V must be made to decrease after some time. To accomplish this, a choice needs to be made as to how long the increases in V can be tolerated before switching to the bounded controller is deemed necessary. In the statement of Theorem 5.14, this user-specified choice is quantified by N which represents the maximum number of times that $\dot{V}$ is allowed by the designer to change sign. This requirement is intended to avoid scenarios where V neither keeps monotonically increasing nor decreasing for all times.

Remark 5.16. Note that the times T_b and T_i are not known a priori. They are implicity defined by the relations of (5.15–5.16), respectively, which are continuously checked on-line by the supervisor. The switching scheme proposed in Theorem 5.14 then is best understood as follows. For illustration purposes we choose $N = 2$. The closed-loop system is initialized within $\Omega(u_{max})$ using MPC, and the temporal evolution of V is monitored on-line. Suppose that V keeps increasing monotonically until, at some time t, (5.15) is satisfied (trajectory hits boundary of $\Omega(u_{max})$). In this case, we have that $T_b = t$ and that T_1, T_2, if exist, must be greater than t. Therefore $\min\{T_b, T_2\} = t$ and, according to the switching law of (5.17), MPC implementation should be terminated and the bounded controller switched in at T_b. Now suppose, instead, that at some time $\bar{t} < t$, (5.16) is satisfied, then in this case we would have $T_1 = \bar{t}$, $T_b = t$. Since T_2 in this case, if exists, is greater than t, we set $\min\{T_b, T_2\} = t$ which implies that MPC should be switched off again at T_b. Finally, consider the case when (5.16) is satisfied at two consecutive times, $t^* < \bar{t}$, prior to which (5.15) is not fulfilled. Then in this case, we have $T_1 = t^*$, $T_2 = \bar{t}$, and T_b, if exists, must be greater than $\bar{t}$. Therefore, we set $\min\{T_b, T_2\} = \bar{t}$ and conclude that MPC should be switched off at $\bar{t}$. To summarize, in all the cases, switching is done either when the trajectory gets "close" to the boundary of the stability region or after $\dot{V}$ has changed sign N times, whichever occurs first.

Remark 5.17. The switching schemes proposed in Theorems 5.7 and 5.14 can be further generalized to allow for multiple switchings between MPC and the bounded controller. In this case, when switched in, the bounded controller need not stay in the closed-loop for all future times. Instead, it is employed only until it brings the closed-loop state trajectory sufficiently closer to the origin (for example, to a pre-determined inner level set of V) at which point MPC is activated once again. This scheme offers the possibility of further enhancement in closed-loop performance (over that obtained from the bounded controller) by implementing MPC for longer periods of time (as permitted by stability considerations). It also allows for the possibility of stabilizing the closed-loop system starting from initial conditions for which a given (fixed-horizon) MPC controller is not stabilizing. This can occur, for example, when the stability region for such a controller is contained within $\Omega(u_{max})$. In this case, only a finite number of switches between the two controllers is needed to steer the closed-loop trajectory to that point, beyond which the MPC controller can stay in the closed-loop for all times (with no further switching).

5.3.4 Switching Using Advanced MPC Formulations

In Theorems 5.7 and 5.14, a classical MPC formulation (with no stability constraints) was used as an example to illustrate the basic idea of controller switching and how it can be used to aid MPC implementation. The value of the hybrid predictive control structure (see Fig. 5.3), however, is not restricted to classical MPC formulations, and extends to any MPC formulation for which a priori knowledge of the set of admissible initial conditions is lacking (or computationally expensive to obtain). The structure can be adapted to more advanced versions of MPC by appropriate design of the switching logic. To illustrate this point, consider the case when advanced MPC algorithms, that employ stability constraints in the optimization formulation, are used to compute the control action. An example of such formulations is the following one:

$$\begin{array}{rcl} & \min\limits_{u(\cdot)} \displaystyle\int_t^{t+T} (x'(s)Qx(s) + u'(s)Ru(s))ds & \\ s.t. \quad \dot{x}(t) & = & Ax(t) + Bu(t), \; x(0) = x_0 \\ u(\cdot) & \in & S \\ x(t+T) & = & 0 \end{array} \tag{5.18}$$

which uses a terminal equality constraint (note that other types of stability constraints, such as control Lyapunov function (CLF)-based inequality constraints of the form $V(x(t+T)) < V(x(t))$ can be treated similarly). When the above formulation is used to compute the MPC control law in our switching scheme, it is clear that monitoring the evolution of the closed-loop trajectory (which was the basis for the implementation of the switching rules in Theorems 5.7 and 5.14) is no longer appropriate as a switching criterion, since, by definition, a closed-loop trajectory that solves the above optimization will not "evolve" unless it stabilizes at the origin. By incorporating stability conditions as part of the constrained optimization problem, asymptotic stability under MPC can be guaranteed provided that the initial optimization yields a feasible solution. However, inclusion of the stability constraints does not, by itself, provide any a priori explicit knowledge of the feasible initial conditions, which must then be identified through simulations. This observation suggests that the hybrid control structure can be used here to safeguard closed-loop stability in the event of MPC infeasibility. Specifically, in lieu of monitoring the growth of V, a different switching logic, based on the feasibility of the optimization in (5.18), needs to be implemented. The switching algorithm in this case can be summarized as follows (see also Fig. 5.5 for a schematic representation):

- Given any initial condition, x_0, within $\Omega(u_{max})$, check, off-line, whether the constrained optimization in (5.18) yields a feasible solution. If a solution exists, no switching is needed and MPC can be implemented for all time.
- If an initial feasible solution is not found, implement the bounded controller instead, i.e., set $u(0) = b(x_0)$.
- While the bounded controller is in the closed-loop, continue to check, off-line, and as frequently as is desired, the feasibility of the optimization in (5.18).
- At the earliest time that a feasible solution is found, switch to the MPC controller, else the bounded controller remains active.

The above discussion underscores an important attribute of the proposed hybrid control structure, and that is the fact that it is not proposed as a substitute for stability constraints (even though it guarantees stability when the MPC formulation is not stabilizing), but as a reliable fall-back mechanism from which any MPC formulation can benefit.

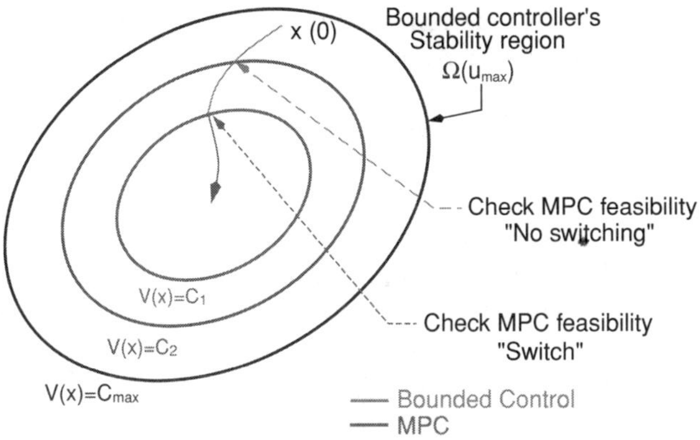

Fig. 5.5. A schematic representation of the implementation of switching when a "stabilizing" MPC formulation is used. Starting at $x(0)$, MPC does not yield a feasible solution and, therefore, the bounded controller is used. When the trajectory crosses the C_2-level set of V, the supervisor checks off-line the feasibility of the MPC and finds that no feasible solution exists. The bounded controller remains in action. Finally, when the trajectory crosses the C_1-level set of V, the supervisor finds that the optimization is feasible and therefore switches to the MPC controller which stays in the closed-loop system for all future times

Remark 5.18. An important issue in the practical implementation of MPC is the selection of the horizon length. It is well known that this selection can have a profound effect on nominal closed-loop stability, and results are available in the literature that establish, under certain assumptions, the existence of "sufficiently large" finite horizons that ensure stability (see, for example, [48,228]). However, a priori knowledge (without closed-loop simulations) of the critical horizon length that guarantees closed-loop stability, from an arbitrary initial condition, is currently not available. Therefore, in practice the horizon length is typically chosen based on available selection criteria, tested through closed-loop simulations, and varied, if necessary, to achieve stability. Note that in all the switching schemes proposed in Sect. 5.2.2, closed-loop stability is maintained independent of the horizon length, and therefore selection of the horizon length can be made solely on the basis of what is computationally practical for the size of the optimization problem, without increasing, unnecessarily, the horizon length (and consequently the computational load) out of concern for stability.

Remark 5.19. The supervisory checks required in the implementation of all the switching schemes developed in this chapter do not incur any additional computational costs beyond that involved in MPC implementation alone. In the event that MPC is infeasible from a given initial condition (due, for example, to insufficient number of control moves), the supervisor checks as frequently as desired the initial feasibility of the MPC by actually solving the optimization problem at each time step and implementing the resulting solution (if found). If a solution is not found at a given time step, the supervisor implements the bounded controller (an inexpensive algebraic computation) and continues, in the mean time, to solve the optimization until it finds a feasible solution to implement.

Remark 5.20. The switching schemes proposed in this chapter differ, both in their objective and implementation, from other MPC formulations involving switching which have appeared earlier in the literature. For example, in dual mode MPC [200] the strategy includes switching from MPC to a locally stabilizing controller once the state is brought near the origin by MPC. The purpose of switching in this approach is to relax the terminal equality constraint whose implementation is computationally burdensome for nonlinear systems. However, the set of initial conditions for which MPC is guaranteed to steer the state close to the origin is not explicitly known a priori. In contrast, switching from MPC to the bounded controller is used here only to prevent any potential closed-loop instability arising from implementing MPC without the a priori knowledge of the admissible initial conditions. So depending on the stability properties of the chosen MPC, switching may or may not occur, and if it occurs, it can take place near or far from the origin. Finally, we note that the notion of "switching" from a predictive to an unconstrained LQR controller, for times beyond a finite control horizon, has been used in [48, 243, 263] in the context of determining a finite horizon that solves the infinite horizon constrained LQR problem.

5.3.5 Simulation Example

Consider the following linear system:

$$\dot{x} = \begin{bmatrix} 0.5 & 0.25 \\ 0.5 & 1 \end{bmatrix} x + \begin{bmatrix} 1 & 0 \\ 0 & 1 \end{bmatrix} u \tag{5.19}$$

where both inputs u_1, u_2 are constrained in the interval $[-5, 5]$. It is straightforward to verify that the open-loop system has an unstable equilibrium point at the origin (A has two positive eigenvalues). We initially use (5.7–5.8) to design the bounded controller and construct its stability region via (5.9–5.10). The matrix P is chosen to be:

$$P = \begin{bmatrix} 1.1362 & 0.8102 \\ 0.8102 & 1.8658 \end{bmatrix} \tag{5.20}$$

For the MPC controller, the parameters in the objective function of (5.4) are chosen as penalty on the states, $Q = qI$, with $q = 1$ and penalty on the control inputs, $R = rI$, with $r = 3$. A higher penalty on u is indicative of the requirement of driving the state to the origin with less control action. We also choose a horizon length of $T = 1$ in implementing the MPC controller. The resulting quadratic program is solved using the MATLAB subroutine QP, and the set of ODEs are integrated using the MATLAB solver ODE45.

As shown by the solid curve in Fig. 5.6, applying the MPC controller from the initial condition $x_0 = [6\ -2]^T$ (which belongs to the stability region of the bounded controller, Ω, represented by the ellipse in Fig. 5.6) leads to closed-loop instability. The corresponding input and output profiles are shown by the solid lines in Fig. 5.7. Using the switching scheme of Theorem 5.7, however, we find that the supervisor detects an increase in V at $t = 1.7$ and therefore immediately switches at this time from the MPC controller to the bounded controller in order to preserve closed-loop stability. As expected, the bounded controller asymptotically stabilizes the plant (see dashed lines in Figs. 5.6 and 5.7). It is important to note that one could try to tune the horizon length further in order to achieve stability using the MPC controller. For example, we see from the dotted lines in Figs. 5.6 and 5.7 that stability can be achieved by increasing the horizon length to $T = 1.5$. However, this conclusion could not be reached a priori, i.e., before running the closed-loop simulation in its entirety to check whether the choice $T = 1.5$ is appropriate. In contrast, closed-loop stability starting from the given initial condition, is guaranteed, a priori, under the switching scheme of Theorem 5.7.

In the next simulation run, we demonstrate how the switching scheme of Theorem 5.14 can be used to enhance closed-loop performance by allowing greater flexibility in implementing the MPC controller. To this end, we again consider the initial condition $x_0 = [6\ -2]^T \in \Omega$. For the MPC controller,

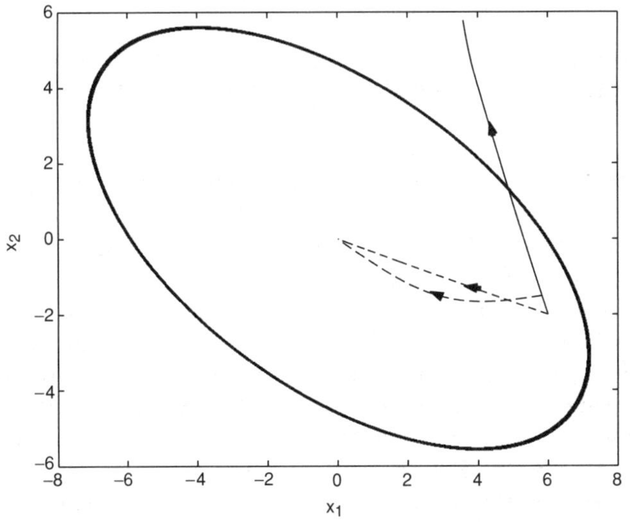

Fig. 5.6. Closed-loop state trajectory under MPC with $T = 1$ (*solid line*), under the switching scheme of Theorem 5.7 with $T = 1$ (*dashed line*), and under MPC with $T = 1.5$ (*dotted line*)

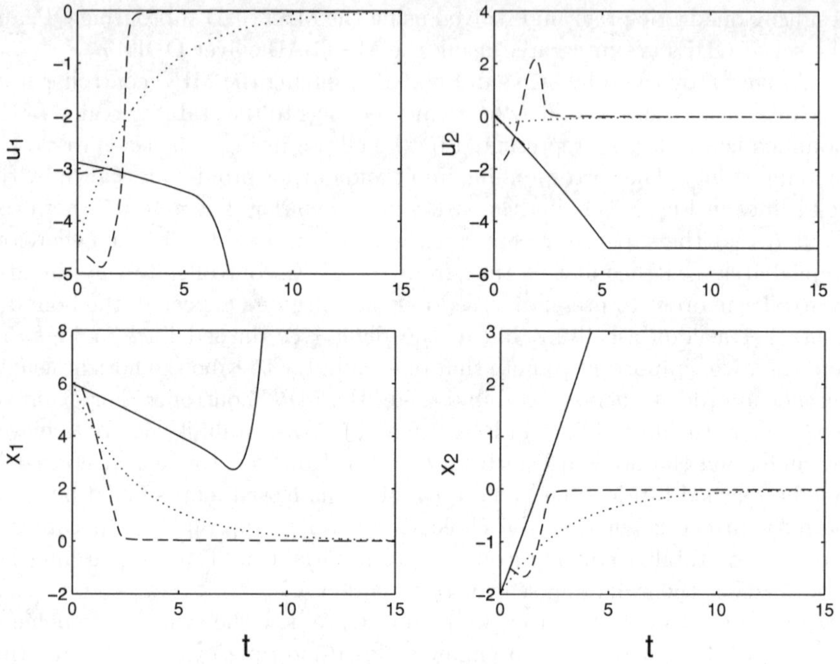

Fig. 5.7. Input and closed-loop state profiles under MPC with $T = 1$ (*solid line*), under the switching scheme of Theorem 5.7 with $T = 1$ (*dashed line*), and under MPC with $T = 1.5$ (*dotted line*)

the objective function includes penalty on the state with $q = .5$ and penalty on the inputs with $r = 1$. We also include a quadratic penalty on the rate of input change weighted by the matrix $S = sI$ with $s = 40$. The horizon length is chosen to be $T = 1.5$. Initially, we implement the MPC controller and use the switching scheme of Theorem 5.7 to stabilize the closed-loop system. In this case, an increase in V is detected immediately (at $t = 0$) by the supervisor which, consequently, switches automatically to the bounded controller. The results for this case are depicted by the solid curves in Figs. 5.8 and 5.9, which reflect the closed-loop performance under the bounded controller. However, when the switching scheme of Theorem 5.14 is used, we find that the MPC controller stays in action, for all times, thus providing the desired optimal performance and, at the same time, asymptotically stabilizing the closed-loop system without the need for switching. The results are shown by the dashed curves in Figs. 5.8 and 5.9. So, while the first switching scheme selects the bounded controller, the second one selects the MPC controller. The reason for the difference between the two scenarios is the fact that the switching law in the second scheme tolerates the initial increase in V (see Fig. 5.10) since this increase does not land the state outside the stability region and is shortly

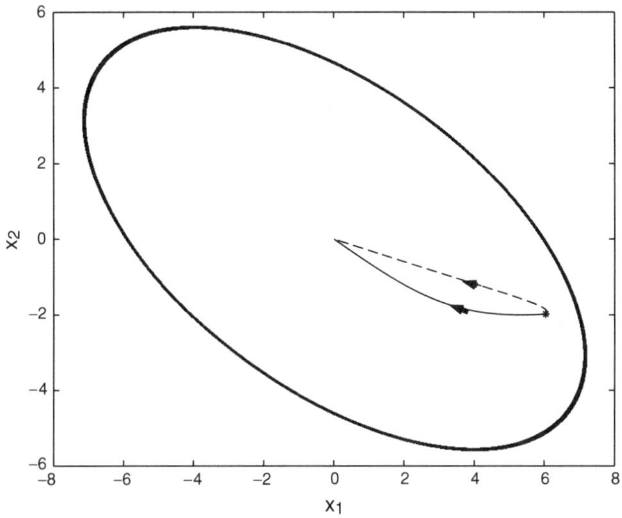

Fig. 5.8. Closed-loop state trajectory under the switching scheme of Theorem 5.7 (*solid line*), and under the switching scheme of Theorem 5.14 (*dashed line*)

followed by continuous decrease, and thus keeps the MPC controller in the closed-loop.

In the following set of simulation runs, we demonstrate an application of the feasibility-based switching scheme, proposed in Sect. 5.3.4, which uses a stable version of MPC, specifically, one employing a terminal equality constraint. For the MPC controller, the parameters in the objective function (5.18) are chosen as $q = 1$ and $r = 1$. The horizon length is chosen to be $T = 1.2$. As shown by the solid lines in Fig. 5.11, starting from an initial condition, $x_0 = [2 \;\; -2]^T$, MPC is found to be feasible and is therefore implemented and kept in the closed-loop for all times driving the trajectory asymptotically to the origin. The corresponding input and output profiles are shown by the solid lines in Fig. 5.12. The scenario, therefore, represents a reasonably well-tuned MPC for initial conditions around the origin. Now suppose that, after settling at the origin, a temporary disturbance drives the closed-loop trajectory to the point $x_i = [5 \;\; -5]^T$ (which is still within the stability region of the bounded controller). After the disturbance disappears, MPC starting from this initial condition is found to be infeasible and, therefore, the bounded controller is employed instead. Implementation of the bounded controller proceeds until the closed-loop trajectory crosses the level set denoted by Ω_2 in Fig. 5.11, which occurs at $t = 0.6$. At this point, feasibility of MPC is checked again and the MPC controller is found to be feasible. Consequently, the bounded controller is terminated immediately and the MPC controller is employed in the closed-loop for the remaining time (inside Ω_2). This scenario is shown by the dashed lines in Figs. 5.11 and 5.12. Once again, note that using

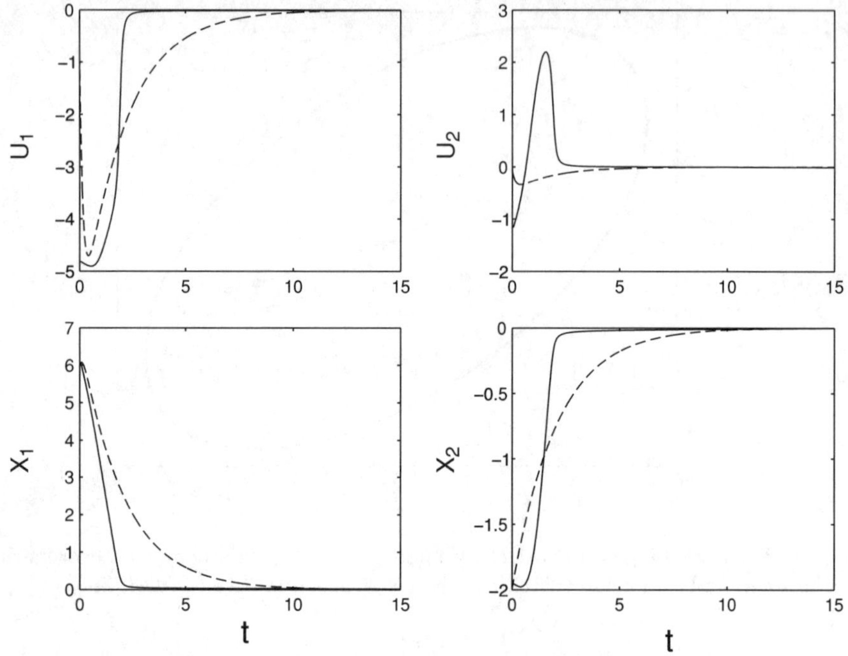

Fig. 5.9. Input and closed-loop state profiles under the switching scheme of Theorem 5.7 (*solid line*) and under the switching scheme of Theorem 5.14 (*dashed line*)

a higher value of the horizon length, $T = 1.5$ can lead to MPC being feasible from the initial condition $x_i = [5\ -5]^T$ (dotted lines in Figs. 5.11 and 5.12). This implies that, once the closed-loop state is driven by the disturbance to x_i, a re-tuning of the horizon length can make MPC feasible. However, a priori knowledge of the value of the horizon length that is guaranteed to work could not be obtained short of testing feasibility at that condition. The prospect of disturbances throwing the system around would therefore necessitate running extensive closed-loop simulations to determine such a value. The feasibility-based switching scheme, on the other hand, guarantees closed-loop stability for any value of the horizon length. In particular, after the disturbance vanishes, the bounded controller drives the closed-loop trajectory to a point at which the original MPC controller (with $T = 1.2$) becomes feasible again and can therefore be implemented. In this example, feasibility was checked only at one level set. The frequency of feasibility checking could be made larger by introducing more level sets, or by checking feasibility at predetermined times.

In the last set of simulation runs, we demonstrate an example of how the switching schemes can be used (or modified) in the case when state measurements are not available continuously, but are rather available only at discrete sampling instants. For simplicity and clarity of presentation, we focus our discussion only on the switching scheme of Theorem 5.14. Recall that the

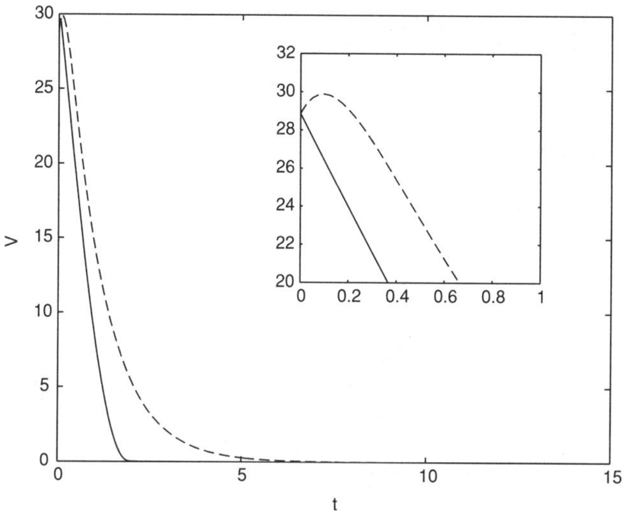

Fig. 5.10. Temporal evolution of the Lyapunov function under the switching scheme of Theorem 5.7 (*solid line*), where the bounded controller is switched in immediately at $t = 0$, and under the switching scheme of Theorem 5.14 (*dashed line*), where the MPC controller stays in action for all time

implementation of the switching rule in Theorem 5.14 requires continuous monitoring of the state trajectory to prevent its potential escape under MPC from the set $\Omega(u_{max})$. The restricted access to state information during the time periods between the sampling instants, however, implies the possibility that, during the first such period after the MPC controller of (5.3–5.4) is implemented, the closed-loop state trajectory could leave the stability region Ω before such an escape is detected, especially if the initial condition is chosen close to the boundary of Ω and/or the sampling period is too large. In this case, closed-loop stability cannot be guaranteed, and any switching to the bounded controller will be too late to recover from the instability of MPC. To guard against this possibility, we slightly modify the switching rule by restricting the implementation of MPC within a subset of Ω. This subset, henceforth denoted by Ω_2 and referred to as the safety zone, is defined as the set of all states starting from where the closed-loop trajectory is guaranteed to remain within Ω after a single sampling period. An estimate of this set can be obtained from:

$$\Omega_2 = \left\{ x \in \mathbb{R}^n : \sqrt{V(x)} \geq \sqrt{c_{max}} \exp\left(-\frac{\alpha\Delta}{2}\right) - \frac{\beta}{\alpha}\left(1 - \exp\left(-\frac{\alpha\Delta}{2}\right)\right)\right\} \tag{5.21}$$

where Δ is the sampling period, $\alpha = \dfrac{\lambda_{max}(PBB'P)}{\lambda_{min}(P)} > 0$, $\beta = 2u_{max}\sqrt{\alpha} > 0$, $\lambda_{max}(\cdot)$ and $\lambda_{min}(\cdot)$ are, respectively, the maximum and minimum eigenvalues

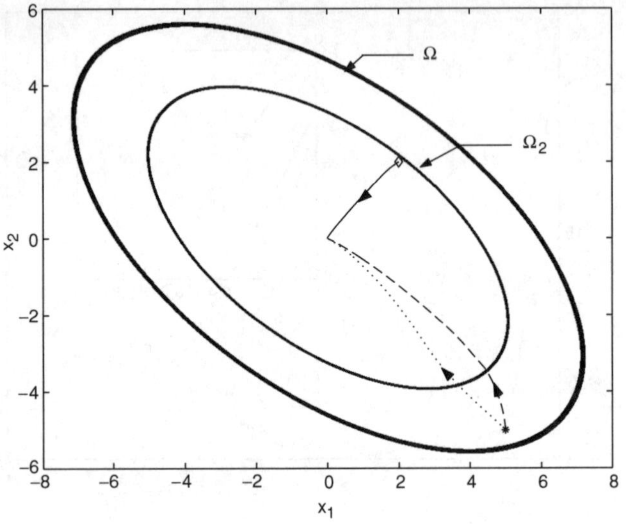

Fig. 5.11. Closed-loop state trajectory starting close to the origin using MPC with $T = 1.2$ (*solid line*) and from a point farther away in state space using the switching scheme with a horizon length $T = 1.2$ (*dashed line*) and using only MPC with $T = 1.5$ (*dotted line*)

of the given matrix. The above estimate is obtained by writing the appropriate Lyapunov dissipation inequality for the system, majorizing the right hand side using appropriate bounds and, finally, integrating both sides of the inequality from $t = 0$ ($V = V(x)$) to $t = \Delta$ ($V = c_{max}$). Since the implementation of MPC is restricted to Ω_2, only the bounded controller is used in the annular region between Ω and Ω_2.

To illustrate the implementation of the resulting switching scheme in the case of measurement sampling, we consider the case when the state measurements arrive every 0.1 time units ($\Delta = 0.1$). This sampling period is found to be sufficiently small for the bounded controller to be stabilizing. For the MPC controller, the parameters in the objective function (5.4) are chosen as $q = 0.5$, $r = 2$ and $s = 40$. The horizon length is chosen to be $T = 1.5$. The safety zone, Ω_2, is constructed using (5.21) to ensure that starting from Ω_2, the state cannot escape Ω in one sampling period. As stated earlier, this ensures that an increase in V under MPC can be detected by the supervisor before the trajectory goes outside the stability region of the bounded controller. Starting from the initial condition, $x_0 = [-3 \ -2.5]$, as shown by the solid lines in Fig. 5.13, the closed-loop state trajectory is clearly unstable under MPC. The corresponding input and state profiles are shown by the solid lines in Fig. 5.14. Note that since the closed-loop is initialized outside Ω_2, the closed-loop trajectory under MPC was able to escape Ω before being detected. To remedy this problem, the bounded controller is employed, starting from

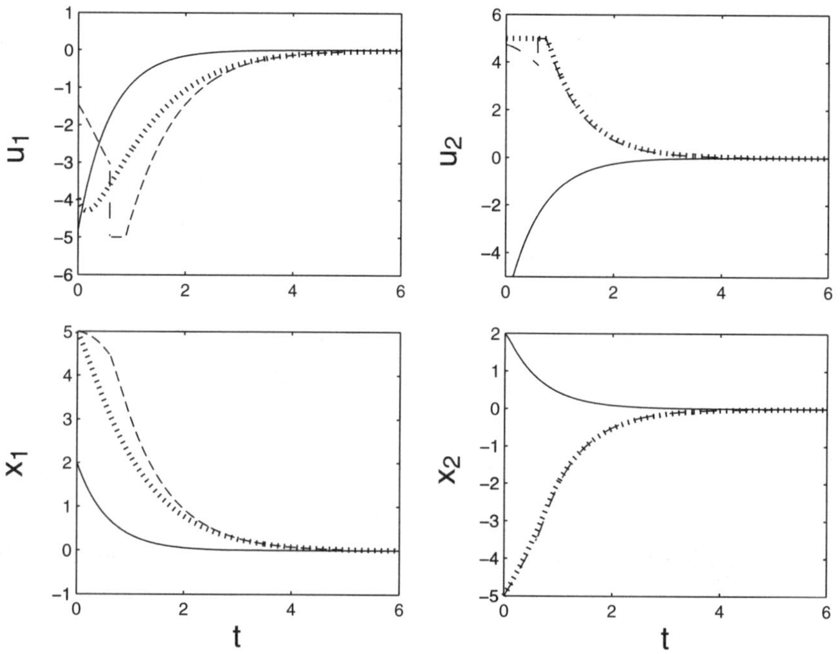

Fig. 5.12. Input and closed-loop state profiles starting close to the origin using MPC with $T = 1.2$ (*solid line*) and from a point farther away in state space using the switching scheme with a horizon length $T = 1.2$ (*dashed line*) and using only MPC with $T = 1.5$ (*dotted line*)

$x_0 = [-3\ -2.5]$, until the closed-loop state is driven inside Ω_2. Once inside the safety zone, the MPC controller is switched in which, as can be seen from the dashed lines in Figs. 5.13 and 5.14, asymptotically stabilizes the closed-loop system. It is important to keep in mind that Ω_2 is not necessarily a stability region for MPC. Therefore, even when the MPC controller is activated only inside Ω_2, it is possible for the closed-loop trajectory to leave the set at some future time. However, if this happens the supervisor will be able, at the next sampling instance, to detect this behavior and switch back to the bounded controller, before the trajectory can escape the stability region of the bounded controller, Ω.

5.4 Hybrid Predictive Output Feedback Control

The hybrid predictive control structure presented in Sect. 5.3 was developed under the assumption of full state feedback. In this section, we address the hybrid predictive control problem under output feedback. Due to the absence of complete state measurements, a state observer is constructed to provide

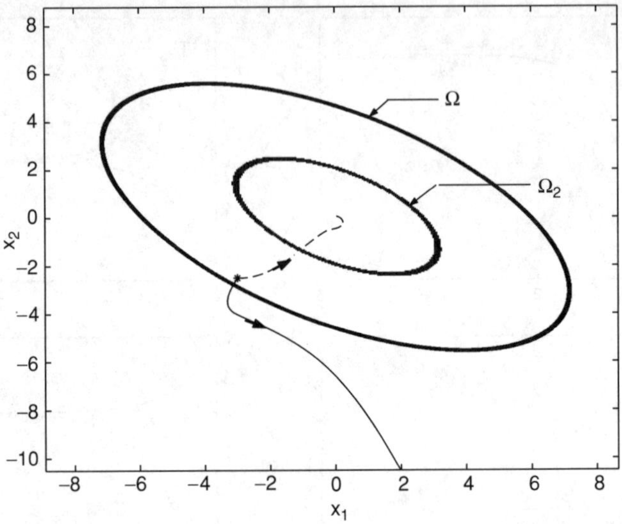

Fig. 5.13. Closed-loop state trajectory under MPC (*solid line*) and under the switching scheme accounting for sampling (*dashed line*)

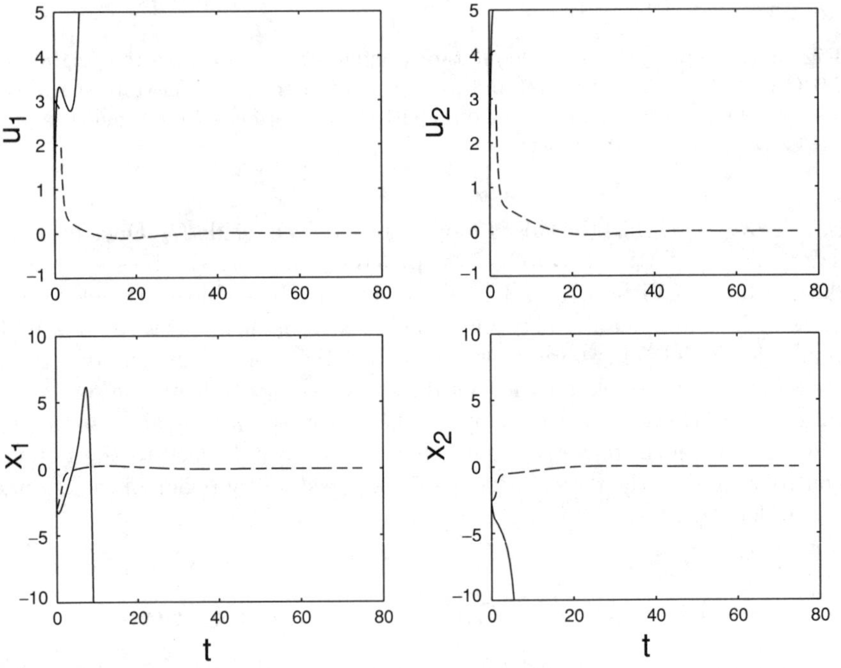

Fig. 5.14. Input and closed-loop state profiles under MPC (*solid line*) and under the switching scheme accounting for sampling (*dashed line*)

the controllers, as well as the supervisor, with appropriate state estimates. The observer is tuned in a way so as to guarantee closed–loop stability for all initial conditions within the bounded controller's output feedback stability region (which can be chosen arbitrarily close in size to its state feedback counterpart, provided that the observer gain is sufficiently large). Switching laws, that monitor the evolution of the state estimates, are then derived to orchestrate the transition between the two controllers in a way that reconciles their respective stability and performance properties and safeguards closed–loop stability in the event of MPC infeasibility or instability.

In addition to the set of switching rules being different from the ones proposed in Sect. 5.3 under state feedback, an important characteristic of the hybrid predictive control strategy under output feedback is the inherent coupling, brought about by the lack of full state measurements, between the tasks of controller design, characterization of the stability region and supervisory switching logic design, on one hand, and the task of observer design, on the other. The rest of this section is organized as follows. In Sect. 5.4.1, we present some preliminaries that describe the class of systems considered and review the design of a Luenberger state observer. Next, in Sect. 5.4.2 we discuss the stability properties, of both MPC and Lyapunov–based bounded control, under output feedback. Then, in Sect. 5.4.3, we present the hybrid predictive output feedback control structure. The theoretical underpinnings and practical implications of the proposed hybrid predictive control structure are highlighted, and possible extensions of the supervisory switching logic, that address a variety of practical implementation issues, are discussed. Finally, in Sect. 5.4.4, two simulation studies are presented to demonstrate the implementation and evaluate the effectiveness of the proposed control strategy, as well as test its robustness with respect to modeling errors and measurement noise.

5.4.1 Preliminaries

We consider the problem of output feedback stabilization of continuous–time linear time–invariant (LTI) systems with input constraints, with the following state–space description:

$$\dot{x} = Ax + Bu \tag{5.22}$$

$$y = Cx \tag{5.23}$$

$$u \in \mathcal{U} \subset \mathbb{R}^m \tag{5.24}$$

where C is a constant $k \times n$ matrix. The pair (A, B) is assumed to be controllable and the pair (C, A) observable.

We will now briefly review the design of a Luenberger state observer for the system of (5.22–5.23) which will be used later in the development of the hybrid predictive control structure (see Remark 5.21 below for a discussion on the use of other possible estimation schemes). Specifically, we consider a standard Luenberger state observer described by:

$$\dot{\hat{x}} = A\hat{x} + Bu + L(y - C\hat{x}) \tag{5.25}$$

where $\hat{x} = [\hat{x}_1 \cdots \hat{x}_n]' \in \mathbb{R}^n$ denotes the vector of estimates of the state variables and L is a constant $n \times k$ matrix that multiplies the discrepancy between the actual and estimated outputs. Under the state observer of (5.25), the estimation error in the closed–loop system, defined as $e = x - \hat{x}$, evolves, independently of the controller, according to the following equation:

$$\dot{e} = (A - LC)e \tag{5.26}$$

The pair $(C,\ A)$ is assumed to be observable in the sense that the observer gain matrix, L, can be chosen such that the norm of the estimation error in (5.26) evolves according to $\|e(t)\| \leq \kappa(\beta)\|e(0)\|\exp(-\beta t)$, where $-\beta < 0$ is the largest eigenvalue of $A - LC$ and $\kappa(\beta)$ is a polynomial function of β.

Remark 5.21. Referring to the state observer of (5.25), it should be noted that the results presented in this section are not restricted to this particular class of observers. Any other observer that allows us to control the rate of decay of the estimation error at will, can be used. Our choice of using this particular observer design is motivated by the fact that it provides a transparent relationship between the temporal evolution of the estimation error bound and the observer parameters. For example, this design guarantees convergence of the state estimates in a way such that for larger values of β, the error decreases faster. As we discuss later (see Sect. 5.4.3), the ability to ensure a sufficiently fast decay of the estimation error is necessary in order to guarantee closed–loop stability under output feedback control. This requirement or constraint on the error dynamics is present even when other estimation schemes, such as moving horizon observers, are used (for example, see [201, 226]) to ensure closed–loop stability. For such observers, however, it is difficult in general to obtain a transparent relationship between the tunable observer parameters and the error decay rate.

Remark 5.22. For large values of β, the estimation error could possibly increase to large values before eventually decaying to zero (a phenomenon known as "peaking"; see [261]). However, this does not pose a problem in our design because the physical constraints on the manipulated inputs prevent transmission of the incorrect estimates to the process (see also Remark 5.28 for a detailed discussion of this issue). In the special case when C is a square matrix of rank n, the observer gain matrix, L, can be chosen as $L = AC^{-1} - RC^{-1}$, where R is a diagonal matrix whose diagonal elements are of the form $R_{ii} = -\beta a_i$, where $a_i \neq a_j \geq 1$. For this choice of L, the evolution of each error state is completely decoupled from the rest of the error states, i.e., each error state evolves according to $\dot{e}_i = -\beta a_i e_i$. In this case, no peaking occurs since each error term decreases monotonically.

5.4.2 Stability Properties Under Output Feedback Control

Model Predictive Control

For the sake of a concrete illustration of the hybrid predictive control strategy, we will consider the following stabilizing MPC formulation [145, 199], which includes a terminal equality constraint in the optimization problem (see Remark 5.23 below for more advanced MPC formulations that can also be used). This model predictive controller is mathematically described by:

$$J_s(x,t,u(\cdot)) = \int_t^{t+T} (x'(s)Qx(s) + u'(s)Ru(s))ds \tag{5.27}$$

$$u(\cdot) = argmin\{J_s(x,t,u(\cdot))|u(\cdot) \in S\} := M_s(x)$$

$$\begin{aligned} s.t. \quad \dot{x} &= Ax + Bu, \quad x(0) = x_0 \\ u(\cdot) &\in S \end{aligned} \tag{5.28}$$

$$x(t+T) = 0$$

Remark 5.23. Recall from the discussion in Sect. 5.3.4 that the hybrid predictive control strategy (whether under state or output feedback) is not restricted to any particular MPC formulation. The results apply to any MPC formulation for which a priori knowledge of the set of admissible initial conditions is lacking (or computationally expensive to obtain). In this sense, the above formulation is really intended as a symbolic example of stabilizing MPC formulations, with the understanding that other advanced MPC formulations that employ less restrictive constraints and/or penalties on the state at the end of the horizon can be used. Examples include formulations with terminal inequality constraints, such as contractive MPC (see, for example, [155, 220]) and CLF–based receding horizon control (see, for example, [221]), as well as formulations that employ a combination of terminal penalties and inequality constraints (see. for example, [45, 179, 264]). The reader is referred to [191] for a more exhaustive list.

Remark 5.24. By incorporating stability conditions directly as part of the optimization problem in (5.28), asymptotic stability under state feedback MPC is guaranteed provided that the initial condition is chosen so that the optimization yields a feasible solution. However, the implicit nature of the MPC control law, obtained through repeated on–line optimization, limits our ability to obtain, a priori, an explicit characterization of the admissible initial conditions starting from where the given model predictive controller (with fixed horizon length) is guaranteed to be feasible and enforce asymptotic stability. This set is a complex function of the constraints and the horizon length. Estimates of sufficiently large horizon lengths that ensure stability (for example, see [48]) are typically conservative and, if used, may lead to a significant computational burden due to the increased size of the optimization problem. Therefore, in practice, the initial conditions and/or horizon lengths are usually chosen using ad hoc criteria and tested through closed–loop simulations which can add to the computational burden in implementing the model predictive controller. The difficulties encountered in characterizing the stability region under state feedback control carry over to the case of output feedback control, where the lack of state measurements requires that the control action be computed using the state estimates. Feasibility of the MPC optimization problem based on the state estimates, however, does not guarantee closed–loop stability or even the continued feasibility of the ensuing state estimates. This motivates implementing MPC within a hybrid control structure that provides a "fall–back" controller for which a region of constrained closed–loop stability can be obtained off–line. Lyapunov–based controller design techniques provide a natural framework for the design of a stabilizing "fall–back" controller for which an explicit characterization of the region of closed–loop stability can be obtained.

Lyapunov–Based Bounded Control

In this section, we couple the bounded state feedback controller of (5.7–5.8) with the state observer of (5.25) to design an output feedback controller and characterize the stability properties of the resulting closed–loop system including the set of initial conditions for which closed–loop stability is guaranteed. When a state estimator of the form of (5.25) is used, the resulting closed–loop system is composed of a cascade, between the error system and the plant, of the form:

$$\begin{aligned} \dot{x} &= Ax + Bu(x - e) \\ \dot{e} &= (A - LC)e \end{aligned} \tag{5.29}$$

Note that the values of the states used in the controller contain errors. The state feedback stability region, therefore, is not exactly preserved under output feedback. However, by exploiting the error dynamics of (5.26), it is possible to recover arbitrarily large compact subsets of the state feedback stability region, provided that the poles of the observer are placed sufficiently far in the left half of the complex plane (which can be accomplished by choosing the observer gain parameter β sufficiently large). This idea is consistent with earlier results on semi–global output feedback stabilization of unconstrained systems using high–gain observers (for example, see [50, 183, 267]).

By viewing the estimation error as a dynamic perturbation to the state feedback problem, the basic idea, in constructing the output feedback controller and characterizing its stability region, is to design the observer in a way that exploits the robustness of the state feedback controller with respect to bounded estimation errors. To this end, given the state feedback stability region, Ω, we initially quantify the robustness margin by deriving a bound on the estimation error, $e_m > 0$, that can be tolerated by the state feedback

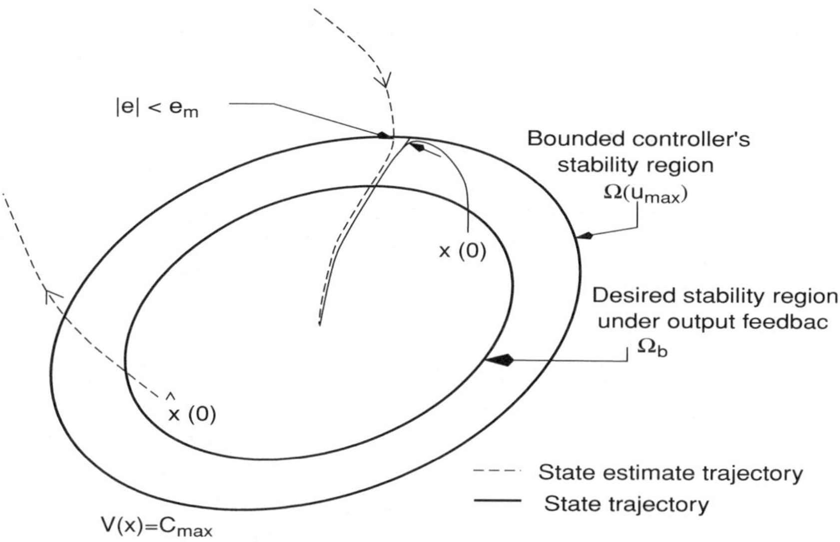

Fig. 5.15. Schematic representation of the evolution of the trajectories of the closed–loop state and state estimate. Starting within Ω_b, the observer gain ensures that, before the state of the closed–loop system reaches the boundary of Ω, the state estimate has converged close enough to the true value of the state, so that the norm of the estimation error is below the allowable error tolerated by the controller within Ω

controller, in the sense that the closed–loop state trajectory does not escape Ω for all $\|e\| \leq e_m$. Once this bound is computed, the idea is to initialize the states and the state estimates within any subset $\Omega_b \in \Omega$ and choose a consistent observer gain matrix L (parameterized by β) to ensure that, before the states reach the boundary of Ω, the norm of the estimation error has decayed to a value less than e_m (see Fig. 5.15 for a pictorial representation of this idea). These arguments are formalized in Propositions 5.25 and 5.27 below (the proofs are given in Appendix C).

Proposition 5.25. *Consider the constrained LTI system of (5.22–5.24) under the bounded control law of (5.7–5.8) with* $u = u(x - e)$, *where* e *is the state measurement error. Then, there exists a positive real number,* e_m, *such that if* $x(0) \in \Omega$ *and* $\|e(t)\| \leq e_m \; \forall \, t \geq 0$, *then* $x(t) \in \Omega \; \forall \, t \geq 0$.

Remark 5.26. The bounded control law of (5.7–5.8) will be used in the next section to illustrate the basic idea of the proposed hybrid control scheme. Our choice of using this particular design is motivated by its explicit structure and well–defined region of stability, which allows for an estimation of the robustness margin of the state feedback controller with respect to bounded measurement errors. However, the results of this work are not restricted to this particular choice of bounded controllers. Any other analytical bounded control law, with an explicit structure and well–defined region of robust stability with respect to bounded measurement errors, can also be used in implementing the proposed control strategy.

Proposition 5.27. *Consider the constrained LTI system of (5.22–5.24), the state observer of (5.25) and the bounded control law of (5.7–5.8). Then, given any positive real number,* δ_b*, such that* $\Omega_b = \{x \in \mathbb{R}^n : \|x\|_P^2 \leq \delta_b\} \subset \Omega$*, there exists a positive real number* β^* *such that if* $\|x(0)\|_P^2 \leq \delta_b$*,* $\|\hat{x}(0)\|_P^2 \leq \delta_b$ *and* $\beta \geq \beta^*$*, the origin of the constrained closed–loop system is asymptotically stable.*

Remark 5.28. Since the only assumption on C is that the pair (C, A) is observable, the observer states may "peak" before they converge to the true state values. This, however, does not pose a problem in our design because: (a) the physical constraints on the manipulated input eliminate the occurrence of instability due to peaking of the state estimates (i.e., they prevent transmission of peaking to the process), and (b) by "stepping back" from the boundary of the state feedback stability region and choosing an appropriate value for β, the design ensures that the system states cannot leave the stability region of the bounded controller before the estimation errors have gone below the permissible value.

Remark 5.29. In principle, the stability region under output feedback, Ω_b, can be chosen as close as desired to Ω by increasing the observer gain parameter β. However, it is well–known that large observer gains can amplify measurement noise and induce poor performance (see the simulation studies section for how this issue is addressed in observer implementation). This points to a fundamental tradeoff that cannot be resolved by simply changing the estimation scheme. For example, while one could replace the high–gain observer design with other observer designs (for example, a moving horizon estimator) to get a better handle on measurement noise, it is difficult in such schemes to obtain an explicit relationship between the observer tuning parameters and the output feedback stability region.

Remark 5.30. Note that, for the bounded controller, initializing x and $\hat{x}$ within Ω_b guarantees that x cannot escape Ω because of the inherent robustness of the bounded controller design with respect to bounded estimation errors (see the proof of Proposition 5.27 in the appendix). For MPC, however, there is no guarantee that x will stay in Ω, even if both x and $\hat{x}$ are initialized within Ω_b, and, therefore, whenever MPC is used, it will be necessary to rely on the evolution of the estimate, $\hat{x}(t)$, to deduce whether $x(t)$ is or is not within Ω. As we discuss in greater detail in the next section, the ability to reliably deduce the position of the true state at any given time is essential to the ability of the bounded controller to "step in" to preserve closed–loop stability in the event of failure of MPC (see Remark 5.35). One way of ascertaining bounds on the position of the true states, by looking only at the values of the estimates, is described in Proposition 5.31 below. The proof of this proposition is given in the appendix.

Proposition 5.31. *Consider the constrained system of (5.22–5.24), the state observer of (5.25) and the bounded control law of (5.7–5.8). Let*

$$T_d^* := \frac{1}{\beta} \ln \left(\frac{\kappa(\beta) e_{max}(0)}{\epsilon \sqrt{c_{max}/\lambda_{max}(P)}} \right)$$

for some $0 < \epsilon < 1$ *where* $e_{max}(0) = \max_{\hat{x}(0), x(0) \in \Omega_b} \|\hat{x}(0) - x(0)\|$. *Then, there exists a positive real number* $\delta_s^* < c_{max}$ *such that for all* $\delta_s \leq \delta_s^*$, *and for all* $t \geq T_d^*$, $\hat{x}'(t) P \hat{x}(t) \leq \delta_s \Longrightarrow x'(t) P x(t) \leq c_{max}$.

Remark 5.32. The above proposition establishes the existence of a set, $\Omega_s := \{x \in \mathbb{R}^m : x'Px \leq \delta_s\}$, that allows us to ascertain bounds on the position of the true state, by looking only at whether the estimate is within this set. Specifically, for a given bound on the norm of the estimation error, at any given time, if the state estimate is within Ω_s, then we can conclude with certainty that the true state cannot be outside of Ω, irrespective of the controller being used (see the proof in the appendix for the mathematical details of this argument). Note also that the size of Ω_s depends on the size of the error, and, therefore, as time evolves and the estimation error keeps decreasing, the size of Ω_s approaches that of Ω.

5.4.3 Hybrid Predictive Controller Design

Consider the LTI system of (5.22–5.24), for which the bounded controller of (5.7–5.8), the state estimator of (5.25) and the model predictive controller of (5.27–5.28) have been designed. The control problem is formulated as the one of designing a set of switching laws that orchestrate the transition between the predictive controller and the bounded controller under output feedback in a way that: (1) respects input constraints, (2) guarantees asymptotic stability of the origin of the closed–loop system starting from any initial condition within arbitrarily large compact subsets of the state feedback stability region of the bounded controller, and (3) accommodates the optimality requirements of MPC whenever possible (i.e., when closed–loop stability is guaranteed). For a precise statement of the problem, we first cast the system of (5.22–5.24) as a switched linear system of the form:

$$\begin{aligned} \dot{x} &= Ax + Bu_{i(t)} \\ y &= Cx \\ \|u_i\| &\leq u_{max} \\ i(t) &\in \{1, 2\} \end{aligned} \tag{5.30}$$

where $i : [0, \infty) \rightarrow \{1, 2\}$ is the switching signal which is assumed to be a piecewise continuous (from the right) function of time, implying that only a finite number of switches is allowed on any finite interval of time. The index $i(t)$, which takes values in the finite set $\{1, 2\}$, represents a discrete state that indexes the control input, $u(\cdot)$, with the understanding that $i(t) = 1$ if and only if $u_i(\hat{x}(t)) = b(\hat{x}(t))$ (i.e., the bounded controller is implemented in the closed–loop system), and $i(t) = 2$ if and only if $u_i(\hat{x}(t)) = M_s(\hat{x}(t))$ (i.e., MPC is implemented in the closed–loop system). Our goal, is to construct a switching law, based on the available state estimates:

$$i(t) = \psi(\hat{x}(t), t) \tag{5.31}$$

that provides the set of switching times that ensure asymptotically stabilizing transitions between the predictive and bounded controllers, in the event that the predictive controller is unable to enforce closed–loop stability for a given initial condition within Ω. In the remainder of this section, we present a switching scheme that addresses this problem. Theorem 5.33 below summarizes the main result (the proof is given in Appendix C).

Theorem 5.33. *Consider the constrained system of (5.30), the bounded controller of (5.7–5.8), the state estimator of (5.25), and the MPC law of (5.27–5.28). Let* $x(0) \in \Omega_b$, $\hat{x}(0) \in \Omega_b$, $\beta \geq \beta^*$, $\delta_s \leq \delta_s^*$, $\Omega_s(T_d^*) = \{x \in \mathbb{R}^n : x'Px \leq \delta_s(T_d^*)\}$, *and* $0 < T_d < T_{min} := \min\{t \geq 0 : V(x(0)) = \delta_b, V(x(t)) = c_{max}, u(t) \in \mathcal{U}\}$, *where* Ω_b, β^* *were defined in Proposition 5.27, and* δ_s^*, T_d^* *were defined in Proposition 5.31. Let* $T_m \geq \max\{T_d, T_d^*\}$ *be the earliest time for which* $\hat{x}(T_m) \in \Omega_s$ *and the MPC law prescribes a feasible solution,* $M_s(\hat{x}(T_m))$. *Also, let* $T_f > T_m$ *be the earliest time for which* $\dot{V}(\hat{x}(T_f)) \geq 0$. *Then, the following switching rule:*

$$i(t) = \begin{cases} 1, & 0 \leq t < T_m \\ 2, & T_m \leq t < T_f \\ 1, & t \geq T_f \end{cases} \tag{5.32}$$

where $i(t) = 1$ *if and only if* $u_i(\hat{x}(t)) = b(\hat{x}(t))$ *and* $i(t) = 2$ *if and only if* $u_i(\hat{x}(t)) = M_s(\hat{x}(t))$, *asymptotically stabilizes the origin of the closed–loop system.*

Remark 5.34. Figure 5.16 depicts a schematic representation of the hybrid predictive control structure proposed in Theorem 5.33. The four main components of this structure include the state estimator, the bounded controller, the predictive controller, and a high–level supervisor that orchestrates the switching between the two controllers. The implementation of the control strategy is best understood through the following stepwise procedure (see Fig. 5.17):

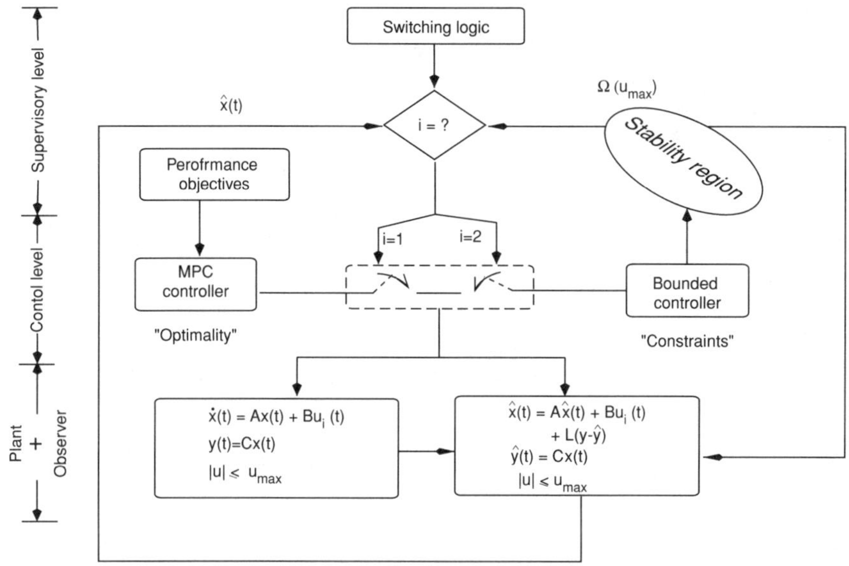

Fig. 5.16. Schematic representation of the hierarchical hybrid predictive control structure merging the bounded and model predictive controllers and a state observer

1. Given the system of (5.22) and the constraints of (5.24), design the bounded controller using (5.7–5.8). Given the performance objective, design the predictive controller.
2. Compute the stability region estimate for the bounded controller under state feedback, Ω, using (5.9–5.10) and, for a choice of the output feedback stability region, $\Omega_b \subset \Omega$, compute e_m, β and T_d defined in Propositions 5.25 and 5.27.
3. Compute the region Ω_s and T_d^* (see Proposition 5.31) which ensure that $\hat{x} \in \Omega_s \Rightarrow x \in \Omega$ for all times greater than T_d^*.
4. Initialize the closed–loop system at any initial condition, $x(0)$, within Ω_b, under the bounded controller using an initial guess for the state, $\hat{x}(0)$, within Ω_b. Keep the bounded controller active for a period of time, $[0, T_m)$.
5. At $t = T_m$ (by which time $\hat{x} \in \Omega_s$), test the feasibility of MPC using values of the estimates generated by the state observer. If MPC yields no feasible solution, keep implementing the bounded controller.
6. If MPC is feasible, disengage the bounded controller from the closed–loop system and implement the predictive controller instead. Keep MPC active for as long as $\hat{x} \in \Omega_s$ and $V(\hat{x})$ keeps decaying.
7. At the first instance that either $V(\hat{x})$ begins to increase under MPC, or MPC runs into infeasibilities, switch back to the bounded controller and implement it for all future times; else keep the predictive controller active.

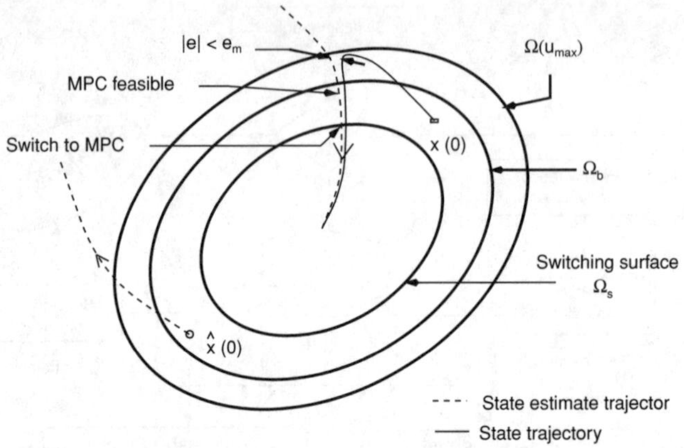

Fig. 5.17. Schematic representation of the evolution of the closed–loop system under the controller switching strategy proposed in Theorem 5.33. Note that MPC is not activated (even if feasible) before $\hat{x}$ enters Ω_s

Remark 5.35. An important feature that distinguishes the switching logic of Theorem 5.33 from the state feedback logic in Sect. 5.3 is the fact that, in the output feedback case, MPC implementation does not commence until some period of time, $[0, T_m)$, has elapsed, even if MPC is initially feasible. The rationale for this delay (during which only bounded control is implemented) is the need to ensure that, by the time MPC is implemented, the norm of the estimation error has already decayed to sufficiently low values that would allow the bounded controller to preserve closed–loop stability in the event that MPC needs to be switched out after its implementation. More specifically, before MPC can be safely activated, the supervisor needs to make sure that: (1) the estimation error has decayed to levels below e_m, and (2) the state, x, will be contained within Ω at any time that MPC could be switched out after its implementation. The first requirement ensures that the bounded controller, when (and if) reactivated, is able to tolerate the estimation error and still be stabilizing within Ω (see Proposition 5.25). This requirement also implies that T_m must be greater than T_d, which is the time beyond which the observer design guarantees $\|e\| \leq e_m$ (see Part 1 of the Proof of Proposition 5.27). The second requirement, on the other hand, implies that MPC cannot be safely switched in before the estimate enters Ω_s and the norm of the error is such that $\hat{x} \in \Omega_s \Longrightarrow x \in \Omega$ (see Remarks 5.30-(5.32)). Recall from Proposition 5.31 that this property holds for all $t \geq T_d^*$; hence T_m must also be greater than T_d^*, i.e., $T_m \geq \max\{T_d, T_d^*\}$. For times greater than T_m, the implementation of MPC can be safely tested without fear of closed–loop instability, since even if $\hat{x}$ tries to escape Ω_s (possibly indicating that x under MPC is about to escape Ω), the switching rule allows the supervisor to switch back immediately to the bounded controller (before $\hat{x}$ leaves Ω_s), thus guaranteeing that at the switching time $\hat{x} \in \Omega_s$, $x \in \Omega$ and $\|e\| \leq e_m$, which altogether imply that the bounded controller, once switched in, will be asymptotically stabilizing (see the Proof of Theorem 5.33 for the mathematical details of this argument).

Remark 5.36. The proposed approach does not require or provide any information as to whether the model predictive controller itself is asymptotically stabilizing or not, starting from any initial condition within Ω_b. In other words, the approach does not turn Ω_b into a stability region for MPC. What the approach does, however, is turn Ω_b into a stability region for the switched closed–loop system. The value of this can be understood in light of the difficulty in obtaining, *a priori*, an analytical characterization of the set of admissible initial conditions that the model predictive controller can steer to the origin in the presence of input constraints. Given this difficulty, by embedding MPC implementation within Ω_b and using the bounded controller as a fall–back controller, the switching scheme of Theorem 5.33 allows us to safely initialize the closed–loop system anywhere within Ω_b with the guarantee that the bounded controller can always intervene to preserve closed–loop stability in case the MPC is infeasible or destabilizing (due, for example, to a poor choice of the initial condition and/or improper tuning of the horizon length).

Remark 5.37. Note that, once MPC is switched in, if $V(\hat{x})$ continues to decrease monotonically, then the predictive controller will be implemented for all $t \geq T_m$. In this case, the optimal performance of the predictive controller is practically recovered. Note also, that in this approach, the state–feedback predictive controller design is not required to be robust with respect to state measurement errors (see [104] for examples when MPC is not robust) because even if it is not robust, closed–loop stability can always be guaranteed by switching back to the bounded controller (within its associated stability region) which provides the desired robustness with respect to the measurement errors.

Remark 5.38. Even though the result of Theorem 5.33 was derived using the MPC formulation of (5.27–5.28), it is important to point out that the same result applies to MPC formulations that employ different types of stability constraints (see Remark 5.23). The basic commonality between the formulation of (5.27–5.28) and other stability–handling formulations is the fact that, in all of these formulations, the implementation of MPC depends on whether the optimization problem, subject to the stability constraints, is feasible or not. This issue of feasibility is therefore accommodated in the switching logic by requiring the supervisor to check the feasibility of MPC in addition to monitoring the evolution of V. On the other hand, when a conventional MPC formulation (i.e., one with no stability constraints) is used, potential infeasibility of the optimization problem is no longer a concern, and, therefore, in this case the supervisory switching logic rests only on monitoring the evolution of V.

Remark 5.39. In conventional MPC formulations (with no stability constraints), the objective function is typically a tuning parameter that can be manipulated, by arbitrarily changing the weighting matrices, to ensure stability. In the hybrid predictive control framework, on the other hand, since closed–loop stability is guaranteed by the bounded controller, the objective function and the weights, instead of being arbitrarily tuned to guarantee stability, can be chosen to reflect actual physical costs and satisfy desirable performance objectives (for example, fast closed–loop response).

5.4.4 Simulation Studies

In this section, two simulation studies are presented to demonstrate the implementation, and evaluate the effectiveness, of the proposed hybrid predictive control strategy as well as test its robustness with respect to modeling errors and measurement noise.

Illustrative Example

In this section, we demonstrate an application of the proposed hybrid predictive control strategy to a three dimensional linear system where only two of the states are measured. Specifically, we consider an exponentially unstable linear system of the form of (5.22–5.23) with

$$A = \begin{bmatrix} 0.55 & 0.15 & 0.05 \\ 0.15 & 0.40 & 0.20 \\ 0.10 & 0.15 & 0.45 \end{bmatrix}, \quad B = \begin{bmatrix} 1 & 0 \\ 0 & 1 \\ 1 & 1 \end{bmatrix}, \quad C = \begin{bmatrix} 1 & 0 & 0 \\ 0 & 0 & 1 \end{bmatrix},$$

where both inputs u_1, u_2 are constrained in the interval $[-1, 1]$. Using (5.7–5.8), we initially designed the bounded controller and constructed its stability region, Ω, via (5.9–5.10). The matrix P was chosen as:

$$P = \begin{bmatrix} 6.5843 & 4.2398 & -3.830 \\ 4.2398 & 3.6091 & -2.667 \\ -3.830 & -2.667 & 2.8033 \end{bmatrix}$$

and the observer gain parameter was chosen to be $\beta = 500$ to ensure closed–loop stability for all initial conditions within a set $\Omega_b \subset \Omega$. For the model predictive controller, the parameters in the objective function of (5.27) were chosen as $Q = qI$, with $q = 1$ and $R = rI$, with $r = 0.1$. We also chose a horizon length of $T = 1.5$ in implementing the predictive controller of (5.25, 5.27–5.28). The resulting quadratic program was solved using the MATLAB subroutine QuadProg, and the set of ODEs integrated using the MATLAB solver ODE45.

In the first simulation run (solid lines in Figs. 5.18–5.19), the states were initialized at $x_0 = [0.75 \ -0.5 \ 1.0]'$, while the observer states were initialized at $\hat{x}_0 = [0 \ 0 \ 0]'$ (which belong to the stability region of the bounded

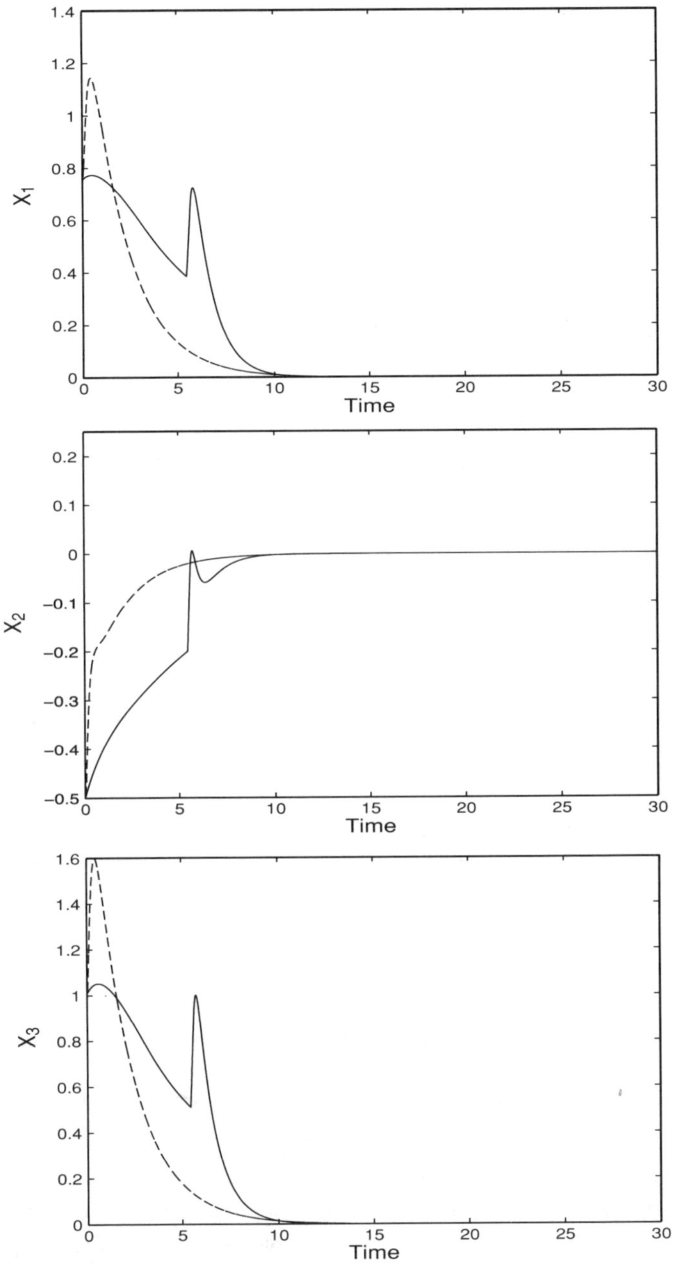

Fig. 5.18. States of the closed–loop system under the proposed hybrid predictive control strategy using two different horizon lengths of $T = 1.5$ (*solid*) and $T = 3.5$ (*dashed*) in MPC implementation, and an observer gain parameter of $\beta = 500$

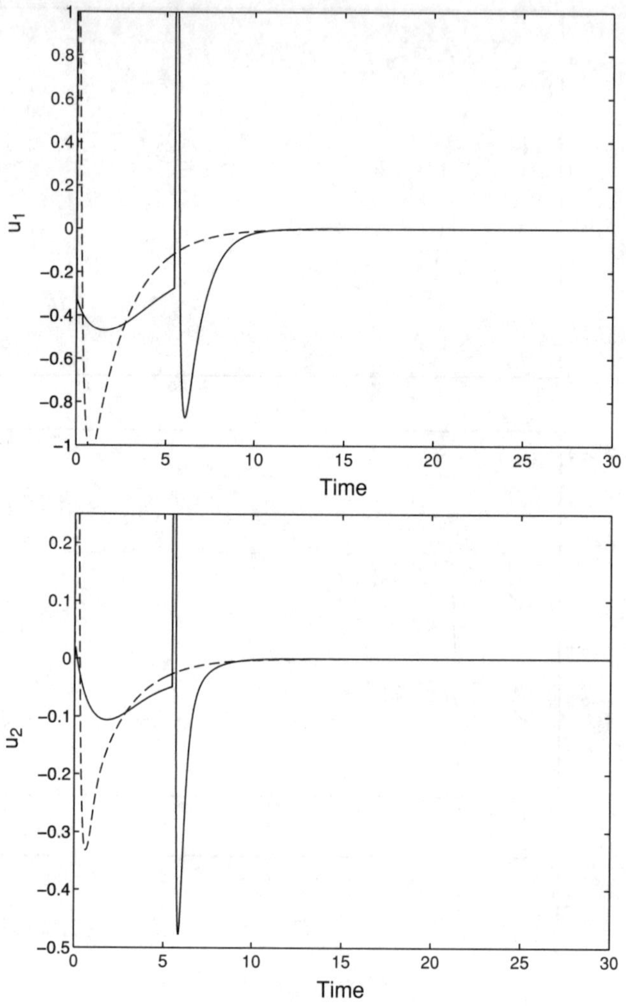

Fig. 5.19. Manipulated input profiles under the proposed hybrid predictive control strategy using two different horizon lengths of $T = 1.5$ (*solid*) and $T = 3.5$ (*dashed*) in MPC implementation, and an observer gain parameter $\beta = 500$

controller, Ω_b). The supervisor employs the bounded controller, while continuously checking the feasibility of MPC. At $t = 5.45$, MPC (based on the state estimates) becomes feasible and is implemented in the closed–loop system to achieve asymptotic stability. The total cost of stabilization achieved by the switching strategy (as measured by (5.27)) is $J = 80.12$. The switching scheme leads to better performance when compared with the scenario where the bounded controller is implemented for all times (the cost in this case is $J = 121.07$), and also better than the model predictive controller, because,

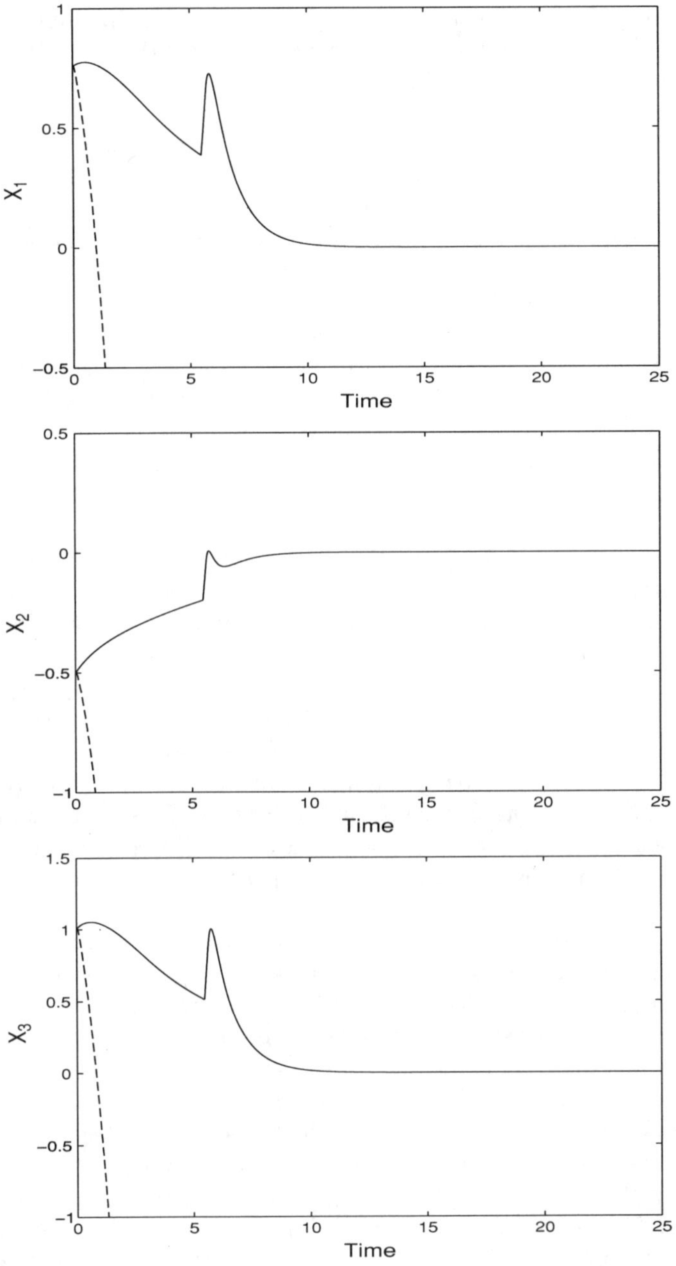

Fig. 5.20. States of the closed–loop system under the proposed hybrid predictive control strategy using two different gain parameters of $\beta = 500$ (*solid*) and $\beta = 0.5$ (*dashed*) in observer implementation, and horizon length of $T = 1.5$ in the predictive controller

starting from this initial condition, the predictive controller prescribes no feasible control move (equivalent to an infinite value for the objective function). Note that even though feasibility of MPC could have been achieved earlier by increasing the horizon length to $T = 3.5$ (dashed lines in Figs. 5.18–5.19), this conclusion could not be reached *a priori*, i.e., before running the closed–loop simulation in its entirety to check whether the choice $T = 3.5$ is appropriate. In the absence of any *a priori* knowledge of the necessary horizon length, for a given initial condition, the proposed hybrid predictive control can be used to guarantee closed–loop stability regardless of the choice of the horizon length.

In the second set of simulation results, we demonstrate the need for a choice of an observer gain consistent with the choice of Ω_b. To this end, we consider an observer design with a low gain ($\beta = 0.5$) placing the observer poles at $-0.5, -1.0, -1.5$. With the low observer gain, the estimates take a long time to converge to the true state values, resulting in the implementation of "incorrect" control action for a large period of time, by the end of which the states have already escaped out of Ω (even though the states and state estimates were initiated within Ω_b), thus resulting in closed–loop instability (dashed lines in Figs. 5.20–5.21). Recall that stability was achieved when the observer gain was chosen to be $\beta = 500$ in the first set of simulation runs (the state and input profiles for this case are reproduced by the solid lines in Figs. 5.20–5.21 for convenience of comparison).

To recover, as closely as possible, the state feedback stability region, large values of the observer gain are needed. However, it is well–known that high observer gains can amplify measurement noise and induce poor performance. These observations point to a fundamental tradeoff that cannot be resolved by simply changing the estimation scheme (see Remark 5.29). For example, if the observer gain consistent with the choice of the output feedback stability region is abandoned, the noise problem may disappear, but then stability cannot be guaranteed. One approach to avoid this problem in practice is to initially use a large observer gain that ensures quick decay of the initial estimation error, and then switch to a low observer gain. In the following simulation, we demonstrate how this idea, in conjunction with switching between the controllers, can be used to mitigate the undesirable effects of measurement noise. To illustrate this point, we switch between the high and low observer gains used in the first two simulation runs and demonstrate the attenuation of noise. Specifically, we consider the nominal system described by (5.22–5.23) (see the first paragraph of this subsection for the values of A, B and C), together with model uncertainty and measurement noise. The model matrix A_m (used for controller and observer design) is assumed to be within five percent error of the system matrix A and the sensors are assumed to introduce noise in the measured outputs as $y(t) = Cx(t) + \delta(t)$ where $\delta(t)$ is a random gaussian noise with zero mean and a variance of 0.01. As seen by the solid lines in Fig. 5.22, starting from the initial condition, $x_0 = [0.75 \ -0.5 \ 1.0]'$, using a high observer gain followed by a switch to a low observer gain at $t = 1.0$, and a switch from bounded control to MPC at $t = 3.5$, the supervisor is still

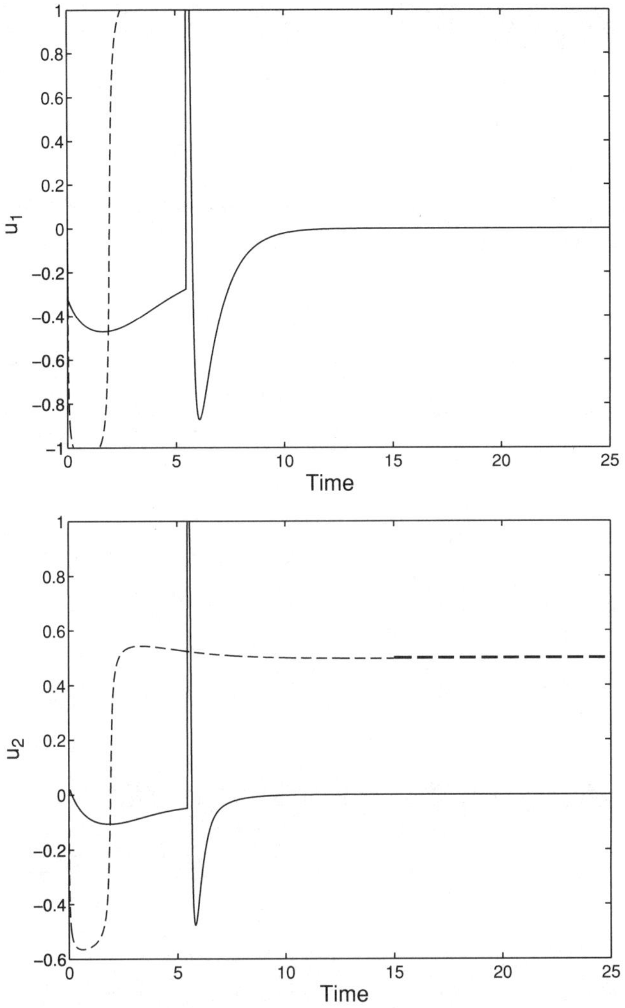

Fig. 5.21. Manipulated input profiles under the proposed hybrid predictive control strategy using two different gain parameters of $\beta = 500$ (*solid*) and $\beta = 0.5$ (*dashed*) in observer implementation, and horizon length of $T = 1.5$ in the predictive controller

able to preserve closed–loop stability, while at the same time resulting in a smoother control action (solid lines in Fig. 5.23) when compared to the case where a high observer gain is used for the entire duration of the simulation (dotted lines in Figs. 5.22–5.23.

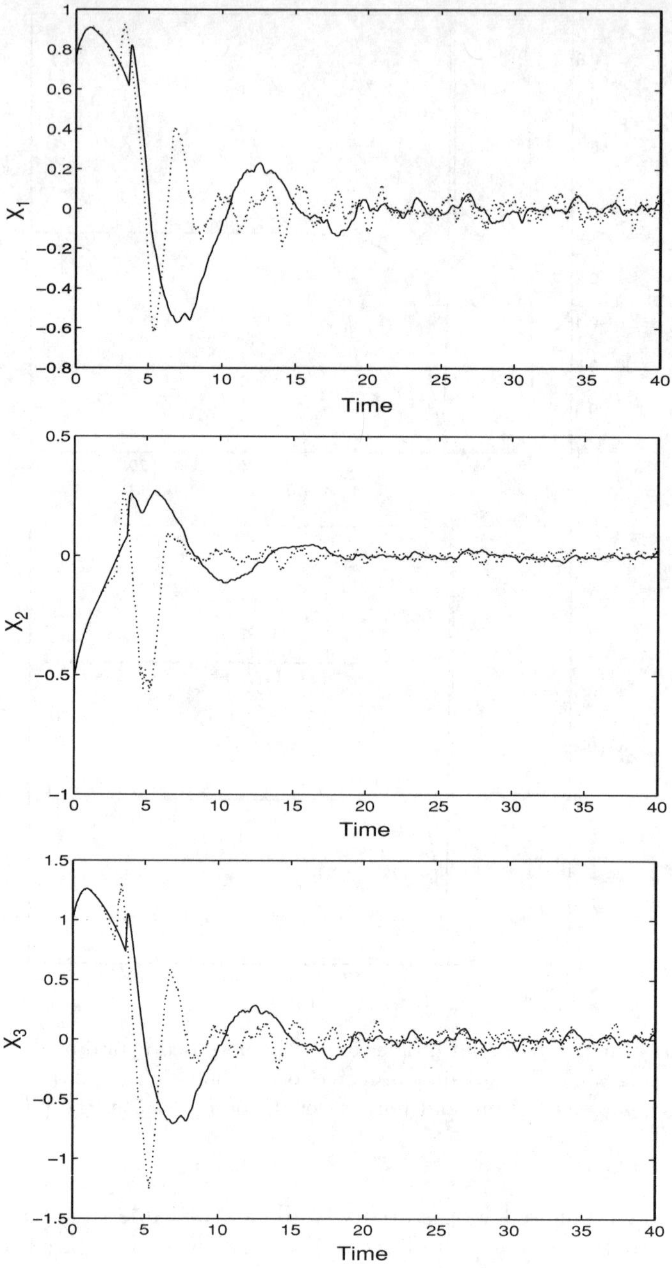

Fig. 5.22. Closed–loop states under hybrid predictive control, in the presence of measurement noise and modeling errors, for the case when the observer is initially implemented using $\beta = 500$ and then switched to $\beta = 0.5$ at $t = 1.0$ (*solid lines*) and the case when the observer is implemented using $\beta = 500$ for all times (*dotted lines*). The horizon length for the predictive controller is $T = 1.5$

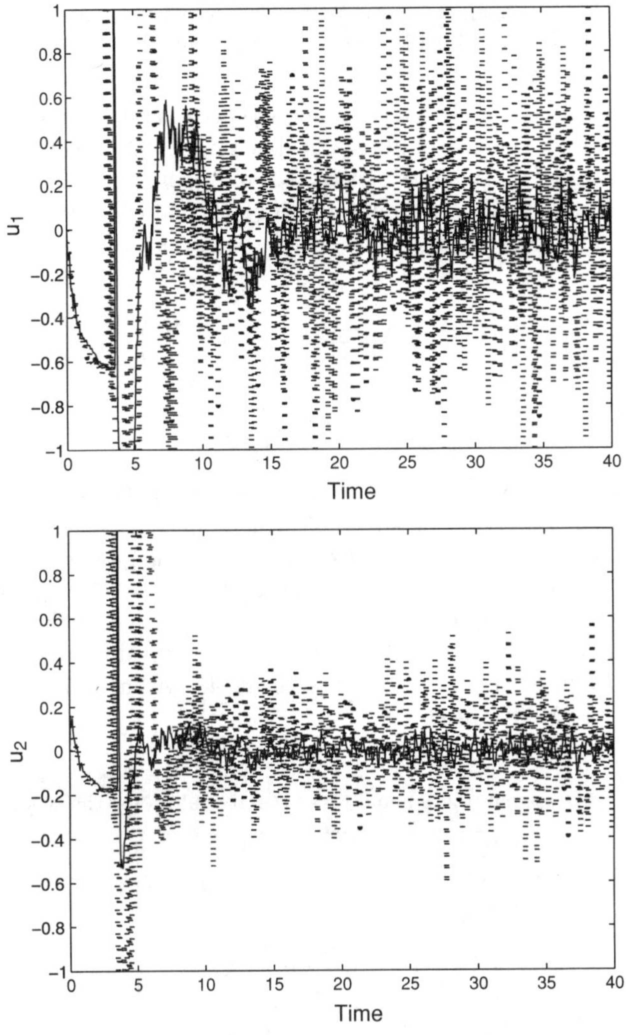

Fig. 5.23. Manipulated input profiles under the proposed hybrid predictive control strategy in the presence of measurement noise and modeling errors. The solid profiles depict the case when the state observer is initially implemented using $\beta = 500$ and then switched to $\beta = 0.5$ at $t = 1.0$. The dotted profiles depict the case when the state observer is implemented using $\beta = 500$ for all times. The horizon length for the model predictive controller is $T = 1.5$

Application to a Chemical Reactor Example

In this subsection, we consider a nonlinear chemical reactor example to test the robustness of the proposed hybrid predictive control structure (designed on the basis of the linearization of the nonlinear model around the operating steady–state) with respect to modeling errors arising due to the presence of nonlinear terms. Specifically, we consider a well–mixed continuous stirred tank reactor where an irreversible elementary exothermic reaction of the form $A \xrightarrow{k} B$, takes place, where A is the reactant species and B is the product. The feed to the reactor consists of pure A at flow rate F, molar concentration C_{A0} and temperature T_{A0}. A jacket is used to remove/provide heat to the reactor. Under standard modelling assumptions, a mathematical model of the process can be derived from material and energy balances and takes the following form:

$$
\begin{aligned}
V\frac{dT_R}{dt} &= F(T_{A0} - T_R) + \frac{(-\Delta H)}{\rho c_p} k_0 \exp\left(\frac{-E}{RT_R}\right) C_A V + \frac{Q}{\rho c_p} \\
V\frac{dC_A}{dt} &= F(C_{A0} - C_A) - k_0 \exp\left(\frac{-E}{RT_R}\right) C_A V
\end{aligned}
\tag{5.33}
$$

where C_A denotes the concentrations of the species A, T_R denotes the temperature of the reactor, Q denotes rate of heat input/removal from the reactor, V denotes the volume of the reactor, ΔH, k, E denote the enthalpy, pre–exponential constant and activation energy of the reaction, c_p and ρ denote the heat capacity and density of the reactor, respectively. The values of the process parameters are given in Table 5.1. It was verified that these conditions correspond to three steady–states, one unstable (given in Table 1) and two locally asymptotically stable.

Table 5.1. Process parameters and steady–state values for the reactor of (5.33)

V	$= 0.1$	m^3
R	$= 8.314$	$kJ/kmol \cdot K$
C_{A0s}	$= 1.0$	$kmol/m^3$
T_{A0s}	$= 310.0$	K
ΔH	$= -4.78 \times 10^4$	$kJ/kmol$
k_0	$= 1.2 \times 10^9$	s^{-1}
E	$= 8.314 \times 10^4$	$kJ/kmol$
c_p	$= 0.239$	$kJ/kg \cdot K$
ρ	$= 1000.0$	kg/m^3
F	$= 1.67 \times 10^{-3}$	m^3/s
T_{Rs}	$= 395.33$	K
C_{As}	$= 0.57$	$kmol/m^3$

The control problem is to regulate both the outlet concentration of the reactant, C_A, and the reactor temperature, T_R, at the unstable steady–state by manipulating the inlet reactant concentration, C_{A0}, and the rate of heat input, Q, provided by the jacket. The control objective is to be accomplished in the presence of constraints on the manipulated inputs ($|Q| \leq 1$ KJ/min,

$|\Delta C_{A0}| = |C_{A0} - C_{A0s}| \leq 1$ mol/L) using measurements of the reactor temperature. For the purpose of implementing the control scheme, the process model of (5.33) was linearized around the unstable steady–state, and the resulting linear model was used for the design of the hybrid predictive control structure. Using (5.7–5.8), we initially designed the bounded controller and constructed its stability region via (5.9–5.10). The matrix P was chosen as:

$$P = \begin{bmatrix} 0.027 & 0.47 \\ 0.47 & 15.13 \end{bmatrix}$$

and the observer gain parameter was chosen to be $\beta = 5$. It should be noted here that, because of the linearization effect, the largest invariant region, Ω, computed using (5.9–5.10) applied to the linear model, includes physically meaningless initial conditions ($C_A < 0$) and, therefore, a smaller level set, $\Omega' \subset \Omega$, that includes only physically meaningful initial conditions, was chosen and used as the stability region estimate for the bounded controller (see Fig. 5.24). For the predictive controller, the parameters in the objective function of (5.27) were chosen as $Q = qI$, with $q = 1$ and $R = rI$, with $r = 1$. We also chose a horizon length of $T = 2$. The resulting quadratic program was solved using the MATLAB subroutine QuadProg, and the set of ODEs integrated using the MATLAB solver ODE45. In all simulations, the controllers, designed on the basis of the linearization around the unstable steady–state, were implemented on the nonlinear system of (5.33).

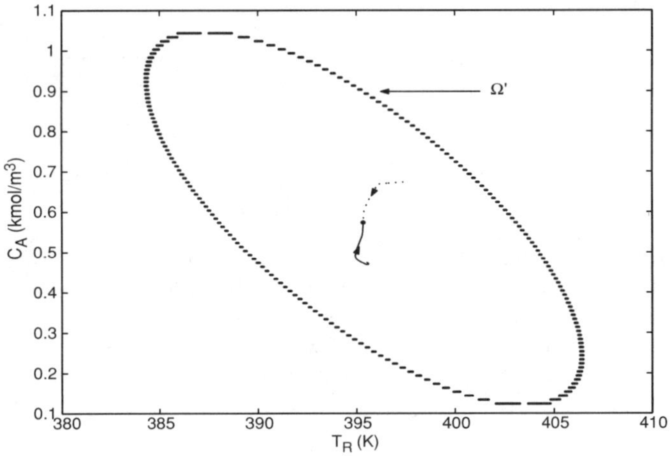

Fig. 5.24. Stability region estimate, Ω', based on the linearization of the nonlinear model of (5.33). The solid line depicts the closed–loop state trajectory starting from an initial condition for which MPC (designed based on the linear model with $T = 2$) is feasible and stabilizes the nonlinear closed–loop system, while the dotted line depicts the state trajectory starting from an initial condition for which MPC is infeasible, and the hybrid predictive control strategy is implemented instead to enforce closed–loop stability

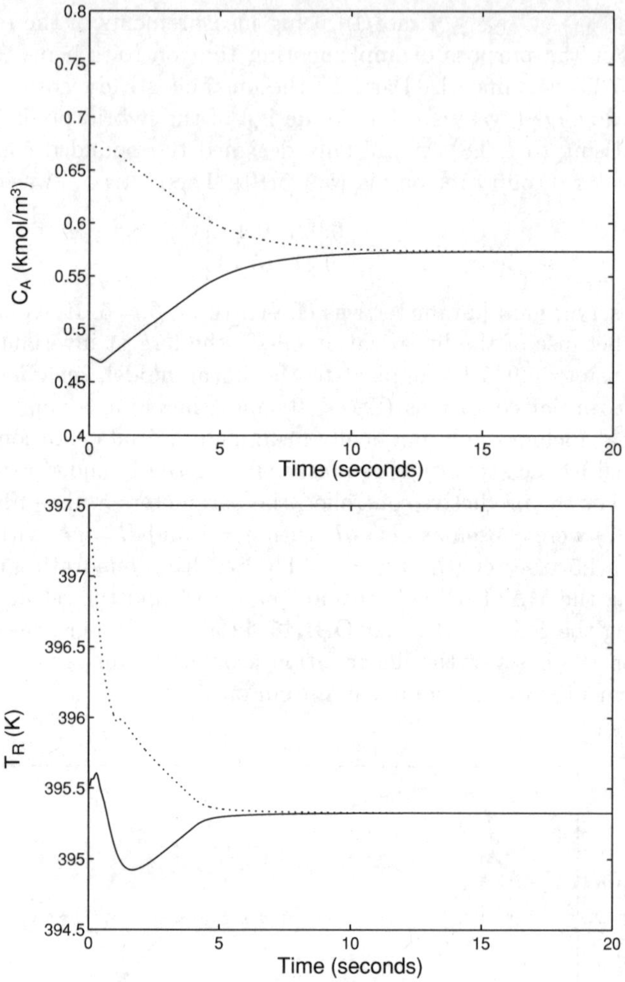

Fig. 5.25. Closed–loop reactant concentration (*top*) and reactor temperature (*bottom*) starting from an initial condition for which MPC (designed based on the linear model with $T = 2$) is feasible and stabilizes the nonlinear closed–loop system (*solid*), and starting from an initial condition for which MPC is infeasible, and the hybrid predictive control strategy is implemented instead to enforce closed–loop stability (*dotted*)

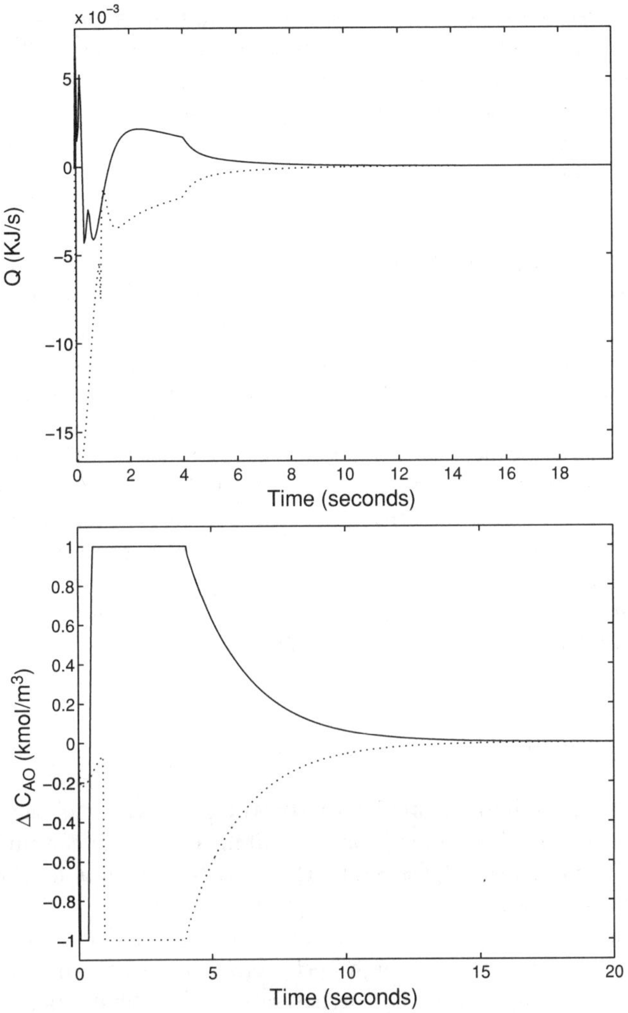

Fig. 5.26. Rate of heat input (*top*) and inlet reactant concentration in deviation variable form (*bottom*) when MPC is implemented (*solid*) and when the hybrid predictive control strategy is implemented (*dotted*).

As seen by the solid lines in Figs. 5.25–5.26, starting from the point (T_R, C_A) = (395.53 K, 0.47 $kmol/m^3$), the model predictive controller is able to stabilize the process at the desired steady–state, (T_{RS}, C_{AS}) = (395.33 K, 0.57 $kmol/m^3$). From the initial condition, (T_R, C_A) = (397.33 K, 0.67 $kmol/m^3$), however, the model predictive controller yields an infeasible solution. Therefore, using the proposed hybrid control strategy (dotted lines in Figs. 5.25–5.26, the supervisor implements the bounded controller in the closed–loop system, while continuously checking the feasibility of MPC. At

$t = 0.9$ seconds, MPC is found to be feasible and the supervisor switches to the model predictive controller, which in turn stabilizes the nonlinear closed–loop system at the desired steady state. These results clearly demonstrate a certain degree of robustness that the proposed hybrid predictive control structure possesses with respect to modeling errors (arising in this case due to neglecting the nonlinearities).

Remark 5.40. In both simulation studies presented above, we have treated robustness as a practical implementation issue and performed simulations that test the performance of the control strategy in the event of some plant–model mismatch to show that the hybrid predictive controller does posses some robustness margins that make it suited for practical implementation. We note that our objective here is not to incorporate robustness explicitly into the controller design, but rather to investigate the robustness margins of the controllers. The issue of incorporating robustness explicitly into the hybrid predictive control strategy is addressed in the next chapter. Furthermore, within predictive control, a common approach to the robust controller design problem is through the min–max type formulations. However, these formulations are well–known to be computationally demanding and, more importantly, suffer from similar difficulties in identifying the set of initial conditions from which robust closed–loop stability can be guaranteed *a priori*.

5.5 Conclusions

In this chapter, a hybrid predictive control structure, uniting MPC and bounded control, was developed for the stabilization of linear time-invariant systems with input constraints. Both the state and output feedback control problems were considered. Under full state feedback, the hybrid predictive control structure consists of three components: a bounded controller with a well-defined stability region, a high-performance model predictive controller, and a higher-level supervisor that orchestrates the switching between the two controllers in a way that: (1) guarantees closed-loop stability for all initial conditions within the stability region of the bounded controller, and (2) reconciles the stability and performance properties of the two controllers. The proposed hybrid control structure was used to construct a number of switching schemes that address a variety of control objectives, such as closed-loop stability (stability-based switching) and closed-loop performance (performance-driven switching). A switching scheme was also developed for more advanced formulations of MPC (feasibility-based switching). In each case, the switching logic was tailored to deal with the specific MPC formulation used and the specific objectives sought.

The hybrid predictive control structure was then extended to address the output feedback stabilization problem, via the addition of an appropriate state observer to provide the controllers, as well as the supervisor, with appropriate state estimates. The observer was tuned in a way so as to guarantee closed–loop stability for all initial conditions within the bounded controller's output feedback stability region (which can be chosen arbitrarily close in size to its state feedback counterpart, provided that the observer gain is sufficiently large). Switching laws, that monitor the evolution of the state estimates, were derived to orchestrate the transition between the two controllers in the event of MPC infeasibility or instability.

The hybrid predictive control structure was shown to provide, irrespective of the specific MPC formulation used, a safety net for the practical implementation of state and output feedback MPC by providing, through off–line computations, a priori knowledge of a large set of initial conditions for which stability of the switched closed–loop system is guaranteed. Finally, simulation studies were presented throughout the chapter to demonstrate the implementation and evaluate the effectiveness of the proposed hybrid predictive control strategy, as well as test its robustness with respect to modeling errors and measurement noise.

6

Hybrid Predictive Control of Nonlinear and Uncertain Systems

6.1 Introduction

In Chap. 5, we developed a hybrid predictive control strategy for the stabilization of constrained linear systems and demonstrated some of its benefits in terms of providing a safety net for the practical implementation of MPC. In this chapter, we focus on nonlinear systems, with and without uncertainty, and show how the hybrid predictive control strategy can be extended in several directions to address the constrained stabilization problem for these systems and aid MPC implementation. To motivate the hybrid predictive control structure for nonlinear and uncertain systems, we note that the practical implementation difficulties of MPC are in general more pronounced for nonlinear and uncertain systems than in the linear case.

For example, when the system is linear, the cost quadratic, and the constraints convex, the MPC optimization problem reduces to a quadratic program for which efficient software exists and, consequently, a number of control-relevant issues have been explored, including issues of closed–loop stability, performance, implementation and constraint satisfaction (see, for example, the tutorial paper [227]). For nonlinear systems, several nonlinear model predictive control (NMPC) schemes have been developed in the literature (see, for example, [45, 155, 179, 190, 200, 262, 264]) that focus on the issues of stability, constraint satisfaction and performance optimization for nonlinear systems. A common idea of these approaches is to enforce stability by means of suitable penalties and constraints on the state at the end of the finite optimization horizon. Because of the system nonlinearities, however, the resulting optimization problem is non-convex and, therefore, much harder to solve, even if the cost functional and constraints are convex. The computational burden is more pronounced for NMPC algorithms that employ terminal stability constraints (e.g., equality constraints) whose enforcement requires intensive computations that typically cannot be performed within a limited time–window.

In addition to the computational difficulties of solving a nonlinear optimization problem at each time step, one of the key challenges that impact on the practical implementation of NMPC is the inherent difficulty of characterizing, *a priori*, the set of initial conditions starting from where a given NMPC controller is guaranteed to stabilize the closed–loop system. For finite horizon MPC, an adequate characterization of the stability region requires an explicit characterization of the complex interplay between several factors, such as the initial condition, the size of the constraints, the horizon length, the penalty weights, etc. Use of conservatively large horizon lengths to address stability only increases the size and complexity of the nonlinear optimization problem and could make it intractable. Furthermore, since feasibility of NMPC is determined through on-line optimization, unless an NMPC controller is exhaustively tested by simulation over the whole range of potential initial states, doubt will always remain as to whether or not a state will be encountered for which an acceptable solution to the finite horizon problem can be found.

The above host of theoretical and computational issues continue to motivate research efforts in this area. Most available predictive control formulations for nonlinear systems, however, either do not explicitly characterize the stability region, or provide estimates of this region based on linearized models, used as part of some scheduling scheme between a set of local predictive controllers. The idea of scheduling of a set of local controllers to enlarge the operating region was proposed earlier in the context of analytic control of nonlinear systems (e.g., see [173, 193]) and requires an estimate of the region of stability for the local controller designed at each scheduling point. Then by designing a set of local controllers with their estimated regions of stability overlapping each other, supervisory scheduling of the local controllers can move the state through the intersections of the estimated regions of stability of different controllers to the desired operating point. Similar ideas were used in [46, 283] for scheduled predictive control of nonlinear systems. All of these approaches require the existence of an equilibrium surface that connects the scheduling points, and the resulting stability region estimate is the union of the local stability regions, which typically forms an envelope around the equilibrium surface. Stability region estimates based on linearization, however, are inherently conservative.

For systems with uncertainty, the MPC problem has been approached either as an analysis problem, where the robustness properties of nominal MPC formulations are analyzed (see, for example, [67, 102, 181]), or as a synthesis problem, where the goal is to develop MPC formulations that explicitly account for uncertainty (see [32,191] for surveys of results in this area). The problem of designing robust predictive controllers for uncertain linear systems has been extensively investigated (see, for example, [31, 43, 62, 168, 170, 240, 284]) and is typically addressed within a min–max optimization framework, where the optimization problem is solved in a way such that the constraints are satisfied for all possible realizations of the bounded uncertainty.

For uncertain nonlinear systems, the problem of robust MPC design continues to be an area of ongoing research (see, for example, [1, 93, 166, 180, 200, 236]). While min-max formulations provide a natural setting within which to address this problem, computational complexity remains a very serious issue. The computational burden stems in part from the nonlinearity of the model which makes the optimization problem non–convex and, therefore, generally hard to solve. Performing the min–max optimization over the non-convex problem further increases the computational complexity of the optimization problem and makes it unsuitable for the purpose of on–line implementation. These implementation difficulties are further compounded by the inherent difficulty of obtaining, a priori (i.e., before controller implementation), an explicit characterization of the set of initial conditions for which closed–loop stability, under any nonlinear MPC formulation (with or without stability conditions, and with or without robustness considerations), can be guaranteed in the presence of uncertainty and constraints. Faced with these difficulties, in current industrial implementation, the robust stability of MPC controllers is usually tested through extensive simulations.

The rest of this chapter is organized as follows. In Sect. 6.2, we develop a hybrid predictive control structure for the stabilization of nonlinear systems with input constraints. The central idea is to use a family of bounded nonlinear controllers, each with an explicitly characterized stability region, as fall–back controllers, and embed the operation of MPC within the union of these regions. In the event that the given predictive controller (which can be linear, nonlinear, or even scheduled) is unable to stabilize the closed–loop system (e.g., due to failure of the optimization algorithm, poor choice of the initial condition, insufficient horizon length, etc.), supervisory switching from MPC to any of the bounded controllers, whose stability region contains the trajectory, guarantees closed–loop stability. Two representative switching schemes that address (with varying degrees of flexibility) stability and performance considerations are described, and possible extensions of the supervisory switching logic, to address a variety of practical implementation issues, are discussed. The efficacy of the proposed approach is demonstrated through applications to chemical reactor and crystallization process examples. In Sect. 6.3, we extend the hybrid predictive control structure to nonlinear systems with input constraints and uncertain variables, and show that the resulting robust hybrid predictive control structure provides a safety net for the implementation of any available MPC formulation, designed with or without taking uncertainty into account. Two representative switching schemes that address robust stability and performance objectives are described to highlight the effect of uncertainty on the design of the switching logic. Finally, the implementation and efficacy of the robust hybrid predictive control strategy are demonstrated through simulations using a chemical reactor example. The results of this chapter were first presented in [88] and [197].

6.2 Hybrid Predictive Control of Nonlinear Systems

6.2.1 Preliminaries

In this section, we consider the problem of asymptotic stabilization of continuous-time nonlinear systems with input constraints, with the following state-space description:

$$\dot{x}(t) = f(x(t)) + g(x(t))u(t) \tag{6.1}$$

$$\|u\| \leq u_{max} \tag{6.2}$$

where $x = [x_1 \cdots x_n]' \in \mathbb{R}^n$ denotes the vector of state variables, $u = [u_1 \cdots u_m]'$ is the vector of manipulated inputs, $u_{max} \geq 0$ denotes the bound on the manipulated inputs, $f(\cdot)$ is a sufficiently smooth $n \times 1$ nonlinear vector function, and $g(\cdot)$ is a sufficiently smooth $n \times m$ nonlinear matrix functions. Without loss of generality, it is assumed that the origin is the equilibrium point of the unforced system (i.e., $f(0) = 0$). Throughout the chapter, the notation, $\|\cdot\|$, will be used to denote the standard Euclidean norm of a vector, while the notation $\|\cdot\|_Q$ refers to the weighted norm, defined by $\|x\|_Q^2 = x'Qx$ for all $x \in \mathbb{R}^n$, where Q is a positive–definite symmetric matrix and x' denotes the transpose of x. In order to provide the necessary background for the main results in Sect. 6.2.2, we will briefly review in the remainder of this section the design procedure for, and the stability properties of, both the bounded and model predictive controllers, which constitute the basic components of our hybrid control scheme. For clarity of presentation, we will focus only on the state feedback problem where measurements of $x(t)$ are assumed to be available for all t (see Remark 6.16 below for a discussion on the issue of measurement sampling and how it can be handled).

Bounded Lyapunov-Based Control

Consider the system of (6.1–6.2), for which a family of control Lyapunov functions (CLFs), $V_k(x)$, $k \in \mathcal{K} \equiv \{1, \ldots, p\}$ has been found (see Remark 6.1 below for a discussion on the construction of CLFs). Using each control Lyapunov function, we construct, using the results in [177] (see also [78,81]), the following continuous bounded control law:

$$u_k(x) = -k_k(x)(L_gV_k)'(x) := b_k(x) \tag{6.3}$$

where

$$k_k(x) = \frac{L_fV_k(x) + \sqrt{(L_fV_k(x))^2 + (u_{max}\|(L_gV_k)'(x)\|)^4}}{\|(L_gV_k)'(x)\|^2 \left[1 + \sqrt{1 + (u_{max}\|(L_gV_k)'(x)\|)^2}\right]} \tag{6.4}$$

when $(L_gV_k)'(x) \neq 0$, and $k_k(x) = 0$ when $(L_gV_k)'(x) = 0$, $L_fV_k(x) = \frac{\partial V_k(x)}{\partial x}f(x)$, $L_gV_k(x) = [L_{g_1}V_k(x) \cdots L_{g_m}V_k(x)]'$ and $g_i(x)$ is the i-th column of the matrix $g(x)$. For the above controller, it can be shown (see, for example, the Proof of Theorem 5.7 in Appendix C), using standard Lyapunov arguments, that whenever the closed–loop state trajectory, x, evolves within the state–space region described by the set:

$$\Phi_k(u_{max}) = \{x \in \mathbb{R}^n : L_fV_k(x) < u_{max}\|(L_gV_k)'(x)\|\} \tag{6.5}$$

then the controller satisfies the constraints, and the time-derivative of the Lyapunov function is negative–definite. Therefore, starting from any initial state within the set, $\Phi_k(u_{max})$, asymptotic stability of the origin of the constrained closed–loop system can be guaranteed, provided that the state trajectory remains within the region described by $\Phi_k(u_{max})$ whenever $x \neq 0$. To ensure this, we consider initial conditions that belong to an invariant subset (preferably the largest), $\Omega_k(u_{max})$. One way to construct such a subset is using the level sets of V_k, i.e.,

$$\Omega_k(u_{max}) = \{x \in \mathbb{R}^n : V_k(x) \leq c_k^{max}\} \tag{6.6}$$

where $c_k^{max} > 0$ is the largest number for which $\Phi_k(u_{max}) \supset \Omega_k(u_{max})\backslash\{0\}$. The union of the invariant regions described by the set:

$$\Omega(u_{max}) = \bigcup_{k=1}^{p} \Omega_k(u_{max}) \tag{6.7}$$

then provides an estimate of the stability region, starting from where the origin of the constrained closed–loop system, under the appropriate control law from the family of (6.3–6.4), is guaranteed to be asymptotically stable.

Remark 6.1. CLF-based stabilization of nonlinear systems has been studied extensively in the nonlinear control literature (see, for example, [17, 97, 177, 245]). The construction of constrained CLFs (i.e., CLFs that take the constraints into account) remains a difficult problem (especially for nonlinear systems) that is the subject of ongoing research. For several classes of nonlinear systems that arise commonly in the modeling of practical systems, systematic and computationally feasible methods are available for constructing unconstrained CLFs (CLFs for the unconstrained system) by exploiting the system structure. Examples include the use of quadratic functions for feedback linearizable systems and the use of back-stepping techniques to construct CLFs for systems in strict feedback form. In this chapter, the bounded controllers in (6.3–6.4) are designed using unconstrained CLFs, which are also used to explicitly characterize the associated regions of stability via (6.5–6.6). While the resulting estimates do not necessarily capture the entire domain of attraction, we will use them throughout the chapter only for a concrete illustration of the basic ideas of the results. It is possible to obtain substantially improved (i.e., less conservative) estimates by using, for example, a combination of several CLFs (see Sect. 6.2.3 for examples).

Model Predictive Control

In this section, we consider model predictive control of the system described by (6.1), subject to the control constraints of (6.2). In the literature, several MPC formulations are currently available, each with its own merits and limitations. While the hybrid predictive control structure is not restricted to any particular MPC formulation (see Remark 6.7 and Chap. 5 for further details on this issue), we will briefly describe here the "traditional" formulation for the purpose of highlighting some of the theoretical and computational issues involved in the nonlinear setting. For this case, MPC at state x and time t is conventionally obtained by solving, on-line, a finite horizon optimal control problem [192] of the form:

$$P(x,t) : \min\{J(x,t,u(\cdot)) | u(\cdot) \in S\} \tag{6.8}$$

$$s.t. \ \dot{x} = f(x) + g(x)u \tag{6.9}$$

where $S = S(t,T)$ is the family of piecewise continuous functions (functions continuous from the right), with period Δ, mapping $[t, t+T]$ into $\mathcal{U} := \{u \in \mathbb{R}^m : \|u\| \leq u_{max}\}$ and T is the specified horizon. Equation (6.9) is a nonlinear model describing the time evolution of the states x. A control $u(\cdot)$ in S is characterized by the sequence $\{u[k]\}$ where $u[k] := u(k\Delta)$. A control $u(\cdot)$ in S satisfies $u(t) = u[k]$ for all $t \in [k\Delta, (k+1)\Delta)$. The performance index is given by:

$$J(x,t,u(\cdot)) = \int_t^{t+T} \left[\|x^u(s;x,t)\|_Q^2 + \|u(s)\|_R^2\right] ds + F(x(t+T)) \tag{6.10}$$

where R and Q are strictly positive–definite, symmetric matrices and $x^u(s;x,t)$ denotes the solution of (6.1), due to control u, with initial state x at time t and $F(\cdot)$ denotes the terminal penalty. The minimizing control $u^0(\cdot) \in S$ is then applied to the plant over the interval $[k\Delta, (k+1)\Delta)$ and the procedure is repeated indefinitely. This defines an implicit model predictive control law:

$$M(x) = argmin(J(x,t,u(\cdot))) = u^0(t;x,t) \tag{6.11}$$

While the use of a nonlinear model as part of the optimization problem is desirable to account for the system's nonlinear behavior, it also raises a number of well-known theoretical and computational issues [192] that impact on the practical implementation of MPC. For example, in the nonlinear setting, the optimization problem is non-convex and, in general, harder to solve than in the linear case. Furthermore, the issue of closed–loop stability is typically addressed by introducing penalties and constraints on the state at the end of the finite optimization horizon (see [191] for a survey of different constraints proposed in the literature). Imposing constraints adds to the computational complexity of the nonlinear optimization problem which must be solved at each time instance.

Even if the optimization problem could be solved in a reasonable time, any guarantee of closed–loop stability remains critically dependent upon making the appropriate choice of the initial condition, which must belong to the predictive controller's region of stability (or feasibility), which, in turn, is a complex function of the constraints, the performance objective, and the horizon length. However, the implicit nature of the nonlinear MPC law, obtained through repeated on-line optimization, limits our ability to obtain, a priori (i.e., before controller implementation), an explicit characterization of the admissible initial conditions starting from where a given MPC controller (with a fixed performance index and horizon length) is guaranteed to asymptotically stabilize the nonlinear closed–loop system. Therefore, the initial conditions and the horizon lengths are usually tested through closed–loop simulations, which can add to the computational burden prior to the implementation of MPC.

6.2.2 Controller Switching Strategies

In this section, we show how to reconcile the bounded nonlinear control and model predictive control approaches by means of switching schemes that provide a safety net for the implementation of MPC to nonlinear systems with input constraints. To this end, consider the constrained nonlinear system of (6.1–6.2), for which the bounded controllers of (6.3–6.4) and the predictive controller of (6.8–6.11) have been designed. The hybrid predictive control problem is formulated as the one of designing a set of switching laws that orchestrate the transition between MPC and the bounded controllers in a way that guarantees asymptotic stability of the origin of the closed–loop system starting from any initial condition in the set, $\Omega(u_{max})$, defined in (6.7), respects input constraints, and accommodates the performance requirements whenever possible. For a precise statement of the problem, the system of (6.1) is first cast as a switched nonlinear system of the form:

$$\begin{aligned} \dot{x} &= f(x) + g(x)u_{i(t)} \\ \|u_i\| &\leq u_{max} \\ i(t) &\in \{1,2\} \end{aligned} \tag{6.12}$$

where $i : [0, \infty) \to \{1, 2\}$ is the switching signal, which is assumed to be a piecewise continuous (from the right) function of time, implying that only a finite number of switches, between the predictive and bounded controllers, is allowed on any finite interval of time. The index, $i(t)$, which takes values in the set $\{1,2\}$, represents a discrete state that indexes the control input $u(\cdot)$, with the understanding that $i(t)=1$ if and only if $u_i(x(t))=M(x(t))$ and $i(t)=2$ if and only if $u_i(x(t))=b_k(x(t))$ for some $k\in\mathcal{K}$. Our goal is to construct a switching law, $i(t)=\psi(x(t),t)$, that provides the set of switching times that ensure stabilizing transitions between the predictive and bounded controllers, in the event that the predictive controller is unable to enforce closed–loop stability. This in turn determines the time-course of the discrete state, $i(t)$.

In the remainder of this section, two switching schemes that address the above problem are presented. The first scheme is given in Theorem 6.2 and focuses primarily on the issue of closed–loop stability, while the second scheme, given in Theorem 6.11, provides more flexible switching rules that guarantee closed–loop stability and, simultaneously, enhance the overall closed–loop performance beyond that obtained from the first scheme. The proofs of both theorems are given in Appendix D.

Stability-Based Controller Switching

Theorem 6.2. *Consider the constrained nonlinear system of (6.12), with any initial condition $x(0):=x_0\in\Omega_k(u_{max})$, for some $k\in\mathcal{K}\equiv\{1,\dots,p\}$, where Ω_k was defined in (6.6), under the model predictive controller of (6.8–6.11). Also, let $\bar{T}\geq 0$ be the earliest time for which either the closed–loop state, under MPC, satisfies:*

$$L_fV_k(x(\bar{T}))+L_gV_k(x(\bar{T}))M(x(\bar{T}))\geq 0 \tag{6.13}$$

or the MPC algorithm fails to prescribe any control move. Then, the switching rule given by:

$$i(t)=\left\{\begin{array}{ll}1, & 0\leq t<\bar{T}\\ \\ 2, & t\geq\bar{T}\end{array}\right\} \tag{6.14}$$

where $i(t)=1\Leftrightarrow u_i(x(t))=M(x(t))$ and $i(t)=2\Leftrightarrow u_i(x(t))=b_k(x(t))$, guarantees that the origin of the switched closed–loop system is asymptotically stable.

Remark 6.3. Theorem 6.2 describes a stability-based switching strategy for control of nonlinear systems with input constraints. The main components of this strategy include the predictive controller, a family of bounded nonlinear controllers, with their estimated regions of constrained stability, and a high–level supervisor that orchestrates the switching between the controllers. A schematic representation of the hybrid control structure is shown in Fig. 6.1. The implementation procedure of this hybrid control strategy is outlined below:

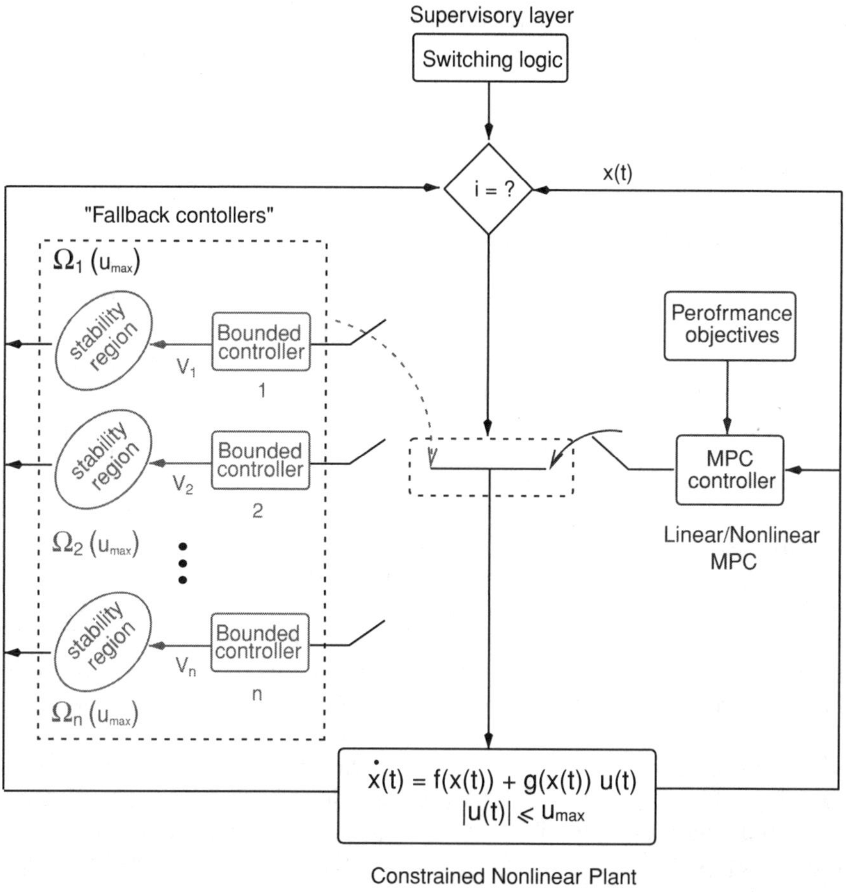

Fig. 6.1. Schematic representation of the hybrid predictive control structure merging MPC and a family of fall–back bounded controllers with their stability regions

- Given the system model of (6.1), the constraints on the input and the family of CLFs, design the bounded controllers using (6.3–6.4). Given the performance objective, set up the MPC optimization problem.
- Compute the stability region estimate for each of the bounded controllers, $\Omega_k(u_{max})$, using (6.5–6.6), for $k = 1, \ldots, p$, and $\Omega(u_{max}) = \bigcup_{k=1}^{p} \Omega_k(u_{max})$.
- Initialize the closed–loop system under MPC at any initial condition, x_0, within Ω, and identify a CLF, $V_k(x)$, for which the initial condition is within the corresponding stability region estimate, Ω_k.
- Monitor the temporal evolution of the closed–loop trajectory (by checking (6.13) at each time) until the earliest time that either (6.13) holds or the MPC algorithm prescribes no solution, $\bar{T}$.

- If such a $\bar{T}$ exists, discontinue MPC implementation, switch to the k-th bounded controller (whose stability region contains x_0) and implement it for all future times.

Remark 6.4. The use of multiple CLFs to design a family of bounded controllers, with their estimated regions of stability, allows us to initialize the closed–loop system from a larger set of initial conditions than in the case when a single CLF is used. Note, however, that once the initial condition is fixed, this determines both the region where MPC operation will be confined (and monitored) and the corresponding fall–back bounded controller to be used in the event of MPC failure. If the initial condition falls within the intersection of several stability regions, then any of the corresponding bounded controllers can be used as the fall–back controller.

Remark 6.5. The relation of (6.13–6.14) represents the switching rule that the supervisor observes when contemplating whether a switch, between MPC and any of the bounded controllers at a given time, is needed. The left hand-side of (6.13) is the rate at which the Lyapunov function grows or decays along the trajectories of the closed–loop system, under MPC, at time $\bar{T}$. By observing this rule, the supervisor tracks the temporal evolution of V_k under MPC such that whenever an increase in V_k is detected after the initial implementation of MPC (or if the MPC algorithm fails to prescribe any control move due, for example, to failure of the optimization algorithm), the predictive controller is disengaged from the closed–loop system, and the appropriate bounded controller is switched in, thus steering the closed–loop trajectory to the origin asymptotically. This switching rule, together with the choice of the initial condition, guarantee that the closed–loop trajectory, under MPC, never escapes $\Omega_k(u_{max})$ before the corresponding bounded controller can be activated.

Remark 6.6. In the case when the condition of (6.13) is never fulfilled (i.e., MPC continues to be feasible for all times and the Lyapunov function continues to decay monotonically for all times ($\bar{T} = \infty$)), the switching rule of (6.14) ensures that only MPC is implemented for all times and that no switching to the fall–back controllers takes place. In this case, MPC is stabilizing and its performance is fully recovered by the hybrid control structure. This particular feature underscores the central objective of the hybrid control structure, which is not to replace or subsume MPC but, instead, to provide a safe environment for the implementation of any predictive control policy for which *a priori* guarantees of stability are not available. Note also that, to the extent that stability under MPC is captured by the given Lyapunov function, the notion of switching, as described in Theorem 6.2, does not result in loss of performance, since the transition to the bounded controller takes place only if MPC is infeasible or destabilizing. Clearly, under these circumstances the issue of performance is not very meaningful for the predictive controller.

Remark 6.7. The fact that closed–loop stability is guaranteed, for all $x_0 \in \Omega$ (through supervisory switching), independently of whether MPC itself is stabilizing or not, allows us to use any desired MPC formulation within the switching scheme (and not just the one mentioned in Theorem 6.2), whether linear or nonlinear, and with or without terminal stability constraints or terminal penalties, without concern for loss of stability. This flexibility of using any desired MPC formulation has important practical implications for reducing the computational complexities that arise in implementing predictive control algorithms to nonlinear systems by allowing, for example, the use of less-computationally demanding NMPC formulations, or even the use of linear MPC algorithms instead (based on linearized models of the plant), with guaranteed stability (see Sect. 6.2.3 for examples). In all cases, by embedding the operation of the chosen MPC algorithm within the large and well-defined stability regions of the bounded controllers, the switching scheme provides a safe fall–back mechanism that can be called upon at any moment to preserve closed–loop stability should MPC become unable to achieve closed–loop stability. Finally, we note that the Lyapunov functions used by the bounded controllers can also be used as a guide to design and tune the MPC in case a Lyapunov-based constraint is used at the end of the prediction horizon.

Remark 6.8. Note that no assumption is made regarding how fast the MPC optimization needs to be solved since, even if this time is relatively significant and the plant dynamics are unstable, the implementation of the switching rule in Theorem 6.2 guarantees an "instantaneous" switch to the appropriate stabilizing bounded controller before such delays can adversely affect closed–loop stability (the times needed to check (6.13–6.14) and compute the control action of the bounded controller are insignificant as they involve only algebraic computations). This feature is valuable when computationally-intensive NMPC formulations fail, in the course of the online optimization, to provide an acceptable solution in a reasonable time. In this case, switching to the bounded controller allows us to safely abort the optimization without loss of stability.

Remark 6.9. One of the important issues in the practical implementation of finite–horizon MPC is the selection of the horizon length. It is well known that this selection can have a profound effect on nominal closed–loop stability as well as the size and complexity of the optimization problem. For NMPC, however, *a priori* knowledge of the shortest horizon length that guarantees closed–loop stability, from a given set of initial conditions (alternatively, the set of feasible initial conditions for a given horizon length) is not available. Therefore, in practice the horizon length is typically chosen using *ad hoc* selection criteria, tested through extensive closed–loop simulations, and varied, if necessary, to achieve stability. In the switching scheme of Theorem 6.2, closed–loop stability is maintained independently of the horizon length. This allows the predictive control designer to choose the horizon length solely on the basis of what is computationally practical for the size of the optimization problem, and without increasing, unnecessarily, the horizon length (and consequently the computational load) out of concern for stability. Furthermore, even if a conservative estimate of the necessary horizon length were to be ascertained for a small set of initial conditions before MPC implementation (say through extensive closed–loop simulations and testing), then if some disturbance were to drive the state outside of this set during the on-line implementation of MPC, this estimate may not be sufficient to stabilize the closed–loop system from the new state. Clearly, in this case, running extensive closed–loop simulations on-line to try to find the new horizon length needed is not a feasible option, considering the computational difficulties of NMPC as well as the significant delays that would be introduced until (and if) the new horizon could be determined. In contrast to the on-line re-tuning of MPC, stability can be preserved by switching to the fall–back controllers (provided that the disturbed state still lies within Ω).

Remark 6.10. When compared with other MPC-based approaches, the proposed hybrid control scheme is conceptually aligned with Lyapunov-based approaches in the sense that it, too, employs a Lyapunov stability condition to guarantee asymptotic closed–loop stability. However, this condition is enforced at the supervisory level, via continuous monitoring of the temporal evolution of V_k and explicitly switching between two controllers, rather than being incorporated in the optimization problem, as is customarily done in Lyapunov-based MPC approaches, whether as a terminal inequality constraint (e.g., contractive MPC [155,220], CLF-based RHC for unconstrained nonlinear systems [221]) or through a CLF-based terminal penalty (e.g., [262,264]). The methods proposed in these works do not provide an explicit characterization of the set of states starting from where feasibility and/or stability is guaranteed *a priori*. Furthermore, the idea of switching to a fall–back controller with a well-characterized stability region, in the event that the MPC controller does not yield a feasible solution, is not considered in these approaches.

Enhancing Closed–Loop Performance

The switching rule in Theorem 6.2 requires monitoring only one of the CLFs (any one for which $x_0 \in \Omega_k$) and does not permit any transient increase in this CLF under MPC (by switching immediately to the appropriate bounded controller). While this condition is sufficient to guarantee closed–loop stability, it does not take full advantage of the behavior of other CLFs at the switching time. For example, even though a given CLF, say V_1, may start increasing at time $\bar{T}$ under MPC, another CLF, say V_2, (for which $x(\bar{T})$ lies inside the corresponding stability region) may still be decreasing (see Fig. 6.2). In this case, it would be desirable to keep MPC in the closed–loop (rather than switch to the bounded controller) and start monitoring the growth of V_2 instead of V_1, because if V_2 continues to decay, then MPC can be kept active for all times

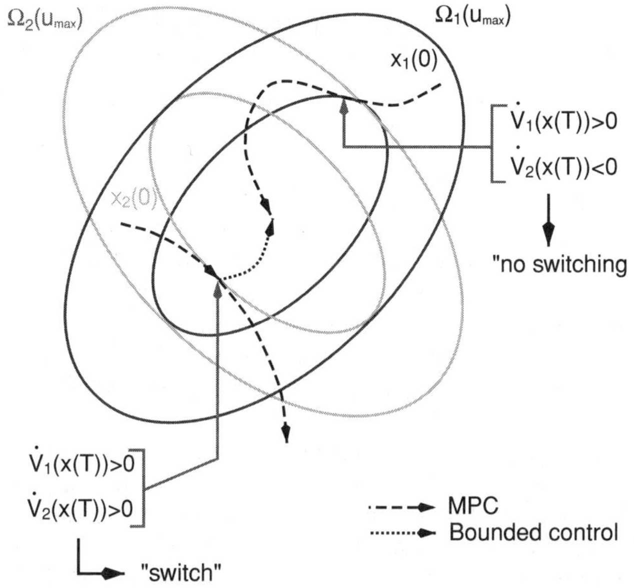

Fig. 6.2. Schematic representation illustrating the main idea of the controller switching scheme proposed in Theorems 6.2 and 6.11

and its performance fully recovered. To allow for such flexibility, we extend in the remainder of this section the switching strategy of Theorem 6.2 by relaxing the switching rule. This is formalized in the following theorem whose proof is given in the Appendix D.

Theorem 6.11. *Consider the constrained nonlinear system of (6.12), with any initial condition* $x(0) := x_0 \in \Omega(u_{max})$*, where* $\Omega(u_{max})$ *was defined in (6.7), under the model predictive controller of (6.8–6.11). Let* $T_k \geq 0$ *be the earliest time for which:*

$$\begin{array}{c} V_k(x(T_k))) \leq c_k^{max} \\ L_f V_k(x(T_k)) + L_g V_k(x(T_k)) M(x(T_k)) \geq 0, \quad k \in \mathcal{K} \end{array} \tag{6.15}$$

and define:

$$T_k^*(t) = \left\{ \begin{array}{c} T_k, \ 0 \leq t \leq T_k \\ \\ 0, \ t > T_k \end{array} \right\} \tag{6.16}$$

Then, the switching rule given by:

$$i(t) = \left\{ \begin{array}{l} 1, \ 0 \leq t < T^* \\ \\ 2, \ t \geq T^* \end{array} \right\} \tag{6.17}$$

where $T^* = \min\{T^i, T^f\}$, $T^i \geq 0$ *is the earliest time for which the MPC algorithm fails to prescribe any control move, and* $T^f = \max_j\{T_j^*\}$ *for all* j *such that* $x(t) \in \Omega_j(u_{max})$, *guarantees that the origin of the switched closed–loop system is asymptotically stable.*

Remark 6.12. The implementation of the switching scheme proposed in Theorem 6.11 can be understood as follows:

- Initialize the closed–loop system using MPC at any initial condition, x_0, within Ω, and start monitoring the growth of all the Lyapunov functions whose corresponding stability regions contain the state.
- The supervisor disregards (i.e., permanently stops monitoring) any CLF whose value ceases to decay (i.e., any CLF for which the condition of (6.15) holds) by setting the corresponding $T_k^*(t)$ to zero for all future times.
- As the closed–loop trajectory continues to evolve, the supervisor includes in the pool of Lyapunov functions it monitors any CLF for which the state enters the corresponding stability region, provided that this CLF has not been discarded at some earlier time.
- Continue implementing MPC as long as at least one of the CLFs being monitored is decreasing and the MPC algorithm yields a solution.
- If at any time, there is no active CLF (this corresponds to T^f), or if the MPC algorithm fails to prescribe any control move (this corresponds to T^i), switch to the bounded controller whose stability region contains the state at this time, else the MPC controller stays in the closed–loop system.

Remark 6.13. The purpose of permanently disregarding a given CLF, once its value begins to increase, is to avoid (or "break") a potential perpetual cycle in which the value of such a CLF increases (at times when other CLFs are decreasing) and then decreases (at times when other CLFs are increasing) and keeps repeating this pattern without decaying to zero. The inclusion of such a CLF among the pool of active CLFs used to decide whether MPC should be kept active, would result in MPC staying in the closed–loop indefinitely, but without actually converging to the origin (i.e., only boundedness of the closed–loop state can be established but not asymptotic stability). Instead of disregarding, for all future times, a CLF that begins to increase at some point, however, an alternative strategy is to consider such a CLF in the supervisory decision-making only if its value, at the current state, falls below the value from which it began to increase. This idea is similar to the one used in the stability analysis of switched systems using multiple Lyapunov functions (MLFs) (see Chap. 7 for further details on this issue). The greater CLF-monitoring flexibility resulting from this policy can increase the likelihood of MPC implementation and enhance closed–loop performance, while still guaranteeing asymptotic stability.

Remark 6.14. The switching schemes proposed in Theorems 6.2 and 6.11 can be further generalized to allow for multiple switchings between MPC and the bounded controllers. For example, if a given MPC is initially found infeasible (due, for example, to some terminal equality constraints), the bounded controller can be activated initially, but need not stay in the closed–loop system for all future times. Instead, it could be employed only until it brings the closed–loop trajectory to a point where MPC becomes feasible, at which time MPC can take over (see Sect. 6.2.3 for illustrations of this scenario). This scheme offers the possibility of further enhancement in the closed–loop performance by implementing MPC for all the times that it is feasible (instead of using the bounded controller). If MPC runs into any feasibility or stability problems, then the bounded controller can be re-activated. Chattering problems, due to back and forth switching, are avoided by allowing only a finite number of switches over any finite time-interval.

Remark 6.15. Note that the implementation of the switching schemes proposed in Theorems 6.2 and 6.11 can be easily adapted to the case when actuator dynamics are not negligible and, consequently, a sudden change in the control action (resulting from switching between controllers) may not be achieved instantaneously. A new estimate of the stability region under the bounded controllers can be generated by sufficiently "stepping back" from the boundary of the original stability region estimate, in order to prevent the escape of the closed–loop trajectory as a result of the implementation of possibly incorrect control action for some time (due to the delay introduced by the actuator dynamics). The switching schemes can then be applied as described in Theorems 6.2 and 6.11, using the revised estimate of the stability region for the family of bounded controllers.

Remark 6.16. When the proposed hybrid control schemes are applied to a process on-line, state measurements are typically available only at discrete sampling instants (and not continuously). In this case, the restricted access that the supervisor has to the state evolution between sampling times can lead to the possibility that the closed–loop state trajectory under MPC may leave the stability region without being detected, particularly if the initial condition is close to the boundary of the stability region and/or the sampling period is too large. In such an event, switching to the bounded controller at the next sampling time may be too late to recover from the instability of MPC. To guard against this possibility, the switching rules in Theorems 6.2 and 6.11 can be modified by restricting the implementation of MPC within a subset of the stability region, computed such that, starting from this subset, the system trajectory is guaranteed to remain within the stability region after one sampling period. Explicit estimates of this subset, which is parameterized by the sampling period, can be readily obtained off-line by computing (or estimating) the time-derivative of the Lyapunov function under the maximum allowable control action and then integrating both sides of the resulting inequality over one sampling period. Note that this computation of a "worst-case" estimate does not require knowledge of the solution of the closed–loop system.

Remark 6.17. The hybrid predictive control schemes proposed in this chapter can be extended to deal with the case when both input and state constraints are present. In one possible extension, state constraints would be incorporated directly as part of the constrained optimization problem that yields the model predictive control law. In addition, estimates of the stability regions for the bounded controllers would be obtained by intersecting the region described by (6.5) with the region described by the state constraints, and computing the largest invariant subset within the intersection. Using these estimates (which now account for both input and state constraints), implementation of the switching schemes can proceed following the same logic outlined for each case.

6.2.3 Applications to Chemical Process Examples

In this section, we present two simulation studies of chemical process examples to demonstrate the implementation of the proposed hybrid predictive control structure and evaluate its effectiveness.

Application to a Chemical Reactor Example

We consider a continuous stirred tank reactor where an irreversible, first-order exothermic reaction of the form $A \xrightarrow{k} B$ takes place. The inlet stream consists of pure A at flow rate F, concentration C_{A0} and temperature T_{A0}. Under standard modeling assumptions, the mathematical model for the process takes the form:

$$\frac{dC_A}{dt} = \frac{F}{V}(C_{A0} - C_A) - k_0 \exp\left(\frac{-E}{RT_R}\right) C_A$$

$$\frac{dT_R}{dt} = \frac{F}{V}(T_{A0} - T_R) + \frac{(-\Delta H)}{\rho c_p} k_0 \exp\left(\frac{-E}{RT_R}\right) C_A + \frac{UA}{\rho c_p V}(T_c - T_R) \tag{6.18}$$

where C_A denotes the concentration of the species A, T_R denotes the temperature of the reactor, T_c is the temperature of the coolant in the surrounding jacket, U is the heat–transfer coefficient, A is the jacket area, V is the volume of the reactor, k_0, E, ΔH are the pre-exponential constant, the activation energy, and the enthalpy of the reaction, c_p and ρ, are the heat capacity and fluid density in the reactor. The values of all process parameters can be found in Table 6.1. At the nominal operating condition of $T_c^{nom} = 302\ K$, the system has three equilibrium points, one of which is unstable. The control objective is to stabilize the reactor at the unstable equilibrium point $(C_A^s, T_R^s) = (0.52, 398.9)$ using the coolant temperature, T_c, as the manipulated input with constraints: $275\ K \leq T_c \leq 370\ K$.

Defining $x = [x_1 \ \ x_2]' = [(C_A - C_A^s) \ \ (T_R - T_R^s)]'$ and $u = T_c - T_c^{nom}$, the process model of (6.18) can be written in the form of (6.1). Defining an

Table 6.1. Process parameters and steady–state values for the reactor of (6.18)

V	$= 100.0$	L
E/R	$= 8000$	K
C_{A0}	$= 1.0$	mol/L
T_{A0}	$= 400.0$	K
ΔH	$= -2.0 \times 10^5$	J/mol
k_0	$= 4.71 \times 10^8$	min^{-1}
c_p	$= 1.0$	$J/g.K$
ρ	$= 1000.0$	g/L
UA	$= 1.0 \times 10^5$	$J/min.K$
F	$= 100.0$	L/min
C_A^s	$= 0.52$	mol/L
T_R^s	$= 398.97$	K
T_c^{nom}	$= 302$	K

auxiliary output, $y = h(x) = x_1$ (for the purpose of designing the controller), and using the invertible coordinate transformation: $\xi = [\xi_1 \; \xi_2]' = T(x) = [x_1 \; f_1(x)]'$, where $f_1(x) = \dot{x}_1$, the system of (6.18) can be transformed into the following partially linear form:

$$\dot{\xi} = A\xi + bl(\xi) + b\alpha(\xi)u \tag{6.19}$$

where $A = \begin{bmatrix} 0 & 1 \\ 0 & 0 \end{bmatrix}$, $b = [0 \; 1]'$, $l(\xi) = L_f^2 h(T^{-1}(\xi))$ is the second-order Lie derivative of $h(\cdot)$ along the vector field $f(\cdot)$, $\alpha(\xi) = L_g L_f h(T^{-1}(\xi))$ is the mixed order Lie derivative. The system of (6.19) will be used to design the bounded controllers and compute their estimated regions of stability. A common choice of CLFs for this system is quadratic functions of the form, $V_k = \xi' P_k \xi$, where the positive–definite matrix P_k is chosen to satisfy the Riccati matrix inequality: $A'P_k + P_k A - P_k bb' P_k < 0$. The following matrices

$$P_1 = \begin{bmatrix} 1.45 & 1.0 \\ 1.0 & 1.45 \end{bmatrix}, \; P_2 = \begin{bmatrix} 0.55 & 0.1 \\ 0.1 & 0.55 \end{bmatrix}, \; P_3 = \begin{bmatrix} 8.02 & 3.16 \\ 3.16 & 2.53 \end{bmatrix} \tag{6.20}$$

were used to construct a family of three CLFs and three bounded controllers, and compute their stability region estimates, Ω_k, $k = 1, 2, 3$, in the ξ-coordinate system. The corresponding stability regions in the (C_A, T_R) coordinate system are then computed using the transformation $\xi = T(x)$ defined earlier. The union of these regions, Ω', is shown in Fig. 6.3.

For the design of the predictive controller, a linear MPC formulation (based on the linearization of the process model around the unstable equilibrium point) with terminal equality constraints, $x(t+T) = 0$, is chosen for the sake of illustration (other MPC formulations that use terminal penalties, instead of terminal equality constraints, could also be used). The parameters in the objective function of (6.10) are chosen as $Q = qI$, with $q = 1$, $R = rI$,

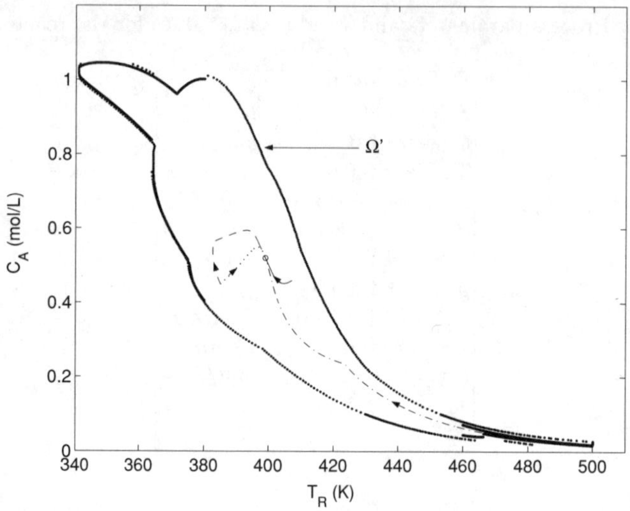

Fig. 6.3. Implementation of the proposed hybrid control structure: Closed–loop state trajectory under MPC with $T = 0.25$ (*solid trajectory*), under the switched MPC/bounded controller(3) with $T = 0.25$ (*dashed trajectory*), under MPC with $T = 0.5$ (*dotted trajectory*), and under the switched bounded controller(2)/MPC with $T = 0.5$ (*dash-dotted line*)

with $r = 1.0$, and $F = 0$. We also choose a horizon length of $T = 0.25$ in implementing the MPC controller. The resulting quadratic program is solved using the MATLAB subroutine QuadProg, and the set of nonlinear ODEs is integrated using the MATLAB solver ODE45.

As shown by the solid trajectory in Fig. 6.3, starting from the initial condition $[C_A(0) \quad T_R(0)]' = [0.46 \quad 407.0]'$, MPC using a horizon length of $T = 0.25$ yields a feasible solution, and when implemented in the closed–loop, stabilizes the nonlinear closed–loop system. The corresponding state and input profiles are shown in Figs. 6.4(a)–(c). Starting from the initial condition $[C_A(0) \quad T_R(0)]' = [0.45 \quad 385.0]'$ (dashed lines in Figs. 6.3–6.4), however, the linear MPC controller is infeasible. Recognizing that the initial condition is within the stability region estimate, Ω_3', the supervisor implements the third bounded controller, while continuously checking feasibility of MPC. At $t = 0.4$, the predictive controller becomes feasible and, therefore, the supervisor switches to MPC and keeps monitoring the evolution of V_3. In this case, the value of V_3 keeps decreasing and MPC stays in the closed–loop for all future times, thus asymptotically stabilizing the nonlinear plant. Note that, from the same initial condition, if the horizon length in the MPC is increased to $T = 0.5$, MPC yields a feasible solution and, when implemented, asymptotically stabilizes the closed–loop system (dotted lines in Figs. 6.3–6.4). However, the initial feasibility, based on the linearized model, does not necessarily imply that the predictive controller will be stabilizing, or for that matter even

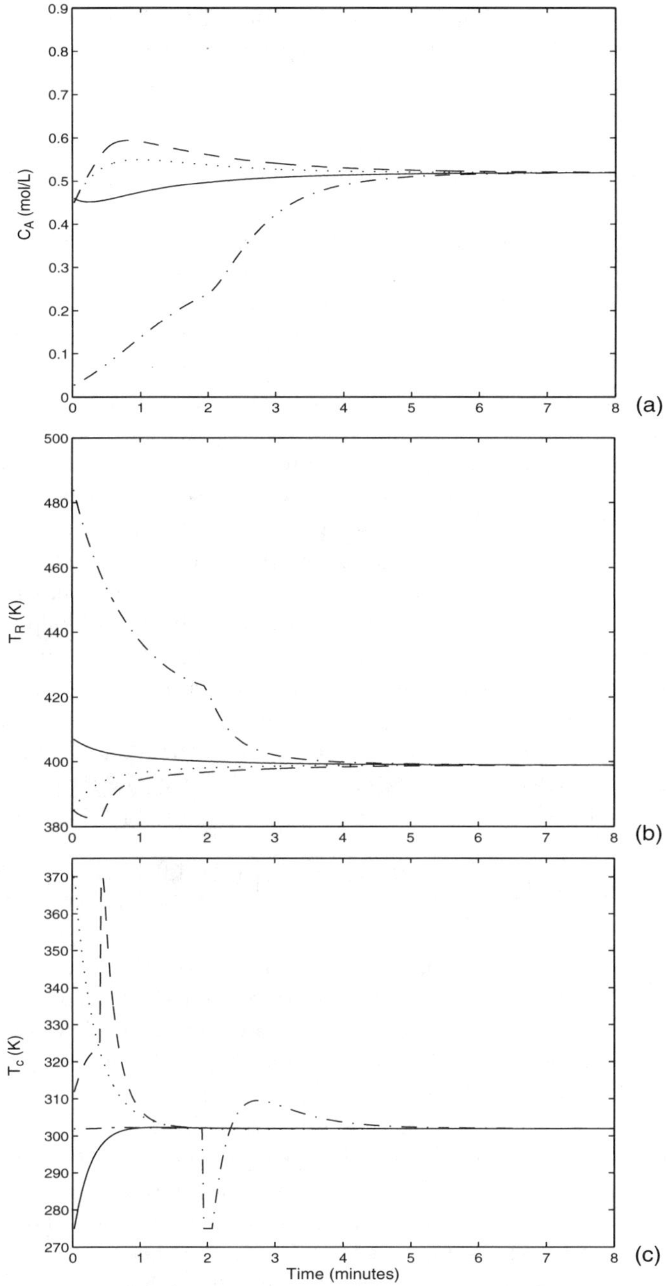

Fig. 6.4. Closed-loop reactant concentration profile (**a**), reactor temperature profile (**b**) and coolant temperature profile (**c**) under MPC with $T = 0.25$ (*solid trajectory*), under the switched MPC/bounded controller(3) with $T = 0.25$ (*dashed trajectory*), under MPC with $T = 0.5$ (*dotted trajectory*), and under the switched bounded controller(2)/MPC with $T = 0.5$ (*dash-dotted line*)

feasible at future times. Furthermore, this value of the horizon length (which yields a feasible solution) could not be determined *a priori* (without actually solving the optimization problem with the given initial condition), and stability of the closed–loop could not be ascertained, *a priori*, without running the closed–loop simulation in its entirety.

While our particular choice of implementing linear MPC (using the linearized model) facilitates implementation by making the optimization problem more easily solvable, the stabilizability of a given initial condition is limited, not only by the possibility of insufficient horizon length, but also by the linearization procedure itself. To demonstrate this, we consider an initial condition, $[C_A(0) \;\; T_R(0)]' = [0.028 \;\; 484.0]'$, belonging to the stability region Ω', (dash-dotted lines in Figs. 6.3–6.4). For this initial condition, linear MPC is found infeasible, no matter how large T is chosen to be, suggesting that this initial condition is outside the feasibility region based on the linear model. Therefore, using the Lyapunov function, V_2, (since the initial condition belongs to Ω_2'), the supervisor activates the second bounded controller which brings the state trajectory closer to the desired equilibrium point, while continuously checking feasibility of linear MPC. At $t = 1.925$, the predictive controller with $T = 0.5$ is found to be feasible and is, therefore, employed in the closed–loop system to asymptotically stabilize the closed–loop system.

To demonstrate some of the performance benefits of using the more flexible switching rules in Theorem 6.11, we consider the same control problem, described above, with relaxed constraints on the manipulated input: $250\ K \le T_c \le 500\ K$. Using these constraints, a new set of four bounded controllers are designed using a family of four CLFs of the form $V_k = \xi' P_k \xi$, $k = 1, 2, 3, 4$, where:

$$P_1 = \begin{bmatrix} 1.03 & 0.32 \\ 0.32 & 3.26 \end{bmatrix}, P_2 = \begin{bmatrix} 0.4 & 0.32 \\ 0.32 & 1.28 \end{bmatrix}, P_3 = \begin{bmatrix} 1.45 & 1.0 \\ 1.0 & 1.45 \end{bmatrix}, P_4 = \begin{bmatrix} 4.78 & 2.24 \\ 2.24 & 2.14 \end{bmatrix} \tag{6.21}$$

The stability region estimates of the controllers, Ω_k, $k = 1, 2, 3, 4$, are depicted in Fig. 6.5. The relaxed input constraints are also incorporated in the design of the predictive controller, using the same MPC formulation employed in the preceding simulations. Starting from the initial condition $[C_A(0) \;\; T_R(0)]' = [0.75 \;\; 361.0]'$ (Figs. 6.6(a)–(c)), the predictive controller, with $T = 0.1$, does not yield a feasible solution and, therefore, the supervisor implements the first bounded controller, using V_1, instead. At $t = 0.95$, however, MPC yields a feasible solution and, therefore, the supervisor switches to MPC. Even though $\dot{V}_1 > 0$ at this time, recognizing the fact that the state at this time ($[0.84 \;\; 363.3]'$, denoted by $\triangle$ in Fig. 6.5) belongs to Ω_2' and that $\dot{V}_2 < 0$, the supervisor continues to implement MPC while monitoring $\dot{V}_2$ (instead of V_1). At $t = 1.1$, the supervisor detects that $\dot{V}_2 > 0$. However, the state at this time ($[0.84 \;\; 378.0]'$, denoted by $\diamondsuit$ in Fig. 6.5) is within Ω_3' where $\dot{V}_3 < 0$. Therefore, the supervisor continues the implementation of MPC while

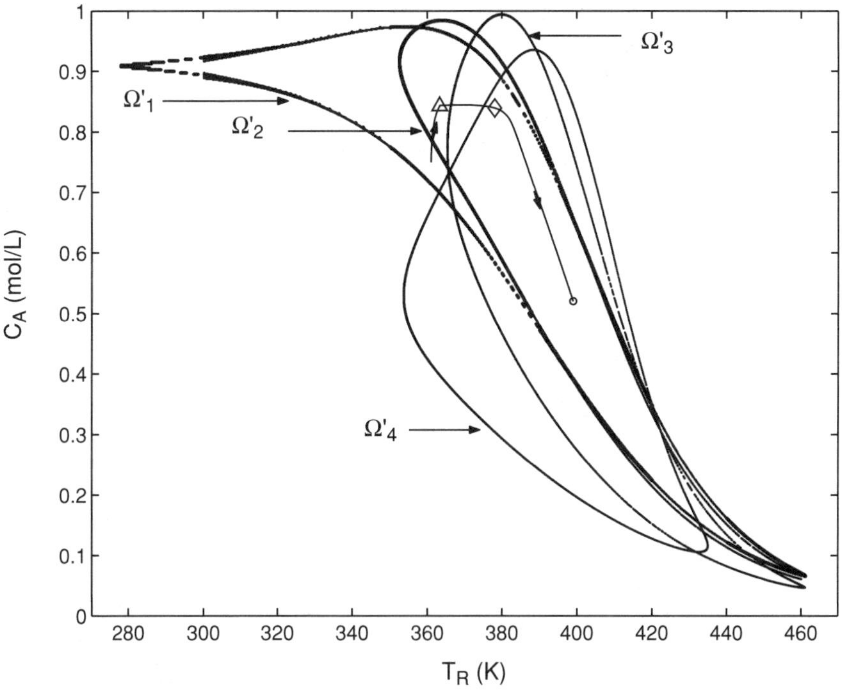

Fig. 6.5. Closed-loop state trajectory under the switching rules of Theorem 6.11 with $T = 0.1$ for MPC

monitoring V_3. From this point onwards, V_3 continues to decay monotonically and MPC is implemented in the closed–loop system for all future times to achieve asymptotic stability. Note that the switching scheme of Theorem 6.2 would have dictated a switch back to the first bounded controller at $t = 0.95$ and would not have allowed for MPC to be implemented in closed–loop, leading to a total cost of $J = 1.81 \times 10^6$ for the objective function of (6.10). The switching rules in Theorem 6.11, on the other hand, allow the implementation of MPC for all the times that it is feasible, leading to a lower total cost of $J = 1.64 \times 10^5$, while guaranteeing, at the same time, closed–loop stability.

Application to a Continuous Crystallizer Example

We consider a continuous crystallizer described by a fifth-order moment model of the following form:

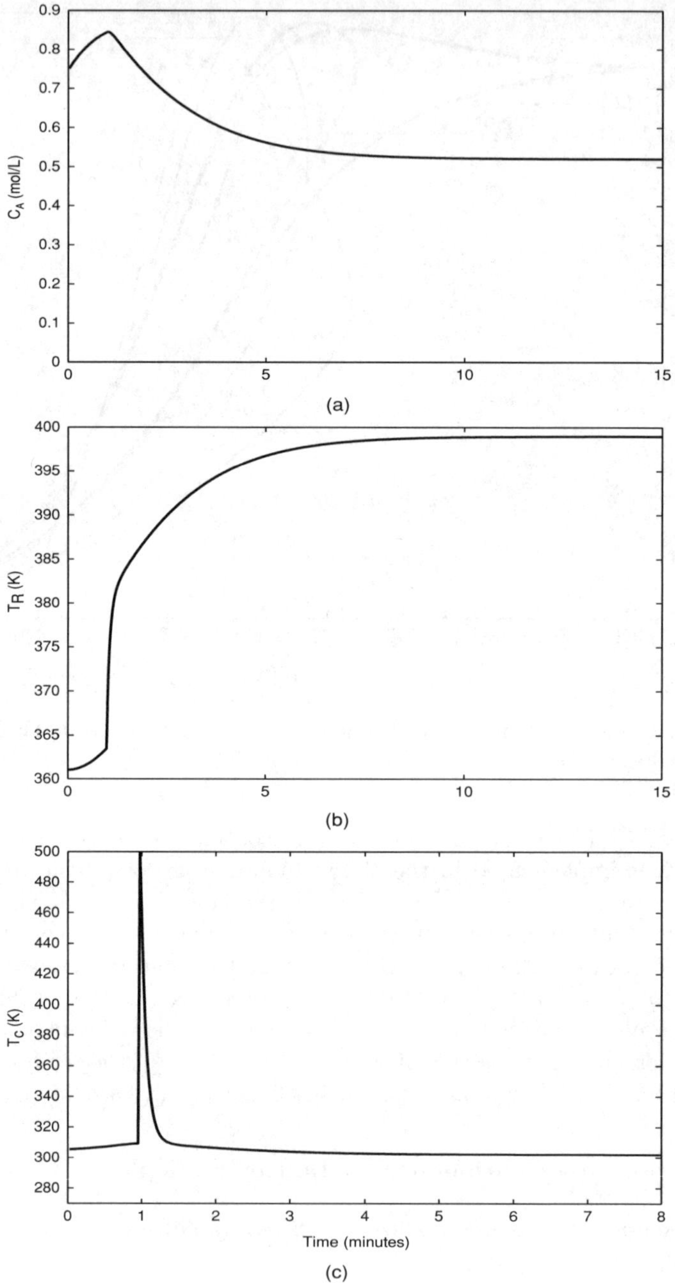

Fig. 6.6. Closed-loop reactant concentration profile (**a**), reactor temperature profile (**b**) and coolant temperature profile (**c**) under the switching scheme of Theorem 6.11 with $T = 0.1$ for MPC

$$
\begin{aligned}
\dot{x}_0 &= -x_0 + (1 - x_3) Da \exp\left(\frac{-F}{y^2}\right) \\
\dot{x}_1 &= -x_1 + y x_0 \\
\dot{x}_2 &= -x_2 + y x_1 \\
\dot{x}_3 &= -x_3 + y x_2 \\
\dot{y} &= \frac{1 - y - (\alpha - y) y x_2}{1 - x_3} + \frac{u}{1 - x_3}
\end{aligned}
\tag{6.22}
$$

where x_i, $i = 0, 1, 2, 3$, are dimensionless moments of the crystal size distribution, y is a dimensionless concentration of the solute in the crystallizer, and u is a dimensionless concentration of the solute in the feed (the reader may refer to [47, 51, 76] for a detailed process description, population balance modeling of the crystal size distribution and derivation of the reduced-order moments model, and to [16, 24, 55, 75, 84] for further results on model reduction of dissipative partial differential equation systems). The values of the dimensionless process parameters are chosen to be: $F = 3.0$, $\alpha = 40.0$ and $Da = 200.0$. For these values, and at the nominal operating condition of $u^{nom} = 0$, the above system has an unstable equilibrium point, surrounded by a stable limit cycle. The control objective is to stabilize the system at the unstable equilibrium point, $x^s = [x_o^s \; x_1^s \; x_2^s \; x_3^s \; y^s]' = [0.0471,\ 0.0283,\ 0.0169,\ 0.0102,\ 0.5996]'$, where the superscript s denotes the desired steady–state, by manipulating the dimensionless solute feed concentration, u, subject to the constraints: $-1 \leq u \leq 1$.

To facilitate the design of the bounded controller, we initially transform the system of (6.22) into the normal form. To this end, we define the auxiliary output variable, $\bar{y} = h(x) = x_0$, and introduce the invertible coordinate transformation: $[\xi' \; \eta']' = T(x) = [x_0 \; f_1(x) \; x_1 \; x_2 \; x_3]'$, where $\xi = [\xi_1 \; \xi_2]' = [x_0 \; f_1(x)]'$, $\bar{y} = \xi_1$, $f_1(x) = -x_0 + (1 - x_3) Da \exp(-F/y^2)$, and $\eta = [\eta_1 \; \eta_2 \; \eta_3]' = [x_1 \; x_2 \; x_3]'$. The state-space description of the system in the transformed coordinates takes the form:

$$
\begin{aligned}
\dot{\xi} &= A\xi + bl(\xi, \eta) + b\alpha(\xi, \eta) u \\
\dot{\eta} &= \Psi(\eta, \xi)
\end{aligned}
\tag{6.23}
$$

where $A = \begin{bmatrix} 0 & 1 \\ 0 & 0 \end{bmatrix}$, $b = [0 \; 1]'$, $l(\xi, \eta) = L_f^2 h(T^{-1}(\xi, \eta))$ is the second-order Lie derivative of the scalar function, $h(\cdot)$, along the vector field $f(\cdot)$, and $\alpha(\xi, \eta) = L_g L_f h(T^{-1}(\xi, \eta))$ is the mixed Lie derivative. The forms of $f(\cdot)$ and $g(\cdot)$ can be obtained by re-writing the system of (6.22) in the form of (6.1), and are omitted for brevity.

The partially-linear ξ-subsystem in (6.23) is used to design a bounded controller that stabilizes the full interconnected system of (6.23) and, consequently, the original system of (6.22). For this purpose, a quadratic function of the form, $V_\xi = \xi' P \xi$, is used as a CLF in the controller synthesis formula of (6.3), where the positive–definite matrix, P, is chosen to satisfy the Riccati matrix equality: $A'P + PA - Pbb'P = -Q$ where Q is a positive–definite matrix. An estimate of the region of constrained closed–loop stability for the full system is obtained by defining a composite Lyapunov function of the form $V_c = V_\xi + V_\eta$, where $V_\eta = \eta' P_\eta \eta$ and P_η is a positive–definite matrix, and choosing a level set of V_c, Ω_c, for which $\dot{V}_c < 0$ for all x in Ω_c. The two-dimensional projections of the stability region are shown in Fig. 6.7 for all possible combinations of the system states.

In designing the predictive controller, a linear MPC formulation, with a terminal equality constraint of the form $x(t+T) = 0$, is chosen (based on the linearization of the process model of (6.22) around the unstable equilibrium point). The parameters in the objective function of (6.10) are taken to be: $Q = qI$, with $q = 1$, $R = rI$, with $r = 1.0$, and $F = 0$. We also choose a horizon length of $T = 0.5$ in implementing the predictive controller. The resulting quadratic program is solved using the MATLAB subroutine QuadProg, and the nonlinear closed–loop system is integrated using the MATLAB solver ODE45.

In the first set of simulation runs, we test the ability of the predictive controller to stabilize the closed–loop system starting from the initial condition, $x(0) = [0.046 \quad 0.0277 \quad 0.0166 \quad 0.01 \quad 0.58]'$. The result is shown by the solid lines in Fig. 6.8(a–e) where it is seen that the predictive controller, with a horizon length of $T = 0.5$, is able to stabilize the closed–loop system at the desired equilibrium point. Starting from the initial condition $x(0) = [0.032 \quad 0.035 \quad 0.010 \quad 0.009 \quad 0.58]'$, however, the predictive controller yields no feasible solution. If the terminal equality constraint is removed, to make MPC yield a feasible solution, we see from the dashed lines in Fig. 6.8(a–e) that the resulting control action cannot stabilize the closed–loop system and sends the system states into a limit cycle. On the other hand, when the switching scheme of Theorem 6.2 is employed, the supervisor immediately switches to the bounded controller which in turn stabilizes the closed–loop system at the desired equilibrium point. This is depicted by the dotted lines in Fig. 6.8(a)–(e). The manipulated input profiles for the three scenarios are shown in Fig. 6.8(f). Finally, we note that the hybrid predictive controller designed here has also been implemented successfully on the population balance model of the crystallizer (see [246]).

6.3 Robust Hybrid Predictive Control of Nonlinear Systems

In the previous section, a hybrid predictive control strategy was developed for nonlinear systems with no uncertainty. In the presence of uncertainty, however, the nominal controllers (designed without taking the uncertainty

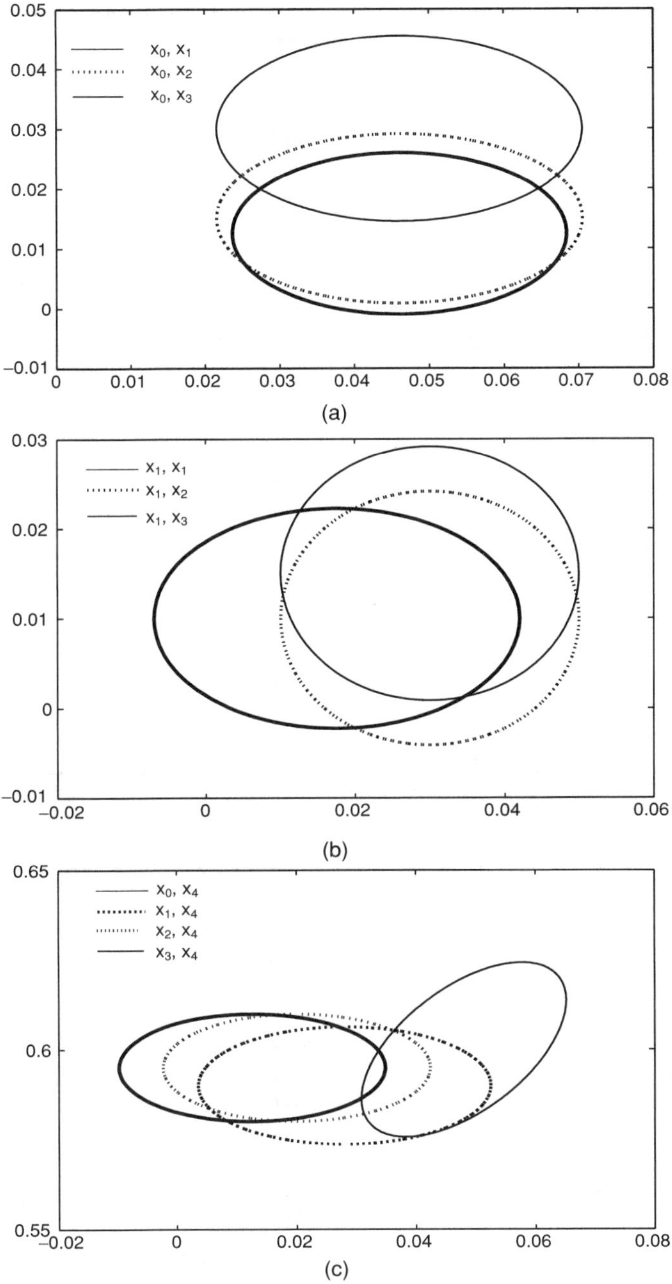

Fig. 6.7. Implementation of the proposed hybrid control structure to a continuous crystallizer: two-dimensional projections of the stability region for the 10 distinct combinations of the process states

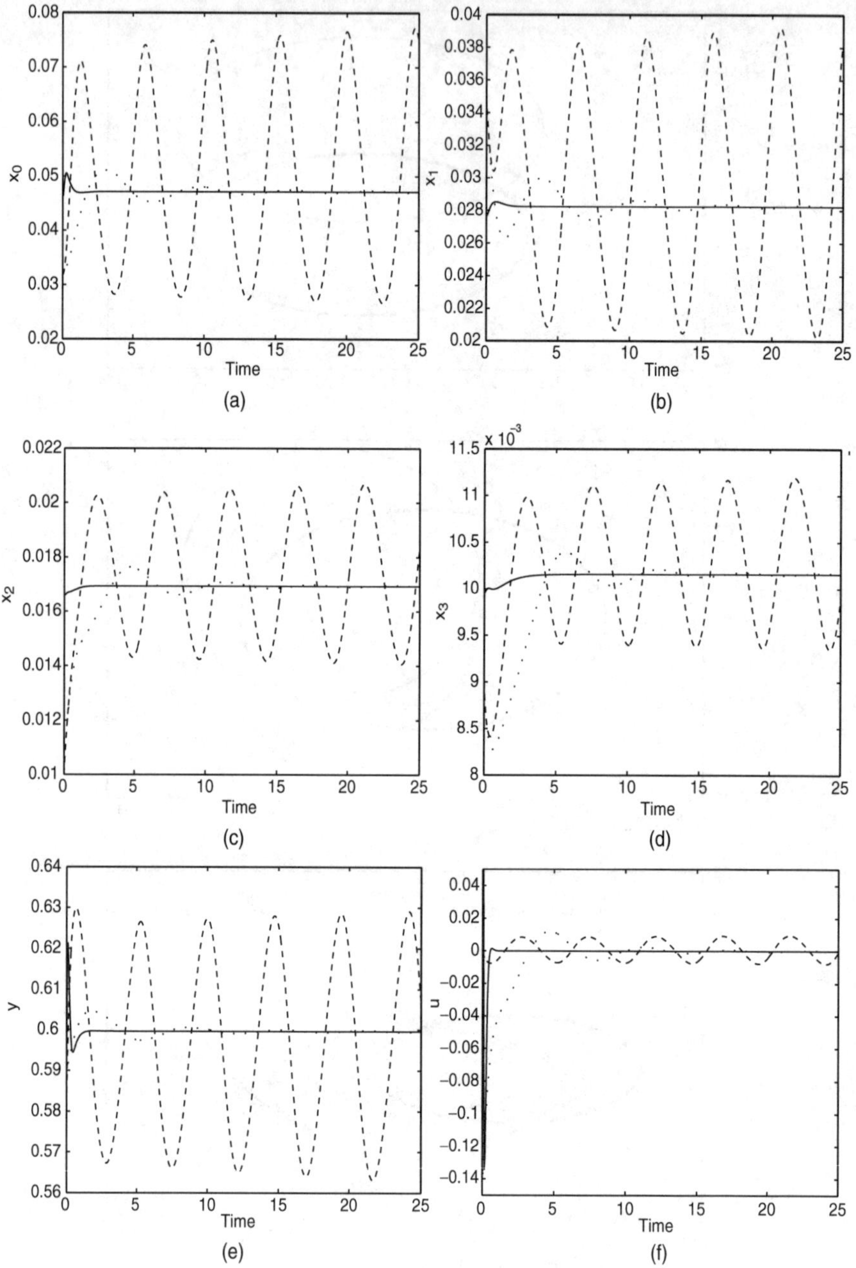

Fig. 6.8. Closed-loop profiles of the dimensionless crystallizer moments (**a**)–(**d**), the solute concentration in the crystallizer (**e**) and the manipulated input (**f**) under MPC with stability constraints (*solid line*), under MPC without terminal constraints (*dashed line*), and using the switching scheme of Theorem 6.2 (*dotted line*)

into account) may not be stabilizing as the uncertainty fundamentally alters the nominal stability region of the closed-loop system. Furthermore, simply replacing the fall–back controller by an appropriate robust controller and implementing the same switching logics proposed in the previous section may lead to a switching scheme that is too conservative, resulting in the implementation of the fall–back controller for almost all times and severely restricting the chances of MPC implementation. Motivated by these considerations, we consider in this section nonlinear systems with input constraints and uncertain variables, and develop a robust hybrid predictive control structure that provides a safety net for the implementation of any available MPC formulation, designed with or without taking uncertainty into account. The key idea is to use a Lyapunov–based robust controller, for which an explicit characterization of the closed–loop robust stability region can be obtained, to provide a stability region within which MPC can be implemented. This is achieved by devising a set of switching laws that orchestrate switching between MPC and the bounded robust controller in a way that exploits the performance of MPC whenever possible, while using the bounded robust controller as a fall–back mechanism that can be switched in to preserve robust closed–loop stability in the event that MPC fails to prescribe a control move (due, for example, to computational problems or infeasibility) or leads to instability (due, for example, to inappropriate penalties and/or horizon length in the objective function).

The rest of this section is organized as follows. In Sect. 6.3.1, some mathematical preliminaries are presented to describe the class of uncertain nonlinear systems under consideration and briefly review how the robust constrained control problem is addressed within the bounded control and MPC frameworks. In Sect. 6.3.2, the controller switching problem is formulated, and two representative switching schemes, that address robust stability and performance objectives, are described. Finally in Sect. 6.3.3, the implementation and efficacy of the robust hybrid predictive control structure are demonstrated through simulations using a chemical reactor example.

6.3.1 Preliminaries

We consider nonlinear systems with uncertain variables and input constraints, with the following state–space description:

$$\begin{aligned} \dot{x} &= f(x) + G(x)u(t) + W(x)\theta(t) \\ u(t) &\in \mathcal{U} \end{aligned} \tag{6.24}$$

where $x \in \mathbb{R}^n$ denotes the vector of state variables, $u(t) = [u^1(t) \cdots u^m(t)]'$ denotes the vector of constrained manipulated inputs, taking values in a nonempty convex subset $\mathcal{U}$ of $\mathbb{R}^m$, where $\mathcal{U} = \{u \in \mathbb{R}^m : \|u\| \leq u_{max}\}$, $\|\cdot\|$ is the Euclidean norm of a vector, $u_{max} \geq 0$ is the magnitude of input constraints, and $\theta(t) = [\theta^1(t) \cdots \theta^q(t)]' \in \Theta \subset \mathbb{R}^q$ denotes the vector of uncertain (possibly time–varying) but bounded variables taking values in a nonempty compact convex subset of $\mathbb{R}^q$. The vector of uncertain variables, $\theta(t)$, may describe time–varying parametric uncertainty and/or exogenous disturbances. It is assumed that the origin is the equilibrium point of the nominal (i.e., $u(t) = \theta(t) \equiv 0$) system of (6.24). The vector function, $f(x)$, the matrices, $G(x) = [g^1(x) \cdots g^m(x)]$ and $W(x) = [w^1(x) \cdots w^q(x)]$, where $g^i(x) \in \mathbb{R}^n$, $i = 1 \cdots m$, and $w^i(x) \in \mathbb{R}^n$, $i = 1 \cdots q$, are assumed to be sufficiently smooth on their domains of definition.

In order to provide the necessary background for the main results in Sect. 6.3.2, we will briefly review in the remainder of this section the design procedure and the stability properties of a Lyapunov–based bounded robust controller and a model predictive controller, which together constitute the two components of the robusy hybrid predictive control structure. For simplicity, we will focus only on the state feedback control problem where measurements of the entire state, $x(t)$, are assumed to be available for all t.

Bounded Robust Lyapunov–Based Control

Consider the nonlinear system of (6.24), where the uncertain variables, $W(x)\theta$, are assumed to be non–vanishing (in the sense that the uncertainty changes the nominal equilibrium point) and a robust control Lyapunov function (RCLF) [97], V, exists. Due to the non–vanishing nature of the uncertainty, asymptotic convergence to the origin cannot be achieved via continuous feedback. However, the states of the closed–loop system can be steered, in finite time, to an arbitrarily small (invariant) neighborhood of the origin by appropriate choice of the controller (practical stability). In order to enforce boundedness of the closed–loop trajectories and achieve an arbitrary degree of attenuation of the effect of the uncertain variables on the states, the existence of known (not necessarily small) bounds that capture the size of the uncertain variables for all times is assumed.

Using the RCLF, V, the following bounded robust state feedback control law can be designed (see Chap. 4 for details on similar controller designs):

$$u = -k(x, \theta_b, u_{max})\,(L_G V)' \tag{6.25}$$

$$k(\cdot) = \left\{ \begin{array}{ll} \dfrac{L_f^* V + \sqrt{\left(L_f^{**} V\right)^2 + (u_{max}\|(L_G V)'\|)^4}}{\|(L_G V)'\|^2 \left[1 + \sqrt{1 + (u_{max}\|(L_G V)'\|)^2}\right]} & , \|(L_G V)'\| \neq 0 \\ 0 & , \|(L_G V)'\| = 0 \end{array} \right\} \tag{6.26}$$

$$L_f^*V = L_fV + (\rho\|x\| + \chi\theta_b\|L_WV\|)\left(\frac{\|x\|}{\|x\|+\phi}\right)$$
$$L_f^{**}V = L_fV + \rho\|x\| + \chi\theta_b\|L_WV\| \tag{6.27}$$

where $L_GV = [L_{g^1}V \ \cdots \ L_{g^m}V]$ is a row vector, θ_b is a positive real number such that $\|\theta(t)\| \leq \theta_b$, for all $t \geq 0$, and ρ, χ and ϕ are adjustable parameters that satisfy $\rho > 0$, $\chi > 1$ and $\phi > 0$. Let Π be the set defined by:

$$\Pi(\theta_b, u_{max}) = \{x \in \mathbb{R}^n : L_f^{**}V \ \leq \ u_{max}\|(L_GV)'\|\} \tag{6.28}$$

and assume that the origin is contained in the interior of Π. Also, let Ω be a subset of Π, defined by:

$$\Omega = \{x \in \mathbb{R}^n : V(x) \leq c^{max}\} \tag{6.29}$$

Then, given any positive real number, d, such that:

$$\mathbb{D} := \{x \in \mathbb{R}^n : \|x\| \leq d\} \ \subset \ \Omega \tag{6.30}$$

and for any initial condition, $x_0 \in \Omega$, it can be shown that there exists a positive real number, ϵ^*, such that if $\phi/(\chi - 1) < \epsilon^*$, the states of the closed–loop system satisfy $x(t) \in \Omega \ \forall \ t \geq 0$ and $\limsup_{t\to\infty} \|x(t)\| \leq d$ (see Appendix B for an analogous proof).

Remark 6.18. Referring to the above controller design, it is important to make the following remarks. First, a general procedure for the construction of RCLFs for nonlinear systems of the form of (6.24) is currently not available. Yet, for several classes of nonlinear systems that arise commonly in the modeling of engineering applications, it is possible to exploit system structure to construct RCLFs. For example, for feedback linearizable systems, quadratic Lyapunov functions can be chosen as candidate RCLFs and can be made RCLFs with appropriate choice of the function parameters based on the process parameters (see, for example, [97]). Also, for nonlinear systems in strict feedback form, backstepping techniques can be employed for the construction of RCLFs [164]. Second, given that an RCLF, V, has been obtained for the system of (6.24), it is important to clarify the essence and scope of the additional assumption that there exists a level set, Ω, of V that is contained in Π. Specifically, the assumption that the set, Π, contains an invariant subset around the origin, is necessary to guarantee the existence of a set of initial conditions for which closed-loop stability is guaranteed (note that even though $\dot{V} < 0$ for all $x \in \Pi \setminus \mathbb{D}$, there is no guarantee that trajectories starting within Π will remain within Π for all times). Moreover, the assumption that Ω is a level set of V is made only to simplify the construction of Ω. This assumption restricts the applicability of the proposed control method because a direct method for the construction of an RCLF with level sets contained in Π is not available. However, the proposed control method remains applicable if the invariant set, Ω, is not a level set of V but can be constructed in some other way (which is a difficult task in general).

Remark 6.19. Regarding the choice of the above controller design, we note that the problem of designing control laws that guarantee stability in the presence of input constraints has been extensively studied (see, for example, [78, 81, 176, 177, 265]). The bounded robust controller design of (6.25–6.27) (proposed in [82] and inspired by the results on bounded control in [177] for systems without uncertainty) is an example of a controller design that (1) guarantees robust stability in the presence of constraints, and (2) allows for an explicit characterization of the closed-loop stability region, which are the essential features that make a given controller a suitable fall–back component in the robust hybrid predictive control structure (to be proposed in Sect. 6.3.2). The results of this section are not limited to this particular choice of controllers, and any other robust controller that satisfies the aforementioned properties, (1) and (2), can be used. Finally, note that when the uncertain variables in (6.24) are vanishing (in the sense that the uncertainty does not change the nominal equilibrium point), then starting from any initial state within Ω, the bounded robust controller of (6.25–6.27) enforces asymptotic closed–loop stability.

Model Predictive Control

For the purpose of illustrating the main results, we describe here a symbolic MPC formulation that incorporates most existing MPC formulations as special cases. This is not a new formulation of MPC; the general description is only intended for the purpose of highlighting the fact that the robust hybrid predictive control structure (to be proposed in the next section) can incorporate any available MPC formulation. In MPC, the control action at time t is conventionally obtained by solving, on–line, a finite horizon optimal control problem. The generic form of the optimization problem can be described as:

$$J_s(x,t,u(\cdot)) = \int_t^{t+T} (x'(s)Qx(s) + u'(s)Ru(s))ds + F(x(t+T)) \tag{6.31}$$

$$\begin{aligned} &u(\cdot) = argmin\{max\{J_s(x,t,u(\cdot))|\theta(\cdot) \in \Theta\}|u(\cdot) \in S\} := M_s(x) \\ &s.t. \quad \dot{x}(t) = f(x(t)) + G(x(t))u(t) + W(x(t))\theta(t), \;\; x(0) = x_0 \\ &x(t+T) \in \Omega_{MPC}(x,t,\theta) \end{aligned} \tag{6.32}$$

where $S = S(t,T)$ is the family of piecewise continuous functions, with period Δ, mapping $[t, t+T]$ into the set of admissible controls, T is the horizon length and θ is the bounded uncertainty assumed to belong to a set Θ. A control $u(\cdot)$ in S is characterized by the sequence $\{u[k]\}$ where $u[k] := u(k\Delta)$ and satisfies $u(t) = u[k]$ for all $t \in [k\Delta, (k+1)\Delta)$. J_s is the performance index, R and Q are strictly positive-definite, symmetric matrices and the function $F(x(t+T))$ represents a penalty on the states at the end of the horizon. The maximization over θ may not be carried out if the MPC version used is not a min-max type formulation. The set $\Omega_{MPC}(x,t,\theta)$ could be a fixed, terminal set or may represent inequality constraints (as in the case of MPC formulations that require some norm of the state, or a Lyapunov function for the system, to decrease at the end of the horizon). This stability constraint may or may not account for uncertainty. The stability guarantees in MPC formulations (with or without explicit stability conditions, and with or without robustness considerations, and whether or not it is a min-max type formulation) are dependent on the assumption of initial feasibility. Obtaining an explicit characterization of the closed–loop stability region of the predictive controller under uncertainty and constraints remains a difficult task.

6.3.2 Robust Hybrid Predictive Controller Design

Consider the nonlinear system of (6.24) with input constraints, $\|u\| \leq u_{max}$, and bounded non–vanishing uncertainty, $\|\theta\| \leq \theta_b$, for which: (1) a predictive controller of the form of (6.31–6.32) and the bounded robust controller of (6.25–6.27) have been designed, and (2) the set, Ω, starting from where the convergence of the closed–loop states to an arbitrarily small (invariant) neighborhood of the origin, $\mathbb{D}$, under the bounded robust controller, is ensured, has been computed using (6.28–6.29). We formulate the controller switching problem as the one of designing a set of switching laws that orchestrate the transition between the predictive controller and the bounded controller in a way that: (1) enforces boundedness of the closed–loop trajectories with an arbitrary degree of uncertainty attenuation, (2) guarantees convergence of the closed–loop states, in finite time, to the set $\mathbb{D}$ (practical stability), starting from any initial condition within Ω, and (3) allows recovering the MPC performance whenever the pertinent stability criteria are met. For a precise statement of the problem, we first cast the system of (6.24) as a switched system of the form:

$$\begin{aligned} \dot{x} &= f(x) + G(x)u_{i(t)} + W(x)\theta(t) \\ \|u_{i(t)}\| &\leq u_{max},\ i(t) \in \{1,2\} \end{aligned} \tag{6.33}$$

where $i : [0,\infty) \rightarrow \{1,2\}$ is the switching signal which is assumed to be a piecewise continuous (from the right) function of time, implying that only a finite number of switches between the two controllers is allowed on any finite–time interval. The index, $i(t)$, represents a discrete state that indexes the control input, u, with the understanding that $i(t) = 1$ if and only if $u_i(x(t)) = M_s(x(t))$ (i.e., MPC is used) and $i(t) = 2$ if and only if $u_i(x(t)) = b(x(t))$ (i.e., bounded control is used). The value of $i(t)$ is determined by a higher–level supervisor responsible for executing the transition between the two controllers. Our goal is to construct a switching law, $i(t) = \psi(x(t),t)$, that provides the supervisor with the switching times that ensure stabilizing

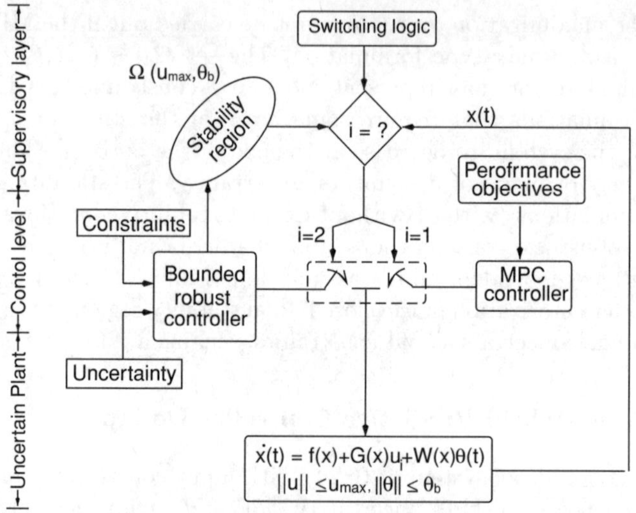

Fig. 6.9. A schematic representation of the robust hybrid predictive control structure

transitions between the two controllers. A schematic representation of the proposed hybrid predictive control structure is depicted in Fig. 6.9.

In the remainder of this section, two switching schemes are presented. The first scheme is given in Theorem 6.20 and focuses primarily on the issue of robust closed–loop stability, while the second scheme, given in Theorem 6.27, provides more flexible switching rules that guarantee robust closed–loop stability and, in addition, enhance the overall closed–loop performance. The proofs of both theorems are given in Appendix D.

Robust Stability-Based Controller Switching

Theorem 6.20. *Consider the switched nonlinear system of (6.33), the model predictive controller of (6.31–6.32) and the bounded controller design of (6.25–6.27). Let* $x(0) := x_0 \in \Omega$*, and initially set* $T_s = T_D = T_{inf} = \infty$*. At the earliest time,* $t \geq 0$*, for which the closed–loop state under MPC satisfies*

$$V(x(t^-)) = c^{max} \tag{6.34}$$

set $T_s = t$*. At the earliest time for which the closed–loop state under MPC satisfies* $\|x(t)\| \leq d$ *where* d *was defined in (6.30), set* $T_D = t$*. Finally, at the earliest time,* t*, that MPC is infeasible, set* $T_{inf} = t$*. Define* $T_{switch} = \min\{T_s, T_D, T_{design}, T_{inf}\}$*, where* $0 \leq T_{design} < \infty$ *is arbitrary. Then, the switching rule:*

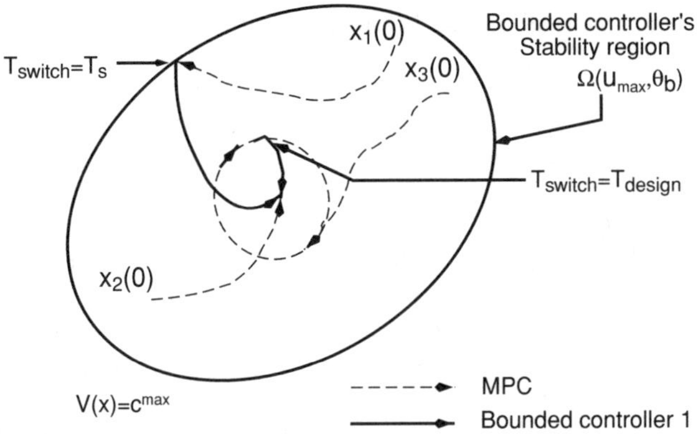

Fig. 6.10. Examples of the evolution of the closed–loop state trajectory under the switching scheme of Theorem 6.20. Starting from $x_1(0)$, the trajectory under MPC hits the boundary of the stability region, upon which the bounded controller is switched in. Starting from $x_2(0)$, MPC is able to achieve practical stability. Starting from $x_3(0)$, the trajectory under MPC converges to a limit cycle and, therefore, after a time T_{design} has elapsed, the bounded controller is switched in to enforce convergence close to the origin

$$i(t) = \left\{ \begin{array}{l} 1, \ \ 0 \leq t < T_{switch} \\ \\ 2, \ \ t \geq T_{switch} \end{array} \right\} \tag{6.35}$$

guarantees that $x(t) \in \Omega$, *for all* $t \geq 0$, *and that* $\limsup_{t \to \infty} \|x(t)\| \leq d$.

Remark 6.21. The design and implementation of the robust hybrid predictive control scheme can be understood algorithmically as follows:

- Given the nonlinear system of (6.24), the bounds on the uncertain variables and the constraints on the manipulated inputs, design the bounded robust controller of (6.25–6.27), and calculate an estimate of its stability region, Ω.
- Design (or pick) an MPC formulation (the MPC formulation could be min–max optimization-based, linear or nonlinear, and with or without stability constraints). For convenience, we refer to the MPC formulation of (6.31–6.32).
- Given any $x_0 \in \Omega$, check the feasibility of the optimization in (6.31–6.32) (for a given horizon length, T) at $t = 0$, and if feasible, start implementing MPC (i.e., set $u(0) = M_s(x_0)$).
- If at any time, MPC becomes infeasible ($t = T_{inf}$), or the states of the closed–loop system approach the boundary of Ω ($t = T_s$; see the scenario starting form $x_1(0)$ in Fig. 6.10), or the closed–loop states enter the set $\mathbb{D}$ ($t = T_D$; see the scenario starting form $x_2(0)$ in Fig. 6.10), then switch to

the bounded controller, else keep MPC active in the closed–loop system until a time T_{design} (see the scenario starting form $x_3(0)$ in Fig. 6.10).
- Switch to the bounded robust controller at T_s, T_D, T_{design}, or T_{inf}, whichever comes earliest, and implement it in the closed–loop system to achieve practical closed–loop stability.

Remark 6.22. The stability guarantees in various MPC formulations is contingent upon the assumption of initial feasibility of the optimization problem, and the set of initial conditions for which the optimization problem is feasible is not explicitly characterized. By using the bounded robust controller as a fall–back controller, the switching scheme allows us to safely initialize the closed–loop system anywhere within Ω using MPC. The use of the fall-back controller turns Ω into a stability region for the switched closed–loop system because the bounded controller can always intervene (through switching) to maintain closed–loop stability in case the closed–loop system is infeasible under MPC or starts to become unstable (due, for example, to using a computationally simple, nominal MPC that does not take uncertainty into account).

Remark 6.23. Even though the MPC framework provides a transparent way of specifying performance objectives, the various MPC formulations found in the literature may, in general, not be optimal, and only approximate the infinite horizon optimal cost to varying degrees of success. Within the robust hybrid predictive control framework, the choice of a particular MPC design can be made entirely on the basis of the desired tradeoff between performance and computational complexity because the stability guarantees of the robust hybrid predictive controller are independent of the specific MPC formulation being used.

Remark 6.24. The purpose of introducing the design parameter, T_{design}, (which is the time beyond which only the bounded robust controller is implemented) is to ensure convergence to $\mathbb{D}$ and avoid possible cases where the closed–loop states, under MPC, could wander forever inside Ω without actually converging to, and staying within, $\mathbb{D}$ (see, for example, the scenario starting form $x_3(0)$ in Fig. 6.10). Other possible means of ensuring convergence to $\mathbb{D}$ include switching to the bounded controller when $\dot{V} \geq 0$ under MPC (see, for example, the switching rules in Chap. 5 and Sect. 6.2.2). However, in the presence of uncertainty, if a nominal MPC design is being used, such a condition might be very restrictive in the sense that it may terminate MPC implementation too early. For example, it could be that the nominal MPC is able to practically stabilize the closed–loop system when implemented in the closed–loop system for all times, yet is unable to maintain negative-definiteness of $\dot{V}$ for all times due to the disturbances (see Sect. 6.3.3 for a demonstration of this point in the context of a chemical process example). If the switching schemes in Sect. 6.2.2 (designed for systems without uncertainty) were to be used, it would dictate an immediate switch to the bounded controller, thus failing to exploit the high performance of MPC. The parameter T_{design}, on the other hand, if chosen large enough, provides ample time for the predictive controller to be implemented and possibly achieve practical stabilization.

Remark 6.25. Note that, while the result of Theorem 6.20 holds for any arbitrary value of the design parameter, T_{design}, the choice of this parameter provides a handle for extracting better performance out of the robust hybrid predictive control structure through the possibility of the extended use of MPC. To achieve this, the value of T_{design} can be suitably altered, depending on the specific MPC formulation being used. For example, if a robust MPC design that guarantees stability for the uncertain nonlinear system is used, T_{design} can be chosen to be practically infinity. In this case, if MPC is feasible, it could be implemented for practically most of the time to stabilize the closed–loop system (note that the stability safeguards provided by the bounded controller would still be required because the stability of any MPC formulation, robust or otherwise, are based on the assumption of initial feasibility, which cannot be verified short of running closed-loop simulations). On the other hand, if a "nominal," computationally inexpensive formulation of MPC is used, it is possible that it drives the states close to the origin, but is unable to enforce convergence to the set $\mathbb{D}$. In this case, T_{design} could be chosen in a way such that after MPC has driven the states close to the origin, the bounded controller is switched in (and MPC switched out) to enforce convergence to $\mathbb{D}$ and achieve the desired degree of uncertainty attenuation.

Remark 6.26. When the uncertainty is vanishing, the same switching scheme proposed in Theorem 6.20 can be applied by simply setting $d = 0$. Recall that, under vanishing uncertainty, the bounded controller guarantees asymptotic closed–loop stability for all $x(0) \in \Omega$ and therefore the set, $\mathbb{D}$, collapses to the origin (see [87] for how the switching scheme is designed in the case of linear systems with vanishing disturbances).

Enhancing Closed–Loop Performance

The switching rule in Theorem 6.20 relies on a single bounded controller for backup, and uses the stability region estimate generated from a single Lyapunov function within which it restricts the evolution of the closed–loop state trajectory under MPC. While this is sufficient to guarantee robust closed–loop stability, the switching logic may limit the chances of MPC implementation (and thus limit the resulting closed-loop performance) in cases when the closed–loop state under MPC might temporarily leave the stability region of the bounded controller, but eventually converge to $\mathbb{D}$. Furthermore, the switching scheme in Theorem 6.20 prescribes only a single switch from MPC to the bounded controller which is then implemented in the closed–loop for all future times. By not allowing for multiple switches, the switching scheme might not be able to take advantage of MPC implementation in cases where MPC is not stabilizing from one initial condition but is stabilizing from another initial condition. In this section, we present a switching scheme that relaxes the switching rules of Theorem 6.20 in two directions, in order to take better advantage of the performance of MPC. In one direction, to expand the region within which MPC implementation is allowed, we use a combination of several Lyapunov functions ($V_k, k = 1, \ldots, l$, where l is the number of bounded controllers), and MPC is allowed to remain in the closed–loop system as long as the closed–loop state trajectory does not escape the union of the stability regions ($\Omega_L = \bigcup_{k=1}^{l} \Omega_k$). In the second direction, to allow for the possibility of multiple switches between the predictive and bounded controllers, at any time the closed–loop state trajectory hits the boundary of the union of the stability regions, the appropriate bounded controller is switched in and implemented only until it brings the state within a pre–determined, smaller subregion where MPC is re–activated once again and confined.

With the availability of $l+1$ controllers (l bounded controllers and MPC) that can be used, we recast the system of (6.24) as a switched system of the form:

$$
\begin{aligned}
\dot{x} &= f(x) + G(x)u_{i(t)} + W(x)\theta(t) \\
\|u_{i(t)}\| &\leq u_{max},\ i(t) \in \{1, 2, \ldots, l+1\}
\end{aligned}
\tag{6.36}
$$

where $i(t) = k$ if and only if $u_i(x(t)) = b_k(x(t))$ (i.e., the k-th bounded controller is used) and $i(t) = l+1$ if and only if $u_i(x(t)) = M_s(x(t))$ (i.e., MPC is used). We also define, for the l bounded controllers, and the robust control Lyapunov functions, $V_k,\ k = 1, \ldots, l,\ 0 < \alpha < 1,\ c_{k,j}^{max} = \alpha^j c_k^{max}$, $\Omega_k^j = \{x \in \mathbb{R}^n : V_k(x) \leq c_{k,j}^{max}\}$ and $\Omega_L^j = \bigcup_{k=1}^{l} \Omega_k^j$, for $j = 0, \cdots, N-1$ where $N < \infty$ is an integer, and $d^{max} = \max_{k=1,\ldots,l} \{d_k\}$, where d_k are arbitrarily small positive numbers such that $\mathbb{D}_k = \{x \in \mathbb{R}^n : \|x\| \leq d_k\} \subset \Omega_k^{N-1}$. The notation, $\partial\Omega$, will also be used to denote the boundary of a closed set, Ω.

We are now in a position to present the relaxed switching scheme which is formalized in Theorem 6.27. The proof of this theorem is given in Appendix D.

Theorem 6.27. *Consider the constrained switched nonlinear system of (6.36), the model predictive controller of (6.31–6.32) and the l bounded controllers of the form of (6.25–6.27), designed using the Lyapunov functions V_k, $k = 1, \ldots, l$. Consider any initial condition, $x(0) := x_0 \in \Omega_L^0$, and initially set $T_0 = 0$, $T_1 = T_2 = \cdots = T_N = T_D = \infty$. At the earliest time, $t \geq 0$, for which the closed–loop state under MPC satisfies:*

$$
\|x(t)\| \leq d^{max}
\tag{6.37}
$$

set $T_D = t$. At the earliest time, $t \geq 0$, under MPC such that:

$$x(t) \in \partial\Omega_L^{j-1},\ x(t^-) \in \Omega_L^{j-1} \text{ and } T_{j-1} < \infty \tag{6.38}$$

for some $j \in \{1, \cdots, N\}$, or MPC is infeasible, set $T_j = t$. Define $T_{switch} = \min\{T_D, T_N, T_{design}\}$, where $0 \leq T_{design} < \infty$ is arbitrary. Then, the switching rule:

$$i(t) = \begin{cases} l+1, & 0 \leq T_j \leq t \leq T_{j+1} \leq T_{switch},\ j = 0, \ldots, N-1 \text{ and } x \in \Omega_L^j \\ a, & 0 \leq T_j \leq t \leq T_{j+1} \leq T_{switch},\ j = 0, \ldots, N-1 \text{ and } x \notin \Omega_L^j \\ f, & t \geq T_{switch} \end{cases} \tag{6.39}$$

for some $a \in \{1, \ldots, l\}$ for which $x(T_j) \in \Omega_a^{j-1}$, and for some $f \in \{1, \ldots, l\}$ for which $x(T_{switch}) \in \Omega_f^0$, guarantees that the state of the switched closed–loop system is bounded and $\limsup_{t \to \infty} \|x(t)\| \leq d^{max}$.

Remark 6.28. The implementation of the switching scheme proposed in Theorem 6.27 can be understood as follows (see also the schematic representation in Fig. 6.11 for an example when $l = N = 2$):

- For each Lyapunov function, V_k, $k = 1, \ldots, l$: construct a bounded controller using (6.25–6.27), a set of N concentric level sets, Ω_k^j, $j = 0, \ldots, N-1$, with $\Omega_k^{N-1} \subset \Omega_k^{N-2} \subset \cdots \subset \Omega_k^1 \subset \Omega_k^0$, and a terminal set, $\mathbb{D}_k$, fully contained in Ω_k^{N-1}.
- Initialize the closed–loop system using MPC at any initial condition, x_0, within Ω_L^0 (the union of the level sets Ω_k^0) and start monitoring the position of the state of the closed–loop system.
- At the earliest time that the closed-loop state approaches the boundary of Ω_L^0, or MPC is infeasible, switch to bounded control (this corresponds to encountering T_1; see Fig. 6.11).
- Implement any of the bounded controllers whose stability region contains the state at T_1 until the state enters Ω_L^1 (the union of the level sets Ω_k^1) after which switch back to MPC and implement it in the closed–loop system (in Fig. 6.11 this happens when the state enters Ω_1^1).
- If the closed–loop state under MPC does not escape the boundary of Ω_L^1 and MPC continues to be feasible, keep MPC in the closed–loop system (this happens in Fig. 6.11(a) where the state continues to stay within Ω_L^1). If the closed–loop state, however, tries to escape the boundary of Ω_L^1 (this happens in Fig. 6.11(b) when the state hits the boundary of Ω_2^1 at $t = T_2$) or MPC is infeasible again, switch to bounded control and implement it until the states are driven inside the next safety zone where MPC is re–activated and confined (note that each time MPC is switched back in, its implementation is confined within a smaller safety zone since $\Omega_L^1 \supset \Omega_L^2 \supset \cdots \supset \Omega_L^{N-1}$). Repeat this process until either the state under MPC is about to escape the N-th safety zone, Ω_L^{N-1}, or the state

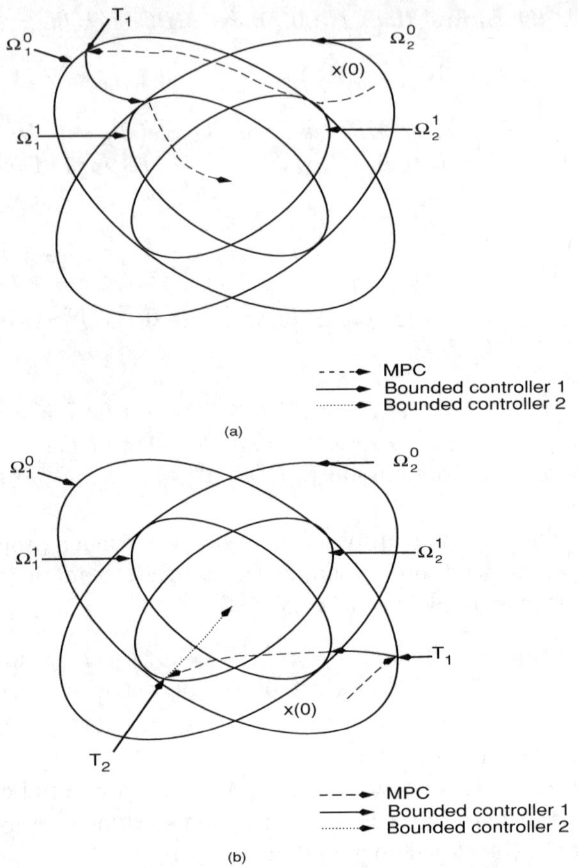

Fig. 6.11. Examples of the evolution of closed–loop state trajectory under the switching scheme of Theorem 6.27: (**a**) starting from $x(0)$, the closed–loop trajectory under MPC hits the boundary of Ω_1^0 (at T_1), upon which the bounded controller 1 is switched in and implemented until the trajectory enters Ω_1^1, following which MPC is switched back in leading to practical stability, (**b**) starting from $x(0)$, the closed–loop trajectory under MPC hits the boundary of Ω_1^0 (at T_1), upon which the bounded controller 1 is switched in and implemented until the trajectory enters Ω_1^1, following which MPC is switched back in. The trajectory under MPC hits the boundary of Ω_2^1 ($t = T_2$) at which time the bounded controller 2 is switched in and implemented for all future times

enters the largest terminal set, $\mathbb{D}_{max}$, or a time greater than the design parameter, T_{design}, has elapsed. At the earliest time that any of these events takes place, switch permanently to any of the bounded controllers whose stability region contains the state at the time to ensure practical stability of the closed–loop system.

Remark 6.29. Even though the switching scheme of Theorem 6.27 allows back and forth switching between MPC and the bounded controllers, chattering is avoided by virtue of the fact that the switching scheme limits the maximum number of switches to N and requires a finite–time interval to elapse in between consecutive switchings of MPC, which is the time it takes the appropriate bounded controller to drive the state from an outer level set to an inner one (note that the level sets of a given V_k are finitely–spaced since α is strictly less than one; see Fig. 6.11 for an illustration). The bounded controller is invoked to bring the state of the closed–loop system closer to the terminal set. Switching back to MPC and trying to implement it in the closed–loop system is valuable because in some cases MPC may be feasible/stabilizing from this new initial condition, which is closer to the terminal set.

Remark 6.30. Both N, the maximum allowable number of switches between MPC and the bounded controllers, and T_{design} serve as design parameters that help take advantage of MPC performance, while still guaranteeing robust stability. Note that while Theorem 6.27 allows for the possibility of N switches between MPC and the bounded controllers, it does not necessarily dictate that N switches take place (see Fig. 6.11(a) for example). If the predictive controller itself is able to achieve practical stability, then back and forth switching between the predictive and bounded controllers is not required and, therefore, not executed.

6.3.3 Application to Robust Stabilization of a Chemical Reactor

Consider a well–mixed continuous stirred tank reactor where an irreversible elementary exothermic reaction of the form $A \xrightarrow{k} B$ takes place. The feed to the reactor consists of pure A at flow rate F, molar concentration C_{A0} and temperature T_{A0}. Under standard modeling assumptions, the process model takes the following form:

$$\begin{aligned} V\frac{dC_A}{dt} &= F(C_{A0} - C_A) - k_0 \exp\left(\frac{-E}{RT}\right) C_A V \\ V\frac{dT}{dt} &= F(T_{A0} - T) + \frac{(-\Delta H)}{\rho c_p} k_0 \exp\left(\frac{-E}{RT}\right) C_A V + \frac{Q}{\rho c_p} \end{aligned} \tag{6.40}$$

where C_A denotes the concentration of species A, T denotes the temperature of the reactor, Q denotes the rate of heat input to the reactor, V denotes the volume of the reactor, k_0, E, ΔH denote the pre–exponential constant, the activation energy, and the enthalpy of the reaction, c_p and ρ denote the heat capacity and density of the fluid in the reactor. The steady–state values and process parameters are given in Table 6.2. For these parameters, it was verified that the given equilibrium point is an unstable one (the system also possesses two other locally asymptotically stable equilibrium points). The control objective is to regulate both the reactor temperature and reactant

Table 6.2. Process parameters and steady–state values for the reactor of (6.40)

V	$= 0.1$	m^3
R	$= 8.314$	$kJ/kmol \cdot K$
C_{A0s}	$= 1.0$	$kmol/m^3$
T_{A0s}	$= 310.0$	K
ΔH	$= -4.78 \times 10^4$	$kJ/kmol$
k_0	$= 1.2 \times 10^9$	s^{-1}
E	$= 8.314 \times 10^4$	$kJ/kmol$
c_p	$= 0.239$	$kJ/kg \cdot K$
ρ	$= 1000.0$	kg/m^3
F	$= 1.67 \times 10^{-3}$	m^3/s
T_{Rs}	$= 395.33$	K
C_{As}	$= 0.57$	$kmol/m^3$

concentration at the (open–loop) unstable equilibrium point by manipulating both the rate of heat input/removal and the inlet reactant concentration. Defining $x_1 = C_A - C_{As}$, $x_2 = T - T_s$, $u_1 = C_{A0} - C_{A0s}$, $u_2 = Q$, $\theta_1(t) = T_{A0} - T_{A0s}$, $\theta_2(t) = \Delta H - \Delta H_{nom}$, where the subscript s denotes the steady–state value and ΔH_{nom} denotes the nominal value of the heat of reaction, the process model of (6.40) can be cast in the form of (6.24). In all simulation runs, the following time–varying function was considered to simulate the effect of exogenous disturbances in the feed temperature: $\theta_1(t) = \theta_0 sin(3t)$, where $\theta_0 = 0.08T_{A0s}$. In addition, a parametric uncertainty of 50% in the heat of reaction was considered, i.e., $\theta_2(t) = 0.5\,(-\Delta H_{nom})$. The upper bounds on the uncertain variables were therefore taken to be $\theta_{b1} = 0.08T_{A0s}$, $\theta_{b2} = 0.5\,|(-\Delta H_{nom})|$. Also, the following constraints were imposed on the manipulated inputs: $|u_1| \leq 1.0\ mol/L$ and $|u_2| \leq 92\ KJ/s$.

A quadratic Lyapunov function of the form $V = (C_A - C_{As})^2 + \frac{1}{T_{Aos}^2}(T - T_s)^2$ was used to design the bounded robust controller of (6.25–6.27) and compute the associated region of guaranteed closed–loop stability, using (6.28–6.29). The following values were used for the tuning parameters: $\chi = 1.01$, $\phi = 0.0001$, $\rho = 0.01$ to guarantee that the closed–loop states satisfy a relation of the form $\limsup_{t \to \infty} \|x(t)\| \leq 0.0005$. The set of nonlinear ODEs was integrated using the MATLAB solver, ODE45, and the optimization problem in MPC was solved using the MATLAB nonlinear constrained optimization solver, *fmincon*.

Figure 6.12 shows the stability region for the nominal bounded controller (Ω), and the bounded robust controller ($\Omega(u_{max}, \theta_b)$), which, as expected, is smaller in size since it accounts for the presence of uncertainty. As can be seen from the solid lines in Figs. 6.12–6.13, when the nominal bounded controller (6.25–6.27) with $\chi = 0$) is implemented, the desired uncertainty attenuation is not achieved. In contrast, from the dashed lines in Figs. 6.12–6.13, it is clear that, starting from the same initial condition, the properly tuned bounded

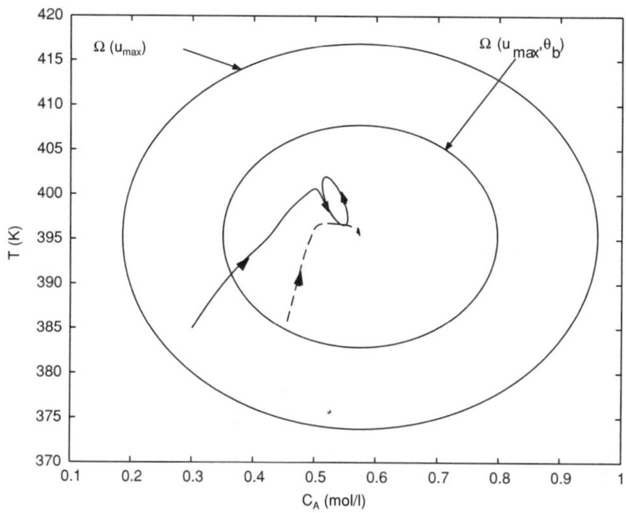

Fig. 6.12. Closed–loop state trajectories under the bounded controller when initialized within the stability region estimated without taking the uncertainty into account (*solid line*) and the stability region that accounts for the uncertainty (*dashed line*)

robust controller successfully drives both states to the desired steady–state, in the presence of input constraints, while simultaneously attenuating the effect of the uncertain variables on the states. We see from this comparison that the uncertain variables have a significant effect on the states and that failure to explicitly account for them in the controller design leads to poor transient performance and the inability to enforce convergence close to the origin.

The next set of simulation runs demonstrate the use of the robust hybrid predictive control strategy of Theorem 6.20. In the first scenario, a nominal nonlinear MPC formulation, without stability constraints, is used as part of the robust predictive control structure (setting $\Omega_{MPC} = \mathbb{R}^n$, $F(x(t+T)) = 0$, $T = 0.02$ minutes in (6.31–6.32) and with the design parameter $T_{design} = 10$ minutes. The solid lines in Figs. 6.14–6.15 represent the states of the closed–loop system initialized from a point within the stability region $\Omega(u_{max}, \theta_b)$. After the initial implementation of MPC, the supervisor detects, at $t = 0.6$ seconds, that the closed–loop states are close to the boundary of $\Omega(u_{max}, \theta_b)$ and therefore switches to the bounded robust controller to stabilize the closed–loop system. In the next scenario, a stabilizing formulation of MPC is used (requiring the states to go to a small invariant set at the end of the horizon), with a horizon length of $T = 0.02$ and a $T_{design} = 20$ minutes. For the initial condition of the trajectory shown by the dashed lines in Figs. 6.14–6.15, the MPC controller yields a feasible solution and drives the states close to the origin. In this case, since the MPC is able to achieve the desired degree

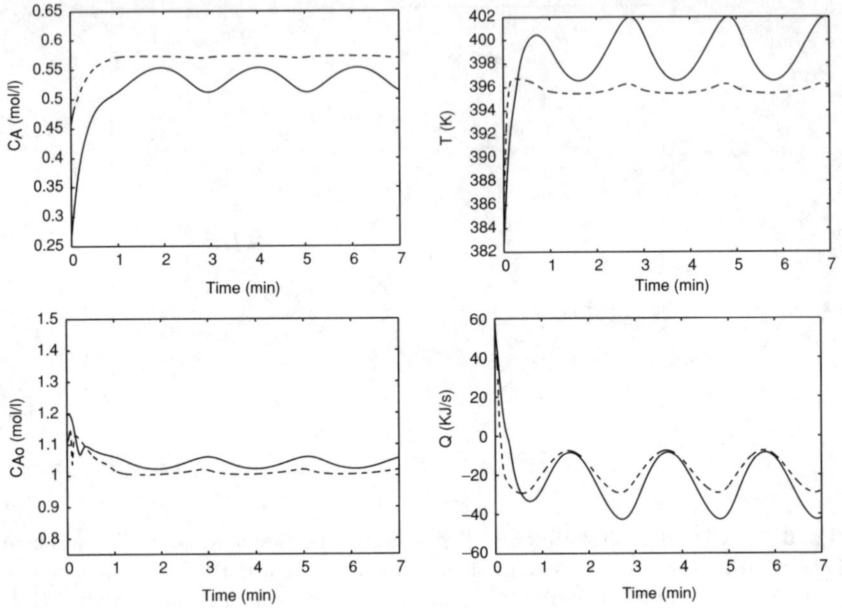

Fig. 6.13. Closed–loop state (*top*) and input (*bottom*) profiles under the bounded controller when initialized within the stability region estimated without taking the uncertainty into account (*solid lines*) and the stability region that accounts for the uncertainty (*dashed lines*)

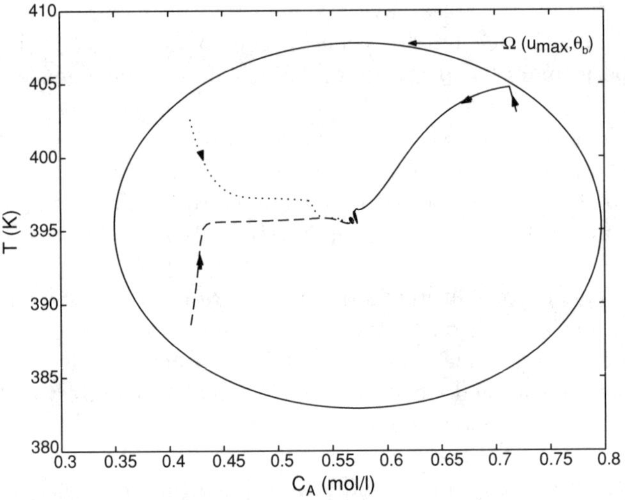

Fig. 6.14. Closed–loop state trajectory: implementation of the robust hybrid predictive controller of Theorem 6.20 using a nominal MPC formulation without stability constraints (*solid line*) and a nonlinear MPC formulation with stability constraints, from two different initial conditions (*dashed and dotted lines*)

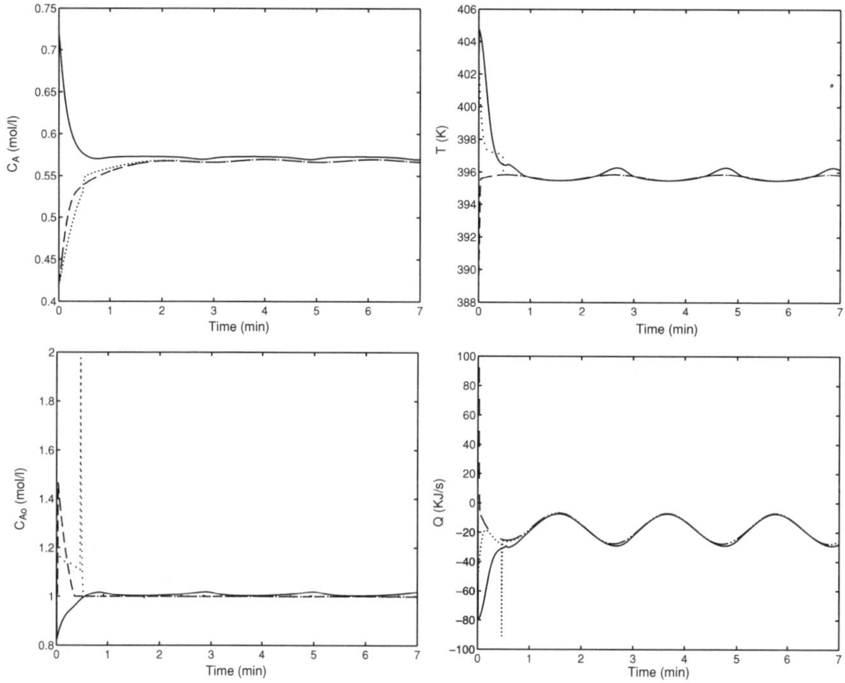

Fig. 6.15. Closed–loop state (*top*) and input (*bottom*) profiles: implementation of the robust hybrid predictive controller of Theorem 6.20 using a nominal MPC formulation without stability constraints (*solid lines*) and a nonlinear MPC formulation with stability constraints, from two different initial conditions (*dashed and dotted lines*)

of uncertainty attenuation, it is implemented for all future times. For the initial condition depicted by the dotted lines in Figs. 6.14–6.15, however, the MPC controller does not yield a feasible solution, and therefore the supervisor initially implements the bounded robust controller. At $t = 0.465$ minutes, the supervisor detects that MPC becomes feasible and, therefore, switches it in (and switches the bounded controller out) leading to closed–loop stability.

Finally, we demonstrate the implementation of the relaxed switching scheme of Theorem 6.27 using two Lyapunov functions. Figure 6.16 shows the stability region for the two bounded controllers used as part of the robust hybrid predictive controller. We use a nominal nonlinear MPC formulation, without stability constraints, as part of the robust predictive control structure (setting $\Omega_{MPC} = \mathbb{R}^n$, $F(x(t+T)) = 0$, $T = 0.02$ minutes in (6.31–6.32) and with the design parameter $T_{design} = 3$ minutes. The solid lines in Figs. 6.16 and 6.17 represent the states of the closed–loop system initialized, under MPC, from a point within the stability region Ω_2. As seen in Fig. 6.16, the switching logic allow MPC to be implemented in the closed–loop even though the states

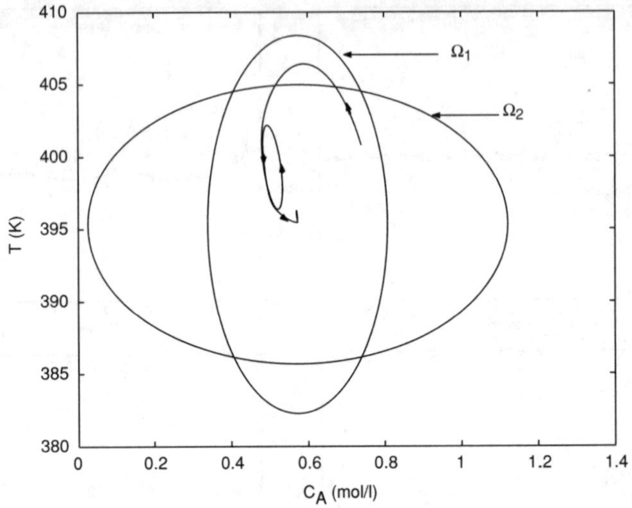

Fig. 6.16. Closed–loop state trajectory:implementation of the robust hybrid predictive controller of Theorem 6.27, with $l = 2$, $N = 1$, using a nominal MPC formulation without stability constraints (*solid line*)

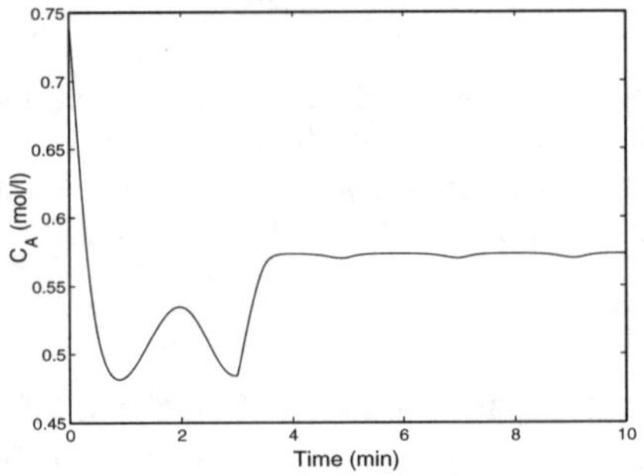

Fig. 6.17. Closed–loop state (*top*) and input (*bottom*) profiles: implementation of the robust hybrid predictive controller of Theorem 6.27, with $l = 2$, $N = 1$, using a nominal MPC formulation without stability constraints (*solid line*)

escape out of Ω_2, since they are still within Ω_1. Note also that in this case the nominal MPC does not enforce the desired degree of uncertainty attenuation (note the oscillations) and, therefore, after the time T_{design} has elapsed, the supervisor implements the bounded controller (associated with Ω_2) in the closed–loop system to achieve practical stabilization.

6.4 Conclusions

In this chapter, we presented hybrid predictive control strategies for the stabilization of input-constrained nonlinear systems with and without uncertainty. In the absence of uncertainty, the strategy involved using a family of bounded nonlinear controllers, each with an explicitly characterized stability region, as fall–back controllers and embedding the operation of MPC within the union of these regions. In the event that the predictive controller was unable to stabilize the closed–loop system, supervisory switching from MPC to any of the bounded controllers, whose stability region contained the trajectory at the switching time, guaranteed closed–loop stability. The strategy was subsequently extended to nonlinear systems with time-varying, bounded uncertain variables. The resulting robust hybrid predictive control strategy involved using a bounded robust controller, with a well-characterized region of robust stability, as a fall–back mechanism and devising a set of switching rules that take the uncertainty into account and orchestrate, accordingly, the transition from MPC to bounded control to maintain robust closed–loop stability in the event that the predictive controller was unable to maintain it. The proposed strategies were shown to provide, through appropriate construction of the switching logic and irrespective of the MPC formulation used, a safety net for the implementation of predictive control algorithms to nonlinear and uncertain systems. The efficacy of the proposed strategies was demonstrated through applications to chemical process examples.

7

Control of Hybrid Nonlinear Systems

7.1 Introduction

In this chapter, we develop hybrid nonlinear control methodologies for various classes of hybrid nonlinear processes. Initially, switched processes whose dynamics are both constrained and uncertain are considered. These are systems that consist of a finite family of continuous uncertain nonlinear dynamical subsystems, subject to hard constraints on their manipulated inputs, together with a higher-level supervisor that governs the transitions between the constituent modes. The key feature of the proposed control methodology is the integrated synthesis, via multiple Lyapunov functions (MLFs), of: (1) a family of lower-level robust bounded nonlinear feedback controllers that enforce robust stability in the constituent uncertain modes, and provide an explicit characterization of the stability region for each mode under uncertainty and constraints, and (2) upper-level robust switching laws that orchestrate safe transitions between the modes in a way that guarantees robust stability in the overall switched closed-loop system. Next, we consider switched nonlinear systems with scheduled mode transitions. These are systems that transit between their constituent modes at some predetermined switching times, following a prescribed switching sequence. For such systems, we develop a Lyapunov-based predictive control strategy that enforces both the required switching schedule and closed-loop stability. The main idea is to design a Lyapunov–based predictive controller for each mode, and incorporate transition constraints in the predictive controller design to ensure that the prescribed transitions between the modes occur in a way that guarantees stability of the switched closed–loop system. The proposed control methods are demonstrated through applications to chemical process examples. Finally, the chapter concludes with a demonstration of how hybrid systems techniques – in particular, the idea of coupling the switching logic to stability regions – can be applied for the analysis and control of mode transitions in biological networks. The results in this chapter were first presented in [82, 85, 196].

7.2 Switched Nonlinear Processes with Uncertain Dynamics

7.2.1 State-Space Description

We consider the class of switched uncertain nonlinear processes described by the following state-space representation:

$$\begin{aligned} \dot{x}(t) &= f_{\sigma(t)}(x(t)) + G_{\sigma(t)}(x(t))u_{\sigma(t)} + W_{\sigma(t)}(x(t))\theta_{\sigma(t)}(t) \\ \sigma(t) &\in \mathcal{I} = \{1, \ldots, N\} \end{aligned} \tag{7.1}$$

where $x(t) \in \mathbb{R}^n$ denotes the vector of continuous process state variables, $u(t) = [u^1(t) \cdots u^m(t)]^T \in \mathcal{U} \subset \mathbb{R}^m$ denotes the vector of control inputs taking values in a nonempty compact convex subset of $\mathbb{R}^m$ that contains the origin in its interior, $\theta(t) = [\theta^1(t) \cdots \theta^q(t)]^T \in \Theta \subset \mathbb{R}^q$ denotes the vector of uncertain (possibly time-varying) – but bounded – variables taking values in a nonempty compact convex subset of $\mathbb{R}^q$. The uncertain variables, $\theta(t)$, may describe time-varying parametric uncertainty and/or exogenous disturbances. $\sigma : [0, \infty) \to \mathcal{I}$ is the switching signal which is assumed to be a piecewise continuous (from the right) function of time, implying that only a finite number of switches is allowed on any finite interval of time. $\sigma(t)$, which takes values in the finite index set $\mathcal{I}$, represents a discrete state that indexes the vector field $f(\cdot)$, the matrices $G(\cdot)$ and $W(\cdot)$, the control input $u(\cdot)$, and the uncertain variable $\theta(\cdot)$, which altogether determine $\dot{x}$. For each value that the discrete state, σ, assumes in $\mathcal{I}$, the temporal evolution of the continuous state, x, is governed by a different set of differential equations. Processes of the form of (7.1) are therefore of variable structure; they consist of a finite family of N continuous-time uncertain nonlinear subsystems (or modes) and some rules for switching between them (see Fig. 7.1 for a graphical representation). These rules define a switching sequence that describes the temporal evolution of the discrete state. Note that, by indexing the vector of uncertain variables, $\theta(t)$, and the matrix $W(x)$ in (7.1) by σ, it is implied that the constituent modes do not necessarily share the same uncertain variables nor are they equally affected by them. The uncertainty is therefore allowed to influence the dynamics of different modes differently.

Throughout the chapter, the notation t_{i_k} and $t_{i'_k}$ is used to denote the k-th times that the i-th subsystem is switched in and out, respectively, i.e. $\sigma(t_{i_k}^+) = \sigma(t_{i'_k}^-) = i$, for all $k \in Z_+$. With this notation, it is understood that, when the i-th mode is active, the continuous state evolves according to $\dot{x} = f_i(x) + G_i(x)u_i + W_i(x)\theta_i$ for $t_{i_k} \leq t < t_{i'_k}$. It is assumed that all entries of the vector functions $f_i(x)$, the $n \times m$ matrices $G_i(x)$, the $n \times q$ matrices $W_i(x)$ are sufficiently smooth on $\mathbb{R}^n$ and, without loss of generality, that the origin is the nominal equilibrium point of each mode, i.e. $f_i(0) = 0$ for all $i \in \mathcal{I}$. We also assume that the state, x, does not jump at the

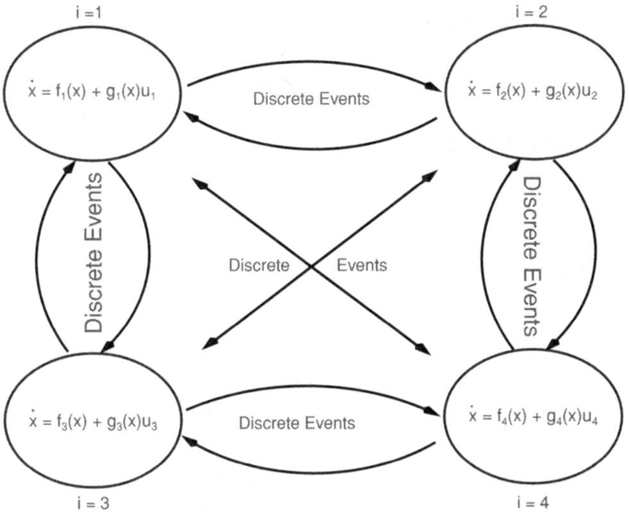

Fig. 7.1. A multi-modal representation of a hybrid process, involving a finite family of continuous subsystems and discrete events governing the transitions between them

switching instants, i.e. the solution, $x(\cdot)$, is everywhere continuous. Note that changes in the discrete state $\sigma(t)$ (i.e., transitions between the continuous dynamical modes) may, in general, be a function of time, state, or both. When changes in $\sigma(t)$ depend only on inherent process characteristics, the switching is referred to as autonomous. However, when $\sigma(t)$ is chosen by some higher process such as a controller or human operator, the switching is referred to as controlled. In this chapter, we focus on controlled switching where mode transitions are decided and executed by some higher-level supervisor. This class of systems arises naturally in the context of coordinated supervisory and feedback control of chemical process systems (see the illustrative example below and the simulation study in Sect. 7.3.5).

7.2.2 Stability Analysis Via Multiple Lyapunov Functions

For purely continuous-time nonlinear processes, Lyapunov techniques provide useful tools for stability analysis as well as nonlinear and robust controller design (e.g., see [78]). The basic conceptual idea behind any Lyapunov design is that of "energy shaping", where an appropriate "energy" function (called a Lyapunov function) is chosen for the system, and the controller is designed in a way that enforces the monotonic decay of this function along the trajectories of the closed-loop system (energy dissipation). Given the view of hybrid processes as a finite collection of continuous-time nonlinear processes with discrete events that govern the transition between them, it is quite intuitive to exploit Lyapunov tools to analyze stability of hybrid processes. In this direction, one of the main tools for analyzing stability is multiple Lyapunov functions (MLFs) (e.g., see [70]). In many respects, the MLF framework extends the classical energy shaping idea of Lyapunov analysis for continuous-time systems to switched systems, but provides, in addition, the tools necessary to account for the hybrid dynamics of such systems. Preparatory for its use later in robust controller design, we will briefly review in this section the main idea of MLF analysis. To this end, consider the switched process of (7.1), with $u_i = \theta_i \equiv 0$, $i = 1, \ldots, N$, and suppose that we can find a family of Lyapunov functions $\{V_i : i \in \mathcal{I}\}$ such that the value of V_i decreases on each interval when the i-th subsystem is active, i.e.,

$$V_i(x(t_{i'_k})) < V_i(x(t_{i_k})) \tag{7.2}$$

for all $i \in \mathcal{I}$, $k \in Z_+$. The key idea here is that, even if there exists such a Lyapunov function for each subsystem, f_i, individually (i.e., each mode is stable), restrictions must be placed on the switching logic to guarantee stability of the overall switched system. The reason is that during the time periods when a particular mode is inactive, its energy might be adversely affected by the evolution of the active mode such that at the next time that the inactive mode is activated, its energy already exceeds the level it had attained during its last period of activity. When this happens, the overall energy of the system may keep increasing indefinitely, as the process keeps switching in and out between the various modes, thus leading to instability. In fact, it is easy to construct examples of globally asymptotically stable systems and a switching rule that sends all trajectories to infinity (see [39] for some classical examples). There are multiple ways of guarding against such instability due to switching. One possibility is to require, in addition to (7.2), that for every $i \in \mathcal{I}$, the value of V_i at the beginning of each interval on which the i-th mode is active exceed (or, at least, be equal to) the value at the beginning of the next such interval (see Fig. 7.2); more precisely:

$$V_{\sigma(t_{i_{k+1}})}(x(t_{i_{k+1}})) \leq V_{\sigma(t_{i_k})}(x(t_{i_k})) \tag{7.3}$$

where $\sigma(t_{i_{k+1}}) = \sigma(t_{i_k}) = i$. This guarantees that the switched system is Lyapunov stable (see [70] and the references therein for alternative switching rules). In Theorem 7.1 below, a stronger switching condition than the one given in (7.3) will be invoked to enforce asymptotic stability in the overall switched closed-loop system (i.e., to enforce both Lyapunov stability and asymptotic convergence to the origin).

7.2.3 Illustrative Example

In this section, we introduce an example of a hybrid nonlinear chemical process with model uncertainty and actuator constraints that will be used throughout the chapter to illustrate the implementation of the hybrid control strategy. To this end, consider a continuous stirred tank reactor where an irreversible first-order exothermic reaction of the form $A \xrightarrow{k} B$ takes place. As shown in Fig.

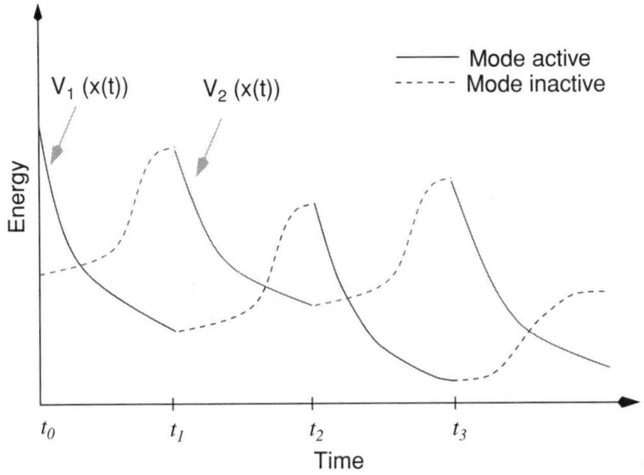

Fig. 7.2. Temporal evolution of Multiple Lyapunov Functions for an asymptotically stable switched system consisting of twomodes

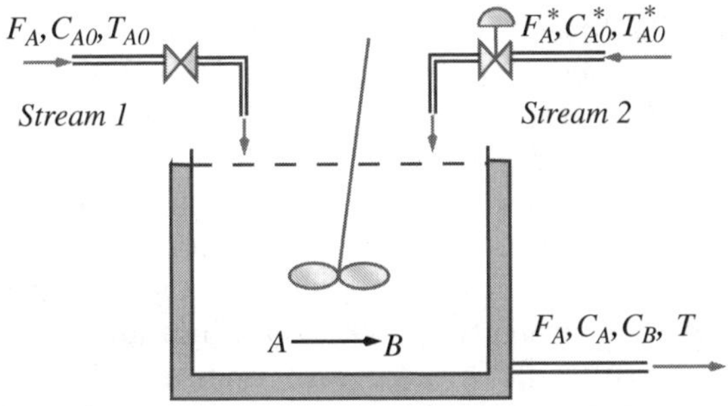

Fig. 7.3. A switched non-isothermal continuous stirred tank reactor

7.3, the reactor has two inlet streams; the first continuously feeds pure A at flow rate F, concentration C_{A0} and temperature T_{A0}, while the second has a control valve that can be turned on or off depending on operational requirements. When the control valve is open, the second stream feeds pure A at flow rate F^*, concentration C_{A0}^* and temperature T_{A0}^*. Under standard modeling assumptions, the mathematical model for the process takes the form:

$$\begin{aligned} \frac{dC_A}{dt} &= \frac{F}{V}(C_{A0} - C_A) + \sigma(t)\frac{F^*}{V}(C_{A0}^* - C_A) - k_0 \exp\left(\frac{-E}{RT}\right) C_A \\ \frac{dT}{dt} &= \frac{F}{V}(T_{A0} - T) + \sigma(t)\frac{F^*}{V}(T_{A0}^* - T) + \frac{(-\Delta H_r)}{\rho c_p} k_0 \exp\left(\frac{-E}{RT}\right) C_A \\ &\quad + \frac{Q}{\rho c_p V} \end{aligned} \tag{7.4}$$

Table 7.1. Process parameters and steady-state values for the reactor of (7.4)

V	$= 0.1$	m^3
R	$= 8.314$	$kJ/kmol.K$
C_{A0}	$= 1.0$	$kmol/m^3$
T_{A0s}	$= 310.0$	K
ΔH_{nom}	$= -400$	$kJ/kmol$
k_0	$= 2.0 \times 10^7$	s^{-1}
E	$= 8.314 \times 10^4$	$kJ/kmol$
c_p	$= 0.002$	$kJ/kg.K$
ρ	$= 1000.0$	kg/m^3
F	$= 2.77 \times 10^{-5}$	m^3/s
$C_{As}(\sigma = 0)$	$= 0.577$	$kmol/m^3$
$T_s(\sigma = 0)$	$= 395.3$	K
F^*	$= 5.56 \times 10^{-5}$	m^3/s
C^*_{A0}	$= 2.0$	$kmol/m^3$
T^*_{A0s}	$= 350.0$	K

where C_A denotes the concentration of A, T denotes the reactor temperature, Q denotes the rate of heat input/removal from the reactor, V denotes the reactor volume, k_0, E, ΔH denote the pre-exponential constant, the activation energy, and the enthalpy of the reaction, c_p and ρ, denote the heat capacity and density of the fluid in the reactor. The process parameters and steady-state values are given in Table 7.1. $\sigma(t)$ is a discrete control variable that takes a value of zero when the control valve, on the second inlet stream, is closed and a value of one when the valve is open. Initially, we assume that the valve is closed (i.e., $\sigma(0) = 0$). During reactor operation, however, it is desired to open this valve and feed through additional reactant material through the second inlet stream (i.e., $\sigma = 1$) in order to enhance the product concentration leaving the reactor.

The above requirement gives rise to two distinct modes of reactor operation, between which switching is desired. These modes correspond to the off($\sigma = 0$)/on($\sigma = 1$) conditions of the control valve on the second inlet stream. Since the initial operating mode ($\sigma = 0$) has an open-loop unstable steady-state that corresponds to $T_s = 395.3$ K, our control objective will be to stabilize the reactor temperature at this point by manipulating the rate of heat input. However, since switching to the second mode ($\sigma = 1$) at some later point in time can potentially destabilize the process, our switching objective will be to carry out the transition between the two modes at the earliest time which does not jeopardize process stability. The control and switching objectives are to be accomplished in the presence of: (1) hard constraints on the manipulated input, $|Q| \leq 80$ KJ/hr, (2) time-varying external disturbances in the feed temperature of both inlet streams, and (3) time-varying parametric uncertainty in the enthalpy of reaction. Note that the disturbances in the feed temperature of the second inlet stream take effect only after switching and, therefore, impact only the second mode, while the disturbances in the first stream's temperature and the parametric uncertainty in the enthalpy influence both modes.

7.3 Coordinating Feedback and Switching for Robust Hybrid Control

7.3.1 Problem Formulation and Solution Overview

Consider the switched nonlinear process of (7.1) where, for each $i \in \mathcal{I}$, a robust control Lyapunov function, V_i, is available, the vector of manipulated inputs, u_i, is constrained by $\|u_i\| \leq u_i^{max}$, and the vector of uncertain variables is bounded by $\|\theta_i\| \leq \theta_{bi}$, where the notation $\|\cdot\|$ denotes the Euclidean norm of a vector. The bounds that capture the size of the uncertainty can be arbitrarily large. Two control problems will be considered. In the first problem, the uncertain variables are assumed to be vanishing, in the sense that $W_i(0)\theta_i = 0$ for any $\theta_i \in \Theta$ (note that this does not require the variable, θ_i, itself to vanish in time). Under this assumption, the origin – which is an equilibrium point for the nominal modes of the hybrid process – is an equilibrium point for the uncertain modes as well. For this case, and given that switching is controlled by a higher-level supervisor, the problem is how to coordinate switching between the constituent modes, and their respective controllers, in a way that respects the constraints and guarantees asymptotic stability of the overall closed-loop system in the presence of uncertainty. To address the problem, we formulate the following objectives. The first is to synthesize, using a family of Lyapunov functions, a family of N bounded robust nonlinear continuous feedback control laws of the general form:

$$u_i = -k_i(V_i, u_i^{max}, \theta_{bi})(L_{G_i}V_i)^T, \quad i = 1, \ldots, N \tag{7.5}$$

that: (1) enforce robust asymptotic stability for their respective closed-loop subsystems, and (2) provide, for each mode, an explicit characterization of the set of admissible initial conditions starting from where a given mode is guaranteed to be stable in the presence of model uncertainty and input constraints. The scalar gain, $k_i(\cdot)$, of the L_GV controller in (7.5) is to be designed so that $\|u_i\| \leq u_i^{max}$ and the energy of the i-th mode – as captured by V_i – is monotonically decreasing whenever that mode is active. The second objective is to construct a set of robust switching laws that supply the supervisor with the set of switching times that guarantee stability of the constrained uncertain switched closed-loop system, which in turn determines the time-course of the discrete state, $\sigma(t)$.

In the second control problem, the uncertain variables are assumed to be non-vanishing, in the sense that $W_i(0)\theta_i \neq 0$ for all $i \in \mathcal{I}$. In this case, the origin is no longer an equilibrium point of the uncertain modes or the overall hybrid process and, therefore, the objective is to coordinate feedback controller synthesis and switching in a way that guarantees boundedness of the states of the hybrid process, with an arbitrary degree of attenuation of the effect of uncertainty.

Having formulated the above robust hybrid control problems, we proceed in Sects. 7.3.2–7.3.3 to present their solutions. The first result, given in Theorem 7.1 below, addresses the problem of vanishing uncertainty, while the second result, given in Theorem 7.11, deals with the problem of non-vanishing uncertainty. For a clear presentation of the main ideas, only the state feedback control problem is considered below. Results on the output feedback control problem for switched systems using MLFs can be found in [90].

7.3.2 Hybrid Control Strategy Under Vanishing Uncertainty

Theorem 7.1 below summarizes the proposed robust hybrid control strategy for the case when the uncertainty does not affect the nominal equilibrium point of the hybrid process. Provided in this theorem are the formulae of the family of bounded robust feedback continuous controllers, together with the appropriate switching rules used by the supervisor to govern the transitions between the various closed-loop modes in a way that guarantees the desired properties in the constrained uncertain hybrid closed-loop system. The proof of this theorem is given in Appendix E.

Theorem 7.1. *Consider the switched uncertain nonlinear process of (7.1), where* $W_i(0)\theta_i = 0$ *for all* $i \in \mathcal{I}$, *under the following family of bounded nonlinear feedback controllers:*

$$u_i = -k_i(V_i, u_i^{max}, \theta_{bi}, \chi_i, \phi_i)(L_{G_i}V_i)^T, \; i = 1, \ldots, N \tag{7.6}$$

where

$$k_i(\cdot) = \begin{cases} \dfrac{L_{f_i}^* V_i + \sqrt{(L_{f_i}^{**} V_i)^2 + (u_i^{max}\|(L_{G_i}V_i)^T\|)^4}}{(\|(L_{G_i}V_i)^T\|)^2[1 + \sqrt{1 + (u_i^{max}\|(L_{G_i}V_i)^T\|)^2}]} & , \|(L_{G_i}V_i)^T\| \neq 0 \\ 0 & , \|(L_{G_i}V_i)^T\| = 0 \end{cases} \tag{7.7}$$

$$\begin{aligned} L_{f_i}^* V_i &= L_{f_i}V_i + \left(\rho_i\|x\| + \chi_i\|(L_{W_i}V_i)^T\|\theta_{bi}\right)\left(\frac{\|x\|}{\|x\| + \phi_i}\right) \\ L_{f_i}^{**} V_i &= L_{f_i}V_i + \rho_i\|x\| + \chi_i\|(L_{W_i}V_i)^T\|\theta_{bi} \end{aligned} \tag{7.8}$$

V_i *is a robust control Lyapunov function for the i-th subsystem and* ρ_i, χ_i, ϕ_i *are tunable parameters that satisfy* $\rho_i > 0$, $\chi_i > 1$ *and* $\phi_i > 0$. *Let*

$\Omega_i^*(u_i^{max}, \theta_{bi})$ be the largest invariant set embedded within the region described by the inequality:

$$L_{f_i}V_i + \rho_i\|x\| + \chi_i\|(L_{W_i}V_i)^T\|\theta_{bi} \leq u_i^{max}\|(L_{G_i}V_i)^T\| \tag{7.9}$$

and assume, without loss of generality, that $x(0) := x_0 \in \Omega_i^*(u_i^{max}, \theta_{bi})$ for some $i \in \mathcal{I}$. If, at any given time T such that:

$$x(T) \in \Omega_j^*(u_j^{max}, \theta_{bj}) \tag{7.10}$$

$$V_j(x(T)) < V_j(x(t_{j'})) \tag{7.11}$$

for some $j \in \mathcal{I}$, $j \neq i$, where $t_{j'}$ is the time when the j-th subsystem was last switched out, i.e. $\sigma(t_{j'}^+) \neq \sigma(t_{j'}^-) = j$, we set $\sigma(T^+) = j$, then there exists a positive real number ϕ_j^* such that for any $\phi_j \leq \phi_j^*$, the origin of the switched closed-loop system is asymptotically stable.

Remark 7.2. The bounded robust feedback controllers given in (7.6–7.8) are synthesized, using multiple Lyapunov functions, by reshaping the scalar nonlinear gain of the L_GV controller proposed originally in [177] (see also Chap. 4), in order to account for the effect of the uncertain variables. As a result, the control action now depends explicitly on both the magnitude of the input constraints, u_i^{max}, and the size of the uncertainty, θ_{bi}. Note that only information about the size of the uncertainty (and not the uncertainty itself) is needed for controller design and that this size can be arbitrarily large. Each controller in (7.6–7.8) is a smooth function of the state away from the origin. It is also continuous at the origin whenever the corresponding control Lyapunov function satisfies the small control property (see [97,177] for details on this issue).

Remark 7.3. Each controller in (7.6–7.8) possesses two tuning parameters, ϕ_i and χ_i, responsible for enforcing robust stability and achieving the desired degree of attenuation of the effect of uncertainty on each of the constituent modes of the hybrid process. A significant degree of attenuation can be achieved by selecting the parameter ϕ_i to be sufficiently small and/or choosing the parameter χ_i to be sufficiently large (see Step 1 in the Proof of Theorem 7.1). Another important feature of the uncertainty compensator (i.e., the term $\chi_i\|(L_{W_i}V_i)^T\|\theta_{bi}$) used in designing the controller gain of (7.7–7.8) is the presence of a scaling function of the form $\dfrac{\|x\|}{\|x\| + \phi_i}$ that multiplies the compensator. Since ϕ_i is a small positive number, the scaling function approaches a value of 1 when $\|x\|$ is large (far from the equilibrium point) and a value of zero when $\|x\|$ is small (close to the equilibrium point). This allows us, as we get closer and closer to the equilibrium point, to use smaller and smaller control effort to cancel the uncertainties. The full weight of the uncertainty compensator is used only when the state is far from the equilibrium point.

Remark 7.4. The use of bounded nonlinear controllers of the form of (7.6–7.8) to robustly stabilize the constituent modes is motivated by the fact that this class of controllers provides an explicit characterization of the limitations imposed by uncertainty and constraints on the region of closed-loop stability. Specifically, each controller in (7.6–7.8) provides an explicit characterization of the set of admissible initial conditions starting from where robust closed-loop stability of the corresponding subsystem is guaranteed with the available control action. This characterization can be obtained from the set of inequalities given in (7.9). For each mode, the corresponding inequality describes a closed region in the state-space (henceforth denoted by $\Phi_i(u_i^{max}, \theta_{bi})$) where the corresponding control law satisfies the constraints and the associated Lyapunov function, V_i, decreases monotonically (see Step 1 in the Proof of Theorem 7.1). Owing to the presence of uncertainty, the size of Φ_i now depends on the size of the uncertainty, in addition to the magnitude of input constraints. The larger the uncertainty and/or the tighter the constraints, the smaller Φ_i is in size. It is important to note that even though a trajectory starting in Φ_i will move from one Lyapunov surface to an inner Lyapunov surface with lower energy (because $\dot{V}_i < 0$), there is no guarantee that the trajectory will remain forever in Φ_i since it is not necessarily a region of invariance. Once the trajectory leaves Φ_i, however, there is no guarantee that $\dot{V}_i < 0$. To guarantee that $\dot{V}_i$ remains negative for all times during which the i-th mode is active, we compute the largest invariant set, $\Omega_i^*(u_i^{max}, \theta_{bi})$, within $\Phi_i(u_i^{max}, \theta_{bi})$ (see [148] for details on how to construct these sets). This set, which is also parameterized by the constraints and the uncertainty bounds, represents an estimate of the stability region associated with each mode. Finally, note that in the absence of any plant-model mismatch (i.e., when $\theta_{bi} \equiv 0$), the controllers of (7.6–7.8), together with the expressions for the stability regions, Ω_i^*, reduce to those developed in [80] under nominal conditions.

Remark 7.5. Each of the inequalities in (7.9) captures the classical tradeoff between stability and performance. To see this, note that the size of the region of guaranteed closed-loop stability, obtained from (7.9), can be enlarged by using small values for the controller tuning parameters χ_i and ρ_i. This enlargement of the stability region, however, comes at the expense of the controller's robust performance, since large values for χ_i and ρ_i are typically required to achieve a significant degree of attenuation of the effect of disturbances and plant-model mismatch on the closed-loop system. Therefore, in selecting the controller tuning parameters, one must strike a balance between the need to stabilize the process from a given initial condition and the requirement of achieving a satisfactory robust performance.

Remark 7.6. The two switching laws of (7.10–7.11) determine, implicitly, the times when switching from mode i to mode j is permissible. The first rule tracks the temporal evolution of the continuous state, x, and requires that, at

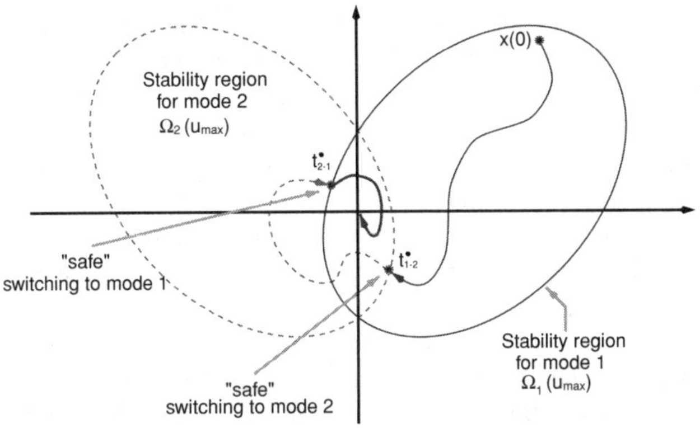

Fig. 7.4. Implementation of the switching law based on monitoring the evolution of the closed-loop trajectory with respect to stability regions

the desired time for switching, the continuous state be within the stability region of the target mode, $\Omega_j^*(u_j^{max}, \theta_{bj})$. A pictorial representation of this idea is shown in Fig. 7.4. This requirement ensures that, once the target mode and its corresponding controller are engaged, the corresponding Lyapunov function continues to decay for as long as the mode remains active. Note that this condition must be enforced at every time that the supervisor considers switching between modes. In contrast, the second switching rule of (7.11) is checked only if the target mode has been activated (at least once) in the past. In this case, (7.11) allows reactivation of mode j provided that its energy at the current "switch in" is less than its energy at the last "switch out". This condition guarantees that, whenever a given mode is activated, the closed-loop state is closer to the origin than it was when the same mode was last activated. Note that if each of the N modes is activated only once during the course of operation (i.e., we never switch back to a mode previously activated), the second condition is automatically satisfied. Furthermore, for the case when only a finite number of switches (over the infinite time-interval) is considered, this condition can be relaxed by allowing switching to take place even when the value of V_j, at the desired switching time, is larger than that when mode j was last switched in, as long as the increase is finite. The reason owes to the fact that these finite increases in V_j (resulting from switching back to mode j) will be overcome when the process eventually settles in the "final" mode whose controller, in turn, forces its Lyapunov function to continue to decay as time tends to infinity, thus asymptotically stabilizing the overall hybrid process.

Remark 7.7. Even though the switching conditions of (7.10–7.11) require knowledge of the temporal evolution of the closed-loop state, $x(t)$, the a priori knowledge of the solution of the constrained closed-loop nonlinear system (which is difficult to obtain in general) is not needed for the practical implementation of the proposed approach. Instead, the supervisor can monitor (on-line) how x evolves in time to determine if and when the switching conditions are satisfied. If the conditions are satisfied, then we conclude that it is "safe" to switch from the current mode/controller combination to the one for which the conditions hold. Otherwise, the current mode is kept active. Note that, absent any failures in the control system, maintaining closed-loop stability does not require switching since the closed-loop system is always initialized within the stability region of at least one of the constituent modes and, therefore, absent switching, the closed-loop trajectory will simply remain in this invariant set and stabilize at the desired equilibrium point. In many practical situations, however, changes in operational conditions and requirements typically motivate switching between various process modes. Under these conditions, the result of Theorem 7.1 provides a strategy for carrying out mode transitions without jeopardizing the stability of the overall process (see the simulation example in Sect. 7.3.5).

Remark 7.8. In many practical situations, the ability of the control system to deal with failure situations requires consideration of multiple control configurations and switching between them to preserve closed-loop stability in the event that the active control configuration fails. For such cases, the switching rules proposed in Theorem 7.1 (based on the stability regions) can be used to explicitly identify the appropriate fall-back control actuator configuration that should be activated to preserve closed-loop stability (see Chap. 8 for how this notion can be applied for the design of fault-tolerant control systems). Clearly, if failure occurs when the state is outside the stability regions of all the available configurations, then closed-loop stability cannot be preserved because of the fundamental limitations imposed by the constraints on the stability regions as well as how many backup configurations are available. However, our approach in this case provides a useful way for analyzing when the control system can or cannot tolerate failures. For example, by analyzing the overlap of the stability regions of the given configurations, one can decide the time periods during which the control system (under a given configuration) cannot tolerate failure (which are the times that the trajectory spends outside the stability region of the other configurations). By relaxing the constraints (i.e., enlarging the stability regions) and/or increasing the available control configurations (this is ultimately limited by system design considerations) one can reduce the possibility of failures taking place outside of the stability regions of all configurations.

Remark 7.9. Note that it is possible for more than one subsystem, j, to satisfy the switching rules given in (7.10–7.11). This occurs when the process state lies within the intersection of several stability regions. In this case, Theorem 7.1 guarantees only that a transition from the current mode to any of these modes is safe, but does not suggest which one to choose since they all guarantee stability. The decision to choose a particular mode to activate is typically made by the supervisor based on the particular operational requirements of the process.

Remark 7.10. Referring to the practical applications of Theorem 7.1, one must initially identify the constituent modes of the hybrid process. A Lyapunov function is then constructed for each mode to synthesize, via (7.6–7.8), a bounded robust controller and construct, with the aid of (7.9), the region of closed-loop stability associated with each mode. Implementation of the control strategy then proceeds by initializing the process within the stability region of the desired initial mode of operation and implementing the corresponding robust controller. Then, the switching laws of (7.10–7.11) are checked on-line, by the supervisor, to determine if it is possible to switch operation to a particular mode at some time. If the conditions are satisfied, then a transition to that mode (and its controller) is executed. If the conditions are not satisfied, then the current operating modes is kept active. A summary of the proposed robust hybrid control methodology is shown schematically in Fig. 7.5.

7.3.3 Hybrid Control Strategy Under Non-Vanishing Uncertainty

In this section, we address the problem of non-vanishing uncertainty, where the uncertainty changes the nominal equilibrium point of the hybrid process. Theorem 7.11 below summarizes the proposed hybrid control strategy for this case and states precisely the resulting closed-loop properties. The proof of this theorem is given in Appendix E.

Theorem 7.11. *Consider the switched uncertain nonlinear process of (7.1), where $W_i(0)\theta_i \neq 0$ for all $i \in \mathcal{I}$, under the family of controllers given in (7.6–7.8). Assume, without loss of generality, that $x(0) \in \Omega_i^*(u_i^{max}, \theta_{bi})$ for some $i \in \mathcal{I}$ and that, for any given $T > 0$, $\sigma(T^+) = j$ only if the conditions of (7.10–7.11) hold for some $j \in \mathcal{I}$, $j \neq i$. Then, given any arbitrarily small real number, $d > 0$, there exist a set of positive real numbers $\{\epsilon_1^*, \dots, \epsilon_N^*\}$ such that if $\epsilon_i \leq \epsilon_i^*$, $i \in \mathcal{I}$, the trajectories of the switched closed-loop system are bounded and satisfy* $\limsup_{t\to\infty}\|x(t)\| \leq d$, *where $\epsilon_i = \phi_i/(\chi_i - 1)$.*

Remark 7.12. Owing to the non-vanishing nature of the uncertainty influencing each mode, asymptotic convergence to the origin is not possible via continuous feedback for any of the constituent modes of the hybrid process. However, in lieu of asymptotic stability, one can show (see the Proof of Theorem 7.11 in Appendix E) that, for every mode i by itself (without switching), the corresponding controller guarantees convergence of the closed-loop trajectory, in finite time, to a small neighborhood of the origin (called a residual or terminal set) such that the trajectory, once inside this set, cannot escape (see Chap. 4 for further details). The size of this set can be made arbitrarily small by choosing the controller tuning parameter, ϕ_i, to be sufficiently

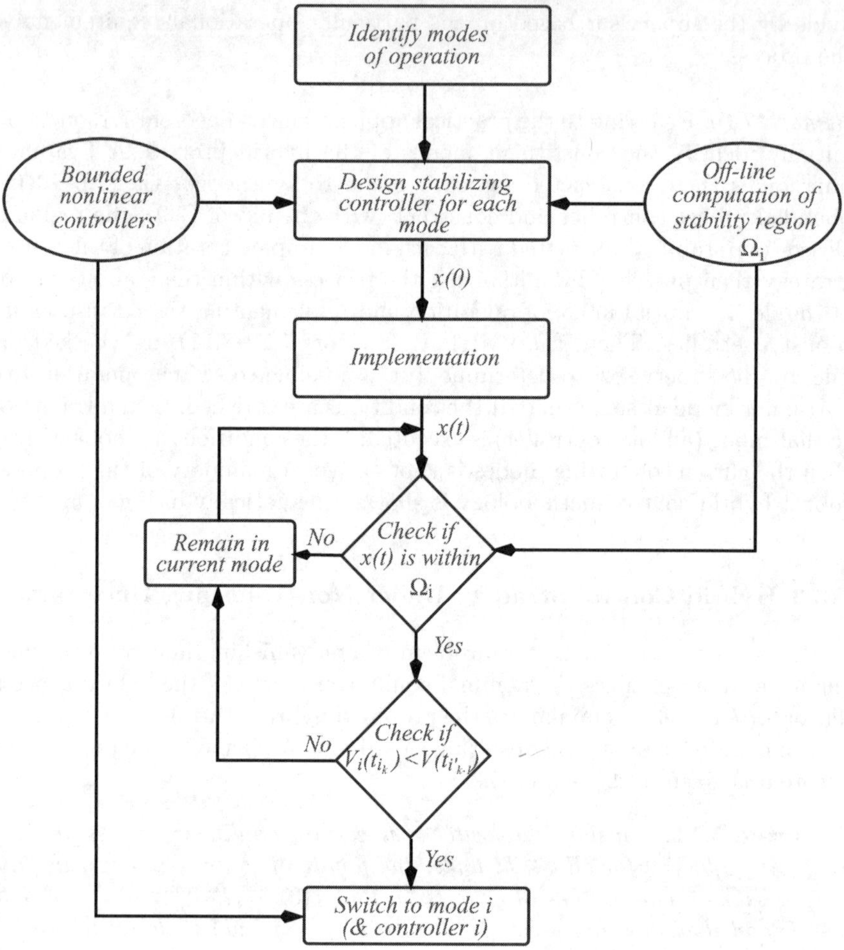

Fig. 7.5. Summary of the proposed hybrid control strategy based on coordinating feedback and switching

small and/or selecting the tuning parameter, χ_i, to be sufficiently large. In this manner, one can achieve an arbitrary degree of robust attenuation of the effect of uncertainty on each mode of the closed-loop system.

Remark 7.13. Boundedness of the state of the constituent modes does not, in and of itself, imply boundedness of the state of the overall hybrid process. Boundedness of the switched closed-loop trajectory can be ensured using the switching rules given in (7.10–7.11), which automatically impose constraints on which mode can be engaged at any given time. These constraints, similar to the case of the vanishing uncertainty, guarantee that whenever a mode is activated (or reactivated), not only is its energy less than what it was before (7.11) but also that this energy continues to decay for as long as the mode remains active (7.10). The only difference in the case of non-vanishing uncertainty is that these rules need only be implemented, by the supervisor, when the state lies outside the residual set. To understand the rationale for this observation, we first observe that since the size of the residual set of each mode can be tuned (by appropriately adjusting ϕ_i and χ_i), a small common residual set can be chosen (for all the modes) that is completely contained within the intersection of the various stability regions. Then, starting from any admissible initial condition, implementation of the switching logic outside this residual set ensures that, for every mode, the closed-loop trajectory moves closer and closer to that set, as we switch in and out of that mode. Since the number of process modes is finite, and only a finite number of switches is allowed over any finite time-interval, then at least one of constituent modes will converge, in finite time, to the residual set. From that time onward, the closed-loop trajectory stays within the common residual set for all times, regardless of any further mode switchings (see Step 2 in the Proof of Theorem 7.11 for the mathematical details).

Remark 7.14. Theorems 7.1 and 7.11 consider, respectively, the cases when the uncertainties affecting the various modes are either all vanishing or all non-vanishing. For the general case when the uncertainty is vanishing for some modes and non-vanishing for others, one can establish only boundedness of the closed-loop trajectory. If, however, only a finite number of switches is allowed over the infinite time-interval, then one can establish asymptotic stability if the uncertainty influencing the final mode – where the hybrid process eventually settles – is vanishing. If such uncertainty, on the other hand, is non-vanishing, then the closed-loop trajectory will instead reach and enter, in finite time, the residual set of the final mode without ever leaving again.

7.3.4 Application to Input/Output Linearizable Processes

An important class of nonlinear processes that has been studied extensively within process control is that of input/output feedback linearizable processes. This class arises frequently in practical problems where the objective is to force the controlled output to follow some reference-input trajectory (rather than stabilize the full state at some nominal equilibrium point). In this section, we illustrate how the coordinated robust feedback and switching methodology proposed in this chapter can be applied when the individual modes of the hybrid process are input/output linearizable. For simplicity, we limit our attention to the single-input single-output case with vanishing uncertainty. Consider the hybrid process:

$$\begin{aligned}\dot{x}(t) &= f_{\sigma(t)}(x(t)) + g_{\sigma(t)}(x(t))u_{\sigma(t)} + w_{\sigma(t)}(x(t))\theta_{\sigma(t)} \\ y &= h(x) \\ \sigma(t) &\in \mathcal{I} = \{1,\ldots,N\}\end{aligned} \tag{7.12}$$

where $y \in \mathbb{R}$ is the controlled output and $h(x)$ is a sufficiently smooth scalar function. Suppose that, for all $i \in \mathcal{I}$, there exists an integer r (this assumption is made only to simplify notation and can be readily relaxed to allow a different relative degree, r_i, for each mode) and a set of coordinates:

$$\begin{bmatrix}\zeta \\ \eta\end{bmatrix} = \begin{bmatrix}\zeta_1 \\ \zeta_2 \\ \vdots \\ \zeta_r \\ \eta_1 \\ \vdots \\ \eta_{n-r}\end{bmatrix} = \mathcal{X}(x) = \begin{bmatrix}h(x) \\ L_{f_i}h(x) \\ \vdots \\ L_{f_i}^{r-1}h(x) \\ T_{1,i}(x) \\ \vdots \\ T_{n-r,i}(x)\end{bmatrix} \tag{7.13}$$

where $T_1(x), \ldots, T_{n-r}(x)$ are nonlinear scalar functions of x, such that the system of (7.12) takes the form:

$$\begin{aligned}\dot{\zeta}_1 &= \zeta_2 \\ &\vdots \\ \dot{\zeta}_{r-1} &= \zeta_r \\ \dot{\zeta}_r &= L_{f_i}^r h(\mathcal{X}^{-1}(\zeta,\eta)) + L_{g_i}L_{f_i}^{r-1}h(\mathcal{X}^{-1}(\zeta,\eta))u_i + L_{w_i}L_{f_i}^{r-1}h(\mathcal{X}^{-1}(\zeta,\eta))\theta_i \\ \dot{\eta}_1 &= \Psi_{1,i}(\zeta,\eta) \\ &\vdots \\ \dot{\eta}_{n-r} &= \Psi_{n-r,i}(\zeta,\eta) \\ y &= \zeta_1)\end{aligned} \tag{7.14}$$

where $L_{g_i}L_{f_i}^{r-1}h(x) \neq 0$ for all $x \in \mathbb{R}^n$, $i \in \mathcal{I}$. Under the assumption that the η-subsystem is input-to-state stable (ISS) with respect to ζ for each $i \in \mathcal{I}$, the controller synthesis task for each mode can be addressed on the basis of the partially linear ζ-subsystem. To this end, upon introducing the notation $e_k = \zeta_k - v^{(k-1)}$, $e = [e_1 \; e_2 \; \cdots \; e_r]^T$, $\bar{v} = [v \; v^{(1)} \; \cdots \; v^{(r-1)}]^T$, where $v^{(k)}$ is the k-th time derivative of the reference input, v, which is assumed to be a smooth function of time, the ζ-subsystem of (7.14) can be further transformed into the following more compact form:

$$\dot{e} = \bar{f}_i(e,\eta,\bar{v}) + \bar{g}_i(e,\eta,\bar{v})u_i + \bar{w}_i(e,\eta,\bar{v})\theta_i, \; i = 1,\ldots,N \tag{7.15}$$

where $\bar{f}_i$, $\bar{g}_i$, $\bar{w}_i$ are $r \times 1$ vector functions given by $\bar{f}_i(e,\eta,\bar{v}) = Ae + bL_{f_i}^r h(\mathcal{X}^{-1}(e,\eta,\bar{v}))$ and $\bar{g}_i(e,\eta,\bar{v}) = bL_{g_i}L_{f_i}^{r-1}h(\mathcal{X}^{-1}(e,\eta,\bar{v}))$, $\bar{w}_i(e,\eta,\bar{v}) = bL_{w_i}L_{f_i}^{r-1}h(\mathcal{X}^{-1}(e,\eta,\bar{v}))$ are $r \times 1$ vector functions, and:

$$A = \begin{bmatrix} 0 & 1 & 0 & \cdots & 0 \\ 0 & 0 & 1 & \cdots & 0 \\ \vdots & & & & \vdots \\ 0 & 0 & 0 & \cdots & 1 \\ 0 & 0 & 0 & \cdots & 0 \end{bmatrix}, \quad b = \begin{bmatrix} 0 \\ 0 \\ \vdots \\ \\ 1 \end{bmatrix} \tag{7.16}$$

are an $r \times r$ matrix and $r \times 1$ vector, respectively. For systems of the form of (7.15–7.16), a simple choice for a robust control Lyapunov function is a quadratic function of the form $\bar{V}_i = e^T P_i e$ where the positive-definite matrix P_i is chosen to satisfy the following Riccati inequality:

$$A^T P_i + P_i A - P_i b b^T P_i < 0 \tag{7.17}$$

Using these quadratic functions, a bounded robust controller can be designed for each mode using (7.6–7.8) applied to the system of (7.15). Using a standard Lyapunov argument, it can then be shown that each controller robustly asymptotically stabilizes the e states in each mode. This result together with the ISS assumption on the η states can then be used to show, via a small gain argument, that the origin of the full closed-loop e-η interconnection, for each individual mode, is asymptotically stable.

Remark 7.15. Note that, since the objective here is output tracking, rather than full state stabilization, the Lyapunov functions used in designing the controllers, $\bar{V}_i$'s, are in general different from the Lyapunov functions, V_i's, used in implementing the switching rules. Owing to the ISS property of the η-subsystem of each mode, only a Lyapunov function for the e-subsystem, namely $\bar{V}_i$, is needed and used to design a controller that robustly stabilizes the full e-η interconnection for each mode. However, when implementing the switching rules (constructing the Ω_i^*'s and verifying (7.11)), we need to track the evolution of x (and hence the evolution of both e and η). Therefore, the Lyapunov functions used in verifying the switching conditions at any given time, V_i, are based on x. From the asymptotic stability of each mode, the existence of these Lyapunov functions is guaranteed by converse Lyapunov theorems (see Chap. 3 in [148] for further details). For systems with relative degree $r = n$, the choice $\bar{V}_i = V_i$ is sufficient.

7.3.5 Application to a Switched Chemical Reactor

In this section, we revisit the switched chemical reactor example, introduced earlier in (7.4), to illustrate, through computer simulations, the application of the proposed hybrid control strategy. Recall that the control objective is to stabilize the reactor temperature at the open-loop unstable steady-state, by manipulating the rate of heat input, while the switching objective is to carry out the transition between the two modes at the earliest time possible without jeopardizing process stability. The control and switching objectives are to be accomplished in the presence of hard constraints on the manipulated input ($|Q| \leq 80$ KJ/hr), time-varying external disturbances in the feed temperature of both inlet streams, and time-varying parametric uncertainty in the enthalpy of reaction. For the purpose of simulating the effect of uncertainty on the process output, we consider time-varying functions of the form $\theta(t) = \theta_b sin(4t)$, where the upper bounds on the feed temperature disturbances are taken to be 10 K for both streams, and the upper bound on the uncertainty in the enthalpy is taken to be 15% of the nominal value. We note that any other bounded and time-varying function can be used to simulate the effect of uncertainty. This choice does not affect the results since it is only the bounds on these functions that are needed for controller and switching law design.

To accommodate both the control and operational objectives, and since the uncertain variables considered are non-vanishing, we follow the strategy proposed in Theorem 7.11. Using two quadratic Lyapunov functions of the form $V_i = \frac{1}{2}c_i(T - T_s)^2$, where $c_i > 0$, we initially use (7.6–7.8) to synthesize two bounded robust controllers (one for each mode) that enforce robust closed-loop stability for their respective modes, and also achieve an arbitrary degree of attenuation of the effect of uncertainty on the reactor temperature. Then, with the aid of (7.9), we compute the region of guaranteed closed-loop stability associated with each mode, which will be used in implementing the necessary switching laws. The following tuning parameters were used for each controller: $c_1 = c_2 = 1$, $\phi_1 = \phi_2 = 0.01$, $\chi_1 = 2$, $\chi_2 = 1.1$, $\rho_1 = \rho_2 = 0.001$, to guarantee that the reactor temperature satisfies a relation of the form $\limsup_{t \to \infty} |T(t) - T_s| \leq 0.01$.

Several closed-loop simulations were performed to evaluate the proposed control strategy. In the first set of simulation runs, the reactor is operated in the first mode (control valve closed, $\sigma = 0$) for all time, with no switching. The feedback controller designed for this mode is consequently implemented to robustly stabilize the reactor temperature, starting from an admissible initial condition. Figure 7.6 (solid lines) depicts the resulting reactor temperature (controlled output) and rate of heat input (manipulated input) profiles. Included in the figure also are the corresponding open-loop profiles with no control (dashed lines). We observe that the controller successfully stabilizes the reactor temperature at the desired steady-state and simultaneously attenuates the effect of disturbances and model uncertainty on the reactor temperature.

In the second set of simulation runs, we seek to accommodate the operational requirement of increasing the product concentration by switching to the second mode (control valve open, $\sigma = 1$) at some point. Note that switching to the second mode (i.e., opening the valve) is accompanied by a switch to the second controller responsible for stabilizing this mode. In the absence of any explicit switching guidelines, suppose that the switching time is randomly set to be as early as $t = 12\ min$. The resulting temperature and heat input profiles in this case are shown in Fig. 7.7. It is clear from the figure that by

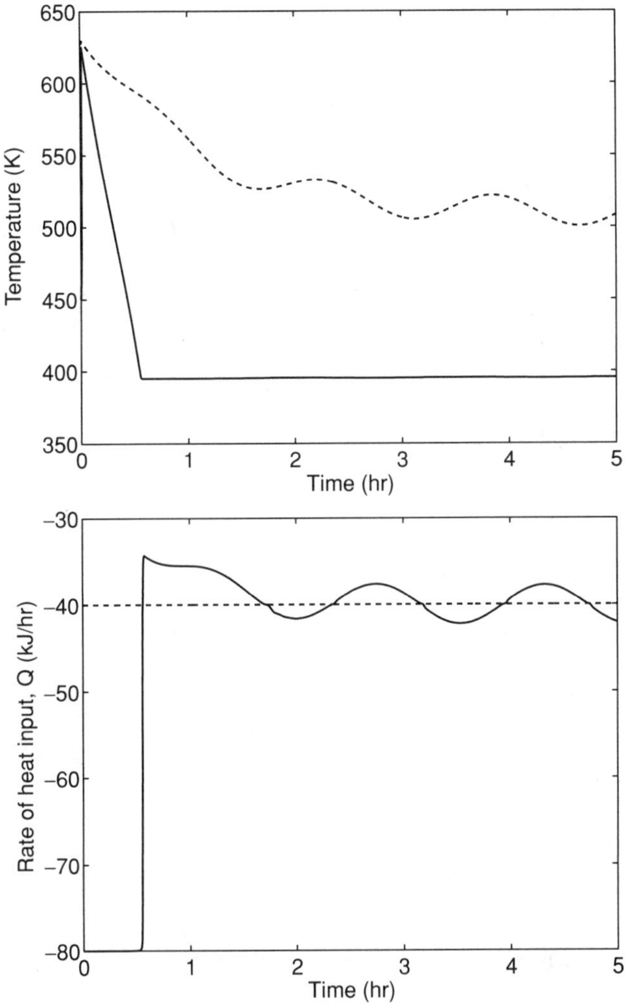

Fig. 7.6. Reactor temperature and rate of heat input profiles under the bounded robust controller of the first mode, i.e., control valve closed and $\sigma = 0$ (*solid lines*) and under open-loop conditions (*dashed lines*)

switching at this arbitrarily chosen time, the controller for the $\sigma = 1$ mode is unable to stabilize the reactor temperature at the desired steady-state nor attenuate the effect of uncertainty on the reactor temperature. The reason – which is reflected in the input profile – is that at this time, the process state lies outside the stability region of the $\sigma = 1$ mode and therefore, the available control action is insufficient to stabilize the temperature.

Now, instead of choosing the switching time arbitrarily, suppose that the nominal switching laws proposed in [77] are used. These laws were used to

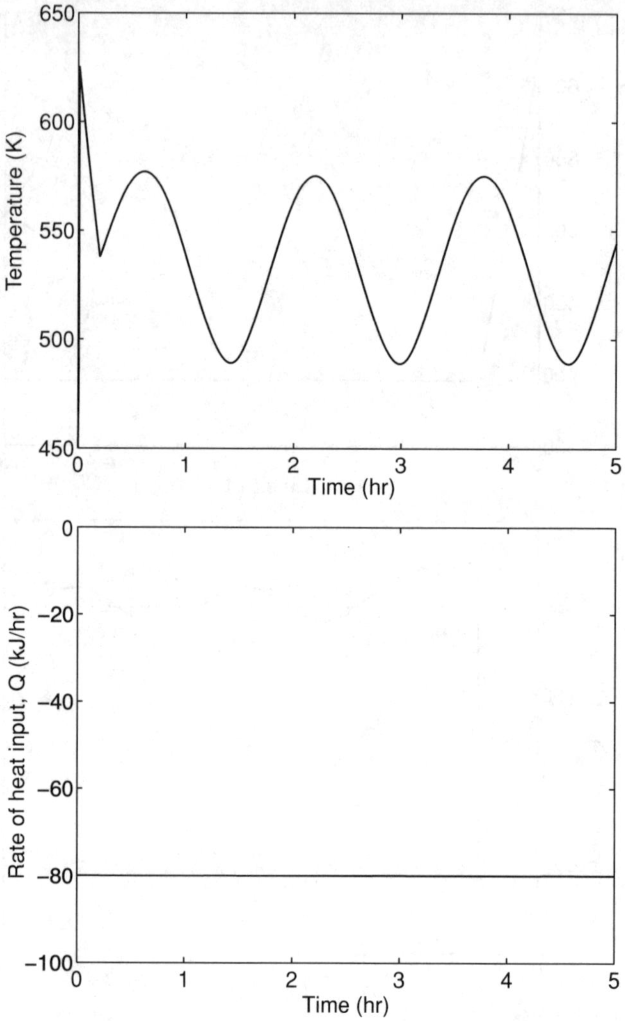

Fig. 7.7. Reactor temperature and rate of heat input profiles when the reactor is initially operated in the first mode (control valve closed, $\sigma = 0$) using the corresponding robust controller, then the control valve is opened ($\sigma = 1$) at $t = 12$ min and the robust controller for the second mode is activated in place of the first controller

address a similar switched reactor control problem under nominal conditions (without uncertainty). This scenario is considered here to study the effect of uncertainty on switching. To this end, consider first the case where no uncertainty is present (i.e., $\theta_i \equiv 0$) and initialize the closed-loop system within the first mode (control valve closed, $\sigma = 0$) at the same admissible initial condition, considered previously, using the first controller with $\rho_1 = \chi_1 = 0$.

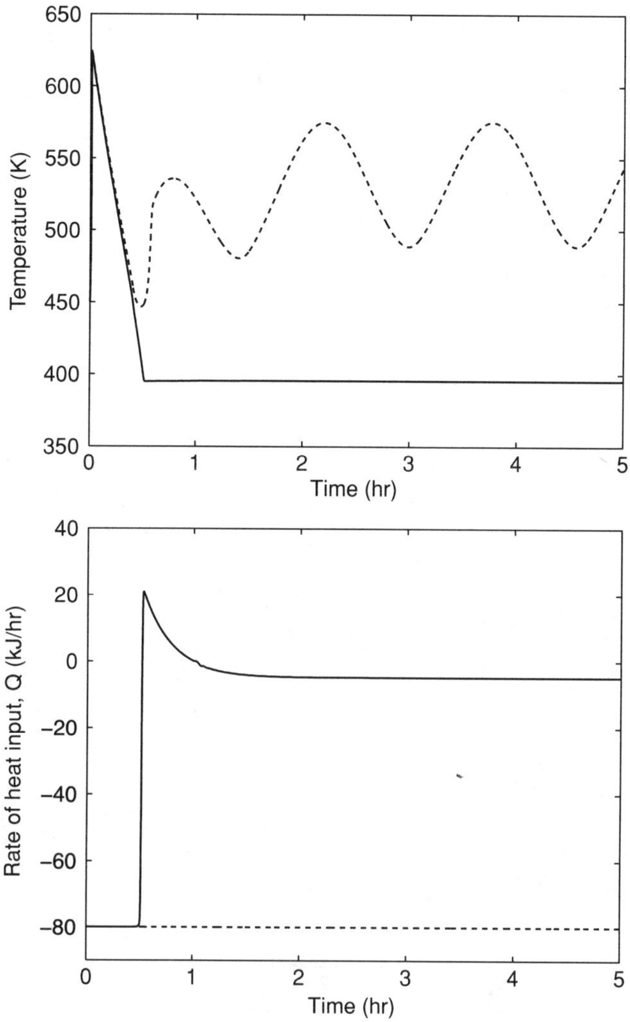

Fig. 7.8. Reactor temperature and rate of heat input profiles for the case when mode switching is carried out at $t = 24$ min with no uncertainty present (*solid lines*) and for the case when uncertainty is present and mode switching (using nominal controllers with $\rho_i = \chi_i = 0$) is carried out at $t = 24$ min (*dashed lines*)

By tracking the closed-loop trajectory in time, it is found that the process state enters the stability region of the second mode at $t = 24\ min$. Consequently, the mode transition (including a switch to the second controller with $\rho_2 = \chi_2 = 0$) is carried out at this time. The resulting controlled output and manipulated input profiles are depicted by the solid lines in Fig. 7.8 which show that the reactor temperature stabilizes at the desired steady-state in the absence of uncertainty.

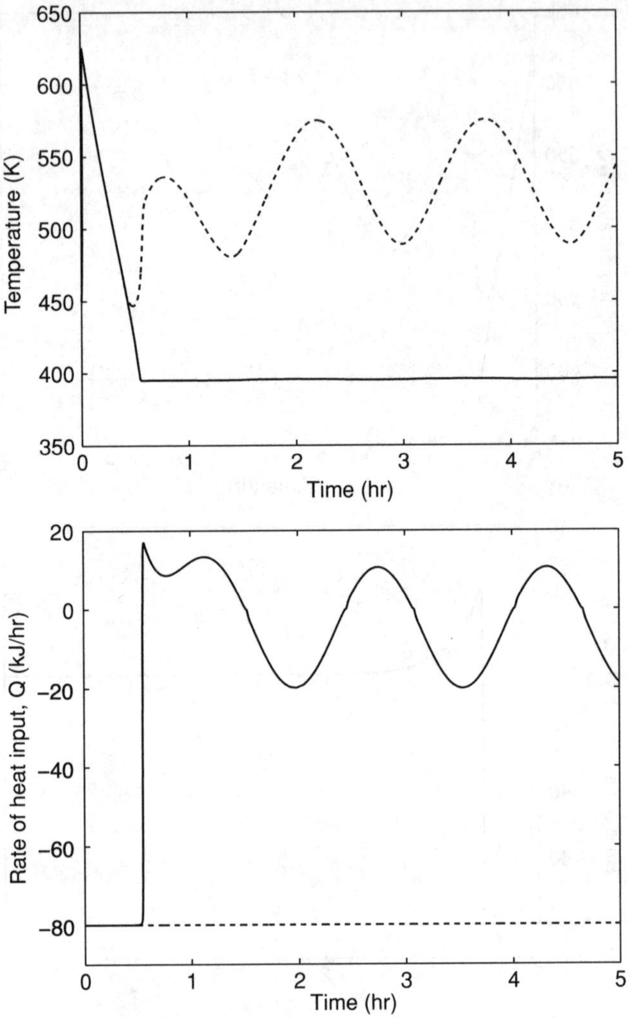

Fig. 7.9. Reactor temperature and rate of heat input profiles for the cases when uncertainty is present and mode switching (using the bounded robust controllers) is carried out at $t = 24$ min (*dashed lines*) and at $t = 30$ min (*solid lines*)

Suppose now that the "nominally-safe" switching time, $t = 24\ min$, is used when model uncertainty is present. The resulting temperature and heat input profiles are shown by the dashed lines in Figs. 7.8 and 7.9, which depict, respectively, the case when the controllers do not compensate for the effect of uncertainty ($\rho_i = \chi_i = 0$) and the case when they compensate for the effect of uncertainty. It is clear from the unstable behavior in both profiles that the effect of uncertainty is significant and that, even when a robust controller is used in each mode, the nominal switching rules (or times) cannot guarantee closed-loop stability when uncertainty is present. As indicated by the input profiles, which remain saturated for all times, the state lies outside the stability region of the second mode at $t = 24\ min$. These results underscore the fact that, for hybrid processes, the impact of model uncertainty transcends the well-known adverse effects on the stability and performance of purely continuous processes (which comprise in this case the lower-level modes of the hybrid process) since uncertainty impacts also on the design of the switching logic. Owing to the limitations imposed by actuator constraints on the stability regions of the constituent modes of the hybrid process, the design of a stabilizing switching scheme requires knowledge of these stability regions, in order to decide when (or where, in the state-space,) a particular mode can be activated. However, when plant-model discrepancies are taken into consideration, the actual stability region of each mode can be quite smaller than the one obtained under nominal conditions (compare, for example, the expression in (7.9) with and without the uncertainty term, θ_{bi}). Consequently, nominal characterizations of stability regions can no longer guarantee safe mode switching, and the switching rules need to be modified.

In the final set of simulation runs, the switching scheme proposed in Theorem 7.11, which is based on stability regions that account for the presence of plant-model mismatch, is implemented. In this case, the reactor is initialized in the $\sigma = 0$ mode, using the corresponding robust controller, and then a switch to the $\sigma = 1$ mode (and its robust controller) is carried out only when the condition in (7.10) is satisfied (note that the condition of (7.11) is not needed since the initial mode is not reactivated). The controlled output and manipulated input profiles for this case are depicted by the solid lines in Fig 7.9 which show that the robust hybrid control strategy successfully drives the reactor temperature to the desired steady-state while attenuating the effect of uncertainty. The transition between the two modes becomes safe after about 30 minutes of reactor startup.

7.4 Predictive Control of Switched Nonlinear Systems

In Sect. 7.3, a framework for coordinating feedback and switching for control of hybrid nonlinear systems with uncertainty and constraints was developed. The key feature of the proposed control methodology is the integrated synthesis of: (1) a family of lower-level bounded nonlinear controllers that stabilize the continuous dynamical modes, and provide an explicit characterization of the stability region associated with each mode, and (2) upper-level switching laws that orchestrate the transition between the modes, on the basis of their stability regions, in a way that ensures stability of the overall switched closed-loop system. While the approach allows one to *determine* whether or not a switch can be made at any given time without loss of stability guarantees, it does not address the problem of *ensuring* that such a switch be made safely at some predetermined time.

Guiding the system through a prescribed switching sequence requires a control algorithm that can incorporate both state and input constraints in the control design. One such method is model predictive control (MPC). As discussed in the previous two chapters, one of the important issues that arise in the context of using predictive control policies for the purpose of stabilization of the individual modes, is the difficulty typically encountered in identifying a priori (i.e., before controller implementation) the set of initial conditions starting from where feasibility and closed-loop stability are guaranteed. This typically results in the need for extensive closed-loop simulations to search over a large set of possible initial conditions, thus adding to the overall computational load. This difficulty is more pronounced when considering MPC of hybrid systems that involve switching between multiple modes. Re-tuning MPC parameters (e.g., horizon length) of each predictive controller on-line, or running extensive closed-loop simulations in the midst of mode transitions, to determine the feasibility of switching, becomes computationally intractable, especially if the hybrid system involves a large number of modes with frequent switches.

For linear systems, the switched system can be transformed into a mixed logical dynamical system, and a mixed-integer linear program can be solved to determine the optimal switching sequence and switching times [33,73]. For nonlinear systems, one can, in principle, set up the mixed integer nonlinear programming problem, where the decision variables (and hence the solution to the optimization problem) include the control action together with the switching schedule. The resulting optimization problem – which in general is non-convex due to the nonlinearity of the system – is harder to solve since it also involves the discrete decision variables that determine mode switchings. The complexity of the optimization problem, and the computation time requirements, limit the applicability of this approach for the purpose of real-time control.

In many systems of practical interest, the switched system is often required to follow a prescribed switching schedule, where the switching times are no longer decision variables, but are prescribed via an operating schedule. Motivated by this practical problem, we present in this section a predictive control framework for the constrained stabilization of switched nonlinear systems that transit between their constituent modes at prescribed switching times. The main idea is to design a Lyapunov–based predictive controller for each mode, and incorporate constraints in the predictive controller design to ensure that the prescribed transitions between the modes occur in a way that guarantees stability of the switched closed–loop system. This is achieved as follows. For each mode, a Lyapunov-based model predictive controller is designed, and an analytic bounded controller that uses the same Lyapunov function is used to explicitly characterize a set of initial conditions for which the predictive controller is guaranteed to be feasible, and hence stabilizing, irrespective of the predictive controller parameters. Then, constraints are incorporated in the MPC design which, upon satisfaction, ensure that: (1) the state of the closed–loop system, at the time of the transition, resides in the stability region of the mode that the system is switched into, and (2) the Lyapunov function for each mode is non-increasing wherever the mode is re-activated, thereby guaranteeing stability. The proposed control method is demonstrated through application to a chemical process example.

7.4.1 Preliminaries

We consider the class of switched nonlinear systems represented by the following state-space description:

$$\begin{aligned}\dot{x}(t) &= f_{\sigma(t)}(x(t)) + G_{\sigma(t)}(x(t))u_{\sigma(t)} \\ u_{\sigma(t)} &\in \mathcal{U}_\sigma \\ \sigma(t) &\in \mathcal{K} := \{1, \ldots, p\}\end{aligned} \tag{7.18}$$

where $x(t) \in \mathbb{R}^n$ denotes the vector of continuous-time state variables, $u_\sigma(t) = [u_\sigma^1(t) \cdots u_\sigma^m(t)]^T \in \mathcal{U}_\sigma \subset \mathbb{R}^m$ denotes the vector of constrained manipulated inputs taking values in a nonempty compact convex set, $\mathcal{U}_\sigma := \{u_\sigma \in \mathbb{R}^m : \|u_\sigma\| \leq u_\sigma^{max}\}$, where $\|\cdot\|$ is the Euclidian norm, $u_\sigma^{max} > 0$ is the magnitude of the constraints, $\sigma : [0, \infty) \rightarrow \mathcal{K}$ is the switching signal which is assumed to be a piecewise continuous (from the right) function of time, i.e., $\sigma(t_k) = \lim\limits_{t \rightarrow t_k^+} \sigma(t)$ for all k, implying that only a finite number of switches is allowed on any finite interval of time. p is the number of modes of the switched system, $\sigma(t)$, which takes different values in the finite index set $\mathcal{K}$, represents a discrete state that indexes the vector field $f(\cdot)$, the matrix $G(\cdot)$, and the control input $u(\cdot)$. Throughout this section, we use the notations $t_{k_r^{in}}$ and $t_{k_r^{out}}$ to denote the times at which the k-th subsystem is switched in and out, respectively, for the r-th time, i.e., $\sigma(t_{k_r^{in}}^+) = \sigma(t_{k_r^{out}}^-) = k$. With this notation, it is understood that the continuous state evolves according to $\dot{x} = f_k(x) + G_k(x)u_k$ for $t_{k_r^{in}} \leq t < t_{k_r^{out}}$.

We denote by $\mathcal{T}_{k,in}$ the set of switching times at which the k-th subsystem is switched in, i.e., $\mathcal{T}_{k,in} = \{t_{k_1^{in}}, t_{k_2^{in}}, \cdots\}$. Similarly, $\mathcal{T}_{k,out}$ denotes the set of switching times at which the k-th subsystem is switched out, i.e., $\mathcal{T}_{k,out} = \{t_{k_1^{out}}, t_{k_2^{out}}, \cdots\}$. It is assumed that all entries of the vector functions $f_k(x)$, and the $n \times m$ matrices $G_k(x)$, are sufficiently smooth on their domains of definition and that $f_k(0) = 0$ for all $k \in \mathcal{K}$. Furthermore, the notation $L_f\bar{h}$ denotes the standard Lie derivative of a scalar function, $\bar{h}(x)$, with respect to the vector function, $f(x)$, i.e., $L_f\bar{h}(x) = \frac{\partial \bar{h}}{\partial x} f(x)$, and $\limsup\limits_{t \rightarrow \infty} f(x(t)) = \lim\limits_{t \rightarrow \infty} \{\sup\limits_{\tau \geq t} f(x(\tau))\}$.

In the remainder of this section, we focus on the problem of stabilizing switched nonlinear systems of the form of (7.18) where mode transitions are decided and executed at prescribed times. In order to provide the necessary background for the main results in Sect. 7.4.4, we will briefly review, in the next subsection, the design procedure for a bounded controller whose stability properties are then exploited for the design of a Lyapunov–based model predictive controller that guarantees stability for an explicitly characterized set of initial conditions. For simplicity, we will focus only on the state feedback control problem where measurements of $x(t)$ are assumed to be available for all t.

7.4.2 Bounded Lyapunov-Based Controller Design

Consider the system of (7.18), for a fixed $\sigma(t) = k$, for some $k \in \mathcal{K}$, for which a control Lyapunov function, V_k, exists. Using the results in [177] (see also Chaps. 4-6, the following bounded control law can be constructed:

$$u_k(x) = -k_k(x) L_{G_k} V_k(x) := b_k(x) \tag{7.19}$$

where

$$k_k(x) = \frac{L_{f_k}^* V_k(x) + \sqrt{\left(L_{f_k}^* V_k(x)\right)^2 + \left(u_k^{max} \|(L_{G_k} V_k)^T(x)\|\right)^4}}{\|(L_{G_k} V_k)^T(x)\|^2 \left[1 + \sqrt{1 + \left(u_k^{max} \|(L_{G_k} V_k)^T(x)\|\right)^2}\right]} \tag{7.20}$$

when $L_{G_k} V_k(x) \neq 0$, and $k_k(x) = 0$ when $L_{G_k} V_k(x) = 0$, $L_{G_k} V_k = [L_{g_k^1} V_k \cdots L_{g_k^m} V_k]$ is a row vector, where g_k^i is the i-th column of G_k, $L_{f_k}^* V_k = L_{f_k} V_k + \rho_k V_k$ and $\rho_k > 0$. For the above controller, one can compute an estimate of the stability region as follows:

$$\Omega_k(u_k^{max}) = \{x \in \mathbb{R}^n : V_k(x) \leq c_k^{max}\} \tag{7.21}$$

where $c_k^{max} > 0$ is the largest number for which $\Omega_k(u_k^{max})$ is completely contained within the set

$$\Phi_k(u_k^{max}) = \{x \in \mathbb{R}^n : L_{f_k}^* V_k(x) \leq u_k^{max} \|(L_{G_k} V_k)^T(x)\|\} \tag{7.22}$$

The bounded controller of (7.19–7.20) possesses a robustness property with respect to measurement errors, that preserves closed–loop stability when the control action is implemented in a discrete (sample and hold) fashion with a sufficiently small hold time (Δ). Specifically, the control law ensures that, for all initial conditions in Ω_k, the closed–loop state remains in Ω_k and eventually converges to some neighborhood of the origin whose size depends on Δ. This robustness property, formalized below in Proposition 7.16 (see Appendix E for the proof), will be exploited in designing an appropriate Lyapunov-based predictive controller in Sect. 7.4.3. For further results on the analysis and control of sampled-data nonlinear systems, the reader may refer to [105, 140, 207, 299].

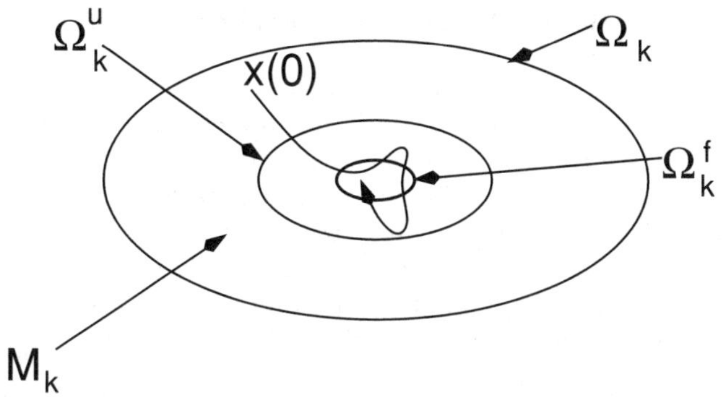

Fig. 7.10. A schematic representation of the stability region and the evolution of the closed–loop state trajectory under discrete implementation of the bounded controller

Proposition 7.16. *Consider the system of (7.18), for a fixed value of* $\sigma(t) = k$*, under the bounded control law of (7.19–7.20) with* $\rho_k > 0$ *and* Ω_k *as the stability region estimate obtained under continuous implementation. Let* $u_k(t) = u_k(j\Delta)$ *for all* $j\Delta \leq t < (j+1)\Delta$ *and* $u_k(j\Delta) = b_k(x(j\Delta))$*,* $j = 0, \cdots, \infty$*. Then, given any positive real number,* d_k*, there exists a positive real number* Δ_k^* *such that if* $x(0) := x_0 \in \Omega_k$ *and* $\Delta \in (0, \Delta_k^*]$*, then* $x(t) \in \Omega_k \; \forall \, t \geq 0$ *and* $\limsup_{t \to \infty} \|x(t)\| \leq d_k$*.*

Remark 7.17. The idea behind Proposition 7.16 can be understood as follows (see also Fig. 7.10). For a given d_k (that defines a ball in the neighborhood of the origin that the closed–loop state is required to converge to), one can find $\delta_k^{'} > 0$ such that if $x \in \Omega_k^u := \{x \in \mathbb{R}^n : V_k(x) \leq \delta_k^{'}\}$, then $\|x\| \leq d_k$. Then, one can find $\Delta_k^* > 0$ and δ_k (that defines the set $\Omega_k^f := \{x \in \mathbb{R}^n : V_k(x) \leq c_k^{max} - \delta_k\}$) such that: (a) for all initial conditions in $\Omega_k \backslash \Omega_k^f := \mathcal{M}_k$, $\dot{V}_k$ stays negative for an implement and hold time of Δ_k^*, and (b) for all initial conditions in Ω_k^f, the closed–loop state cannot escape Ω_k^u under any admissible control action before a time Δ_k^* has elapsed. Note that for all initial conditions in Ω_k^f, the value of V_k may increase during one hold time. For all $x_0 \in \mathcal{M}_k$ and $0 < \Delta \leq \Delta_k^*$, however, the value of V_k continues to decrease under the bounded control law until the closed–loop state enters Ω_k^f (and therefore Ω_k^u). From this point onward, for any state $x \in \Omega_k^f$, $x(t) \in \Omega_k^u$ for $t \in [0, \Delta]$ (by definition of the sets Ω_k^f, Ω_k^u). For any state $x \in \Omega_k^u \backslash \Omega_k^f$, $\dot{V}_k < 0$ for $t \in [0, \Delta]$ and hence $x(t) \in \Omega_k^u$ for $t \in [0, \Delta]$ (since Ω_k^u is a level set of V_k). The state of the closed–loop system, therefore, continues to evolve in Ω_k^u for all times after it has entered Ω_k^u, and hence we have that $\limsup_{t \to \infty} \|x(t)\| \leq d_k$. Note that if $\Delta = 0$, the discrete implementation reduces to continuous implementation, under which asymptotic stability is guaranteed for all initial conditions in Ω_k.

7.4.3 Lyapunov-Based Predictive Controller Design

In this section, we consider model predictive control of the system of (7.18), for a fixed $\sigma(t) = k$ for some $k \in \mathcal{K}$. We present here a Lyapunov–based design of MPC (see Remark 7.19 for a discussion on this formulation and its relationship to other Lyapunov-based formulations) that guarantees feasibility of the optimization problem and hence constrained stabilization of the closed–loop system from an explicitly characterized set of initial conditions. For this MPC design, the control action at state x and time t is obtained by solving, on-line, a finite horizon optimal control problem of the form:

$$P(x,t) : \min\{J(x,t,u_k(\cdot))|u_k(\cdot) \in S_k\} \tag{7.23}$$

$$s.t.\;\; \dot{x} = f_k(x) + G_k(x)u_k \tag{7.24}$$

$$\dot{V}_k(x(\tau)) \leq -\epsilon_k \;\; if \;\; V_k(x(t)) > \delta_k^{'},\; \tau \in [t, t+\Delta) \tag{7.25}$$

$$V_k(x(\tau)) \leq \delta_k^{'} \;\; if \;\; V_k(x(t)) \leq \delta_k^{'},\; \tau \in [t, t+\Delta) \tag{7.26}$$

where ϵ_k is a sufficiently small positive real number (see the Proof of Proposition 7.16 in Appendix E), $\delta_k^{'}$ is a positive real number for which $V_k(x) \leq \delta_k^{'}$ implies $\|x\| \leq d_k$ (see Remark 7.17), $S_k = S_k(t,T)$ is the family of piece-wise continuous functions (functions continuous from the right), with period Δ, mapping $[t, t+T]$ into $\mathcal{U}_k$, T is the specified horizon and V_k is the Lyapunov function used in the bounded controller design. A control $u_k(\cdot)$ in S_k is characterized by the sequence $\{u_k[j]\}$ where $u_k[j] := u_k(j\Delta)$ and satisfies $u_k(t) = u_k[j]$ for all $t \in [j\Delta, (j+1)\Delta)$. The performance index is given by

$$J(x,t,u_k(\cdot)) = \int_t^{t+T} \left[\|x^u(s;x,t)\|_Q^2 + \|u_k(s)\|_R^2\right] ds \tag{7.27}$$

where Q is a positive semi-definite symmetric matrix, R is a strictly positive-definite symmetric matrix, and $x^u(s;x,t)$ denotes the solution of (7.18), due to control u, with initial state x at time t. The minimizing control $u_k^0(\cdot) \in S_k$ is then applied to the plant over the interval $[t, t+\Delta)$ and the procedure is repeated indefinitely. This defines an implicit model predictive control law:

$$M_k(x) := argmin(J(x,t,u_k(\cdot))) := u_k^0(t;x,t) \tag{7.28}$$

The closed–loop stability properties under the Lyapunov–based predictive controller of (7.23–7.28) are inherited from the bounded controller of (7.19–7.20) under discrete implementation, and are formalized in Proposition 7.18 below (see Appendix E for the proof).

Proposition 7.18. *Consider the constrained system of (7.18), for a fixed value of $\sigma(t) = k$, under the MPC law of (7.23–7.28), designed using a control Lyapunov function V_k that yields a stability region Ω_k under continuous implementation of the bounded controller of (7.19–7.20) with a fixed $\rho_k > 0$.*

Then, given any positive real number d_k, *there exist positive real numbers* Δ_k^* *and* δ_k', *such that if* $x(0) \in \Omega_k$ *and* $\Delta \in (0, \Delta_k^*]$, *then* $x(t) \in \Omega_k \ \forall \ t \geq 0$ *and* $\limsup_{t \to \infty} \|x(t)\| \leq d_k$.

Remark 7.19. Note that, to achieve closed–loop stability, the predictive controller formulation of (7.23–7.28) requires that the value of the Lyapunov function decrease during the first time step only. This is due to the receding-horizon nature of controller implementation which dictates that only the first move of the set of calculated control moves be implemented and that the problem be re-solved at the next time step. Other Lyapunov-based predictive control approaches (see, for example, [155, 221]) typically incorporate a similar Lyapunov function decay constraint, albeit requiring that the constraint of (7.25) hold at the *end* of the prediction horizon, and assume initial feasibility of this constraint. In contrast, the requirement that V_k decrease during the first time step only in (7.25) allows the use of the bounded controller as an auxiliary controller to explicitly characterize the set of initial conditions stating from where the predictive controller is guaranteed to be feasible. Extensions of the Lyapnov–based controller design of (7.23–7.28) to the case when both input and state constraints are present can be found in [198].

Remark 7.20. Initial feasibility of the optimization problem is guaranteed because of: (1) the use of the same V_k that was used to design the bounded controller of (7.19–7.20), (2) initializing the closed–loop system within the stability region of the bounded controller, Ω_k, and (3) using a hold time $0 < \Delta \leq \Delta_k^*$. Note that for any initial condition, x_0, such that $x_0 \in \Omega_k \backslash \Omega_k^u$, we have $x_0 \in \mathcal{M}_k$, and that for any $x_0 \in \mathcal{M}_k$, negative-definiteness of $\dot{V}_k$ for a time duration less than Δ_k^* can be achieved with a control action (e.g., the one provided by the bounded controller) that respects the input constraints. For MPC implementation, therefore, and for sufficiently small Δ ($0 < \Delta \leq \Delta_k^*$), the constraint of (7.25) is guaranteed to be satisfied (the control action computed by the bounded controller design can also be used to provide a feasible initial guess to the optimization problem). Furthermore, since the state is initialized in Ω_k, which is a level set of V_k, the decay of V_k during any given time step ensures that the closed–loop state remains within Ω_k, thereby guaranteeing feasibility at future times. Once the state of the closed–loop system enters Ω_k^u, it may not be possible to drive the state closer to the origin, but it is possible to restrict the state to be within Ω_k^u. In this case, the constraint of (7.26) is used to ensure that the state of the closed–loop system does not leave Ω_k^u. Note, once again, that the control action prescribed by the bounded controller provides a feasible initial guess for the satisfaction of this constraint.

Remark 7.21. The fact that only practical stability is achieved is not a limitation of the specific MPC formulation used, but is rather due to the discrete implementation of the controller. Even if the bounded controller were used instead under the same implement-and-hold time of Δ, it would still only guarantee convergence of the closed–loop state to the set Ω_k^u, the size of which is limited by the value of the hold time, Δ (in the limit as Δ goes to zero, i.e., continuous implementation, the bounded controller and the predictive controller enforce asymptotic stability). Note also that any other Lyapunov-based analytic control design that provides an explicit characterization of the stability region and is robust with respect to discrete implementation can be used as an auxiliary controller.

Remark 7.22. As discussed in Chap. 6, one of the key challenges that impact on the practical implementation of NMPC is the inherent difficulty of characterizing, *a priori*, the set of initial conditions starting from where a given NMPC controller is guaranteed to stabilize the closed–loop system, or for a given set of initial conditions, to identify the value of the prediction horizon for which the optimization problem will be feasible. Use of conservatively large horizon lengths to address stability only increases the size and complexity of the nonlinear optimization problem and could make it intractable. Owing to the fact the closed–loop stability is guaranteed by the Lyapunov-based predictive controller from an explicitly characterized set of initial conditions irrespective of the prediction horizon, the time required for the computation of the control action can be made smaller, if so desired, by reducing the prediction horizon without loss of stability.

Remark 7.23. In the event that full state measurements are not available continuously, but are available only at sampling times $\Delta_s > \Delta_k^*$, i.e., greater than what a given bounded control design can tolerate (and, therefore, greater than the maximum allowable discretization for the Lyapunov-based predictive controller), it is necessary to redesign the bounded controller to increase the robustness margin. A larger value of Δ_k^* may be achieved by increasing the value of the parameter ρ_k in the design of the bounded controller (see Proof of Proposition 7.16 in Appendix E). If the value of the sampling time is reasonable, an increase in the value of the parameter ρ_k, while leading to a shrinkage in the stability region estimate for the Lyapunov-based predictive controller, can increase Δ_k^* to a value greater than Δ_s and preserve the desired feasibility and stability guarantees of the Lyapunov-based predictive controller.

7.4.4 Predictive Control with Scheduled Mode Transitions

Consider now the switched nonlinear system of (7.18), with a prescribed switching sequence (including the switching times) defined by the sets $\mathcal{T}_{k,in} = \{t_{k_1^{in}}, t_{k_2^{in}}, \cdots\}$ and $\mathcal{T}_{k,out} = \{t_{k_1^{out}}, t_{k_2^{out}}, \cdots\}$. Also, assume that for each mode of the switched system, a Lyapunov–based predictive controller of the form of (7.23–7.28) has been designed and an estimate of the stability region generated. The control problem is formulated as the one of designing a Lyapunov-based predictive controller that guides the closed–loop system trajectory in a way that the schedule described by the switching times is followed and stability of the closed–loop system is achieved. A predictive control algorithm that addresses this problem is presented in Theorem 7.24 below. The proof of the theorem is given in Appendix E.

Theorem 7.24. *Consider the constrained nonlinear system of (7.18), the control Lyapunov functions* V_k, $k = 1, \dots, p$, *and the stability region estimates,* Ω_k, *obtained under continuous implementation of the bounded controller of (7.19–7.20) with fixed* $\rho_k > 0$, $k = 1, \dots, p$. *Let* $0 < T_{design} < \infty$ *be a design parameter, and let* t *be such that* $t_{k_r^{in}} \leq t < t_{k_r^{out}}$ *and* $t_{m_j^{in}} = t_{k_r^{out}}$ *for some* $m, k \in \mathcal{K}$. *Consider the following optimization problem:*

$$P(x,t) : \min\{J(x,t,u_k(\cdot)) | u_k(\cdot) \in S_k\} \tag{7.29}$$

$$J(x,t,u_k(\cdot)) = \int_t^{t+T} \left[\|x^u(s;x,t)\|_Q^2 + \|u_k(s)\|_R^2 \right] ds \tag{7.30}$$

where T *is the prediction horizon given by* $T = t_{k_r^{out}} - t$, *if* $t_{k_r^{out}} < \infty$, *or* $T = T_{design}$ *if* $t_{k_r^{out}} = \infty$, *subject to the following constraints:*

$$\dot{x} = f_k(x) + G_k(x)u_k \tag{7.31}$$

$$\dot{V}_k(x(\tau)) \leq -\epsilon_k \ \ if \ \ V_k(x(t)) > \delta_k^{'}, \ \tau \in [t, t+\Delta) \tag{7.32}$$

$$V_k(x(\tau)) \leq \delta_k^{'} \ \ if \ \ V_k(x(t)) \leq \delta_k^{'}, \ \tau \in [t, t+\Delta) \tag{7.33}$$

and, if $t_{k_r^{out}} = t_{m_j^{in}} < \infty$,

$$V_m(x(t_{m_j^{in}})) \leq \left\{ \begin{array}{ll} V_m(x(t_{m_{j-1}^{in}})) - \epsilon^* & , j > 1, V_m(x(t_{m_{j-1}^{in}})) > \delta_m^{'} \\ \delta_m^{'} & , j > 1, V_m(x(t_{m_{j-1}^{in}})) \leq \delta_m^{'} \\ c_m^{max} & , \qquad\qquad j = 1 \end{array} \right\} \tag{7.34}$$

where ϵ^* *is a positive real number. Then, given a positive real number* d^{max}, *there exist positive real numbers* Δ^* *and* $\delta_k^{'}$, $k = 1, \dots, p$ *such that if the optimization problem of (7.29–7.34) is feasible at all times, the minimizing control is applied to the system over the interval* $[t, t + \Delta_{k_r}]$, *where* $\Delta_{k_r} \in (0, \Delta^*]$ *and* $t_{k_r^{out}} - t_{k_r^{in}} = l_{k_r}\Delta_{k_r}$ *for some integer* $l_{k_r} > 0$, *and the procedure is repeated, then* $\limsup_{t\to\infty} \|x(t)\| \leq d^{max}$.

Remark 7.25. Note that feasibility of the constraint of (7.32) between mode transitions is guaranteed, provided that the

system is initialized within the stability region, and does not need to be assumed. This stability constraint ensures that the value of the Lyapunov function of the currently active mode keeps decreasing (recall that one of the criteria in the MLF stability analysis is that the individual modes of the switched system be stable). The constraint of (7.34) expresses

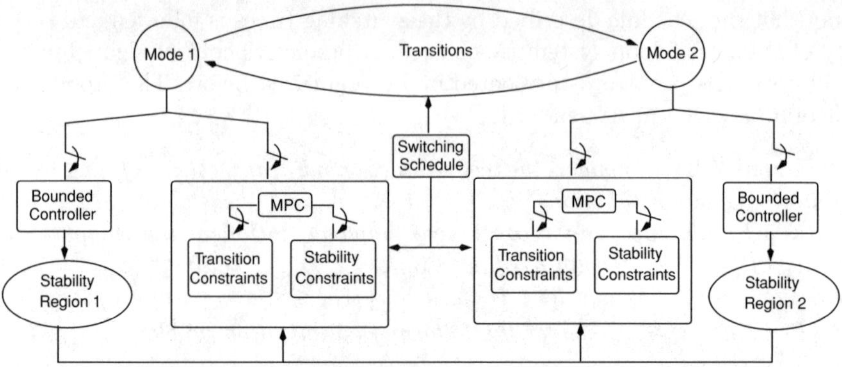

Fig. 7.11. A schematic representation of the predictive control structure comprised of the predictive and bounded controllers for the constituent modes, together with transition constraints

two transition requirements simultaneously: (1) the MLF constraint that requires that the value of the Lyapunov function be less than what it was the last time the system was switched into that mode (required when the switching sequence is infinite, see [39] for details), and (2) the stability region constraint that requires that the state of the process reside within the stability region of the target mode at the time of the switch; since the stability regions of the modes are expressed as level sets of the Lyapunov functions, the MLF-constraint also expresses the stability region constraint. The understanding that the chosen switching schedule is a reasonable one (i.e., one that does not result in closed–loop instability) motivates the assumption on the feasibility of the transition constraints for all times. Note that feasibility of the transition constraints can also be used to validate the given switching schedule, and can be used to abort the switching schedule (i.e., to decide that the remaining switches should not be carried out) in the interest of preserving closed–loop stability.

Remark 7.26. Figure 7.11 is a schematic that depicts the main features of the predictive controller of (7.29–7.34), for representative switching between two modes, where the switching schedule dictates the transitions and is accounted for in the predictive control design. A representative implementation of the predictive controller is also depicted schematically by the solid lines in Fig. 7.12. The algorithm below explains the implementation of the predictive controller of (7.29–7.34), including the course of action that needs to be taken if the predictive controller does not yield a feasible solution:

1. Given the system model of (7.18) and the constraints on the input, design the bounded controller of (7.19–7.20) for each mode and compute the stability region estimate, $\Omega_k(u_k^{max})$, for the bounded controller using (7.21–7.22).

2. Given the size of the ball that the state is required to converge to, d^{max}, compute Δ_k^*, $k = 1, \ldots, p$ for the predictive controller of (7.23–7.28) such that under single mode operation for each mode, $\limsup_{t\to\infty} \|x(t)\| \le d^{max}$. Compute $\Delta^* = \min_{k=1}^{p}\{\Delta_k^*\}$, and use a $\Delta_{k_r} \in (0, \Delta^*]$, $t_{k_r^{out}} - t_{k_r^{in}} = l_{k_r}\Delta_{k_r}$ for some integer $l_{k_r} > 0$, for which for the purpose of MPC implementation.
3. Consider the time $t_{k_r^{in}}$, which designates the time that the closed–loop system is in the k-th mode for the r-th time, and the state belongs to the stability region of the k-th mode (for the purpose of initialization, i.e., at $t = 0$, this would correspond to $t_{1_1^{in}}$, and the state belonging to the stability region $\Omega_1(u_1^{max})$).
4. From the set of prescribed switching times, pick $t_{m_j^{in}} = t_{k_r^{out}}$ ($t_{m_j^{in}}$, therefore, is the time that the next switch takes place, and that the system, upon exiting from the current mode enters mode m for the j-th time).
5. Consider the predictive controller of Theorem 7.24. The constraint that $V_m(t_{m_j^{in}}) \le V_m(t_{m_{j-1}^{in}}) - \epsilon^*$ requires that when the closed–loop system enters the mode m, the value of V_m is less than what it was at the time that the system last entered mode m. If the system has never entered mode m before, i.e., for $j = 1$, set $V_m(t_{m_{j-1}^{in}}) = c_m^{max}$ (this requires that the state belongs to the stability region corresponding to mode m). If the closed–loop state has already entered the desired ball around the origin, implement $V_m(t_{m_j^{in}}) \le \delta_m^{'}$, that ensures that the state stays within the ball, $\|x\| \le d^{max}$.
6. If at any time, the predictive controller of (7.29–7.34) does not yield a feasible solution, or the switching schedule does not prescribe another switch, go to step 7; else implement the predictive controller up-to time $t_{k_r^{out}}$, i.e., until the time that the system switches into mode m and go back to step 4 to proceed with the rest of the switching sequence.
7. Implement the Lyapunov-based predictive controller of (7.23–7.28) for the current mode to stabilize the closed–loop system.

Remark 7.27. Since the switching times are fixed, the prediction of the states in the controller needs to be carried out from the current time up-to the time of the next switch only. The predictive controller is therefore implemented with a shrinking horizon between successive switching times. Note, however, that the value of the horizon is *not* a decision variable; it is obtained simply by evaluating the difference between the next switching time and the current time. In the case that the switching schedule terminates, then after the last switch has been made and the system is evolving in the terminal mode, the horizon is fixed and set equal to a preset design parameter T_{design}. From this point on, the controller design of (7.29–7.33) reduces to the Lyapunov–based predictive controller of (7.23–7.27) for the terminal mode, and guarantees practical stability of the closed–loop system for any value of prediction horizon, including for $T = T_{design}$. Note also that picking the hold time, Δ_{k_r}, during the time interval between $t_{k_r^{in}}$ and $t_{k_r^{out}}$, such that the time interval between switches is an integer multiple of the sampling time is made to ensure that the system does not go through a transition while the final control move in a given mode is being implemented. To see the reasoning behind this, note first that, due to the discrete nature of controller implementation, the final control in the pre-switching mode is implemented for Δ_{k_r} time. Therefore, if the system undergoes a transition during that time (i.e., if $t + \Delta_{k_r}$ is greater than $t_{k_r^{out}}$), then the control could cause the closed-loop state to leave the stability region of the target mode after switching has been executed. Picking Δ_{k_r} as described above precludes such a possibility.

Remark 7.28. The constraints of (7.32–7.33) require that the closed–loop state trajectory evolve in such a way that it enters the stability region of the target mode (due to (7.34)) while, up-to the time of the switch, continuing to evolve in the stability region of the current mode. Owing to this, if at any time the transition constraints become infeasible, and the switching sequence is aborted, the Lyapunov-based predictive controller for the then-current mode is guaranteed to stabilize the closed-loop system. While this condition safeguards against instability, it may be restrictive in the sense that it could hamper the feasibility of the optimization problem, especially if the boundaries of the stability regions of the constituent modes do not have a significant overlap. When dealing with a finite switching sequence, the constraints of (7.32–7.33) can be relaxed for all times before the final switch takes place. Also, a relaxed version of the constraint of (7.34) may be used up to the time of the terminal switch, requiring only that $V_m(t_{m_j^{in}}) \leq c_m^{max}$ (i.e., the closed–loop state resides in the stability region of the target mode; see dashed lines in Fig. 7.12).

Remark 7.29. For purely continuous systems, the problem of implementing MPC with guaranteed stability regions was addressed in Chap. 5 for linear systems and in Chap. 6 for nonlinear systems, by means of a hybrid control structure that unites bounded control and MPC. The hybrid structure was used to embed the implementation of MPC within the stability region of a Lyapunov-based bounded controller which serves as a fall-back component that can be switched to in the event of infeasibility or instability of the predictive controller. In this section, the bounded controller design is not used as a fall–back controller, but rather for the purpose of providing an estimate of the stability region for the Lyapunov-based predictive controller, and feasible initial guesses for the control moves (the decision variables in the optimization problem). Note that the Lyapunov-based predictive controller of (7.23–7.28) is guaranteed to be feasible and stabilizing from an explicitly characterized set of initial conditions.

Remark 7.30. The use of a predictive control frameowrk is both beneficial and essential to the problem of implementing a prescribed switching schedule using the proposed approach. In particular, while the bounded controller

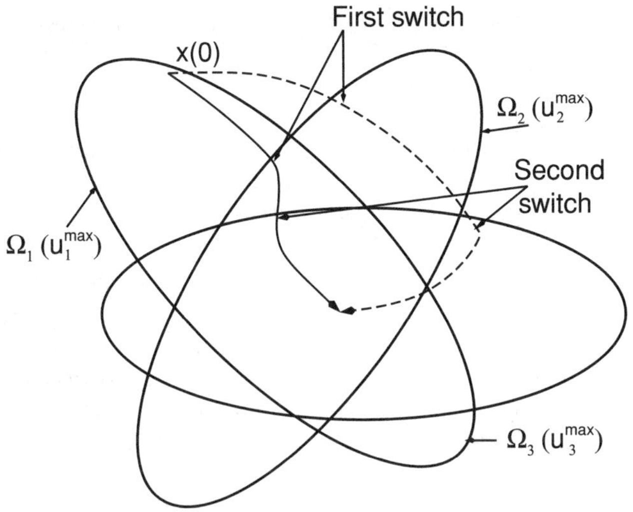

Fig. 7.12. Schematic representation of the closed–loop trajectory of a switched system under the implementation of the predictive control algorithm of (7.29–7.34) (*solid line*) and using the relaxed constraints of Remark 7.28 (*dotted line*)

can achieve stabilization under single mode operation, the bounded controller framework simply does not address performance considerations. More importantly, it does not allow for incorporating transition constraints, and there is no guarantee that the bounded controller (or even the Lyapunov-based predictive controller, if the transition constraints are not incorporated) can stabilize the closed–loop system when following a prescribed switching schedule (see the simulation example in Sect. 7.4.5 for a demonstration). In contrast, the predictive controller approach provides a natural framework for specifying appropriate transitions constraints whose feasibility ensures closed–loop stability.

Remark 7.31. For a nonlinear switched system of the form of (7.18), while one can in principle set up the mixed integer nonlinear programming problem, where the decision variables (and hence the solution to the optimization problem) include the control action together with the switching schedule, it is not possible to use it for the purpose of online implementation. In many systems of practical interest, however, the switched system may be only required to follow a prescribed switching schedule. The predictive controller algorithm provides an implementable controller for switched nonlinear systems with a prescribed switching sequence by incorporating appropriate constraints on the implemented control action in each of the individual modes. Note also that, if it is possible to solve the mixed integer optimization problem off-line, the solution can be used to provide a set of optimal switching times, and an associated switching sequence, which can then be implemented online using the proposed approach.

Remark 7.32. Note that the predictive controller algorithm presented in Theorem 7.24 can be adapted to account for possible uncertainty in the switching schedule. For example, in the case that the switching sequence is known, but only upper and lower bounds on the switching times are available, i.e., we know that $t_{k_r^{in,\ min}} \leq t_{k_r^{in}} \leq t_{k_r^{in,\ max}}$, the constraint of (7.34) can be modified to require that $V_k(t_{k_r^{in,min}}) < V_k(t_{k_{r-1}^{in}})$, i.e., to require that the closed–loop state enters the stability region of the target mode at $t_{k_r^{in,min}}$. Additionally, a constraint that requires the closed–loop system to evolve in the intersection of the stability regions of the current and target modes between $t_{k_r^{in,min}}$ and $t_{k_r^{in,max}}$, can be added so that any time between $t_{k_r^{in,min}}$ and $t_{k_r^{in,max}}$ that the switch occurs, the closed–loop state resides in the stability region of the target mode.

7.4.5 Application to a Chemical Process Example

We consider a continuous stirred tank reactor where an irreversible, first-order exothermic reaction of the form $A \xrightarrow{k} B$ takes place. The operation schedule requires switching between two available inlet streams consisting of pure A at flow rates F_1, F_2, concentrations C_{A1}, C_{A2}, and temperatures T_{A1}, T_{A2}, respectively. For each mode of operation, the mathematical model for the process takes the form:

$$\begin{aligned}\frac{dC_A}{dt} &= \frac{F_\sigma}{V}(C_{A\sigma} - C_A) - k_0 \exp\left(\frac{-E}{RT_R}\right) C_A \\ \frac{dT_R}{dt} &= \frac{F_\sigma}{V}(T_{A\sigma} - T_R) + \frac{(-\Delta H)}{\rho c_p} k_0 \exp\left(\frac{-E}{RT_R}\right) C_A + \frac{Q_\sigma}{\rho c_p V}\end{aligned} \tag{7.35}$$

where C_A denotes the concentration of the species A, T_R denotes the temperature of the reactor, Q_σ is the heat removed from the reactor, V is the volume of the reactor, k_0, E, ΔH are the pre-exponential constant, the activation energy, and the enthalpy of the reaction, c_p and ρ, are the heat capacity and fluid density in the reactor and $\sigma(t) \in \{1,\ 2\}$ is the discrete variable. The values of all process parameters can be found in Table 7.2. The control objective is to stabilize the reactor at the unstable equilibrium point, $(C_A^s,\ T_R^s) = (0.57, 395.3)$, using the rate of heat input, Q_σ, and the change in inlet reactant concentration, $\Delta C_{A\sigma} = C_{A\sigma} - C_{A\sigma_s}$ as manipulated inputs with constraints: $|Q_\sigma| \leq 1KJ/hr$ and $|\Delta C_{A\sigma}| \leq 1mol/l$, $\sigma = 1, 2$. For each mode, we considered a quadratic Lyapunov function of the form $V_k(x) = x^T P_k x$ where $x = [C_A - C_A^s \ \ T_R - T_R^s]^T$, with $P_1 = \begin{bmatrix} 33.24 & 1.32 \\ 1.32 & 0.06 \end{bmatrix}$

Table 7.2. Process parameters and steady–state values for the reactor of (7.35)

V	$= 0.1$	m^3
R	$= 8.314$	$kJ/kmol \cdot K$
C_{A1_s}	$= 0.79$	$kmol/m^3$
C_{A2_s}	$= 1.0$	$kmol/m^3$
T_{A1}	$= 352.6$	K
T_{A2}	$= 310.0$	K
Q_{1_s}	$= 0.0$	KJ/hr
Q_{2_s}	$= 0.0$	KJ/hr
ΔH	$= -4.78 \times 10^4$	$kJ/kmol$
k_0	$= 1.2 \times 10^9$	s^{-1}
E	$= 8.314 \times 10^4$	$kJ/kmol$
c_p	$= 0.239$	$kJ/kg \cdot K$
ρ	$= 1000.0$	kg/m^3
F_1	$= 3.34 \times 10^{-3}$	m^3/s
F_2	$= 1.67 \times 10^{-3}$	m^3/s
T_{Rs}	$= 395.33$	K
C_{As}	$= 0.57$	$kmol/m^3$

and $P_2 = \begin{bmatrix} 7.39 & 0.32 \\ 0.32 & 0.016 \end{bmatrix}$, and used these in the Lyapunov-based controller design to compute the stability regions for the two modes, Ω_1 and Ω_2, shown in Fig. 7.13. The matrices P_k were computed by solving a Riccatti inequality using the linearized system matrices. The computation of the stability regions (using (7.21–7.22)), however, was done using the nonlinear system dynamics. The parameters in the objective function of (7.27) are chosen as $Q = qI$, with $q = 1$, and $R = rI$, with $r = 1.0$. The constrained nonlinear optimization problem is solved using the MATLAB subroutine fmincon, and the set of ODEs is integrated using the MATLAB solver ODE45.

We first demonstrate the implementation of the Lyapunov-based predictive controller to a single mode operation of the chemical reactor, i.e., one in which the process is operated for all times in mode 1. To this end, we consider an initial condition that belongs to the stability region of the predictive controller for mode 1. As shown by the solid line in Fig. 7.13, starting from the initial condition $(C_A,\ T_R) = (0.14,\ 404.9)$, which belongs to the stability region of the predictive controller for mode 1, the controller successfully stabilizes the closed–loop system. The corresponding closed–loop state and manipulated input profiles are shown in Fig. 7.14.

To demonstrate the need to implement the algorithm proposed in Theorem 7.24 for stabilization when switching is involved, we choose a schedule involving a switch from inlet stream 1 (mode 1) to inlet stream 2 (mode 2) at time $t = 0.1\ hr$. Once again, the system is initialized within the stability region of mode 1, and the predictive controller for mode 1 is implemented (without any transition constraints). Up until $t = 0.1\ hr$, the state of the closed–loop

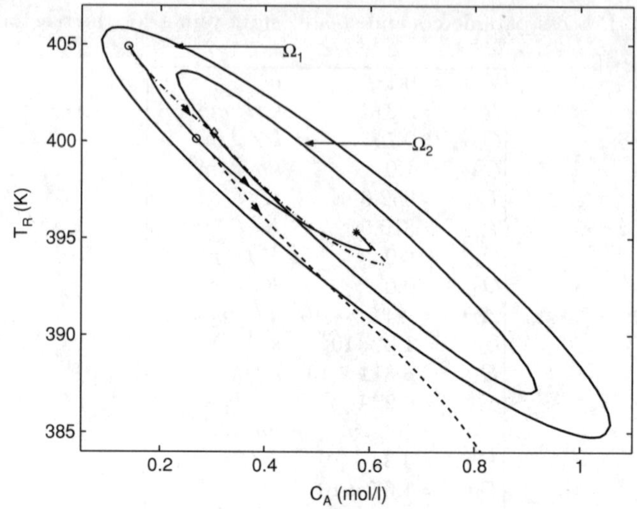

Fig. 7.13. Closed–loop state trajectory when the reactor is operated in mode 1 for all times under the stabilizing MPC formulation of (7.23–7.28) (*solid line*), when the reactor operation involves switching from mode 1 to mode 2 at $t = 0.1\ hr$, under the predictive controller design of (7.23–7.28) without transition constraints (*dashed line*), and when the reactor operation involves switching from mode 1 to mode 2 at $t = 0.1\ hr$, under the predictive controller of Theorem 7.24 with transitions constraints (*dashed–dotted line*)

system moves towards the desired steady-state (as seen from the dashed lines in Fig. 7.13); however, when the system switches to mode 2, the MPC controller, designed for the stabilization of the process in mode 2, does not yield a feasible solution. If the bounded controller for mode 2 is implemented, the resulting control action is not able to stabilize the closed–loop system (dashed lines in Figs. 7.13–7.14). This happens because at the time of the transition, the state of the closed–loop system (marked by *o* in Fig. 7.13) does not belong to the stability region of mode 2. Note also that while the predictive formulation of (7.23–7.28) guarantees stabilization for all initial conditions belonging to the stability region of mode 1, it does not incorporate constraints ensuring a safe transition to mode 2.

Finally, the predictive control algorithm of Theorem 7.24 (which incorporates constraints that account for switching) is implemented (dash-dotted lines in Figs. 7.13–7.14). The MPC controller of mode 1 is designed to drive the state of the closed–loop system such that the state belongs to the stability region of mode 2 at the switching time. Consequently, when the system switches to mode 2 at $t = 0.1\ hr$, the closed–loop system state at the switching time (marked by the $\diamondsuit$ in Fig. 7.13) belongs to the stability region of the MPC designed for mode 2. At this time, when the process switches to mode 2 and

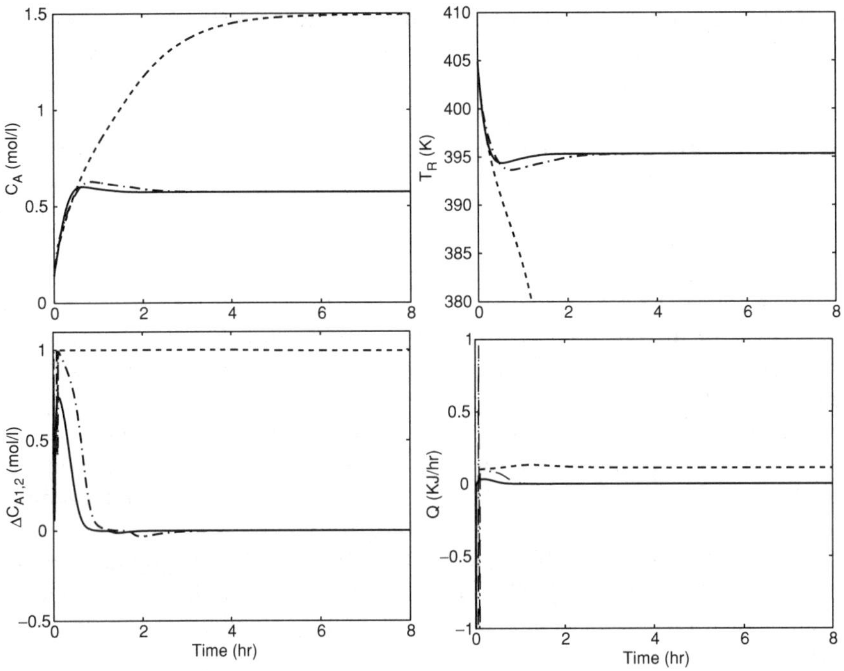

Fig. 7.14. Closed-loop state (*top plots*) and manipulated input (*bottom plots*) profiles when the reactor is operated in mode 1 for all times under the stabilizing MPC formulation of (7.23–7.28) (*solid line*), when the reactor operation involves switching from mode 1 to mode 2 at $t = 0.1$ hr under the predictive controller design of (7.23–7.28) without transition constraints (*dashed line*), and when the reactor operation involves switching from mode 1 to mode 2 at $t = 0.1$ hr under the predictive controller of Theorem 7.24 with transition constraints (*dashed–dotted line*)

the corresponding predictive controller is implemented, closed–loop stability is achieved.

7.5 Applications of Hybrid Systems Tools to Biological Networks

Hybrid system models are increasingly being used not only for the modeling of engineering systems, such as automotive and chemical process control systems, but also for the modeling of a diverse array of biological networks that implement and control important functions in biological cells, such as metabolism, DNA synthesis, movement and information processing. The dynamics of biological networks often involve switching between many qualitatively different modes of behavior. At the molecular level, for example, the fundamental process of inhibitor proteins turning off the transcription of genes by RNA polymerase reflects a switch between two continuous processes. An example of this is the classic genetic switch observed in the bacteriophage λ (e.g., see [114,222,237]), where two distinct behaviors, lysis and lysogeny, each with different mathematical models, are seen. Also, at the cellular level, the cell growth and division in a eukaryotic cell is usually described as a sequence of four processes, each being a continuous process that is triggered by a set of conditions or events (e.g., see [115, 175, 273]). At the inter-cellular level, cell differentiation can also be viewed as a switched system [103].

In addition to naturally occurring switches, switched dynamics can be the result of external intervention that attempts to re-engineer a given network by turning on or off, for example, certain pathways. In all of these examples, the overall behavior of the network is more appropriately viewed as a switched system, i.e., intervals of continuous dynamics interspersed by discrete transitions, and, therefore, a hybrid approach that combines elements of discrete and continuous dynamics is necessary, not only for the modeling, simulation and analysis (e.g., see [5,6]), but also for controlling and modifying the network behavior. Given the similarity that many biological networks exhibit to switched systems encountered in engineering (e.g., involving feedback mechanisms and switching), it is instructive to investigate how these tools can be applied to model, analyze and possibly modify the dynamics of biological networks.

Changes in network dynamics can result from alterations in local conditions (e.g., temperature, nutrient and energy source, light, cell density) and/or changes in the molecular environment of individual regulatory components (e.g. intra-cellular concentrations of transcription factors). Often, the network can be switched between different modes by changes in parameter values. These parameter typically include rate constants and total enzyme concentrations that are under genetic control. Changing the expression of certain genes will change the parameter values of the model and move the network across bifurcation boundaries into regions of qualitatively different behavior (e.g., transitions from limit cycles to single and multiple steady-states). Understanding and analyzing the nature of these qualitatively different modes of behavior typically involves bifurcation analysis which determines how the attractors of the vector field depend on parameter values, leading to a characterization of the regions in parameter space where the different behaviors are observed. The boundaries of these regions represent the bifurcation boundaries.

An important question, however, that is not addressed by bifurcation analysis is that of when, or where in state space, is a transition from one mode to another feasible. For example, bifurcations can predict that a change in a certain parameter is required for the network to move from an oscillatory mode (stable limit cycle) to a multi-stable mode (multiple stable steady-states) but cannot tell us when, or which, of the new steady-states will be observed upon switching. This is an important consideration when one tries to manipulate the network behavior to achieve a certain desirable behavior or steady-state. To address this question, bifurcations must be complemented by a dynamical analysis of the transient behavior of the constituent modes of the overall network. Intuitively, one expects that the newly switched mode will exhibit the desired steady-state if, at the time of switching, the network state is in the vicinity of that steady-state. A precise concept from nonlinear dynamical systems theory that quantifies this closeness is that of the domain of attraction, which is the set of all points in the state space, starting from where the trajectories of the dynamical system converge to a given equilibrium state.

In this section, we present a methodology for the dynamic analysis of mode transitions in biological networks. The proposed approach is based on the notion of coupling the switching logic to the domains of attraction (stability regions) of the constituent modes, which was introduced earlier in this chapter. To this end, we initially model the overall network as a switched nonlinear system that dwells in multiple modes, each governed by a set of continuous-time differential equations. The transition between the continuous modes are triggered by discrete events (changes in model parameters that correspond to alterations in physiological conditions). Then, following the characterization of the steady-state behavior of each mode, Lyapunov techniques are used to characterize the domains of attraction of the steady-states. Finally, by analyzing how the stability regions of the various modes overlap with one other, it is possible to determine when, and if, a given steady-state behavior, for a given mode transition, is feasible or not. The proposed method is demonstrated using a biological network model arising in cell cycle regulation.

7.5.1 A Switched System Representation of Biological Networks

We consider biological networks modeled by systems of nonlinear ordinary differential equations of the general form:

$$\begin{aligned} \frac{dx(t)}{dt} &= f_{i(t)}(x(t), p_{i(t)}) \\ i(t) \in \mathcal{I} &= \{1, \cdots, N\} \end{aligned} \tag{7.36}$$

where $x = [x_1 \ x_2 \ \cdots \ x_n]^T \in \mathbb{R}^n$ is the vector of continuous state variables (e.g., concentrations of the various network components such as proteins, genes, metabolites, etc.), $f_i(\cdot)$ is a smooth nonlinear function, p_i is a vector of network parameters (e.g., kinetic constants, total enzyme concentrations) that are typically under genetic control, $i : [0, \infty) \rightarrow \mathcal{I}$ is the switching signal which is assumed to be a piecewise continuous (from the right) function of time, i.e., $i(t_k) = \lim_{t \rightarrow t_k^+} i(t)$ for all $t_k \geq 0$, $k \in Z_+$, where Z_+ is the set of positive integers and t_k is the k-th switching time, implying that only a finite number of switches occurs on any finite interval of time. N is the number of modes of the switched system, $i(t)$, which takes different values in the finite index set, $\mathcal{I}$, represents a discrete state that indexes the vector field $f(\cdot)$ which determines $\dot{x}$. For each value that i takes in $\mathcal{I}$, the temporal evolution of the continuous state is governed by a different set of differential equations. The system of (7.36) is therefore a switched (multi-modal) system that consists of a finite family of continuous nonlinear subsystems (modes) and a switching rule that orchestrates the transitions between them. In biological networks, mode transitions can be the result of a fundamental change in the vector field itself (e.g., different modes having different f_i's) or, more commonly, a change in network parameter values due to changes in levels of gene expression and enzyme activities (which can occur spontaneously or be induced externally).

The basic problem that we address in this section is that of determining when (or where in the state-space) can a transition from one mode to another produce a certain desired behavior that exists in the target mode (e.g., a desired steady-state). From an analysis point of view, the answer to this question sheds light on why certain naturally-occurring mode transitions seem to always favor a certain steady-state behavior. From a control point of view, on the other hand, the answer provides insight into how and when the designer should enforce the transition in order to bring about a desired steady-state behavior. In the next subsection, we outline a methodology that addresses these questions.

7.5.2 Methodology for Analysis of Mode Transitions

The methodology proposed here is based on the idea of designing the switching logic on the basis of the stability regions of the constituent modes, which was introduced in Sects. 7.3–7.4 in the context of constrained control of switched nonlinear systems. However, unlike the results in Sects. 7.3–7.4, where the presence of the stability regions was a consequence of the constraints imposed on the manipulated inputs of each mode, the stability regions considered here are directly linked to the intrinsic dynamic behavior of the constituent modes, which is dictated by the dependence of the attractors of the vector field on the network parameters. For example, the presence of multiple equilibrium points in a given mode gives rise to multiple stability regions, or domains of attraction, whose union covers the entire state-space. Clearly, which equilibrium state is attained depends on which region contains the system state at the switching time. Below is the proposed methodology:

1. Identify the different modes of the network, where each mode is characterized either by a different set of differential equations or by the same set of equations but with different parameters.
2. Compute the steady-state(s) of each mode by solving:

$$0 = f_i(x_s, p_i) \tag{7.37}$$

where x_s is an admissible steady-state solution. Depending on the values of p, each mode might possess a limit cycle, a single steady-state, or multiple steady-states.

3. Characterize (or estimate) the domain of attraction (stability region) of each steady-state in each mode. For a given steady-state, x_s, the domain of attraction, $\Omega(x_s)$, consists of the set of all states starting from where the system trajectories converge to that steady-state. Estimates of the domain of attraction can be obtained using Lyapunov techniques [148]. For example, consider the case of isolated equilibrium points and let V_i be a Lyapunov function candidate, i.e., $V_i(x_s) = 0$ and $V_i(x) > 0$ for all $x \neq x_s$. Consider also the set $\Pi(x_s) = \{x \in \mathbb{R}^n : \dot{V}_i(x) < 0\}$. Then the level set, $\Omega(x_s) = \{x \in \mathbb{R}^n : V_i(x) \leq c_i^{max}\}$, where $c_i^{max} > 0$ is the largest constant for which Ω is fully contained in Π, provides an estimate of the domain of attraction of x_s. Due to the possible conservatism of the resulting estimates, Lyapunov techniques are usually coupled with other methods in order to obtain larger estimates (e.g., multiple Lyapunov functions; see Chap. 4 in [148] for details).
4. Analyze how the domains of attraction of a given mode overlap with those of another mode. Suppose, for example, that the network is initialized within mode k and let T be the transition time from mode k to mode j. Also, let x_s be an admissible steady-state (among several others) of the j-th mode. Then, if
$$x(T) \in \Omega_j(x_s) \tag{7.38}$$
and $i(t) = j \; \forall \, t \geq T^+$ (i.e., no further switches take place), then we will have $\lim_{t\to\infty} x(t) = x_s$, i.e., the x_s steady-state will be observed following switching. The switching rule of (7.38) requires monitoring the temporal evolution of the state evolution in order to locate where the state is at the switching time, with respect to the domains of attraction of the mode to be activated.

Remark 7.33. Referring to the computation of the steady-states of a biological network, we note that it is, in general, difficult to compute all the steady-state solutions of a system of nonlinear ordinary differential equations (ODEs). For an arbitrary system of nonlinear ODEs, where the right-hand side does not possess any kind of structure, one can resort to general search algorithms, such as Newton-type methods, to solve (7.37). These methods are usually local in character and thus may require an extensive search over all possible initial guesses in order to find all possible solutions. For biological systems, the search complexity can be reduced somewhat by taking advantage of the natural limits on the values of the state variables in order to bracket the region in the state-space where the system is expected to operate and where the search needs to be carried out. More importantly, the dynamic models of biological systems often exhibit specific types of structure that arise from physical considerations and can thus be exploited in the computation of all the steady-states using computational algorithms that have been developed in the literature. For example, if each component on the right-hand side of the system of ODEs in (7.36), f_i, involves linear combinations of rational functions of variables and parameters, then the algorithm developed in [304] can be used to find all the steady-states (the algorithm converts the steady-state equations into a system of polynomial equations and uses a globally convergent homotopy method to find all the roots of the system of polynomials). Most biological models of molecular networks have linear combinations of rational functions for the right-hand side of their system of ODEs (see the cell-cycle and λ-switch models studied in the next two sections for examples). In fact, the right-hand sides are usually even more restricted to mass action and Michaelis-Menten type kinetics. Mass action kinetics have the form $k * S_1 * S_2 * \cdots * S_n$; where k is a rate constant (parameter) and S_i represents the concentration of a protein (variable). Michaelis- Menten kinetics have the form $k * S * E/(K_m + S)$; where k is a rate constant (parameter), K_m is a Michaelis constant (parameter), S is the substrate concentration (variable), and E is the enzyme concentration (variable). Clearly, these kinetics are rational functions. Once the target steady-states are identified, the domains of attraction for each steady-state can be computed. Then, the switching rule of (7.38) ensures a priori where the system will end up upon switching at a given point in the state-space, provided that this point is within the domain of attraction of a stable steady-state. Finally, it should be noted that even in the rare case that a structure cannot be identified – and subsequently not all of the steady-states can be found – the proposed method still provides useful information regarding the feasibility of switching into any of the known steady-states by verifying whether the state at any given given time is contained within its domain of attraction.

Remark 7.34. The issue of robustness of the proposed approach with respect to model uncertainty can be explicitly handled by modifying the computation of the domains of attraction following the results presented in Chap. 4 to account for the presence of parametric model uncertainty in the computation of the domain of attraction using bounds on the variation of the model parameters.

Remark 7.35. The Lyapunov function-based approach that we follow for the construction of the domains of attraction for the individual stable steady-states yields a domain of attraction estimate that is dependent upon the specific Lyapunov function used. To improve upon the obtained estimate, one can use a group of different Lyapunov functions to come up with a larger estimate of the domain of attraction. Other methods for the construction of the Lyapunov function, such as Zubov's method (e.g., [74]) and the sum of squares decomposition approach [216], can also be used. Acceptability of the computed estimates should ultimately be judged with respect to the size of the expected operating regime. Once the domain of attraction estimates are obtained, the switching rule of (7.38) ensures that the system will go to a certain stable steady-state if the switching occurs at a point which is within the domain of attraction of this steady-state. Finally, we note that the case of multiple mode switchings can be handled in a sequential fashion – the same way that the first mode switch is handled – by tracking where the state is at the time of each switch.

Remark 7.36. It should be noted that the proposed approach is not limited by the dimensionality of the system under consideration, and applies to systems of any dimension. The estimation of the domain of attraction utilizes only simple algebraic computations and does not incur prohibitive computational costs with increasing dimensionality. In the simulation study presented below, the domains of attraction are plotted for the sake of a visual demonstration. However, a plot of the domain of attraction is not *required* for the implementation of the switching rule, and, therefore, poses no limitation when considering systems of higher dimensions. The knowledge of the domain of attraction is contained completely in the value of the level set, c_i, obtained when computing the estimate of the domain of attraction. At the time of implementation, to ascertain whether the state is within the domain of attraction requires only evaluating the Lyapunov function and verifying if $V_i(x(T)) \leq c_i$. To reduce the possible conservatism of the resulting estimate, it is often desirable to find the largest value of c_i for which the estimate $\Omega_{c_i} = \{x : V_i(x) \leq c_i\}$ is fully contained within Π_i. For this purpose, an iterative procedure to recompute (and enlarge) the estimate of the domain of attraction can be employed whereby the value of c_i is increased gradually in each iteration until a value, c_i^{max}, is reached where for any $c_i > c_i^{max}$, Ω_{c_i} is no longer fully contained in Π_i. The level set $\Omega_{c_i^{max}}$ then is the largest estimate of the domain of attraction that can be obtained using the level sets of the given Lyapunov function. Note that, for a given value of c_i in each iteration, the determination of whether Ω_{c_i} is fully contained in Π_i involves only algebraic computations and thus this iterative procedure does not incur prohibitive computational costs as the dimensionality of the system increases. The same procedure also applies when a family of Lyapunov functions is used to estimate the domain of attraction of a given steady-state. Finally, it should be noted that how close the obtained estimate is to the actual domain of attraction depends on the particular system structure as well as the method used to compute this estimate (in this case the particular Lyapunov functions chosen). In general, it is expected that the estimate will not capture the entire domain of attraction which implies that the union of all the estimates of the domains of attraction of all the steady-states will not cover the entire state-space. An implication of this, for the case when switching of the network is controlled externally and a priori stability guarantees are sought, is that switching should be delayed until the state trajectory enters the computed estimate of the domain of attraction of the desired target steady-state. The "gaps" between the different estimates (and hence the conservatism of the switching policy) can be reduced either with the help of dynamic simulations or by augmenting the individual estimates using any of the methods cited in Remark 7.35.

Remark 7.37. The proposed approach models biological networks using deterministic differential equations and does not account for possible network stochastic behavior. Such stochasticity can be modeled as uncertainty in the model parameters, and therefore be handled directly by modifying the computation of the domains of attraction in a way that accounts explicitly for the effect of parameter model uncertainty following the results presented in Chap. 4.

In the next subsection, we demonstrate, through computer simulations, an applications of this methodology to the analysis of mode transitions in a biological network that arises in eukaryotic cell cycle regulation.

7.5.3 Application to Eukaryotic Cell Cycle Regulation

We consider here an example network of biochemical reactions, based on cyclin-dependent kinases and their associated proteins, which are involved in cell cycle control in frog egg development. A detailed description of this network is given in [210] where the authors use standard principles of biochemical kinetics and rate equations to construct a nonlinear dynamic model of the network that describes the time evolution of the key species including free cyclin, the M-phase promoting factor (MPF), and other regulatory enzymes. The model parameters have either been estimated from kinetic experiments in frog egg extracts or assigned values consistent with experimental observations. For illustration purposes, we will consider below the simplified network model derived by the authors (focusing only on the positive-feedback loops in the network) which captures the basic stages of frog egg development. The model is given by:

$$\begin{aligned} \frac{du}{dt} &= \frac{k_1'}{G} - \left(k_2' + k_2'' u^2 + k_{wee}\right) u + \left(k_{25}' + k_{25}'' u^2\right) \left(\frac{v}{G} - u\right) \\ \frac{dv}{dt} &= k_1' - (k_2' + k_2'' u^2) v \end{aligned} \tag{7.39}$$

where $G = 1 + \frac{k_{INH}}{k_{CAK}}$, k_{INH} is the rate constant for inhibition of INH, a protein that negatively regulates MPF, k_{CAK} is the rate constant for activation of CAK, a cdc2-activating kinase, u is a dimensionless concentration of active MPF and v is a dimensionless concentration of total cyclin, k_2' and k_2'' are rate constants for the low-activity and high activity forms, respectively, of cyclin degradation; k_{25}' and k_{25}'' are rate constants for the low-activity and high activity forms, respectively, of tyrosine dephosphorylation of MPF; k_1' is a rate constant for cyclin synthesis, k_{wee} is the rate constant for inhibition of Wee1, an enzyme responsible for the tyrosine phosphorylation of MPF (which inhibits MPF activity) (see [210] for model derivation from the molecular mechanism, and Tables 7.3–7.4 for the parameter and steady-state values). Bifurcation and phase-plane analysis of the above model [210] shows that, by changing the values of k_2', k_2'' and k_{wee}, the following four modes of behavior are predicted:

- A G2-arrested state (blocked before the G2-M transition) characterized by high cyclin concentration and little MPF activity. This corresponds to a

Table 7.3. Parameter values for the cell cycle model in (7.39).

k_1'	$= 0.01$
k_{25}'	$= 0.04$
k_{25}''	$= 100$
k_{INH}	$= 0.1$
k_{CAK}	$= 1$

Table 7.4. Steady-state values (u_s, v_s) for the cell cycle model for different values of k_2', k_2'', k_{wee}.

k_2'	k_2''	k_{wee}	Mode	M-arrest state	$G2$-arrest state	Reference
0.01	10	3.5	G2-arrest	n/a	$(0.016, 0.802)$	
0.01	0.5	2.0	M-arrest	$(0.202, 0.329)$	n/a	
0.015	0.1	3.5	Bi-stable	$(0.276, 0.442)$	$(0.012, 0.666)$	
0.01	10	2.0	Oscillatory	n/a	n/a	Fig. 7.16b

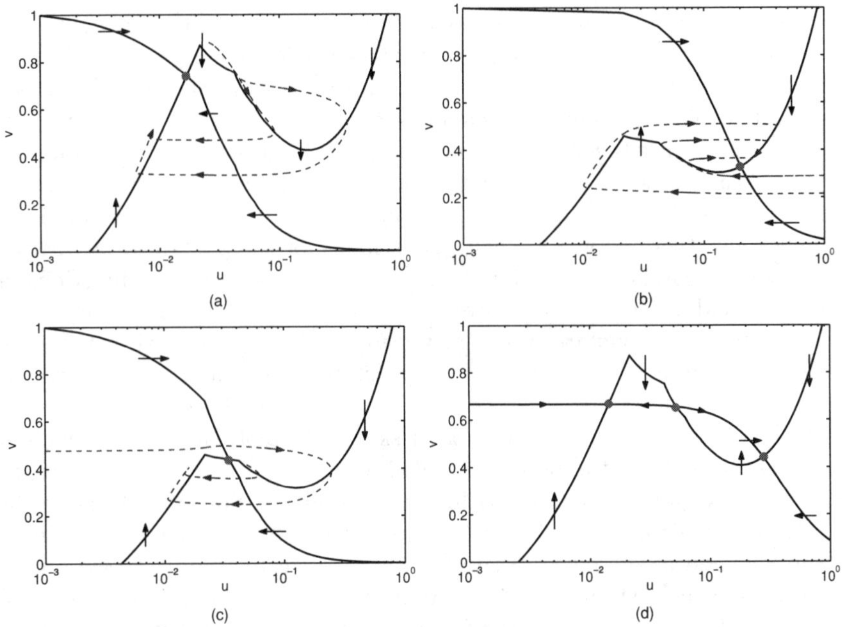

Fig. 7.15. Phase-plane portraits of the system of (7.39), for different values of k_2', k_2'', and k_{wee}, showing: (**a**) Stable steady-state with most MPF inactive, (**b**) Stable steady-state with most MPF active, (**c**) Unstable steady-state surrounded by a limit cycle, and (**d**) Bi-stability: two stable steady-states separated by an unstable saddle point

unique, asymptotically stable steady-state ($k'_2 = 0.01$, $k''_2 = 10$, $k_{wee} = 3.5$; see Fig. 7.15a).

- An M-arrested state (blocked before the meta- to anaphase transition) state with lots of active MPF. This corresponds to a unique, asymptotically stable steady-state ($k'_2 = 0.01$, $k''_2 = 0.5$, $k_{wee} = 2.0$; see Fig. 7.15b).
- An oscillatory state (alternating phases of DNA synthesis and mitosis) exhibiting sustained, periodic fluctuation of MPF activity and total cyclin protein. This corresponds to a stable limit cycle surrounding an unstable equilibrium point ($k'_2 = 0.01$, $k''_2 = 10$, $k_{wee} = 2.0$; see Fig. 7.15c).
- Co-existing stable steady-states of G2 arrest and M arrest. This corresponds to three steady-states; one unstable and two locally asymptotically stable ($k'_2 = 0.015$, $k''_2 = 0.1$, $k_{wee} = 3.5$; see Fig. 7.15d).

The above analysis predicts that slight increases in k'_2, k_{wee}, accompanied by a significant drop in k''_2 (which could be driven, for example, by down-regulation of cyclin degradation) can induce a transition from the oscillatory mode of MPF activity (early embryo stage) to the bi-stable mode. However, it is not clear from this analysis alone whether the cell will end up in a G2- or an M-arrested state upon switching. To address this question, we initially compute the domains of attraction of both steady-states in the bi-stable mode. This is done using a Lyapunov function of the form $V = (u - u_s)^4 + 10(v - v_s)^2$, where u_s and v_s are the steady-state values. The basic idea here is to compute, for each steady-state, the region in the (u, v) space where the time-derivative of V is negative-definite along the trajectories of the dynamical system of (7.39), and then use this region to obtain an estimate of the domain of attraction. While several candidate functions could be used, this particular function was found to yield acceptable estimates of the stability regions. The stability regions for both steady-states are depicted in Fig. 7.16a. The entire area above the dashed curve (the separatrix) is the stability region of the M-arrested state while the area below is the stability region of the G2-arrested state. Both steady-states are denoted by asterisks on the plot. By plotting the limit cycle (obtained from the oscillatory mode) on the same plot, we see that a portion of the limit cycle lies within the stability region of the M-arrested steady-state (segment A in Fig. 7.16a) while the rest is completely within the stability region of the steady-state for the G2-arrested mode. Based on this analysis, we conclude that switching from the oscillatory mode to the bi-stable mode would move the cell to the $G2$-arrested state only if the transition occurs at times when the state is not on segment A, while it would end up in the M-arrested state if switching were to occur on segment A. This conclusion is verified by the dotted and dash-dotted state trajectories, respectively, shown in Fig. 7.16b. The corresponding plots of the time-evolution of the states in both switching scenarios are given in Fig. 7.17 for two representative switching times. Note that because of the periodic nature of the solution in the oscillatory mode, there are many time-intervals, between $t = 0$ and $t = 333.5$ min, when the limit cycle trajectory is on segment A. These intervals are separated

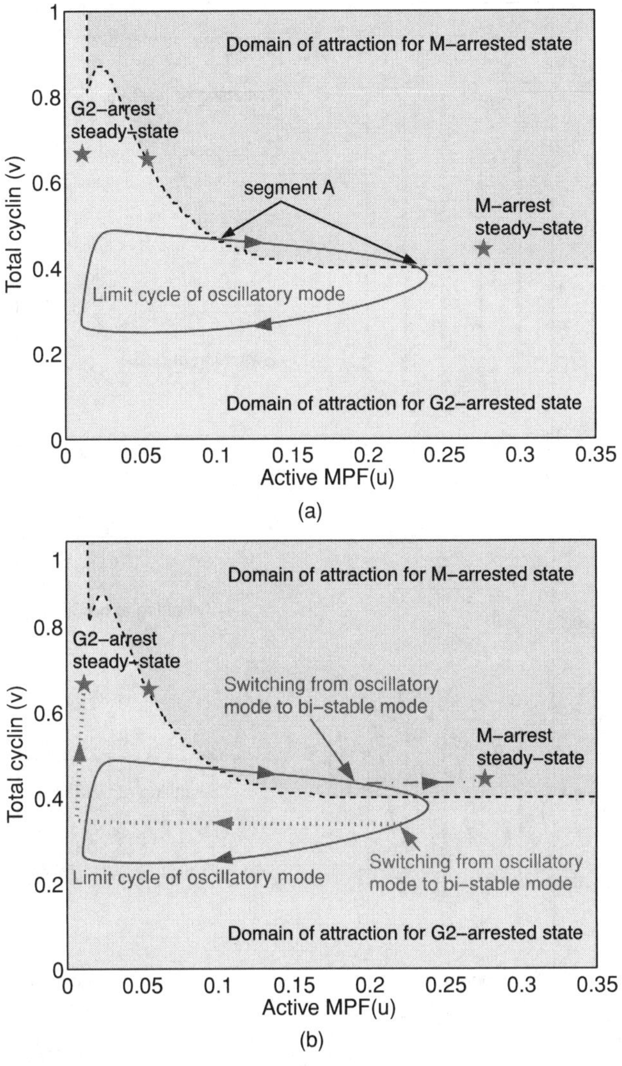

Fig. 7.16. (**a**) A plot showing the overlap of the limit cycle of the oscillatory mode with the domains of attraction for the M-arrested steady-state (*entire area above dashed curve*) and for the $G2$-arrested steady-state (*entire area below the dashed curve*), (b) A plot showing that switching from the oscillatory to the bi-stable mode moves the system to different steady-states depending on where switching takes place. In both cases, the oscillatory mode is fixed at $k_2' = 0.01$, $k_2'' = 10$, $k_{wee} = 2.0$

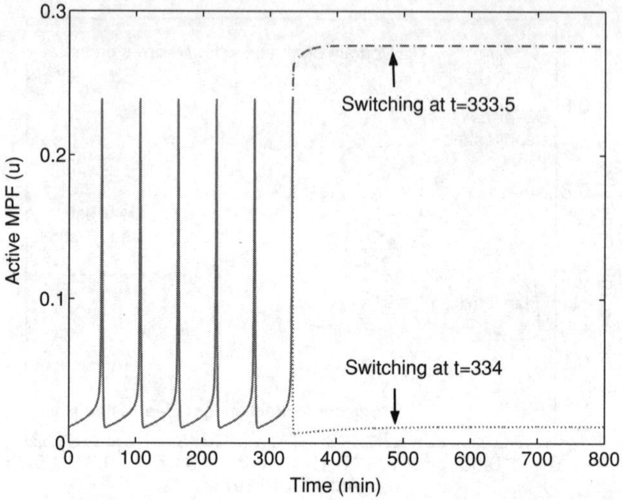

(a)

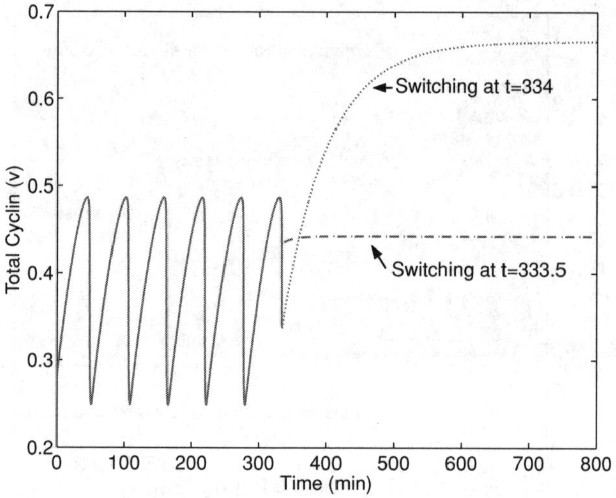

(b)

Fig. 7.17. The time evolution plots of (**a**) active MPF, and (**b**) total cyclin upon switching from the oscillatory to the bi-stable mode at two representative switching times. At $t = 333.5$ min, the state trajectory lies on segment A (see Fig. 7.16a) and therefore switching lands the state in the M-arrested steady-state (*dash-dotted line*), while at $t = 334$ min, switching lands the state in the $G2$-arrested steady-state (*dotted line*). In both cases, the oscillatory mode is fixed at $k_2' = 0.01$, $k_2'' = 10$, $k_{wee} = 2.0$

by one period of the limit cycle. Switching during any of these intervals to the bi-stable mode moves the system to the M-arrested state. Similarly, there are many time-intervals when the trajectory is not on segment A. Switching during any of those intervals will land the system at the $G2$-arrested state.

7.6 Conclusions

In this chapter, a number of hybrid control strategies were developed for broad classes of hybrid nonlinear processes. Initially, hybrid processes with actuator constraints and model uncertainty were considered, and a control strategy that coordinates, via multiple Lyapunov functions, the tasks of feedback controller synthesis and logic-based switching between the constituent modes was developed. A family of feedback controllers were designed to enforce robust stability within the constituent modes and provide an explicit characterization of the constrained region of robust stability for each mode. The stability regions were then used to derive stabilizing switching rules that orchestrate the transition between the various modes (and their controllers) in a way that guarantees robust stability for the overall hybrid process. Hybrid systems with scheduled mode transitions were considered next, and a Lyapunov-based predictive control strategy, that enforces both the switching schedule and closed-loop stability, was presented. The main idea was to design a Lyapunov–based predictive controller for each mode, and incorporate constraints in the predictive controller design to ensure that the prescribed transitions between the modes occur in a way that guarantees stability of the switched closed–loop system. The proposed control methods were demonstrated through applications to chemical process examples. Finally, the chapter was concluded with a demonstration of how hybrid systems techniques – in particular, the idea of coupling the switching logic to the stability regions – can be applied for the analysis of mode transitions in biological networks.

8

Fault-Tolerant Control of Process Systems

8.1 Introduction

Safety and reliability are primary goals in the operation of industrial chemical plants. An important national need currently exists for enhancing the safety and reliability of chemical plants in ways that reduce their vulnerability to serious failures. Increasingly faced with the requirements of operational flexibility under tight performance specifications and other economic drivers, plant operation is relying extensively on highly automated process control systems. Automation, however, tends to increase vulnerability of the plant to faults, such as defects/malfunctions in process equipment, sensors and actuators, failures in the controllers or in the control loops, which, if not appropriately handled in the control system design, can potentially cause a host of undesired economic, environmental, and safety problems that seriously degrade the operating efficiency of the plant. These considerations provide a strong motivation for the development of systematic methods and strategies for the design of fault–tolerant control systems and have motivated many research studies in this area (see, for example, [26, 287, 295] and [38, 182, 218] for references).

Given the complex dynamics of chemical processes (due, for example, to the presence of nonlinearities and constraints) and the geographically distributed, interconnected nature of plant units, as well as the large number of distributed sensors and actuators typically involved, the success of any fault–tolerant control strategy requires an integrated approach that brings together several essential elements, including: (1) the design of advanced feedback control algorithms that handle complex dynamics effectively, (2) the design of supervisory switching schemes that orchestrate the transition from the failed control configuration to available well–functioning fall–back configurations to ensure fault–tolerance, and (3) the efficient exchange of information and communication between the different plant units through a high–level supervisor that coordinates the overall plant response in failure situations and minimizes the effects of failure propagation.

The realization of such an approach is increasingly aided by a confluence of recent, and ongoing, advances in several areas of process control research, including advances in nonlinear controller designs for chemical processes (e.g., [78,81,138,141,275]) and advances in the analysis and control of hybrid process systems leading to the development of a systematic framework for the integration of feedback and supervisory control [80,82]. A hybrid systems framework provides a natural setting for the analysis and design of fault–tolerant control systems since the occurrence of failure and subsequent switching to fall–back control configurations induce discrete transitions superimposed on the underlying continuous dynamics. Hybrid control techniques have been useful in dealing with a wide range of problems that cannot be addressed using classical control approaches, including fault–tolerant control of spatially–distributed systems (e.g., [84]), control of processes with switched dynamics (e.g., [32,82]), and the design of hybrid predictive control structures that overcome some of the limitations of classical predictive control algorithms (e.g., [88]). In addition to control studies, research work on hybrid systems spans a diverse set of problems ranging from the modeling (e.g., [28,294]) and simulation (e.g., [28,100]) to the optimization (e.g., [106,110]) and stability analysis (e..g, [70,120]) of several classes of hybrid systems.

In addition to the above fundamental advances, recent innovations in actuator/sensor and communication technologies are increasingly enabling the integration of communication and control domains [297]. For example, the use of communication networks as media to interconnect the different components in an industrial control system is rapidly increasing and expected to replace the more costly point–to–point connection schemes currently employed in distributed control systems. Figure 8.1 shows the basic networked control architecture for (a) a single–unit plant with few actuators and sensors (centralized structure) and (b) a larger plant with several interconnected processing units and larger number of actuators and sensors (distributed hierarchical structure). Currently, networked control systems is an active area of research within control engineering (e.g., see [203, 271, 282, 293] for some recent results and references in this area). In addition to the advantages of reduced system wiring (reduced installation, maintenance time and costs) in this architecture, the increased flexibility and ease of maintenance of a system using a network to transfer information is an appealing goal. In the context of fault–tolerant control in particular, systems designed in this manner allow for easy modification of the control strategy by rerouting signals, having redundant systems that can be activated automatically when component failure occurs, and in general they allow having a high–level supervisor control over the entire plant. The appealing features of communication networks motivate investigating ways for integrating them in the design of fault–tolerant control systems to ensure a timely and coordinated response of the plant in ways that minimize the effects of failure propagation between plant units. This entails devising strategies to deal with some of the fundamental issues introduced by the network, including issues of bandwidth limitations, quantization effects,

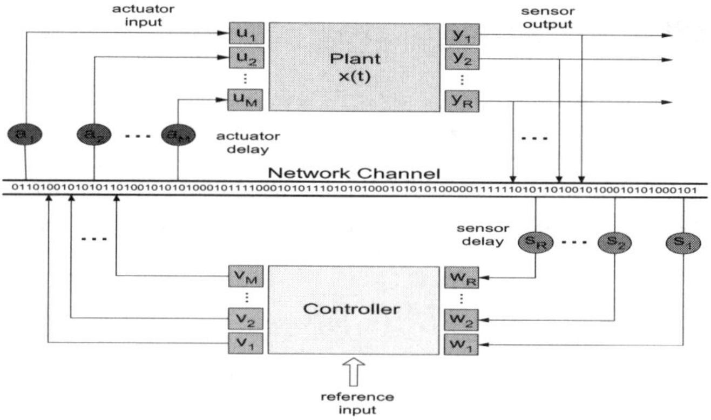

Fig. 8.1. Block diagrams of: (**a**) a centralized networked control system for a single–unit plant, and (**b**) a hierarchical distributed networked control architecture for a multi–unit plant

network scheduling, and communication delays, which continue to be topics of active research.

Motivated by the above considerations, we develop in this chapter a fault–tolerant control system design methodology, for plants with multiple (distributed) interconnected processing units, that accounts explicitly for the inherent complexities in supervisory control and communication tasks resulting from the distributed interconnected nature of plant units. The approach brings together tools from Lyapunov–based control and hybrid systems theory and is based on a hierarchical distributed architecture that integrates lower–level feedback control of the individual units with upper–level logic–based supervisory control over communication networks. The local control systems consist each of a family of feedback control configurations together with a local supervisor that communicates with actuators and sensors, via a local communication network, to orchestrate the transition between control configurations, on the basis of their fault–recovery regions, in the event of failures. The local supervisors communicate, through a plant–wide communication network, with a plant supervisor responsible for monitoring the different units and coordinating their responses in a way that minimizes the propagation of failure effects. The communication logic is designed to ensure efficient transmission of information between units while also respecting the inherent limitations in network resources by minimizing unnecessary network usage and accounting explicitly for the effects of possible delays due to fault–detection, control computations, network communication, and actuator activation. The proposed approach provides explicit guidelines for managing the interplays between the coupled tasks of feedback control, fault–tolerance and communication. The efficacy of the proposed approach is demonstrated through chemical process examples. The results of this chapter were first presented in [86].

8.2 Preliminaries

8.2.1 System Description

We consider a plant composed of l connected processing units, each of which is modeled by a continuous–time multivariable nonlinear system with constraints on the manipulated inputs, and represented by the following state–space description:

$$\begin{aligned}
\dot{x}_1 &= f_1^{k_1}(x_1) + G_1^{k_1}(x_1)u_1^{k_1} \\
\dot{x}_2 &= f_2^{k_2}(x_2) + G_2^{k_2}(x_2)u_2^{k_2} + W_{2,1}^{k_2}(x_2)x_1 \\
&\vdots \\
\dot{x}_l &= f_l^{k_l}(x_l) + G_l^{k_l}(x_l)u_l^{k_l} + \sum_{p=1}^{l-1} W_{l,p}^{k_l}(x_l)x_p \\
\|u_i^{k_i}\| &\leq u_{i,max}^{k_i} \\
k_i(t) &\in \mathcal{K}_i := \{1,\ldots,N_i\}, \quad N_i < \infty, \quad i = 1,\ldots,l
\end{aligned} \tag{8.1}$$

where $x_i := [x_i^{(1)} \; x_i^{(2)} \; \cdots \; x_i^{(n_i)}]^T \in \mathbb{R}^{n_i}$ denotes the vector of process state variables associated with the i–th processing unit, $u_i^{k_i} := [u_{i,1}^{k_i} u_{i,2}^{k_i} \cdots u_{i,m_i}^{k_i}]^T \in \mathbb{R}^{m_i}$ denotes the vector of constrained manipulated inputs associated with the k_i–th control configuration in the i–th processing unit, $u_{i,max}^{k_i}$ is a positive real number that captures the maximum size of the vector of manipulated inputs dictated by the constraints, $\|\cdot\|$ denotes the Euclidean norm of a vector, and N_i is the number of different control configurations that can be used to control the i–th processing unit. The index, $k_i(t)$, which takes values in the finite set $\mathcal{K}_i$, represents a discrete state that indexes the right–hand side of the set of differential equations in (8.1). For each value that k_i assumes in $\mathcal{K}_i$, the i–th processing unit is controlled via a different set of manipulated inputs which define a given control configuration. For each unit, switching between the available N_i control configurations is controlled by a local supervisor that monitors the operation of the unit and orchestrates, accordingly, the transition between the different control configurations in the event of control system failures. This in turn determines the temporal evolution of the discrete state, $k_i(t)$, which takes the form of a piecewise constant function of time. The local supervisor ensures that only one control configuration is active at any given time, and allows only a finite number of switches over any finite interval of time.

Without loss of generality, it is assumed that $x_i = 0$ is an equilibrium point of the uncontrolled i–th processing unit (i.e., with $u_i^{k_i} = 0$) and that the vector functions, $f_i^{k_i}(\cdot)$, and the matrix functions, $G_i^{k_i}(\cdot)$ and $W_{j,p}^{k_j}(\cdot)$, are sufficiently smooth on their domains of definition, for all $k_i \in \mathcal{K}_i$, $i = 1, \ldots, l$, $j = 2, \ldots, l$, $p = 1, \ldots, l-1$. For the j–th processing unit, the term, $W_{j,p}^{k_j}(x_j)x_p$, represents the connection that this unit has with the p–th unit upstream. Note from the summation notation in (8.1) that each processing unit can in general be connected to all the units upstream from it. Our nominal control objective (i.e., in the absence of control system failures) is to design, for each processing unit, a stabilizing feedback controller that enforces asymptotic stability of the origin of the closed–loop system in the presence of control actuator constraints. To simplify the presentation of our results, we will focus only on the state feedback control problem where measurements of all process states are available for all times.

8.2.2 Problem Statement and Solution Overview

Consider the plant of (8.1) where, for each processing unit, a stabilizing feedback control system has been designed and implemented. Given some catastrophic fault – that has been detected and isolated – in the actuators of one of the control systems, our objective is to develop a plant–wide fault–tolerant control strategy that: (1) preserves closed–loop stability of the failing unit, if possible, and (2) minimizes the negative impact of this failure on the closed–loop stability of the remaining processing units downstream. To accomplish both of these objectives, we construct a hierarchical control structure that integrates lower–level feedback control of the individual units with upper–level logic–based supervisory control over communication networks. The local control system for each unit consists of a family of control configurations for each of which a stabilizing feedback controller is designed and the stability region is explicitly characterized. The actuators and sensors of each configuration are connected, via a local communication network, to a local supervisor that orchestrates switching between the constituent configurations, on the basis of the stability regions, in the event of failures. The local supervisors communicate, through a plant–wide communication network, with a plant supervisor responsible for monitoring the different units and coordinating their responses in a way that minimizes the propagation of failure effects. The basic problem under investigation is how to coordinate the tasks of feedback, control system reconfiguration and communication, both at the local (processing unit) and plant–wide levels in a way that ensures timely recovery in the event of failure and preserves closed–loop stability.

Remark 8.1. In the design of any fault–tolerant control system, an important task that precedes the control system reconfiguration is the task of fault–detection and isolation (FDI). There is an extensive body of literature on this topic including, for example, the design of fault–detection and isolation schemes based on fundamental process models (e.g., [69, 95]) and statistical/pattern recognition and fault diagnosis techniques (e.g., [12, 66, 113, 209,

278,286]). In this work, we focus mainly on the interplay between the communication network and the control system reconfiguration task. To this end, we assume that the FDI tasks take place at a time scale that is very fast compared to the time constant of the overall process dynamics and the time needed for the control system reconfiguration, and thus can be treated separately from the control system reconfiguration (we note that the time needed for FDI is accounted for in the control system reconfiguration through a time–delay; see the Sects. 8.3 and 8.4 for details). In the context of process control applications, this sequential and decoupled treatment of FDI and control system reconfiguration is further justified by the overall slow dynamics of chemical plants.

8.2.3 Motivating Example

In this section, we introduce a simple benchmark example that will be revisited later to illustrate the design and implementation aspects of the fault–tolerant control design methodology to be proposed in Sect. 8.3. While the discussion will center around this example, we note that the proposed framework can be applied to more complex plants involving more complex arrangements of processing units as shown in (8.1). To this end, consider two well–mixed, non–isothermal continuous stirred tank reactors (CSTRs) in series, where three parallel irreversible elementary exothermic reactions of the form $A \overset{k_1}{\to} B$, $A \overset{k_2}{\to} U$ and $A \overset{k_3}{\to} R$ take place, where A is the reactant species, B is the desired product and U, R are undesired byproducts. The feed to CSTR 1 consists of pure A at flow rate F_0, molar concentration C_{A0} and temperature T_0, and the feed to CSTR 2 consists of the output of CSTR 1 and an additional fresh stream feeding pure A at flow rate F_3, molar concentration C_{A03} and temperature T_{03}. Due to the non–isothermal nature of the reactions, a jacket is used to remove/provide heat to both reactors. Under standard modeling assumptions, a mathematical model of the plant can be derived from material and energy balances and takes the following form:

$$
\begin{aligned}
\frac{dT_1}{dt} &= \frac{F_0}{V_1}(T_0 - T_1) + \sum_{i=1}^{3} \frac{(-\Delta H_i)}{\rho c_p} R_i(C_{A1}, T_1) + \frac{Q_1}{\rho c_p V_1} \\
\frac{dC_{A1}}{dt} &= \frac{F_0}{V_1}(C_{A0} - C_{A1}) - \sum_{i=1}^{3} R_i(C_{A1}, T_1) \\
\frac{dT_2}{dt} &= \frac{F_1}{V_2}(T_1 - T_2) + \frac{F_3}{V_2}(T_{03} - T_2) + \sum_{i=1}^{3} \frac{(-\Delta H_i)}{\rho c_p} R_i(C_{A2}, T_2) + \frac{Q_2}{\rho c_p V_2} \\
\frac{dC_{A2}}{dt} &= \frac{F_1}{V_2}(C_{A1} - C_{A2}) + \frac{F_3}{V_2}(C_{A03} - C_{A2}) - \sum_{i=1}^{3} R_i(C_{A2}, T_2)
\end{aligned}
\tag{8.2}
$$

where $R_i(C_{Aj}, T_j) = k_{i0} \exp\left(\frac{-E_i}{RT_j}\right) C_{Aj}$, for $j = 1, 2$. T, C_A, Q, and V denote the temperature of the reactor, the concentration of species A, the rate of heat input/removal from the reactor, and the volume of reactor, respectively, with subscript 1 denoting CSTR 1 and subscript 2 denoting CSTR 2. ΔH_i, k_i, E_i, $i = 1, 2, 3$, denote the enthalpies, pre–exponential constants and activation energies of the three reactions, respectively, c_p and ρ denote the heat capacity and density of the fluid in the reactor. Using typical values for the process parameters (see Table 8.1), CSTR 1, with $Q_1 = 0$, has three steady–states: two locally asymptotically stable and one unstable at $(T_1^s, C_{A1}^s) = (388.57\ K, 3.59\ kmol/m^3)$. The unstable steady–state of CSTR 1 corresponds to three steady–states for CSTR 2 (with $Q_2 = 0$), one of which is unstable at $(T_2^s, C_{A2}^s) = (429.24\ K, 2.55\ kmol/m^3)$.

Table 8.1. Process parameters and steady–state values for the reactors of (8.2)

F_0	$= 4.998$	m^3/hr
F_1	$= 4.998$	m^3/hr
F_3	$= 30.0$	m^3/hr
V_1	$= 1.0$	m^3
V_2	$= 3.0$	m^3
R	$= 8.314$	$KJ/kmol \cdot K$
T_0	$= 300.0$	K
T_{03}	$= 300.0$	K
C_{A0}	$= 4.0$	$kmol/m^3$
C_{A03}^s	$= 2.0$	$kmol/m^3$
ΔH_1	$= -5.0 \times 10^4$	$KJ/kmol$
ΔH_2	$= -5.2 \times 10^4$	$KJ/kmol$
ΔH_3	$= -5.4 \times 10^4$	$KJ/kmol$
k_{10}	$= 3.0 \times 10^6$	hr^{-1}
k_{20}	$= 3.0 \times 10^5$	hr^{-1}
k_{30}	$= 3.0 \times 10^5$	hr^{-1}
E_1	$= 5.0 \times 10^4$	$KJ/kmol$
E_2	$= 7.53 \times 10^4$	$KJ/kmol$
E_3	$= 7.53 \times 10^4$	$KJ/kmol$
ρ	$= 1000.0$	kg/m^3
c_p	$= 0.231$	$KJ/kg \cdot K$
T_1^s	$= 388.57$	K
C_{A1}^s	$= 3.59$	$kmol/m^3$
T_2^s	$= 429.24$	K
C_{A2}^s	$= 2.55$	$kmol/m^3$

The control objective is to stabilize both reactors at the (open–loop) unstable steady–states. Operation at these points is typically sought to avoid high temperatures, while simultaneously achieving reasonable conversion. To accomplish the control objective under normal conditions (with no failures),

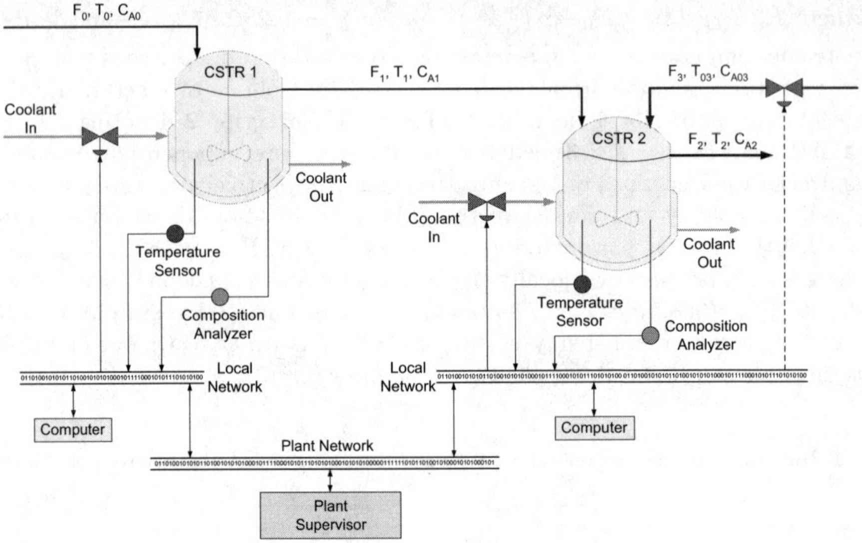

Fig. 8.2. Process flow diagram of two CSTR units in series

we choose as manipulated inputs the rates of heat input, $u_1^1 = Q_1$, subject to the constraint $|Q_1| \leq u_{max}^{Q_1} = 2.7 \times 10^6 \ KJ/hr$ and $u_1^2 = Q_2$, subject to the constraint $|Q_2| \leq u_{max}^{Q_2} = 2.8 \times 10^6 \ KJ/hr$.

As shown in Fig. 8.2, each unit has a local control system with its sensors and actuators connected through a communication network. The local control systems in turn communicate with the plant supervisor (and with each other) through a plant–wide communication network. Note that in designing each control system, only measurements of the local process variables are used (for example, the controller for the second unit uses only measurements of T_2 and C_{A2}). This decentralized architecture is intended to minimize unnecessary communication costs incurred by continuously sending measurement data from the first to the second unit over the network. We note that while this issue may not be a pressing one for the small plant considered here (where a centralized structure can in fact be easily designed), real plants nonetheless involve a far more complex arrangement of units with thousands of actuators and sensors, which makes the complexity of a centralized structure as well as the cost of using the network to share measurements between units quite significant. For this reason, we choose the distributed structure in Fig. 8.2 in order to highlight some of the manifestations of the inherent interplays between the control and communication tasks.

The fault–tolerant control problem under consideration involves a total failure in both control systems after some time of startup, with the failure in the first unit being permanent. Our objective will be to preserve closed–loop stability of CSTR 2 by switching to an alternative control configuration involving, as manipulated variables, the rate of heat input, $u_2^1 = Q_2$, subject to the same constraint, and the inlet reactant concentration, $u_2^2 = C_{A03} - C_{A03}^s$, subject to the constraint $|C_{A03} - C_{A03}^s| \leq u_{max}^{C_{A03}} = 0.4\ kmol/m^3$ where $C_{A03}^s = 3.0\ kmol/m^3$. The main question, which we address in the next section, is how to devise the switching and network communication logics in a way that ensures fault–tolerance in the second unit and, simultaneously, accounts for the inherent limitations in network resources and possible delays in fault–detection, communication and actuator activation.

8.3 Fault–Tolerant Control System Design Methodology

In this section, we outline the main steps involved in the fault–tolerant control system design procedure. These include: (1) the synthesis of a stabilizing feedback controller for each of the available fall–back control configurations, (2) the explicit characterization of the stability region for each configuration which characterize the operating conditions for which fault–recovery can be guaranteed, (3) the design of a switching law that orchestrates the re–configuration of the failing control system in a way that safeguards closed–loop stability in the event of failures, and (4) the design of the network communication logic in a way that minimizes the propagation of failure effects between plant units while also accounting for bandwidth constraints and delays. A major feature of the design methodology is the inherent coupling between the aforementioned tasks whereby each task affects how the rest are carried out. Below is a more detailed description of each step and a discussion on how the tradeoffs between the different steps are managed.

8.3.1 Constrained Feedback Controller Synthesis

Referring to the system of (8.1), consider first the case when no failures take place anywhere in the plant. Under such conditions, our objective is to design, for each processing unit, a "nominal" feedback controller that enforces asymptotic closed–loop stability and provides an explicit characterization of the stability region under actuator constraints. One way to do this is to use Lyapunov–based control techniques. Specifically, consider the nonlinear system describing the i–th processing unit under the k_i–th control configuration, for which a control Lyapunov function, $V_i^{k_i}$, is available. Using this function, one can construct the following bounded nonlinear control law (see [78, 177]):

$$u_i^{k_i} = -r(x_i, u_{i,max}^{k_i})\beta^T(x_i) \tag{8.3}$$

where

$$r(x_i, u_{i,max}^{k_i}) = \frac{\alpha^*(x_i) + \sqrt{(\alpha^*(x_i))^2 + \left(u_{i,max}^{k_i}\|\beta^T(x_i)\|\right)^4}}{\|\beta^T(x_i)\|^2 \left[1 + \sqrt{1 + (u_{i,max}^{k_i}\|\beta^T(x_i)\|)^2}\right]} \tag{8.4}$$

$\alpha^*(x_i) = \alpha(x_i) + \rho_i^{k_i}\|x_i\|^2$, $\rho_i^{k_i} > 0$ is a real number, $\alpha(x_i) = L_{f_i^{k_i}} V_i^{k_i}(x_i)$, $\beta^T(x_i) = (L_{G_i^{k_i}} V_i^{k_i})^T(x_i)$, the notation $L_{f_i^{k_i}} V_i^{k_i}$ is used to denote the Lie derivative of the scalar function, $V_i^{k_i}$, with respect to the vector field, $f_i^{k_i}$, and $L_{G_i^{k_i}} V_i^{k_i}$ is a row vector whose constituent components are the Lie derivatives of $V_i^{k_i}$ along the column vectors of the matrix $G_i^{k_i}$. Note that the control law of (8.3–8.4) requires measurements of the local process state variables, x_i, only and not measurements from other plant units upstream. This fully decentralized design is motivated by the desire to minimize unnecessary communication costs which would be incurred when sharing measurement data between the different units over the communication network. By disregarding the interconnections between the units in the controller design, however, closed–loop stability for a given unit rests on the stability properties of the upstream units. In particular, using a combination of Lyapunov and small–gain theorem type arguments, one can show that, starting from any invariant subset (e.g, a level–set of $V_i^{k_i}$) of the region described by:

$$\Phi_i(u_{i,max}^{k_i}) := \{x_i \in \mathbb{R}^{n_i} : \alpha(x_i) + \rho_i^{k_i}\|x_i\|^2 \leq u_{i,max}^{k_i}\|\beta^T(x_i)\|\} \quad (8.5)$$

the control law of (8.3–8.4) asymptotically stabilizes the i–th unit, under the k_i–th control configuration, at the origin provided that the closed–loop states of the upstream units, $x_1, x_2, \cdots, x_{i-1}$, converge asymptotically to the origin. In this case, and because of the way the various units are connected (see (8.1)), the closed–loop states of the upstream units can be viewed as bounded vanishing perturbations that affect the i–th unit and, therefore, a control law that asymptotically stabilizes the unperturbed i–th unit (i.e., disregarding the upstream states) also stabilizes the closed–loop system when the perturbations (connections) are added.

Having designed the nominal feedback control systems, we now proceed to consider the effect of control actuator failure on the feedback controller design for each unit. To this end, let us consider a total failure in the actuators of the k_i–th control configuration in the i–th control system. This failure, if not addressed properly, can lead to closed–loop instabilities both within the i–th processing unit itself (where the failure has occurred) and within all the remaining units downstream. Minimizing the effects of failure propagation throughout the plant can be achieved in one of two ways. The first involves reconfiguring the local control system of the i–th unit – once the failure is detected and isolated – by appropriately switching from the malfunctioning control configuration to some well–functioning fall–back configuration (recall that each processing unit has a family of control configurations). If this is feasible and can be done sufficiently fast, then the inherent fault–tolerance of the local control system is sufficient to preserve closed–loop stability not only for the i-th unit with the failing control system but also for the other units downstream without having to reconfigure their control systems. However, if local fault–recovery is not possible (this can happen, for example, in cases when the failure occurs at times that the state lies outside the stability regions of all the available fall–back control configurations; see Sect. 8.3.2 for details), then it becomes necessary to communicate the failure information to the control systems downstream and reconfigure them in order to preserve their closed–loop stability.

The main issue here is how to design the feedback control law for a given fall–back configuration in the units downstream in a way that respects the actuators' constraints and guarantees closed–loop stability despite the failure in the control system of some upstream unit. The choice of the feedback law depends on our choice of the communication policy. To explain this interdependence, we first note that a total failure in the control system of the i–th unit will cause its state, x_i, to move away from the origin (possibly settling at some other steady–state). Therefore, unless the nominal feedback controllers for the downstream units, $i+1,\ i+2,\ \cdots, l$, are re–designed to account for this incoming "disturbance", the evolution of their states, $x_{i+1},\ x_{i+2},\ \cdots, x_l$, will be adversely affected driving them away from the origin. To account for the disturbance caused by the upstream control system failure, one option is to send available measurements of x_i, through the communication network, to the affected units and redesign their controllers accordingly. From a communications cost point of view, however, this option may be costly since it requires continued usage of the network resources after the failure, which can adversely affect the performance of other units sharing the same communication medium due to bandwidth limitations and overall delays.

To reduce unnecessary network usage, we propose an alternative approach where the failure in the i–th processing unit is viewed as a bounded non–vanishing disturbance affecting units $i+1,\ i+2,\ \cdots, l$, and use the available process models of these units to capture, or estimate, the size of this disturbance (by comparing, for example, the evolution of the process variables for the i–th unit under the failed and well–functioning control configurations through simulations). In this formulation, state measurements from the i–th unit need not be shared with the other units; instead, only bounds on the disturbance size are transmitted to the downstream units. This approach involves using the network only once at the failure time and not continuously thereafter. The disturbance information can then be used to design an appropriate robust controller for each downstream unit to attenuate the effect of the incoming disturbance and enforce robust closed–loop stability. To illustrate how this can be done, let us assume that the failure in the control system of unit i occurs at $t = T_f$ and that the failure is detected immediately (the effect of possible delays in fault–detection and how to account for them are discussed in Sect. 8.3.4). Consider some unit, j, downstream from the i–th unit, that is described by the following model:

$$\dot{x}_j = f_j^{k_j}(x_j) + G_j^{k_j}(x_j)u_j^{k_j} + \delta_i \sum_{p=1}^{i-1} W_{j,p}^{k_j}(x_j)x_p + \sum_{p=i}^{j-1} W_{j,p}^{k_j}(x_j)\theta_p \tag{8.6}$$

for $i = 1, \ldots, l-1$, $j = i+1, \cdots, l$, where $\delta_i = 0$ for $i = 1$, and $\delta_i = 1$ for $i = 2, \ldots, l-1$. The third term on the right–hand side of (8.6) describes the input from all the units upstream of unit i. The θ_p's are time–varying, but bounded, functions of time that describe the evolution of the states of the i–th unit and all the units downstream from unit i but upstream from unit j (i.e., $\theta_p(t) = x_p(t)$, $p = i, \ldots, j-1$). The choice of using the notation θ_p, instead of x_p, for units $i, \cdots, j-1$ is intended to distinguish the effect of these units (where the failure originates and propagates downstream) as non–vanishing disturbances to the j–th unit, compared with the units upstream from unit i which are unaffected by the failure. Note that for unit $j = i+1$, which immediately follows the failing unit, the only source of disturbances that should be accounted for in its controller design is that coming from the i–th unit with the failing control system. However, for units that lie further downstream, i.e., for $j = i+2, \cdots, l$, the controller design needs to account for the additional disturbances resulting from the effect of the failure on the intermediate units separating units i and j.

For a system of the form of (8.6), one possible choice of a stabilizing controller is the following bounded robust Lyapunov–based control law proposed in Chap. 7 (see also Chap. 4 which has the general form:

$$u_j^{k_j} = -r_j(x_j, u_{j,max}^{k_j}, \theta_b)\beta^T(x_j), \tag{8.7}$$

where

$$r_j(x_j, u_{j,max}^{k_j}, \theta_b) = \frac{\alpha_1(x_j) + \sqrt{(\alpha_2(x_j))^2 + \left(u_{j,max}^{k_j}\|\beta^T(x_j)\|\right)^4}}{(\|\beta^T(x_j)\|)^2\left[1 + \sqrt{1 + (u_{j,max}^{k_j}\|\beta^T(x_j)\|)^2}\right]} \tag{8.8}$$

$$\alpha_1(x_j) = \alpha(x_j) + \left(\rho_j^{k_j}\|x_j\| + \sum_{p=i}^{j-1}\chi_j^{k_j}\theta_b^p(T_f)\|\omega_p^T(x_j)\|\right)\left(\frac{\|x_j\|}{\|x_j\| + \phi_j^{k_j}}\right) \tag{8.9}$$

$$\alpha_2(x_j) = \alpha(x_j) + \rho_j^{k_j}\|x_j\| + \sum_{p=i}^{j-1}\chi_j^{k_j}\theta_b^p(T_f)\|\omega_p^T(x_j)\| \tag{8.10}$$

$\theta_b^p(T_f) := \max\limits_{t \geq T_f}\|x_p(t)\|$, $p = i, \ldots, j-1$ are positive real numbers the capture the size of the disturbances, originating from the failure in the control system of the i–th unit and propagating downstream, $\omega_p(x_j) = (L_{W_{j,p}^{k_j}}V_j^{k_j})(x_j)$ is a row vector whose constituent components are the Lie derivatives of $V_j^{k_j}$ along the column vectors of the matrix $W_{j,p}^{k_j}$, $V_j^{k_j}$ is a robust control Lyapunov function for the j–th system under the k_j–th control configuration, and $\rho_j^{k_j} > 0$, $\chi_j^{k_j} > 1$, $\phi_j^{k_j} > 0$ are tuning parameters. Estimates of the disturbance bounds, θ_b^p, can be obtained by comparing, through simulations for example, the responses of the p–th unit under the pre– and post–failure configurations (see Sect. 8.4 for an example). It should be noted that since all the incoming disturbances to unit j take effect only after T_f, the controller of (8.7–8.10) is implemented only for $t \geq T_f$. For $t < T_f$, the nominal controllers of (8.3–8.4) are used.

Remark 8.2. When compared with the nominal controller of (8.3–8.4), we observe that the nonlinear gain function for the fall–back controller, $r_j(\cdot)$ in (8.7–8.10), depends not only on the size of actuator constraints, $u_{j,max}^{k_j}$, and the particular fall–back control configuration being used, k_j, but also on the size of the disturbances caused by the occurrence of failure, θ_b^p. This gain re–shaping procedure is carried out in order to guarantee constraint satisfaction and enforce robust closed–loop stability, with an arbitrary degree of attenuation of the effect of the failure on the j–th unit downstream. Note that, owing to the assumption of a persistent failure in the i–th unit (i.e., a non–vanishing disturbance), asymptotic closed–loop stability cannot be achieved for any of the units downstream. Instead, practical stability can be enforced whereby the states of each unit are driven, in finite–time, to a neighborhood of the origin whose size can be made arbitrarily small by selecting the controller tuning parameters ($\rho_j^{k_j}$, $\chi_j^{k_j}$, $\phi_j^{k_j}$) appropriately (see [81] for a detailed proof). These closed–loop properties are enforced within a well–defined state–space region that is explicitly characterized in Sect. 8.3.2.

Remark 8.3. Note that since the processing units upstream of unit i are not affected by its failing control system, the nominal controllers designed for these units (see (8.3–8.4)) will asymptotically stabilize their states, x_p, $p = 1, \ldots, i-1$, at the origin regardless of the failure; hence these state can be viewed as bounded vanishing inputs to the j–th unit and thus need not be accounted for in the controller design. The terms describing the intermediate units, $p = i+1, \ldots, j-1$ cannot however be treated as vanishing inputs. The reason is that even if the control systems of these units are immediately and appropriately re–configured to suppress the effect of the failure, their controllers, as discussed above, will at best be able to drive the states of these units, in finite time, only near the origin without achieving asymptotic convergence.

Remark 8.4. It should be noted that the fault–tolerant control system design methodology presented in this section is not restricted to the use of the bounded controller designs given in (8.3–8.4) (for the nominal case) and in (8.7–8.10) (for the case with failure). Any other stabilizing controller design that accounts for the constraints, enforces the desired robustness properties under failure, and provides an explicit characterization of the stability region can be used.

Remark 8.5. The treatment of failure in the control system of unit i as a bounded disturbance is rooted in the assumption that x_i, while moving away from the origin after failure, will eventually settle at some other (undesirable) steady–state (recall that this is how the disturbance bound is computed). In the case that the i–th processing unit has only a single steady–state in the post–failure configuration, however, the failure event cannot be treated as a bounded disturbance since x_i will simply grow unbounded after the failure, and will not settle at any point. In such a case, unless the control system of unit i is repaired in time, a shutdown of the plant will be unavoidable.

8.3.2 Characterization of Fault–Recovery Regions

Consider once again the j–th processing unit described by the model of (8.6). In Sect. 8.3.1, we outlined how to design, for a given fall–back control configuration, $k_j \in \mathcal{K}_j$, a robust feedback controller that, when implemented, can preserve closed–loop stability for this unit in the event of control system failure in some upstream unit, i. Given that actuator constraints place fundamental limitations on the ability of the controller to steer the closed–loop dynamics at will, it is important for the control system designer to explicitly characterize these limitations by identifying, or estimating, the set of admissible states starting from where the controller of (8.7–8.10) is guaranteed to robustly stabilize the closed–loop system for unit j (region of robust closed–loop stability). Since suppression of the upstream failure effects on unit j is formulated as a robust stabilization problem, we shall refer to the robust stability region associated with any of the fall–back configurations also as the fault–recovery region. As discussed in Sect. 8.3.3, the characterization of this region plays a central role in devising the appropriate switching policy that reconfigures the control system and ensures fault–recovery.

For the class of robust control laws given in (8.7–8.10), using a Lyapunov argument one can show that the set, $\Pi_j^{k_j}(u_{j,max}^{k_j}, \theta_b(T_f)) :=$

$$\{x_j \in \mathbb{R}^{n_j} : \alpha(x_j) + \rho_j^{k_j}\|x_j\| + \sum_{p=i}^{j-1} \chi_j^{k_j}\theta_b^p(T_f)\|\omega_p^T(x_j)\| \leq u_{j,max}^{k_j}\|\beta^T(x_j)\|\} \tag{8.11}$$

describes a region in the state–space where the control action satisfies the constraints and the Lyapunov function decays monotonically along the trajectories of the closed–loop system (see [81] for the detailed mathematical analysis). Note that the size of this set depends both on the magnitude of the constraints and the size of the disturbance (which in turn depends on the failure time, T_f). In particular, as the constraints become tighter and/or the disturbances greater, the set gets smaller. For a given control configuration, one can use the above inequality to estimate the fault–recovery region associated with this configuration by constructing, for example, the largest invariant subset of Π_j, which we denote by $\Omega_j^{k_j}(u_{j,max}^{k_j}, \theta_b(T_f))$. For a given fall–back configuration, k_j, implementation of the controller of (8.7–8.10) at any time that the state is within $\Omega_j^{k_j}$ ensures that the closed–loop trajectory stays within the region defined by Π_j – and hence $V_j^{k_j}$ continues to decay monotonically – for all the times that the given fall–back control configuration is active.

Remark 8.6. Note that, unlike the nominal stability regions associated with the nominal controllers of (8.3–8.4) and obtained from (8.5), the fault–recovery region of any downstream unit, j, cannot be computed a priori (i.e., before plant startup) since this region, as can be seen from (8.11), depends on the failure time which is unknown prior to startup. However, once the failure occurs, estimates of the disturbance bounds can be computed by the local supervisors of the upstream units, $i, \cdots, j-1$ (through on–line simulations of each unit's response under the pre– and post–failure configurations) and then transmitted, through the communication network, to unit j which in turn uses these bounds to construct, on–line, both the controller and the fault–recovery region (see Sect. 8.3.4 for a discussion on how the resulting computational delays can be handled).

8.3.3 Supervisory Switching Logic Design

Having designed the robust feedback control law and characterized the fault–recovery region associated with each fall–back configuration, the third step in our design methodology is to derive the switching policy that the local supervisor of the downstream unit, j, needs to follow in reconfiguring the local control system (i.e., activating/deactivating the appropriate fall–back configurations) in the event of the upstream failure. In the general case, when more than one fall–back control configuration is available for the unit under consideration, the question is how to decide which of these configurations can and should be activated at the time of failure in order to preserve closed–loop stability. The key idea here is that, because of the limitations imposed by constraints on the fault–recovery region of each configuration, the local supervisor can only activate the configuration whose fault–recovery region contains the closed–loop state at the time of the failure. Without loss of generality, let the active control configuration in the j–th unit, prior to the occurrence of failure in unit i, be $k_j(T_f^-) = \mu$ for some $\mu \in \mathcal{K}_j$, where $k_j(T_f^-) = \lim_{t \to T_f^-} k_j(t)$ and T_f is the time that the control system of unit i fails, then the switching rule given by

$$k_j(T_f^+) = \nu \;\; if \;\; x_j(T_f) \in \Omega_j^{\nu}(u_{j,max}^{\nu}, \theta_b(T_f)) \tag{8.12}$$

for some $\nu \in \mathcal{K}_j$, $\nu \neq \mu$, guarantees that the closed–loop system of the j–th unit is stable. The implementation of the above switching law requires monitoring, by the local supervisor, of the evolution of the closed–loop state trajectory with respect to the fault–recovery regions associated with the various control actuator configurations. Another way to look at the above switching logic is that it implicitly determines, for a fixed fall–back configuration, the times that the control system of the j–th unit can tolerate upstream failures by switching to this configuration. If failure occurs at times when x_j lies outside the fault–recovery region of all available configurations, this analysis suggests that either the constraints should be relaxed – to enlarge the fault–recovery region of the given configurations – or additional fall–back control loops must be introduced. The second option, however, is ultimately limited by the maximum allowable number of control loops that can be designed for the given processing unit. If neither option is feasible, a shutdown could be unavoidable. The proposition of constructing the switching logic on the basis of the stability regions was first proposed in [80] for the control of switched nonlinear systems.

8.3.4 Design of the Communication Logic

Given the distributed interconnected nature of the plant units – and thus the potential for failure effects propagating from one unit to another – an essential element in the design of the fault–tolerant control system is the use of a communication medium that ensures fast and efficient transmission of information during failure events. As discussed in the introduction, communication networks offer such a medium that is both fast (relative to the typically slow dynamics of chemical processes) and inexpensive (relative to current point–to–point connection schemes which require extensive cabling and higher maintenance time and costs). The ability of the network to fulfill this role, however, requires that we devise the communication policy in a way that respects the inherent limitations in network resources, such as bandwidth constraints and overall delays, by minimizing unnecessary usage of the network.

In Sect. 8.3.1, we have already discussed how the bandwidth constraint issue can be handled by formulating the problem as a robust control problem, where the failure in the control system of the i–th processing unit and the consequent effects on units $i+1, \cdots, j-1$ are treated as a bounded non–vanishing disturbances that affect unit j downstream. The communication policy requires that the local supervisors of units $i, \cdots, j-1$ perform the following tasks: (1) compute the disturbance bounds using the process model of each unit, and (2) send this information, together with other relevant information such as the failure type, the failure time and operating conditions, to the plant supervisor. The plant supervisor in turn forwards the information to the local supervisor of unit j utilizing the plant–wide communication network (see Fig. 8.1b). This policy avoids unnecessary overloading of the network (which could result when measurements from the upstream units are sent continuously to unit j) while also guaranteeing fault–tolerance in the downstream units. The idea of using knowledge of the plant dynamics to balance the tradeoff between bandwidth limitations (which favor reduced communication of measurements) and optimum control performance (which favors increased communication of measurements) is conceptually aligned with the notion of minimum attention control (e.g., see [40,203]). In our work, however, this idea is utilized in the context of fault–tolerant control.

The second consideration in devising the communication logic is the issue of time–delays which typically result from the time sharing of the communication medium as well as the computing time required for the physical signal coding and communication processing. The characteristics of these time–delays depend on the network protocols adopted as well as the hardware chosen. For our purposes here, we consider an overall fixed time–delay (which we denote by τ^j_{max}) that combines the contribution of several delays, including: (1) delays in fault–detection, (2) the time that the local supervisors of units $i,\ldots,j-1$ take to compute the effective disturbance bounds (through simulations comparing the pre– and post–failure state evolutions in each unit), (3) the time that the local supervisors of units $i,\ldots,j-1$ take to send the information to the plant supervisor, (4) the time that it takes the plant supervisor to forward the information to the local supervisor of unit j, (5) the time that it takes the local supervisor for unit j to compute the fault–recovery region for the given fall–back configurations using the information arriving from the upstream units and the time that it takes for the supervisor's decision to reach and activate the appropriate fall–back configuration, and (6) the inherent actuator/sensor dead–times.

Failure to take such delays into account can result in activating the wrong control configuration and subsequent instability. For example, even though the upstream failure may take place at $t = T_f$, the fall–back configuration in the control system of unit j will not be switched in before $t = T_f + \tau^j_{max}$. If the delay is significant, then the switching rule in Sect. 8.3.3 should be modified such that the local supervisor for unit j activates configuration, $k_j = \nu$, for which $x_j(T_f + \tau^j_{max}) \in \Omega^{\nu}_j(u^{\nu}_{j,max}, \theta_b)$. This modification is yet another manifestation of the inherent coupling between the switching and communication logics. The implementation of the modified switching rule that accounts for delays requires that the local supervisor of unit j be able to predict where the state trajectory will be at $t = T_f + \tau^j_{max}$ (e.g., through simulations using the process model) and check whether the state at this time is within the fault–recovery region of a given fall–back configuration. If not, then either an alternative fall–back configuration, for which the fault–recovery region contains the state at the end of the delay, should be activated or a shutdown maybe unavoidable. The availability of several fall–back control loops, however, is limited by process design considerations which dictate, for example, how many variables can be used for control. Figure 8.3 summarizes the overall fault–tolerant control strategy for a two–unit plant.

8.4 Simulation Studies

In this section, we present two simulation studies that demonstrate the application of the proposed fault–tolerant control system design methodology

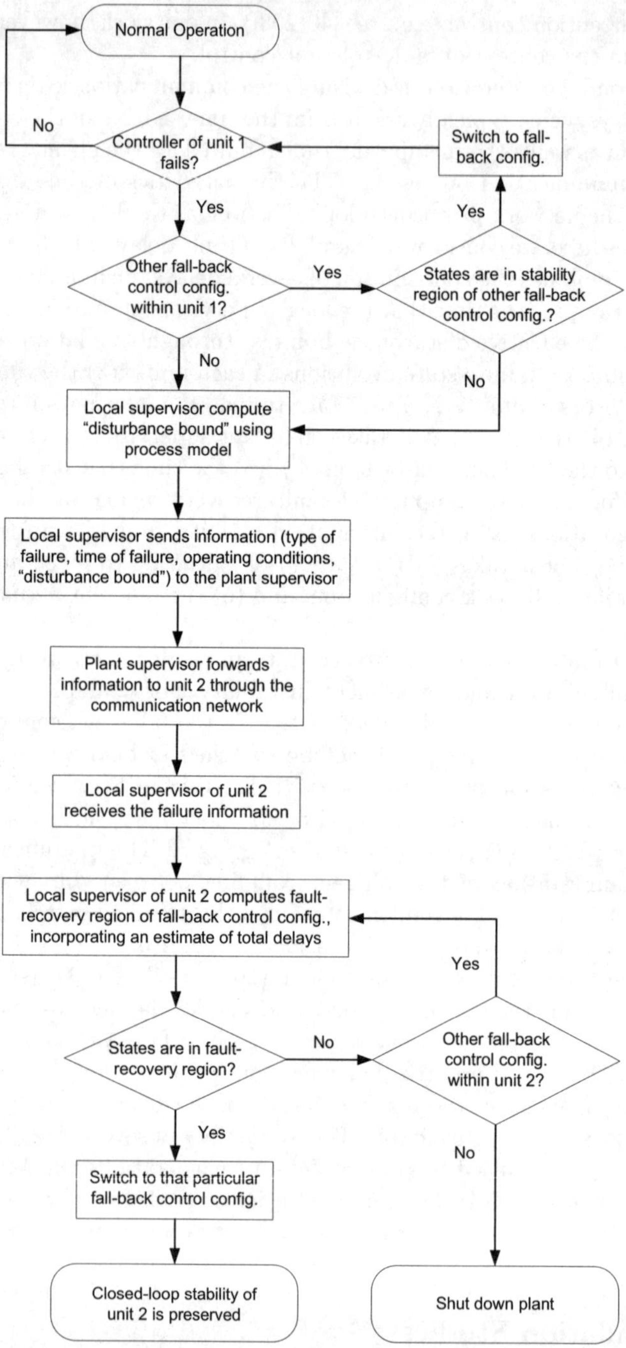

Fig. 8.3. Summary of the fault–tolerant control strategy, for a two–unit plant, using communication networks

to two chemical processes. In the first application, a single chemical reactor example is considered to demonstrate the idea of re–configuring the local control system in the event of failures on the basis of the stability regions of the constituent control configurations, and how overall communication delays impact the re–configuration logic. In the second application, a cascade of two chemical reactors in series is considered to demonstrate how the issue of failure propagation between a multi–unit plant is handled within the proposed methodology, and how the various interplays between the feedback, supervisory control and communication tasks are handled in the multi–unit setting.

8.4.1 Application to a Single Chemical Reactor

Consider a well–mixed, non–isothermal continuous stirred tank reactor where three parallel irreversible elementary exothermic reactions of the form $A \xrightarrow{k_1} B$, $A \xrightarrow{k_2} U$ and $A \xrightarrow{k_3} R$ take place, where A is the reactant species, B is the desired product and U, R are undesired byproducts. The feed to the reactor consists of pure A at flow rate F, molar concentration C_{A0} and temperature T_{A0}. Due to the non–isothermal nature of the reactions, a jacket is used to remove/provide heat to the reactor. Under standard modeling assumptions, a mathematical model of the process can be derived from material and energy balances and takes the following form:

$$
\begin{aligned}
\frac{dT}{dt} &= \frac{F}{V}(T_{A0} - T) + \sum_{i=1}^{3} \frac{(-\Delta H_i)}{\rho c_p} R_i(C_A, T) + \frac{Q}{\rho c_p V} \\
\frac{dC_A}{dt} &= \frac{F}{V}(C_{A0} - C_A) - \sum_{i=1}^{3} R_i(C_A, T) \\
\frac{dC_B}{dt} &= -\frac{F}{V} C_B + R_1(C_A, T)
\end{aligned}
\tag{8.13}
$$

where $R_i(C_A, T) = k_{i0} \exp\left(\frac{-E_i}{RT}\right) C_A$, C_A and C_B denote the concentrations of the species A and B, respectively, T denotes the temperature of the reactor, Q denotes the rate of heat input to the reactor, V denotes the volume of the reactor, ΔH_i, k_i, E_i, $i = 1, 2, 3$, denote the enthalpies, pre–exponential constants and activation energies of the three reactions, respectively, c_p and ρ denote the heat capacity and density of the fluid in the reactor. The values of the process parameters and the corresponding steady–state values are given in Table 8.2. It was verified that under these conditions, the process model of (8.13) has three steady–states: two locally asymptotically stable and one unstable at $(T^s, C_A^s, C_B^s) = (388\ K, 3.59\ kmol/m^3, 0.41\ kmol/m^3)$.

The control objective is to stabilize the reactor at the (open–loop) unstable steady–state. Operation at this point is typically sought to avoid high temperatures while, simultaneously, achieving reasonable reactant conversion. To

Table 8.2. Process parameters and steady–state values for the reactor of (8.13)

F	$= 4.998$	m^3/hr
V	$= 1.0$	m^3
R	$= 8.314$	$KJ/kmol \cdot K$
T_{A0}	$= 300.0$	K
C_{A0}	$= 4.0$	$kmol/m^3$
C_{B0}	$= 0.0$	$kmol/m^3$
ΔH_1	$= -5.0 \times 10^4$	$KJ/kmol$
ΔH_2	$= -5.2 \times 10^4$	$KJ/kmol$
ΔH_3	$= -5.4 \times 10^4$	$KJ/kmol$
k_{10}	$= 3.0 \times 10^6$	hr^{-1}
k_{20}	$= 3.0 \times 10^5$	hr^{-1}
k_{30}	$= 3.0 \times 10^5$	hr^{-1}
E_1	$= 5.0 \times 10^4$	$KJ/kmol$
E_2	$= 7.53 \times 10^4$	$KJ/kmol$
E_3	$= 7.53 \times 10^4$	$KJ/kmol$
ρ	$= 1000.0$	kg/m^3
c_p	$= 0.231$	$KJ/kg \cdot K$
T^s	$= 388.57$	K
C_A^s	$= 3.59$	$kmol/m^3$
C_B^s	$= 0.41$	$kmol/m^3$

accomplish this objective in the presence of control system failures, we consider the following manipulated input candidates (see Fig. 8.4):

1. Rate of heat input, $u_1 = Q$, subject to the constraint $|Q| \leq u^1_{max} = 2.7 \times 10^6 \ KJ/hr$.
2. Inlet stream temperature, $u_2 = T_{A0} - T^s_{A0}$, subject to the constraint $|u_2| \leq u^2_{max} = 100 \ K$.
3. Inlet reactant concentration, $u_3 = C_{A0} - C^s_{A0}$, subject to the constraint $|u_3| \leq u^3_{max} = 4 \ kmol/m^3$.

Each of the above manipulated inputs represents a unique control configuration (or control–loop) that, by itself, can stabilize the reactor using available measurements of the reactor temperature, reactant and product concentrations provided by the sensors. The sensors and control actuators of each configuration are connected to the unit supervisor (e.g., a distant control room) over a communication network (see Fig. 8.5). The first loop involving the heat input, Q, as the manipulated variable will be considered as the primary control configuration. In the event of a total failure in this configuration, however, the supervisor will have to activate one of the other two fall–back configurations in order to maintain closed–loop stability. The main question that we address in this simulation study is how can the supervisor determine which control loop to activate once failure is detected in the active configuration and how overall communication delays influence this decision.

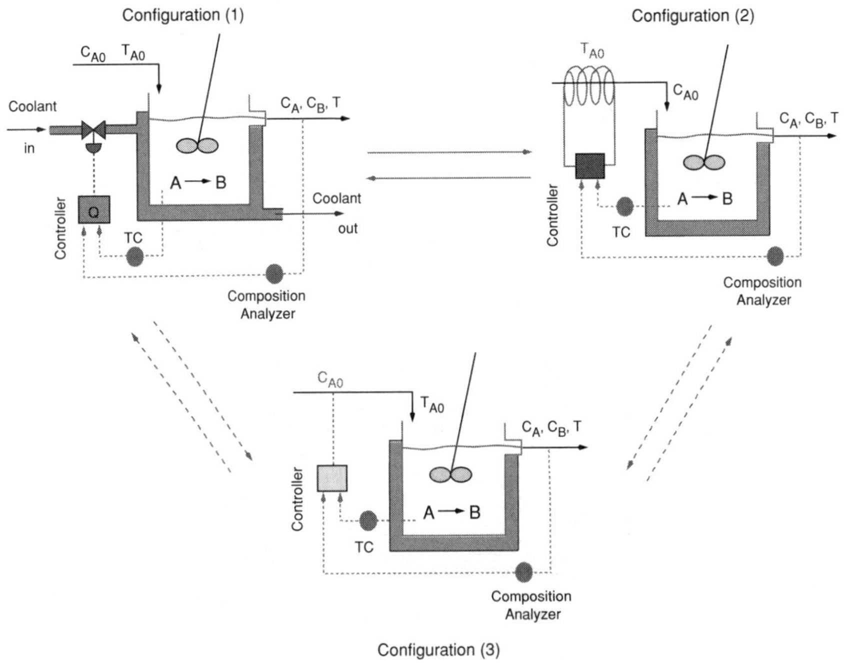

Fig. 8.4. Switching between multiple control configurations, each characterized by a different manipulated input, provides a mechanism for fault-tolerant control

Following the proposed methodology, we initially synthesize, for each control configuration, a feedback controller that enforces asymptotic closed–loop stability in the presence of actuator constraints. This task is carried out on the basis of the process input/output dynamics. While our control objective is to achieve full–state stabilization, auxiliary process outputs are introduced here to facilitate transforming the system of (8.13) into a form more suitable for explicit controller synthesis. In the case of the process of (8.13), a further simplification can be obtained by noting that C_B does not affect the evolution of either T or C_A and, therefore, the controller design can be addressed on the basis of the T and C_A equations only. A controller that stabilizes the (T, C_A) subsystem also stabilizes the entire closed–loop system. For the first configuration with $u_1 = Q$, we consider the output $y_1 = (C_A - C_A^s)/C_A^s$. This choice yields a relative degree of $r_1 = 2$ for the output with respect to the manipulated input. The coordinate transformation (in error variables form) takes the form: $e_1 = (C_A - C_A^s)/C_A^s$, $e_2 = \frac{F}{V}(C_{A0} - C_A)/C_A^s - \sum_{i=1}^{3} k_{i0} \exp\left(\frac{-E_i}{RT}\right) C_A/C_A^s$. For the second configuration with $u_2 = T_{A0} - T_{A0}^s$, we choose the output $y_2 = (C_A - C_A^s)/C_A^s$ which yields the same relative degree as in the first configuration, $r_2 = 2$, and the same coordinate transformation. For the third configuration, with $u_3 = C_{A0} - C_{A0}^s$, we choose the output $y_3 = (T - T^s)/T^s$ which

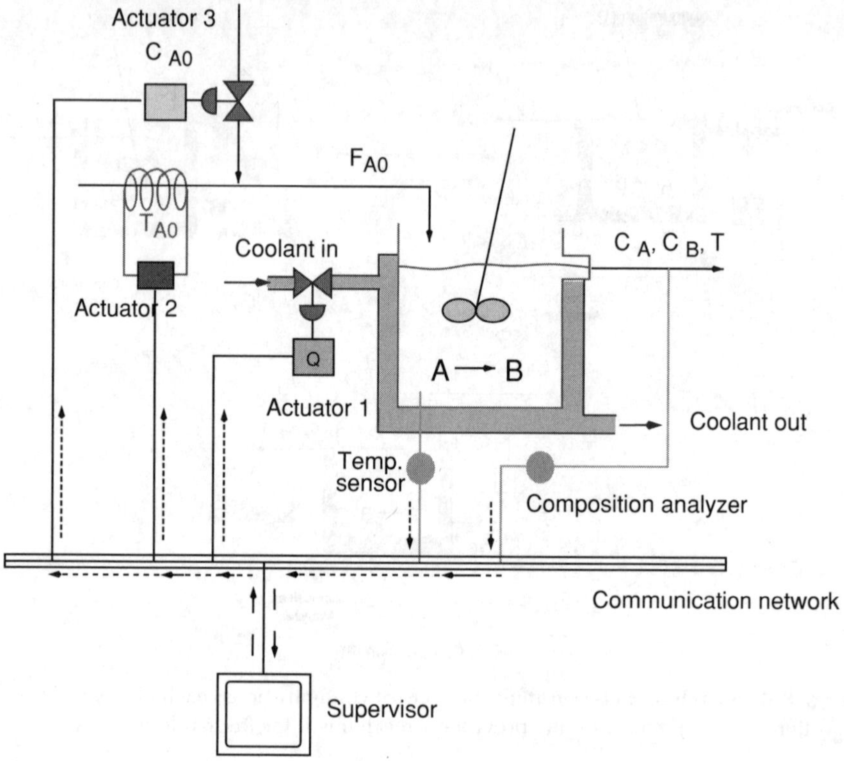

Fig. 8.5. Fault–tolerant control structure for a single unit operation, integrating supervisory and feedback control over a communication network

yields a relative degree of $r_3 = 2$ and a coordinate transformation of the form: $e_1 = (T - T^s)/T_s$, $e_2 = \frac{F}{V}(T_{A0} - T)/T^s + \sum_{i=1}^{3} \frac{(-\Delta H_i)}{\rho c_p T_s} R_i(C_A, T) + \frac{Q}{\rho c_p V T_s}$.

Note that since our objective is full–state stabilization, the choice of the output in each case is really arbitrary. However, to facilitate the controller design and subsequent stability analysis, we have chosen in each case an output that produces a system of relative degree 2. For each configuration, the corresponding state transformation yields a system, describing the input/output dynamics, of the following form:

$$\begin{aligned} \dot{e} &= Ae + l_k(e) + b\alpha_k u_k \\ &:= \bar{f}_k(e) + \bar{g}_k(e)u_k, \quad k = 1, 2, 3 \end{aligned} \tag{8.14}$$

where $A = \begin{bmatrix} 0 & 1 \\ 0 & 0 \end{bmatrix}$, $b = \begin{bmatrix} 0 \\ 1 \end{bmatrix}$, $l_k(\cdot) = L_{f_k}^2 h_k(x)$, $\alpha_k(\cdot) = L_{g_k} L_{f_k} h_k(x)$, $h_k(x) = y_k$ is the output associated with the k–th configuration, $x = [x_1 \; x_2]^T$ with $x_1 = (T - T^s)/T^s$, $x_2 = (C_A - C_A^s)/C_A^s$, and the functions $f_k(\cdot)$ and $g_k(\cdot)$ can be obtained by re–writing the (T, C_A) model equations in (8.13) in the form of (8.1). The explicit forms of these functions are omitted for brevity. Using a quadratic Lyapunov function of the form $V_k = e^T P_k e$, where P_k is a positive–definite symmetric matrix that satisfies the Riccati inequality $A^T P_k + P_k A - P_k bb^T P_k < 0$, we synthesize, for each control–loop, a bounded nonlinear feedback control law of the form of (8.3–8.4) and characterize the associated stability region with the aid of (8.5). Figure 8.6 depicts the stability region, in the (T, C_A) space, for each configuration. The stability region of configuration 1 includes the entire area of the plot. The stability region of configuration 2 is the entire area to the left of the solid line, while the stability region of configuration 3 covers the area to the right of the dashed vertical line. The desired steady–state is depicted with an asterisk that lies in the intersection of the three stability regions.

We consider first the case when no time–delays are involved and the supervisor can switch immediately between the different control loops in the event of failures. To this end, the reactor is initialized at $T(0) = 300\ K$, $C_A(0) = 4.0\ kmol/m^3$, $C_B(0) = 0.0\ kmol/m^3$, using the Q—control configuration, and the supervisor proceeds to monitor the evolution of the closed–loop trajectory. As shown by the solid parts of the closed–loop trajectory in Fig. 8.6, the state profiles in Fig. 8.7 and the rate of heat input profile in Fig. 8.8, the controller proceeds to drive the closed–loop trajectory towards the desired steady–state until the control actuators Q–configuration experiences a total failure after 2.0 hr of startup (simulated by fixing $Q = 0$ for all $t \geq 2.0$ hr).

From the solid part of the trajectory in Fig. 8.6, it is clear that the failure of the primary control configuration occurs when the closed–loop trajectory is

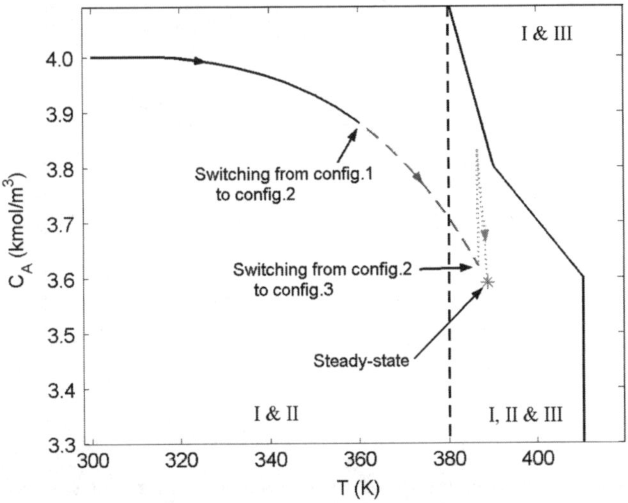

Fig. 8.6. Stability regions of the three control configurations (I, II, III) considered for the chemical reactor example of (8.13)

within the stability region of the second control configuration, and outside the stability region of the third control configuration. Therefore, on the basis of the switching logic, the supervisor immediately activates the second configuration, with T_{A0} as the manipulated input. The result is shown by the dashed parts of the closed–loop trajectory in Fig. 8.6, the state profiles in Fig. 8.7 and the inlet stream temperature profile in Fig. 8.8 where it is seen that, upon switching

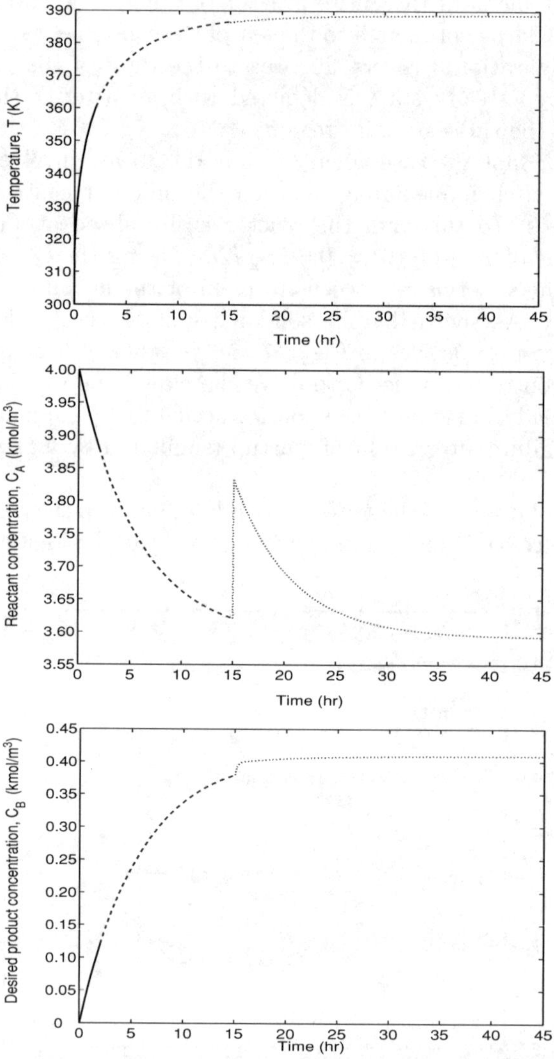

Fig. 8.7. Evolution of the closed–loop state profiles under repeated control system failures and subsequent switching by the supervisor from configuration 1 (*solid lines*) to configuration 2 (*dashed lines*) to configuration 3 (*dotted lines*)

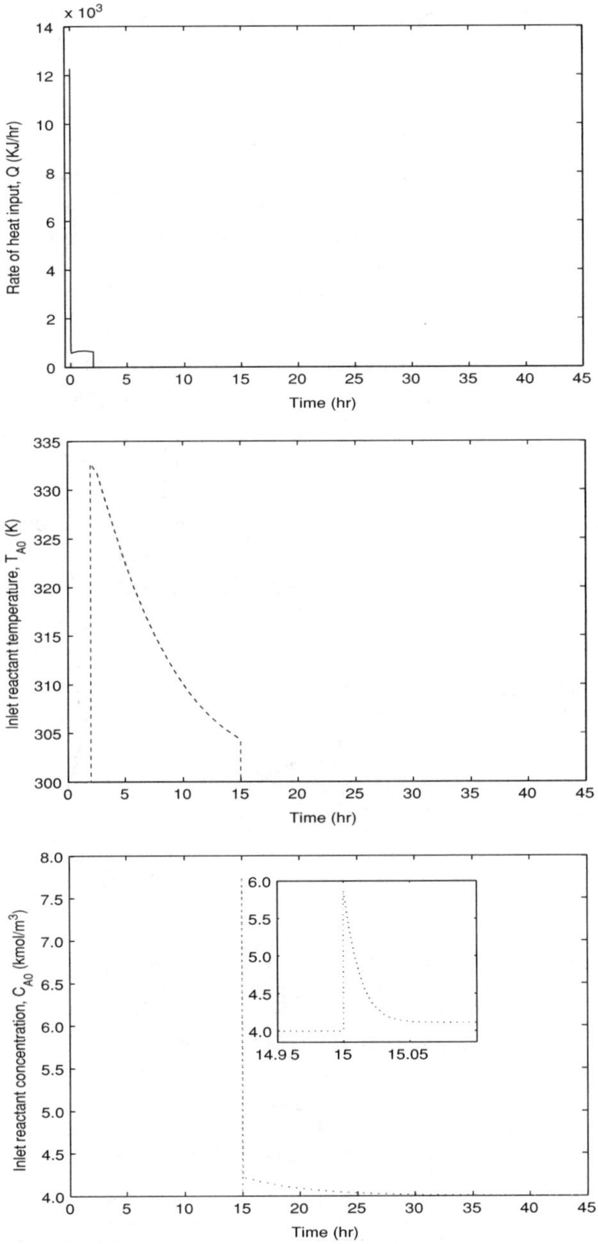

Fig. 8.8. Manipulated input profiles for each control configuration as the supervisor switches from configuration 1 to configuration 2 at $t = 2$ hr and from configuration 2 to configuration 3 at $t = 15$ hr

to the T_{A0}–configuration, the corresponding controller continues to drive the closed–loop trajectory closer to the desired steady–state. At $t = 15.0$ hr, we consider another total failure in the control actuators of the T_{A0}–configuration (simulated by fixing T_{A0} for all $t \geq 15.0$ hr). From the dashed part of the trajectory in Fig. 8.6, it is clear that this failure occurs when the closed–loop trajectory is within the stability region of the third configuration. Therefore, the supervisor immediately activates the third control configuration, with C_{A0} as the manipulated input, which then successfully stabilizes the reactor at the desired steady–state (see the dotted parts of the closed–loop trajectory in Fig. 8.6, the state profiles in Fig. 8.7 and the inlet reactant concentration in Fig. 8.8).

To demonstrate the effect of delays on the implementation of the switching logic, we consider an overall delay, between the supervisor and the constituent control configurations, of $\tau_{max} = 8.0$ min (accounting for possible delays in fault–detection, control computations, network transmission and actuator activation). In this case, the reactor is initialized at $T(0) = 300\ K$, $C_A(0) = 4.0\ kmol/m^3$, $C_B(0) = 0\ kmol/m^3$ under the first control configuration (with Q as the manipulated input). The actual failure of this configuration occurs at $t = 10$ hr which, as can be seen from Fig. 8.9, is a time when the

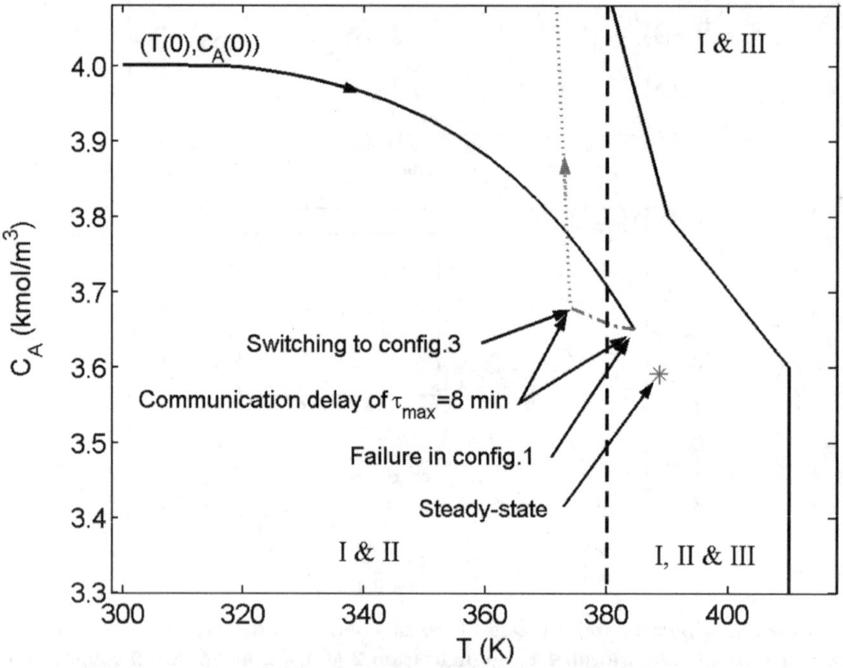

Fig. 8.9. A phase plot showing the closed–loop state trajectory leaving the intersection zone (I,II & III) during the delay period (*dashed–dotted trajectory*) rendering configuration 3 destabilizing (*dotted trajectory*)

closed–loop state trajectory is within the intersection of all three stability regions.

In the absence of delays, this suggests that switching to either configuration 2 or 3 should preserve closed–loop stability. We observe, however, from Fig. 8.10 that, when the delay is present, activation of configuration 3 leads to instability (dotted profile) while activation of configuration 2 achieves stabilization at the desired steady–state (dashed profiles). The reason is the fact that, for the time period between the actual failure ($t = 10$ hr) and the activation of the backup configuration ($t = 10.13$ hr), the process evolves in an open–loop fashion leading the trajectory to move out of the intersection zone such that at $t = 10.13$ hr the state is within the stability region of configuration 2 and outside that of configuration 3. This is shown in Fig. 8.9. The corresponding manipulated input profiles are shown in Fig. 8.11. To activate the correct configuration in this case, the supervisor needs to predict where the state trajectory will be at the end of the communication delay period.

8.4.2 Application to Two Chemical Reactors in Series

In this section, we revisit the two chemical reactors in series of (8.2), introduced earlier in Sect. 8.2.3, to illustrate the implementation of the proposed fault–tolerant control methodology. To this end, the reactors are initialized at $(T_1(0), C_{A1}(0)) = (300\ K, 4.0\ kmol/m^3)$, and $(T_2(0), C_{A2}(0)) = (440\ K, 4.0\ kmol/m^3)$. Under normal operating conditions (with no failures), each reactor is controlled by manipulating the rate of heat input, using a bounded nonlinear control law of the form of (8.3–8.4). To simplify computations, we design each controller on the basis of the temperature equation only. Specifically, a quadratic function of the form $V_2 = \frac{1}{2}a_2(x_2^{(1)})^2$, where $x_2^{(1)} = (T_2 - T_2^s)/T_2^s$, is used to design the controller and estimate the resulting stability region using (8.5). The values of the controller tuning parameters are chosen to be $a_2 = 0.5$ and $\rho_2 = 0.0001$. Figure 8.12 (solid profiles) and Fig. 8.13 show the resulting closed–loop state and manipulated input profiles when the controllers are implemented without failure for both reactors. We observe that each controller successfully stabilizes the corresponding reactor at the desired steady–state.

Consider now a total failure in the actuators of both control systems (Q_1 and Q_2) at $T_f = 5$ min. In this case, both reactors revert to their open–loop mode of behavior and, consequently, if no fall–back control configuration is activated, the states move away from the desired steady–state, as shown by the dashed lines in Fig. 8.12 for the first reactor, and Fig. 8.14 for the second reactor (note that C_{A03} remains fixed for all times since it is not used as a manipulated variable in the pre–failure configuration). As stated in Sect. 8.2.3, we assume that the controller failure in the first reactor is permanent; and our objective is to prevent the propagation of this effect to the second

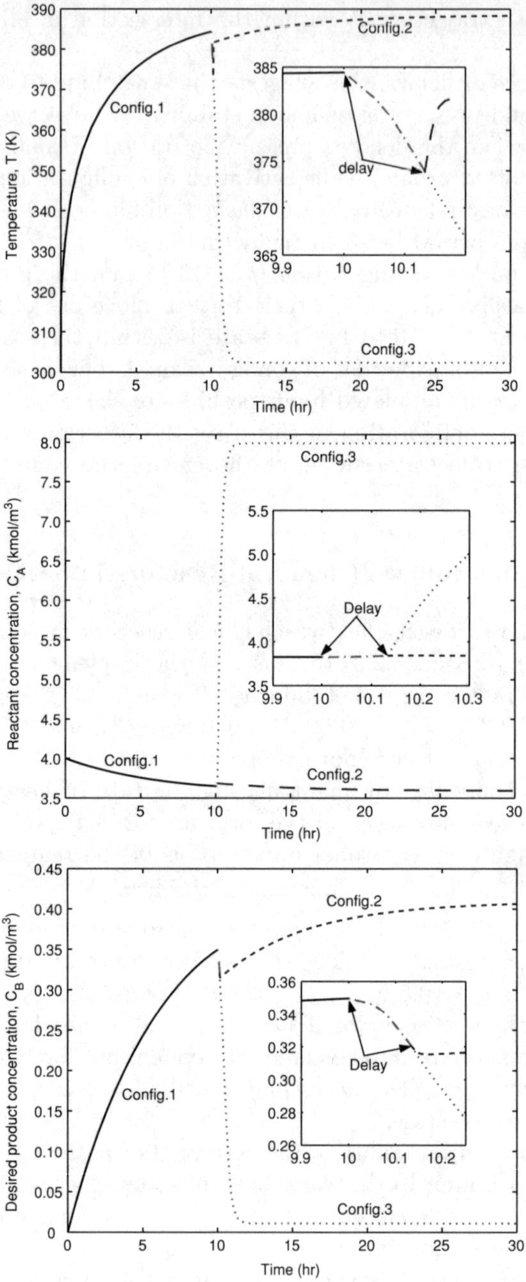

Fig. 8.10. Evolution of the closed–loop state profiles when configuration 1 (*solid lines*) fails at $t = 10$ hr and an overall delay of $\tau_{max} = 8.0$ min elapses before the backup configuration is activated. Activation of configuration 2 preserves closed–loop stability (*dashed lines*) while activation of configuration 3 leads to instability (*dotted lines*)

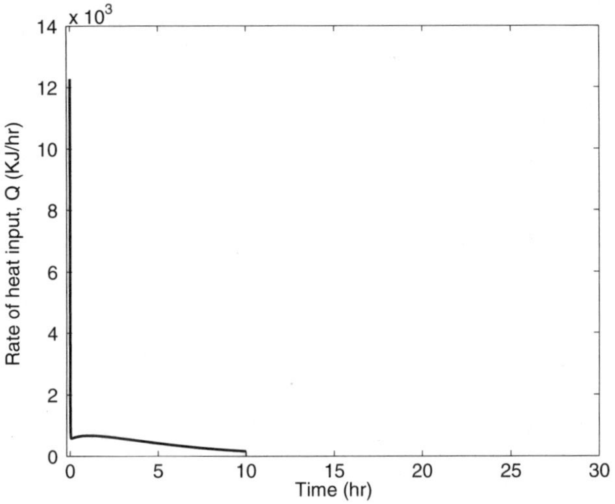

Fig. 8.11. Manipulated input profiles when configuration 1 fails at $t = 10$ hr and an overall delay of $\tau_{max} = 8.0$ min elapses before the backup configuration is activated

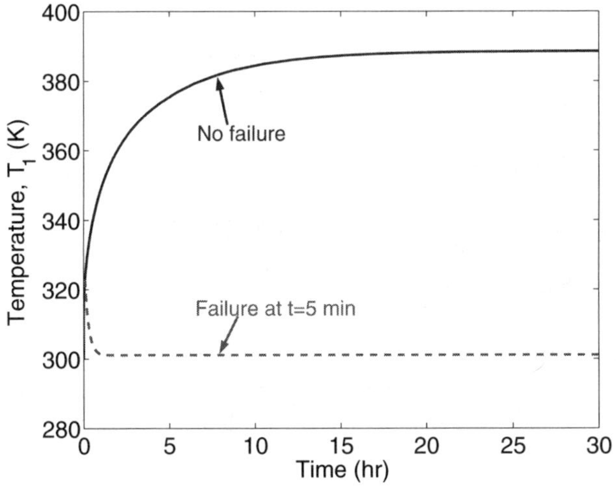

Fig. 8.12. Evolution of the closed–loop state and manipulated input profiles for CSTR 1 under a well–functioning control system (*solid*) and when the control actuator fail at $t = 5$ min (*dashed lines*)

reactor. A permanent failure in the first unit could be the result of lack of sufficient fall–back configurations or because failure occurs at a time when the state is outside the stability regions of all the available configurations for this unit.

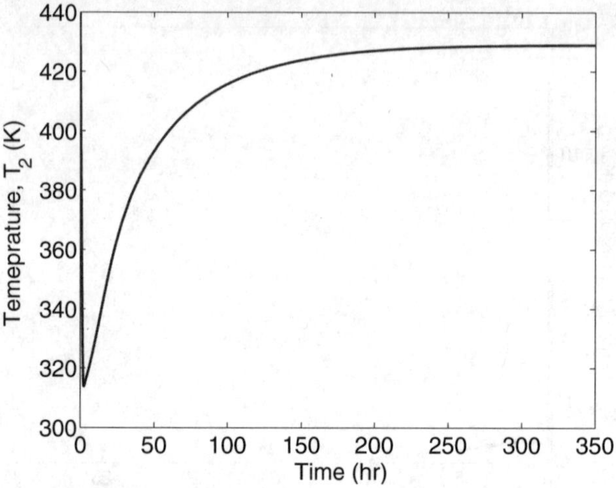

Fig. 8.13. Evolution of the closed–loop state and manipulated input profiles for CSTR 2 under a well–functioning control system

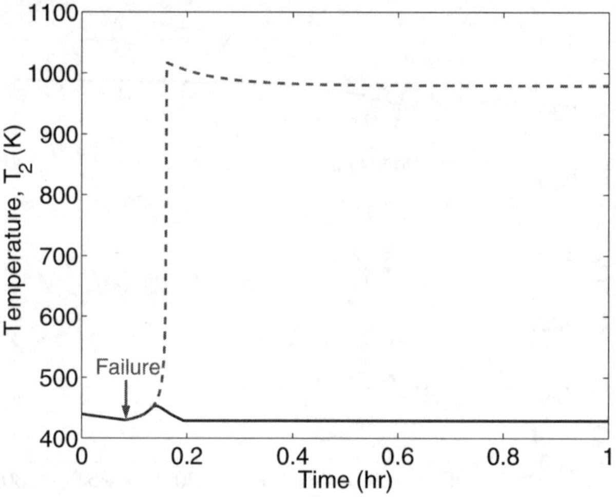

Fig. 8.14. Evolution of the closed–loop state and manipulated input profiles for CSTR 2 when the controller of the fall–back configuration (Q_2, C_{A03}) is activated immediately after the failure (*solid lines*), and the open–loop state and input profiles resulting when the fall–back configuration is not activated after the failure (*dashed lines*)

Using the proposed methodology, the supervisor of CSTR 1, at the failure time, runs both open–loop and closed–loop simulations using the process model of CSTR 1 to estimate the size of the disturbance affecting CSTR 2, and transmits this information to the local supervisor of CSTR 2 through the communication network. The maximum disturbance size is proportional to the largest discrepancy (after the failure time) between the values of C_{A1}, T_1 in the well–functioning (solid lines in Fig. 8.12) and in the failed (dashed lines in Fig. 8.12) modes. Using this information, the local supervisor of CSTR 2 designs a robust control law of the form of (8.7–8.10) to stabilize CSTR 2, using the available fall–back configuration with (Q_2, C_{A03}) as the manipulated inputs, and constructs the associated fault–recovery region for this configuration. The controller design procedure involves re–writing the process model of CSTR 2 in (8.2) in the form of (8.6), using the dimensionless variables, $x_i^{(1)} = (T_i - T_i^s)/T_i^s$, $x_i^{(2)} = (C_{Ai} - C_{Ai}^s)/C_{Ai}^s$, $i = 1, 2$, and with the states of CSTR 1 re–defined as the disturbance variables, $\theta_1(t) = [\theta_1^{(1)}(t)\ \theta_1^{(2)}(t)]^T$, where $\theta_1^{(1)}(t) = (F_1 T_1^s / V_2 T_2^s)(x_1^{(1)}(t) + 1)$ and $\theta_1^{(2)}(t) = (F_1 C_{A1}^s / V_2 C_{A2}^s)(x_1^{(2)}(t) + 1)$, for all $t \geq T_f$. Then, using a quadratic function of the form $V_2 = \frac{1}{2}a_2(x_2^{(1)})^2 + \frac{1}{2}a_2(x_2^{(2)})^2$, the controller of (8.7–8.10) is constructed and its fault–recovery region is computed with the aid of (8.11). The disturbance bound is computed as $\theta_b^1 = \sup_{t \geq T_f} \|\theta_1(t)\|$. The values of the controller tuning parameters are selected to be $a_2 = 0.5$, $\rho_2 = 0.0001$, $\chi_2 = 2.0001$ and $\phi_2 = 0.0001$. The fault–recovery region is depicted by the shaded area in Fig. 8.15.

From Fig. 8.15, we observe that the failure occurs when the states of CSTR 2 are within the fault–recovery region. Therefore, assuming no delays in the fault–detection, computations and communication processing (i.e., instanta-

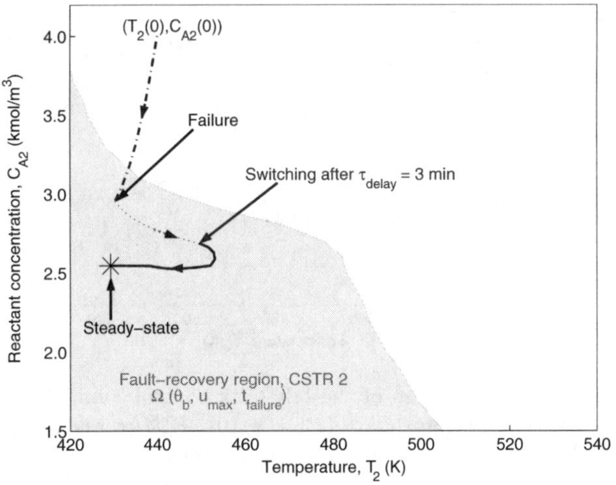

Fig. 8.15. Fault–recovery region of the fall–back control configuration (Q_2, C_{A03}) for CSTR 2, with constraints $|Q_2| \leq 2.8 \times 10^6$ KJ/hr and $|C_{A03} - C_{A03}^s| \leq 0.4$ $kmol/m^3$. when failure occurs at $T_f = 5$ min. Activation of the fall–back configuration after a 3 min delay preserves closed–loop stability (*top plot*), while activation after 4.1 min delay fails to ensure fault–tolerance (*bottom plot*)

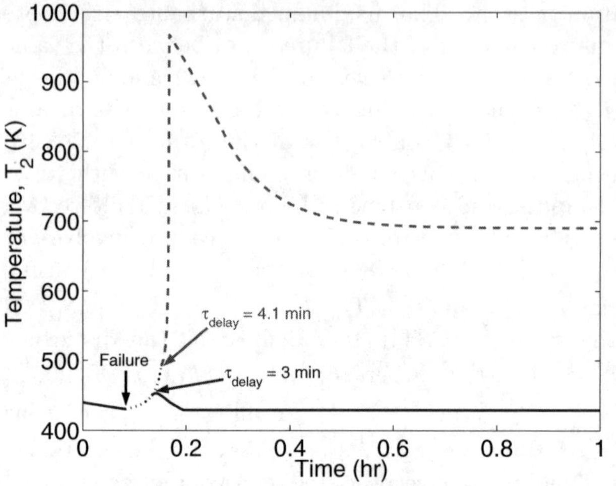

Fig. 8.16. Evolution of the closed–loop state and input profiles when the failure occurs at $T_f = 5$ min and the fall–back configuration (Q_2, C_{A03}), with constraints $|Q_2| \leq 2.8 \times 10^6$ KJ/hr and $|C_{A03} - C_{A03}^s| \leq 0.4$ $kmol/m^3$ is activated after a total delay of 3 min (*solid lines*) and after a total delay of 4.1 min. (*dashed lines*)

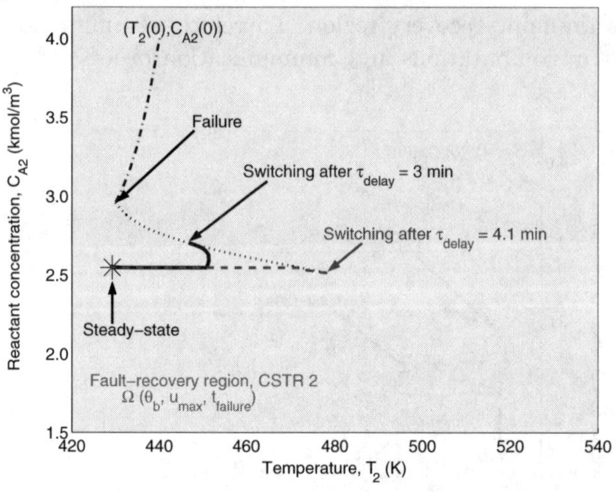

Fig. 8.17. Fault–recovery region of the fall–back control configuration (Q_2, C_{A03}) for CSTR 2, with constraints $|Q_2| \leq 1.4 \times 10^7$ KJ/hr and $|C_{A03} - C_{A03}^s| \leq 2.0$ $kmol/m^3$. when failure occurs at $T_f = 5$ min. Activation of the fall–back configuration after a delay of either 3 min or 4.1 min ensures fault–tolerance

neous switching), when the fall–back controllers are activated, closed–loop stability is preserved and the closed–loop states converge close to the desired steady–state as shown by the solid lines in Fig. 8.14.

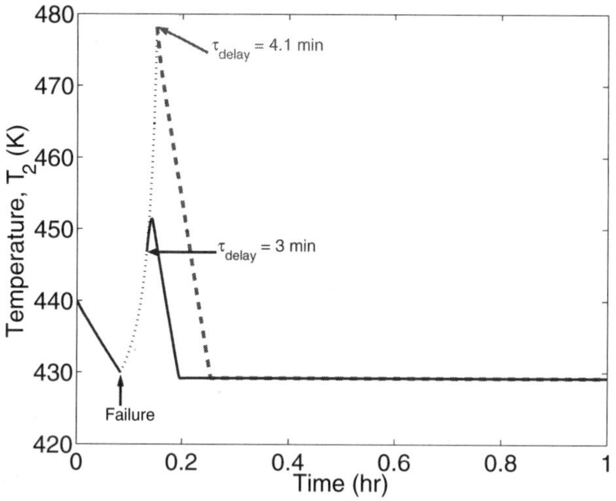

Fig. 8.18. Evolution of the closed–loop state and manipulated input profiles when the failure occurs at $T_f = 5$ min and the fall–back configuration (Q_2, C_{A03}), with constraints $|Q_2| \leq 1.4 \times 10^7\ KJ/hr$ and $|C_{A03} - C^s_{A03}| \leq 2.0\ kmol/m^3$ is activated after a total delay of 3 min (*solid lines*) and after a total delay of 4.1 min. (*dashed lines*)

When delay effects are taken into account, we see from Fig. 8.15 (top plot) that if an overall delay of 3 min (accounting for delays in fault–detection, controller computations, information transmission and actuator activation) elapses between the failure and the activation of the (Q_2, C_{A03}) configuration – during this delay, CSTR 2 evolves in an open–loop mode as indicated by the dotted line – the state at the end of the delay still resides within the fault–recovery region and, therefore, closed–loop stability is preserved by switching to the (Q_2, C^s_{A0}) configuration at the end of the delay. The corresponding state and input profiles are shown by the solid lines in Figs. 8.15–8.16. By contrast, we see from the bottom plot in Fig. 8.15 that when an overall delay of 4.1 min is considered, the state at the end of the delay lies outside the fault–recovery region; hence the fall–back configuration cannot stabilize the system at the desired steady–state, as can be seen from the dashed lines in Figs. 8.15–8.16.

Examination of Fig. 8.15 provides useful insights into how the tradeoff between the controller design, switching and communication logics can be managed to ensure fault–tolerance. For example, the picture suggests that with a larger fault–recovery region, even large delays maybe tolerated by switching to this particular configuration. A larger region can be obtained by relaxing the constraints. Figure 8.17 shows the resulting fault–recovery region for the (Q_2, C_{A03}) configuration when the constraints are relaxed to $|Q_2| \leq u^{Q_2}_{max} = 1.4 \times 10^7\ KJ/hr$ and $|C_{A03} - C_{A03s}| \leq u^{C_{A03}}_{max} = 2.0\ kmol/m^3$. In this case, the fault–recovery region includes the entire area of the plot. As a result, activation of the fall–back configuration, whether after 3 min or 4.1 min from the failure time, stabilizes the reactor since the state at the end of the delay in both cases is contained within the fault–recovery region. Figure 8.18 shows the corresponding closed–loop state and input profiles of CSTR 2 for both scenarios.

8.5 Conclusions

In this chapter, a methodology for the design of fault–tolerant control systems for chemical plants with distributed interconnected processing units was developed. Bringing together tools from Lyapunov–based nonlinear control and hybrid systems theory, the approach is based on a hierarchical architecture that integrates lower–level feedback control of the individual units with upper–level logic–based supervisory control over communication networks. The local control system for each unit consists of a family of control configurations for each of which a stabilizing feedback controller is designed and the stability region is explicitly characterized. The actuators and sensors of each configuration are connected, via a local communication network, to a local supervisor that orchestrates switching between the constituent configurations, on the basis of the stability regions, in the event of failures. The local supervisors communicate, through a plant–wide communication network, with a plant supervisor responsible for monitoring the different units and coordinating their responses in a way that minimizes the propagation of failure effects. The communication logic is designed to ensure efficient transmission of information between units while also respecting the inherent limitations in network resources by minimizing unnecessary network usage and accounting explicitly for the effects of possible delays due to fault–detection, control computations, network communication and actuator activation. Explicit guidelines for managing the various interplays between the coupled tasks of feedback control, fault–tolerance and communication were provided. The efficacy of the proposed approach was demonstrated through chemical process examples.

9

Control of Nonlinear Systems with Time Delays

9.1 Introduction

The dynamic models of many chemical engineering processes involve severe nonlinearities and significant time delays and are naturally described by nonlinear differential difference equation (DDE) systems. Nonlinearities usually arise from complex reaction mechanisms and Arrhenius dependence of reaction rates on temperature, while time delays often occur due to transportation lag such as in flow through pipes, dead times associated with measurement sensors (measurement delays) and control actuators (manipulated input delays), and approximation of high-order dynamics. Typical examples of processes which involve nonlinearities and time delays include chemical reactors with recycle loops, fluidized catalytic cracking units, distillation columns, chemical vapor deposition processes to name a few.

The conventional approach to control linear/nonlinear DDE systems is to neglect the presence of time delays and address the controller design problem on the basis of the resulting linear/nonlinear ordinary differential equation (ODE) systems, employing standard control methods for ODE systems. However, it is well-known (see, for example, [258]) that such an approach may pose unacceptable limitations on the achievable control quality and cause serious problems in the behavior of the closed-loop system including poor performance (e.g., sluggish response, oscillations) and instability.

Motivated by the above, significant research efforts have focused on the development of control methods for linear DDE systems that compensate for the effect of time delays. Research initially focused on linear systems with a single manipulated input delay which are described by transfer function models, in which the presence of the time delay prevents the use of large controller gains (i.e., the proportional gain of a proportional-integral controller should be sufficiently small in order to avoid destabilization of the closed-loop system), thereby leading to sluggish closed-loop response. To overcome this problem, O. J. M. Smith [250] proposed a control structure, known as Smith predictor, which completely eliminates the time delay from the characteristic polynomial of the closed-loop system, allowing the use of larger controller gains. Since then many researchers have proposed alternatives or modifications of the Smith predictor structure (e.g., [279]), extensions to control structures for linear systems including inferential control [41] and internal model control [98], other predictor structures such as the analytical predictor [204, 285] and established connections of the Smith predictor with other predictors [288]. The Smith predictor structure has also been extended to linear multivariable systems with multiple input and output delays which are described by transfer function models, leading to multi-delay compensators [131, 211]. Excellent reviews of results on Smith and other predictor structures can be found in [131, 288].

Even though the above works provided powerful methods for dealing with control actuator and measurement dead time in linear systems, they do not explicitly account for the effect of time delays in the process state variables. This motivates research on the design of controllers for DDE systems with state delays. In this direction, the application of classical optimal control approaches to DDE systems in order to design optimal distributed parameter (infinite-dimensional) controllers was initially studied (e.g., [229, 251]). Then, the distributed parameter nature of the developed controllers motivated research on the problem of model reduction of linear DDE systems. This problem is the one of finding a linear low-dimensional ODE system that accurately reproduces the solutions of a linear DDE system. Approaches to address this problem include balanced approximation based on controllability and observability gramians [178], frequency response analysis [107], and approximations using Fourier-Laguerre models [217, 281] to name a few. An alternative approach for the synthesis of controllers for linear DDE systems with state delays that stabilize the closed-loop system independently of the size of the delays is based on the method of Lyapunov functionals [108]. The central idea of this approach is to synthesize a linear controller so that the time-derivative of an appropriate Lyapunov functional calculated along the trajectories of the closed-loop DDE system is negative definite, independently of the size of the delays. The method of Lyapunov functionals has been used in the design of linear stabilizing controllers for linear DDE systems in [206, 233].

Despite the abundance of results on control of linear DDE systems, most of the research on nonlinear DDE systems has focused on the derivation of conditions for existence and uniqueness of solutions, the understanding of qualitative and geometric properties of the solutions (see the book [108] for results and reference lists), and stability analysis through Razumikhin-type theorems and nonlinear small-gain theorem techniques (e.g., [269, 292]). Very few results are available on control of nonlinear DDE systems, with the exception of optimal control methods [150, 229]. For nonlinear systems represented by input-output models, extensions of the Smith predictor have been proposed in [27]. Within a state-space framework, the only available results on controller synthesis are for single-input single-output nonlinear systems with a single manipulated input delay, for which a nonlinear Smith predictor structure was proposed in [162] under the assumption of open-loop stability, and further extended to open-loop unstable systems in [117]. At this stage, there is no rigorous, yet practical, method for the design of nonlinear controllers for *nonlinear* DDE systems with state, manipulated input and measured output delays.

This chapter proposes a methodology for the synthesis of nonlinear output feedback controllers for single-input single-output nonlinear DDE systems which include time delays in the states, the control actuator and the measurement sensor. Initially, DDE systems which only include state delays are considered and a novel combination of geometric and Lyapunov-based techniques is employed for the synthesis of nonlinear state feedback controllers that guarantee stability and enforce output tracking in the closed-loop system, independently of the size of the state delays. Then, the problem of designing nonlinear distributed state observers, which reconstruct the state of the DDE system while guaranteeing that the discrepancy between the actual and the estimated state tends exponentially to zero, is addressed and solved by using spectral decomposition techniques for DDE systems. The state feedback controllers and the distributed state observers are combined to yield distributed output feedback controllers that enforce stability and output tracking in the closed-loop system, independently of the size of the state delays. For DDE systems with state, control actuator and measurement delays, distributed output feedback controllers are synthesized on the basis of an auxiliary output constructed within a Smith-predictor framework. The results of this chapter were first presented in [9].

9.2 Differential Difference Equation Systems

9.2.1 Description of Nonlinear DDE Systems

We consider single-input single-output systems of nonlinear differential difference equations with the following state-space description:

$$\begin{aligned}
\dot{\bar{x}} = {} & A\bar{x}(\bar{t}) + \sum_{\kappa=1}^{q} B_{\kappa}\bar{x}(\bar{t}-\bar{\alpha}_{\kappa}) + f(\bar{x}(\bar{t}), \bar{x}(\bar{t}-\bar{\alpha}_1), \ldots, \bar{x}(\bar{t}-\bar{\alpha}_q)) \\
& + g(\bar{x}(\bar{t}), \bar{x}(\bar{t}-\bar{\alpha}_1), \ldots, \bar{x}(\bar{t}-\bar{\alpha}_q))u(\bar{t}-\hat{\alpha}_1), \\
& \bar{x}(\xi) = \bar{\eta}(\xi),\ \xi \in [-\bar{\alpha}, 0),\ \ \bar{x}(0) = \bar{\eta}_0 \\
y = {} & h(\bar{x}(\bar{t}-\hat{\alpha}_2))
\end{aligned} \tag{9.1}$$

where $\bar{x} \in \mathbb{R}^n$ denotes the vector of state variables, $u \in \mathbb{R}$ denotes the manipulated input, and $y \in \mathbb{R}$ denotes the controlled output (whose on-line measurements are assumed to be available). $\bar{\alpha}_{\kappa}$, $\kappa = 1, \ldots, q$, denotes the κ−th state delay, $\bar{\alpha} = \max\{\bar{\alpha}_1, \ldots, \bar{\alpha}_q\}$, $\hat{\alpha}_1$ denotes the manipulated input delay (control actuator dead time) and $\hat{\alpha}_2$ denotes the measured output delay (measurement sensor dead time). $A, B_1, \ldots, B_q$ are constant matrices of dimension $n \times n$, $f(\bar{x}(\bar{t}), \bar{x}(\bar{t}-\bar{\alpha}_1), \ldots, \bar{x}(\bar{t}-\bar{\alpha}_q))$, $g(\bar{x}(\bar{t}), \bar{x}(\bar{t}-\bar{\alpha}_1), \ldots, \bar{x}(\bar{t}-\bar{\alpha}_q))$ are locally Lipschitz nonlinear vector functions, $h(\bar{x}(\bar{t}-\hat{\alpha}_2))$ is a locally Lipschitz nonlinear scalar function, $\bar{\eta}(\xi)$ is a smooth vector function defined in the interval $[-\bar{\alpha}, 0)$ and $\bar{\eta}_0$ is a constant vector. We will assume that the vector function $f(\bar{x}(\bar{t}), \bar{x}(\bar{t}-\bar{\alpha}_1), \ldots, \bar{x}(\bar{t}-\bar{\alpha}_q))$ includes only nonlinear terms and satisfies $f(0, 0, \ldots, 0) = 0$ which implies that $x(t) \equiv 0$ is an equilibrium solution for the open-loop (i.e., $u(\bar{t}-\hat{\alpha}_1) \equiv 0$) system of (9.1).

There are many chemical engineering processes whose dynamic models involve time delays in the state variables and are naturally described by nonlinear DDE systems of the form of (9.1). Example of such processes include chemical reactors with recycle loops (where the state delays occur due to transportation lag in the recycle loops), fluidized catalytic cracking reactors (where the state delays occur due to dead time in pipes transferring material from the regenerator to the reactor and vice versa) and distillation columns (where the state delays occur due to dead time in reboiler and condenser recycle loops). Furthermore, control actuator and measurement sensor dead times are also very common sources of time delays in chemical process control, and they are explicitly accounted for in the DDE system of (9.1). The linear appearance of the manipulated input u in the system of (9.1) is also typical in most practical applications, where inlet flow rates, inlet temperatures and concentrations are typically chosen as manipulated inputs. Finally, the assumption that the controlled output is identical to the measured output is done in order to simplify the notation of this work and can be readily relaxed (i.e., the extension of the proposed theory to systems in which the controlled output is different from the measured output is conceptually straightforward)

To simplify the presentation of the results of this work, we will transform the DDE system of (9.1) into an equivalent DDE system which includes state and measurement delays and does not include manipulated input delay. We will also focus on DDE systems with a single state delay (the generalization of the results of this chapter to the case of DDE systems with multiple state delays is conceptually straightforward and will not be presented here for reasons of brevity). To this end, we set $\tilde{\alpha} = \hat{\alpha}_1 + \hat{\alpha}_2$, $t = \bar{t} - \hat{\alpha}_1$, $q = 1$, $B_1 = B$, $\bar{\alpha} = \bar{\alpha}_1$, $\alpha = \hat{\alpha}_1 + \bar{\alpha}$, $x(t) = \bar{x}(t + \hat{\alpha}_1)$ and obtain the following system (which will be used in our development):

$$
\begin{aligned}
\dot{x} &= Ax(t) + Bx(t-\alpha) + f(x(t), x(t-\alpha)) + g(x(t), x(t-\alpha))u(t), \\
&\quad x(\xi) = \bar{\eta}(\xi),\ \xi \in [-\alpha, 0),\ \ x(0) = \bar{\eta}_0 \qquad (9.2) \\
y &= h(x(t-\tilde{\alpha}))
\end{aligned}
$$

Remark 9.1. In order to compare the proposed approach for control of nonlinear DDE systems with existing approaches for control of linear DDE systems, we set $f(x(t), x(t-\alpha)) = 0$, $g(x(t), x(t-\alpha)) = c$ and $h(x(t-\tilde{\alpha})) = wx(t-\tilde{\alpha})$, where c, w are constant vectors, in the system of (9.2) to derive the following linear DDE system:

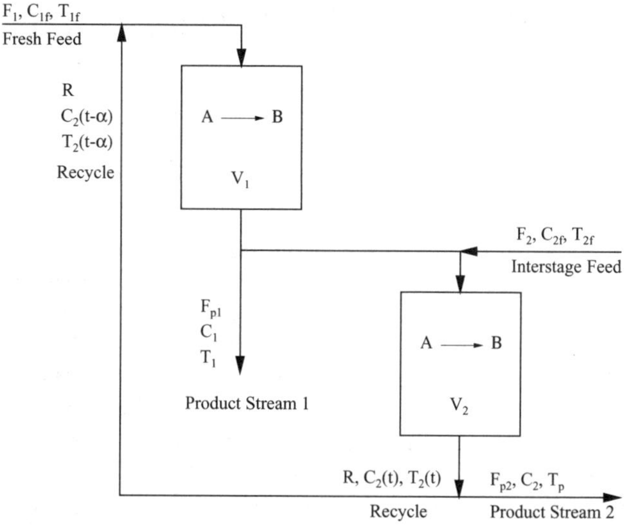

Fig. 9.1. Two chemical reactors with recycle loop

$$\begin{aligned} \dot{x} &= Ax(t) + Bx(t-\alpha) + cu(t), \quad x(\xi) = \bar{\eta}(\xi), \; \xi \in [-\alpha, 0), \quad x(0) = \bar{\eta}_0 \\ y &= wx(t-\tilde{\alpha}) \end{aligned} \tag{9.3}$$

which will be used to synthesize linear state feedback controllers and state observers.

In the next subsection, a typical chemical process example [211] is given in order to illustrate modeling of a chemical process in the form of (9.2).

9.2.2 Example of a Chemical Process Modeled by a Nonlinear DDE System

Consider the cascade of two perfectly-mixed chemical reactors with recycle loop, which is shown in Fig. 9.1. A first-order irreversible reaction of the form $A \to B$ takes place in the reactors. The process possesses an inherent state delay due to the transportation lag in the recycle loop. Under standard modeling assumptions, the dynamic model of the process can be derived from mass and energy balances and consists of the following system of four nonlinear differential difference equations:

$$\begin{aligned} \frac{dC_1}{dt} &= \frac{F_1}{V_1}C_{1f} - \frac{(F_1+R)}{V_1}C_1 + \frac{R}{V_1}C_2(t-\alpha) - k_0 \exp\left(-\frac{E}{RT_1}\right) C_1 \\ \frac{dT_1}{dt} &= \frac{F_1}{V_1}T_{1f} - \frac{(F_1+R)}{V_1}T_1 + \frac{R}{V_1}T_2(t-\alpha) + \frac{(-\Delta H)}{\rho c_p} k_0 \exp\left(-\frac{E}{RT_1}\right) C_1 \end{aligned}$$

$$
\begin{aligned}
\frac{dC_2}{dt} &= \frac{F_2}{V_2}C_{2f} - \frac{(F_{p2}+R)}{V_2}C_2 + \frac{(F_1+R-F_{p1})}{V_2}C_1 - k_0 \exp\left(-\frac{E}{RT_2}\right)C_2 \\
\frac{dT_2}{dt} &= \frac{F_2}{V_2}T_{2f} - \frac{(F_{p2}+R)}{V_2}T_2 + \frac{(F_1+R-F_{p1})}{V_2}T_1 \\
&+ \frac{(-\Delta H)}{\rho c_p}k_0 \exp\left(-\frac{E}{RT_2}\right)C_2
\end{aligned}
\tag{9.4}
$$

where F_1, F_2 denote the flow rates of the inlet streams to the two reactors, V_1, V_2 denote the volumes of the two reactors, R denotes the recycle (from the second to the first reactor) flow rate, F_{p1}, F_{p2} denote the flow rate of the product streams from the two reactors, C_{1f}, C_{2f} denote the concentration of species A in the inlet streams to the reactors, C_1, C_2 denote the concentration of species A in the reactors, T_{1f}, T_{2f} denote the temperature of the inlet streams to the two reactors, T_1, T_2 denote the temperature in the two reactors, k_0, E, ΔH denotes the pre-exponential constant, the activation energy and the enthalpy of the reaction, c_p, ρ denote the heat capacity and the density of the reacting liquid, and α denotes the recycle loop dead time.

A typical control problem for this process can be formulated as the one of regulating the concentration of species A in the first reactor, C_1, by manipulating the feed concentration of A in the first reactor, C_{1f}. Setting:

$$
x_1 = C_1, \;\; x_2 = C_2, \; x_3 = T_1, \;\; x_4 = T_2, \;\; u = C_{1f}, \;\; y = C_1
$$

the original set of equations can be put in the form of (9.2).

9.3 Mathematical Properties of DDE Systems

The objective of this section is to present the basic mathematical properties of DDE systems that will be used in our development. We will begin with the spectral properties of DDE systems, and we will continue with stability concepts and results.

9.3.1 Spectral Properties

In this subsection, we formulate the system of (9.2) as an infinite dimensional system in an appropriate Banach space and provide the statement and solution of the eigenvalue problem for the linear delay operator (see (9.7) below). The solution of the eigenvalue problem will be utilized in the design of nonlinear distributed state observers in Sect. 9.7. We formulate the system of (9.2) in the Banach space $\mathcal{C}([-\alpha,0],\mathbb{R}^n)$ of continuous n-vector valued functions defined in the interval $[-\alpha,0]$ with inner product and norm:

$$
\begin{aligned}
(\tilde{\omega}_1,\tilde{\omega}_2) &= \tilde{\omega}_1(0)\tilde{\omega}_2(0) + \int_{-\alpha}^{0} \tilde{\omega}_1(z+\alpha)B\tilde{\omega}_2(z)dz, \\
\|\tilde{\omega}_1\|_2 &= (\tilde{\omega}_1,\tilde{\omega}_1)^{\frac{1}{2}}
\end{aligned}
\tag{9.5}
$$

where $\tilde{\omega}_1$ is an element of $\mathcal{C}^*([-\alpha,0],\mathbb{R}^{n^*})$, $\mathbb{R}^{n^*}$ is the n-dimensional vector space of row vectors, and $\tilde{\omega}_2 \in \mathcal{C}$. On $\mathcal{C}$, the state function x of the system of (9.2) is defined as:

$$x_t(\xi) = x(t+\xi), \;\; t \geq 0, \;\; \xi \in [-\alpha, 0] \, , \tag{9.6}$$

the operator $\mathcal{A}$ as:

$$\mathcal{A}\phi(\xi) = \left\{ \begin{array}{ll} \dfrac{d\phi(\xi)}{d\xi}, & \xi \in [-\alpha, 0) \\ \\ A\phi(0) + B\phi(-\alpha), & \xi = 0 \end{array} \right\} \tag{9.7}$$

$$\phi(\xi) \in D(\mathcal{A}) = \left\{ \phi \in \mathcal{C}^*([-\alpha,0],\mathbb{R}^{n^*}) : \dot{\phi} \in \mathcal{C}, \dot{\phi}(0) = A\phi(0) + B\phi(-\alpha) \right\} \tag{9.8}$$

and the output as:

$$y(t) = h(Px_t) \tag{9.9}$$

where $P : \mathcal{C} \to \mathbb{R}^n$ is defined by $Px_t = x_t(0)$. $D(\mathcal{A})$ is a dense subset of $\mathcal{C}$. If $\eta \in D(\mathcal{A})$, then the system of (9.2) can be equivalently written as:

$$\begin{aligned} \frac{dx_t}{dt} &= \mathcal{A}x_t + f(Px_t, Qx_t) + g(Px_t, Qx_t)u, \;\; x_0(\xi) = \bar{\eta}, \;\; x_0(0) = \bar{\eta}_0 \\ y(t) &= h(Px_t) \end{aligned} \tag{9.10}$$

where $Qx_t = x_t(-\alpha)$.

The eigenvalue problem for the operator $\mathcal{A}$ is defined as:

$$\mathcal{A}\phi_j = \lambda_j \phi_j, \;\; j = 1, \ldots, \infty \tag{9.11}$$

where λ_j denotes an eigenvalue and ϕ_j denotes an eigenfunction (note that ϕ_j is a vector of dimension n); the eigenspectrum of $\mathcal{A}$, $\sigma(\mathcal{A})$, is defined as the set of all eigenvalues of $\mathcal{A}$, i.e., $\sigma(\mathcal{A}) = \{\lambda_1, \lambda_2, \ldots, \}$ and is given by [108]:

$$\sigma(\mathcal{A}) = \left\{ \lambda : det(\lambda I - A - Be^{-\lambda\alpha}) = 0 \right\} \tag{9.12}$$

The eigenfunctions can be directly computed from the formula $\phi_\lambda = e^{\lambda\xi}\phi_\lambda(0)$, where $\phi_\lambda(0)$ satisfies the equation $(\lambda I - A - Be^{-\lambda\alpha})\phi_\lambda(0) = 0$. The adjoint of operator $\bar{\mathcal{A}}\psi$ of $\mathcal{A}$ is defined from the relation $(\mathcal{A}\phi, \psi) = (\phi, \bar{\mathcal{A}}\psi)$, where $(,)$ denotes the inner product of (9.5), and its eigenspectrum, $\sigma(\bar{\mathcal{A}})$, satisfies $\sigma(\bar{\mathcal{A}}) = \sigma(\mathcal{A})$.

Remark 9.2. Regarding the properties of the eigenspectrum of $\mathcal{A}$, several comments are in order [108, 185]: a) the eigenspectrum $\sigma(\mathcal{A})$ is a point spectrum consisting of eigenvalues, $\lambda \in \sigma(\mathcal{A})$, of finite multiplicity, $\kappa(\lambda)$, b) the number of eigenvalues of $\sigma(\mathcal{A})$ which have positive real part (i.e., they are located in the right-half of the complex plane) is always finite, c) the real parts of all the eigenvalues are bounded from above (i.e., there exists a positive real number β such that $|Re\ \lambda_i| \leq \beta$ for all $i = 1, \ldots, \infty$), and d) the eigenvalues are asymptotically distributed along nearly vertical asymptotes in the complex plane.

Remark 9.3. To illustrate the formulation of a DDE system in an infinite dimensional Banach space, and the formulation and solution of the corresponding eigenvalue problem for the delay operator, we consider the following numerical example:

$$\dot{x} = \begin{bmatrix} -2.0 & 3.5 \\ 3.0 & -3.0 \end{bmatrix} x(t) + \begin{bmatrix} -2.0 & 0.0 \\ 0.0 & 0.0 \end{bmatrix} x(t-3) + \begin{bmatrix} 4x_1x_2 - 3x_1^2 - x_2^2 \\ 0 \end{bmatrix} \tag{9.13}$$

For the above system, the solution $(x_1, x_2) = (0, 0)$ is the unique equilibrium solution, the Banach space is $\mathcal{C}([-3, 0], \mathbb{R}^2)$ and the operator $\mathcal{A}$ takes the form:

$$\mathcal{A}\phi(\xi) = \left\{ \begin{array}{c} \dfrac{d\phi(\xi)}{d\xi}, \quad \xi \in [-3, 0) \\ \\ \begin{bmatrix} -2.0 & 3.5 \\ 3.0 & -3.0 \end{bmatrix} \phi(0) + \begin{bmatrix} -2.0 & 0.0 \\ 0.0 & 0.0 \end{bmatrix} \phi(-3), \quad \xi = 0 \end{array} \right\} \tag{9.14}$$

$$\begin{aligned} \phi(\xi) \in D(\mathcal{A}) = \Big\{ & \phi \in \mathcal{C}^*([-3, 0], \mathbb{R}^{2^*}) : \dot{\phi} \in \mathcal{C}, \\ & \dot{\phi}(0) = \begin{bmatrix} -2.0 & 3.5 \\ 3.0 & -3.0 \end{bmatrix} \phi(0) + \begin{bmatrix} -2.0 & 0.0 \\ 0.0 & 0.0 \end{bmatrix} \phi(-3) \Big\} \end{aligned} \tag{9.15}$$

The eigenvalue problem for $\mathcal{A}$ was solved numerically by using the mathematical software MAPLE and was found that $\sigma(\mathcal{A})$ includes two unstable eigenvalues $\lambda_1 = 0.58$, $\lambda_2 = 0.21$ and infinitely many stable eigenvalues; this implies that the equilibrium solution $(0, 0)$ of the system of (9.13) is unstable.

We finally note that even though the solution $(x_1, x_2) = (0, 0)$ of the DDE system of (9.13) is unstable, the origin of the *undelayed* system:

$$\dot{x} = \begin{bmatrix} -4.0 & 3.5 \\ 3.0 & -3.0 \end{bmatrix} x(t) + \begin{bmatrix} 4x_1x_2 - 3x_1^2 - x_2^2 \\ 0 \end{bmatrix} \tag{9.16}$$

is globally asymptotically stable (the linearization of the above system around the origin possesses two stable eigenvalues: $\lambda_1 = -6.78$, $\lambda_2 = -0.22$ and $(x_1, x_2) = (0, 0)$ is the unique equilibrium point).

9.3.2 Stability Concepts and Results

From the analysis of the previous subsection, it is evident that DDE systems of the form of (9.2) possess fundamentally different properties from ODE systems. The main difference is that the state-space of a DDE system is infinite dimensional, while the state-space of an ODE system is finite dimensional. Therefore, the rigorous analysis of the stability properties of the system of (9.2) requires the use of stability concepts and results for DDE systems. In what follows, we review the definitions of asymptotic stability and input-to-state stability for DDE systems as well as a basic theorem that provides sufficient conditions for assessing asymptotic stability. The reader may refer to the classic book [108] for a complete treatment of stability issues for nonlinear differential difference equations.

Consider the following system of nonlinear differential difference equations:

$$\dot{x}(t) = f(x(t), x(t-\alpha), \theta(t), \theta(t-\alpha)), \quad x(\xi) = \bar{\eta}(\xi), \; \xi \in [-\alpha, 0), \quad x(0) = \bar{\eta}_0 \tag{9.17}$$

where $x \in \mathbb{R}^n$, $\theta \in \mathbb{R}^m$, and suppose that $f(0,0,0,0) = 0$. Now, given a function $\theta : [-\alpha, \infty) \to \mathbb{R}^m$ and $t \in [0, \infty)$, $\theta_t(\xi)$ represents a function from $[-\alpha, 0]$ to $\mathbb{R}^m$ defined by $\theta_t(\xi) := \theta(t+\xi)$. We also define $\|\theta_t(\xi)\| := \max_{t-\alpha \le \xi \le t} \|\theta_t(\xi)\|$, $\|\theta_t\|^s := \sup_{s \ge 0} \|\theta_s(\xi)\|$ and $\|\theta_t\|_T^s := \sup_{0 \le s \le T} \|\theta_s(\xi)\|$. Finally, $\|\cdot\|$ denotes the standard Euclidean norm in $\mathbb{R}^n$. Definition 9.4 that follows provides a rigorous statement of the concept of input-to-state stability for the system of (9.17).

Definition 9.4. *[269] Let γ be a function of class-$\mathcal{Q}$ (see definition of class-$\mathcal{Q}$ function in Appendix F) and δ_x, δ_θ be positive real numbers. The zero solution of (9.17) is said to be input-to-state stable if $\|x_0(\xi)\| \le \delta_x$ and $\|\theta_t\|^s \le \delta_\theta$ imply that the solution of the system of (9.17) is defined for all times and satisfies:*

$$\|x_t(\xi)\| \le \beta(\|x_0(\xi)\|, t) + \gamma(\|\theta_t\|^s), \quad \forall\, t \ge 0 \tag{9.18}$$

The above definition, when $\theta(t) \equiv 0$, $\forall\; t \ge 0$ reduces to the definition of asymptotic stability for the zero solution of the DDE system of (9.17). Furthermore, when $\alpha = 0$, Definition 9.4 reduces to the standard definition of input-to-state stable for nonlinear ODE systems with external inputs (see, for example, [148]). Finally, we note that from the definition of $\|x_t(\xi)\|$ and (9.18), it follows that $\|x(t)\| \le \|x_t(\xi)\| \le \beta(\|x_0(\xi)\|, t) + \gamma(\|\theta_t\|^s)$, $\forall\, t \ge 0$.

The following theorem provides sufficient conditions for the stability of the zero solution of the system of (9.17), expressed in terms of a suitable functional, and consists a natural generalization of the direct method of Lyapunov for ordinary differential equations. The result of this theorem will be directly used in the solution of the state feedback control problem in Sect. 9.5 below.

Theorem 9.5. *[108] Consider the system of (9.17) with $\theta(t) \equiv 0$ and let $\gamma_1, \gamma_2, \gamma_3$ be functions of class Q. If $\gamma_1(s), \gamma_2(s) > 0$ for $s > 0$ and there is a continuous functional $V : \mathcal{C} \to R_{\ge 0}$ such that:*

$$\begin{aligned} \gamma_1(\|x(t)\|) \le V(x_t(\xi)) &\le \gamma_2(\|x_t(\xi)\|) \\ \dot{V}(x_t(\xi)) &\le -\gamma_3(\|x(t)\|) \end{aligned} \tag{9.19}$$

then the solution $x = 0$ is stable. If $\gamma_3(s) > 0$ for $s > 0$, then the solution $x = 0$ is locally asymptotically stable. If $\gamma_1(s) \to \infty$ as $s \to \infty$ and $\gamma_3(s) > 0$ for $s > 0$, then the solution $x = 0$ is globally asymptotically stable.

Remark 9.6. Even though, at this stage, there is no systematic way for selecting the form of the functional $V(x_t(\xi))$ which is suitable for a particular application, a choice for $V(x_t(\xi))$, which is frequently used to show local exponential stability of a DDE system of the form of (9.17) via Theorem 9.5, is:

$$V(x_t(\xi)) = x(t)^T Cx(t) + a^2 \int_{t-\alpha}^{t} x(s)^T Ex(s)ds \tag{9.20}$$

where E, C are symmetric positive definite matrices and a is a positive real number. Clearly, the functional of (9.20) satisfies $K_1\|x(t)\|^2 \leq V(x_t(\xi)) \leq K_2\|x_t(\xi)\|^2$ for some positive K_1, K_2.

9.4 Nonlinear Control of DDE Systems with Small Time Delays

In order to motivate the need for accounting for the presence of time delays in the controller design, we will now establish that a direct application of any nonlinear control method to DDE systems without accounting for the presence of time delays, will lead to the design of controllers that enforce stability and output tracking in the closed-loop system, provided that the time delays are sufficiently small.

We assume that there exists a nonlinear output feedback controller of the form:

$$\begin{aligned} \dot{\omega} &= \mathcal{F}(\omega, y, v) \\ u &= \mathcal{P}(\omega, y, t) \end{aligned} \tag{9.21}$$

where $\mathcal{F}(\omega, y, v)$ is a vector function, $\mathcal{P}(\omega, y, t)$ is a scalar function, and v is the set point, which has been designed on the basis of the system:

$$\begin{aligned} \dot{x} &= Ax(t) + Bx(t) + f(x(t), x(t)) + g(x(t), x(t))u(t), \quad x(0) = x_0 \\ y &= h(x(t)) \end{aligned} \tag{9.22}$$

so that the closed-loop system:

$$\begin{aligned} \dot{\omega} &= \mathcal{F}(\omega, y, v), \quad \omega(0) = \omega_0 \\ \dot{x} &= Ax(t) + Bx(t) + f(x(t), x(t)) + g(x(t), x(t))\mathcal{P}(\omega, y, t), \quad x(0) = x_0 \\ y &= h(x(t)) \end{aligned} \tag{9.23}$$

is locally exponentially stable (see [148] for stability definitions for ODE systems) and the discrepancy between y and v is asymptotically zero (i.e., $\lim_{t\to\infty} \|y(t) - v(t)\| = 0$).

Theorem 9.7 that follows establishes that if the controller of (9.21) enforces local exponential stability and asymptotic output tracking in the closed-loop system of (9.23), then it also enforces these properties in the closed-loop system of (9.2–9.21), provided that the state and measurement delays are sufficiently small (the proof of the theorem can be found in Appendix F).

Theorem 9.7. *If the ODE system of (9.23) is locally exponentially stable, then the nonlinear DDE system of (9.2) under the nonlinear output feedback controller of (9.21) is locally exponentially stable and the discrepancy between* y *and* v *tends asymptotically to zero (i.e.,* $\lim_{t \to \infty} \|y(t) - v(t)\| = 0$*), provided that* α *and* $\tilde{\alpha}$ *are sufficiently small.*

Remark 9.8. Even though the result of the theorem clearly indicates the need for accounting for the presence of time delays in the controller design, it is worth mentioning that Theorem 9.7 establishes a very important robustness property of any nonlinear controller with respect to small time delays. This property may also be useful in many practical applications, because it could lead to simplifications in the controller design task, which, for DDE systems with small time delays, could be addressed on the basis of an ODE system instead of a DDE one.

9.5 Nonlinear Control of DDE Systems with Large Time Delays

9.5.1 Problem Statement

Since the application of controller design methods which do not account for the presence of time delays is limited to DDE systems with small time delays, we address, in this paper, the problem of synthesizing nonlinear output feedback controllers of the general form:

$$\begin{aligned}\dot{\omega} &= A\omega(t) + B\omega(t-\alpha) + f(\omega(t), \omega(t-\alpha)) \\ &\quad + g(\omega(t), \omega(t-\alpha))\mathcal{A}(\omega(t), \bar{v}(t), \omega(t-\alpha), \bar{v}(t-\alpha)) + \mathcal{I}(y(t) - h(\omega(t))) \\ u &= \mathcal{A}(\omega(t), \bar{v}(t), \omega(t-\alpha), \bar{v}(t-\alpha))\end{aligned} \tag{9.24}$$

where $\mathcal{I}(\cdot)$ is a nonlinear integral operator, $\mathcal{A}(\omega(t), \bar{v}(t), \omega(t-\alpha), \bar{v}(t-\alpha))$ is a nonlinear scalar function, $\bar{v}(s) = [v(s)\ v^{(1)}(s)\ \cdots\ v^{(r-1)}(s)]^T$, $s \in [t-\alpha, t]$ and $v^{(k)}$ denotes the $k-$th time-derivative of the reference input $v \in \mathbb{R}$, that enforce exponential stability, asymptotic output tracking and time-delay compensation in the closed-loop system, *independently of the size of the state and measurement delays.*

9.5.2 Methodological Framework

To develop a comprehensive method for the synthesis of controllers of the form of (9.24) that enforce the requested properties in the closed-loop system, we will employ a methodology which involves the following steps:

1. Synthesis of nonlinear state feedback controllers that enforce stability and output tracking in the closed-loop system, independently of the size of the state delay.
2. Design of nonlinear distributed state observers (i.e., the observer itself is a system of nonlinear integro-differential equations) that produce estimates of the unknown state variables of the process with guaranteed asymptotic convergence of the error between the actual and estimated states to zero.
3. Synthesis of distributed nonlinear output feedback controllers through combination of the developed state feedback controllers with the distributed state observers.

Initially, nonlinear state feedback controllers will be synthesized for DDE systems which only include state delays by employing a novel combination of geometric control concepts with the method of Lyapunov functionals (i.e., the controllers will be synthesized in such a way so that the time-derivative of an appropriate Lyapunov-functional calculated along the trajectories of the closed-loop system is negative definite). Then, nonlinear distributed state observers will be constructed for DDE systems which only include state delays by using spectral decomposition techniques. Finally, for DDE systems with state and measurement delays, the output feedback controller will be synthesized on the basis of an auxiliary output constructed within a Smith-predictor framework.

9.6 Nonlinear State Feedback Control for DDE Systems with State Delays

In this section, we consider systems of the form of (9.2) without measurement delay, i.e., $\tilde{\alpha} = 0$, (this assumption will be removed below) and assume that measurements of the states are available (i.e., $x(s)$ for $s \in [t-\alpha, t]$, $t \in [0, \infty)$ is known). For these systems, we address the problem of synthesizing nonlinear static state feedback control laws of the general form:

$$u = \mathcal{A}(x(t), \bar{v}(t), x(t-\alpha), \bar{v}(t-\alpha)) \tag{9.25}$$

where $\mathcal{A}(x(t), \bar{v}(t), x(t-\alpha), \bar{v}(t-\alpha))$ is a nonlinear scalar function, that: a) guarantee local exponential stability, b) force the output to asymptotically follow the reference input (i.e., ensure that $\lim\limits_{t\to\infty} \|y(t) - v(t)\| = 0$), and c) compensate for the effect of the time delay on the output, in the closed-loop system. The structure of the control law of (9.25) is motivated by available results on stabilization of linear DDE systems (e.g., [206, 233]) and the requirement of output tracking.

In order to proceed with the explicit synthesis of the control law of (9.25), we will need to make certain assumptions on the structure and stability properties of the system of (9.2). To simplify the statement of these assumptions, we introduce the notation $f(x(t), x(t-\alpha)) = f_1(x(t)) + f_2(x(t), x(t-\alpha))$, $\tilde{f}(x(t)) = Ax(t) + f_1(x(t))$, and $\bar{p}(x(t), x(t-\alpha)) = Bx(t-\alpha) + f_2(x(t), x(t-\alpha))$, which allows us to rewrite the system of (9.2) in the following form:

$$\begin{aligned} \dot{x} &= \tilde{f}(x(t)) + g(x(t), x(t-\alpha))u + \bar{p}(x(t), x(t-\alpha)) \\ y &= h(x) \end{aligned} \tag{9.26}$$

The first assumption is motivated by the requirement of output tracking and will play a crucial role in the synthesis of the controller.

Assumption 9.1 *Referring to the system of (9.26), there exists an integer r and a change of variables:*

$$\begin{bmatrix} \zeta(s) \\ \eta(s) \end{bmatrix} = \begin{bmatrix} \zeta_1(s) \\ \zeta_2(s) \\ \vdots \\ \zeta_r(s) \\ \eta_1(s) \\ \vdots \\ \eta_{n-r}(s) \end{bmatrix} = \mathcal{X}(x(s)) = \begin{bmatrix} h(x) \\ L_{\tilde{f}} h(x(s)) \\ \vdots \\ L_{\tilde{f}}^{r-1} h(x(s)) \\ \chi_1(x(s)) \\ \vdots \\ \chi_{n-r}(x(s)) \end{bmatrix} \tag{9.27}$$

where $s \in [t-\alpha, t]$ and $\chi_1(x(s)), \ldots, \chi_{n-r}(x(s))$ are scalar functions such that the system of (9.26) takes the form:

$$\begin{aligned} \dot{\zeta}_1 &= \zeta_2(t) + p_1(\zeta(t), \eta(t), \zeta(t-\alpha), \eta(t-\alpha)) \\ &\vdots \\ \dot{\zeta}_{r-1} &= \zeta_r(t) + p_{r-1}(\zeta(t), \eta(t), \zeta(t-\alpha), \eta(t-\alpha)) \\ \dot{\zeta}_r &= L_{\tilde{f}}^r h(\mathcal{X}^{-1}(\zeta(t), \eta(t))) + L_g L_{\tilde{f}}^{r-1} h(\mathcal{X}^{-1}(\zeta(t), \eta(t)))u \\ &\quad + p_r(\zeta(t), \eta(t), \zeta(t-\alpha), \eta(t-\alpha)) \\ \dot{\eta}_1 &= \Psi_1(\zeta(t), \eta(t), \zeta(t-\alpha), \eta(t-\alpha)) \\ &\vdots \\ \dot{\eta}_{n-r} &= \Psi_{n-r}(\zeta(t), \eta(t), \zeta(t-\alpha), \eta(t-\alpha))) \\ y &= \zeta_1 \end{aligned} \tag{9.28}$$

where $p_1(\zeta(t), \eta(t), \zeta(t-\alpha), \eta(t-\alpha)), \ldots, p_r(\zeta(t), \eta(t), \zeta(t-\alpha), \eta(t-\alpha))$, $\Psi_1(\zeta(t), \eta(t), \zeta(t-\alpha), \eta(t-\alpha)), \ldots, \Psi_{n-r}(\zeta(t), \eta(t), \zeta(t-\alpha), \eta(t-\alpha))$ are nonlinear Lipschitz functions and $L_g L_{\tilde{f}}^{r-1} h(x) \neq 0$ for all $x(s) \in \mathbb{R}^n$ and $s \in [t-\alpha, t]$.

Assumption 9.1 provides the explicit form of a coordinate change (which is independent of the state delay present in the system of (9.2)) that transforms the nonlinear DDE system of (9.2) into an interconnection of two subsystems, the ζ-subsystem which describes the input/output dynamics of the system of (9.2) and the η-subsystem which includes the dynamics of the system of (9.2) which are unobservable from the output. Specifically, the interconnection of (9.28) is obtained by considering the change of variables of (9.27) with $s = t$, differentiating it with respect to time, and using that $x(t) = \mathcal{X}^{-1}(\zeta(t), \eta(t))$ and $x(t-\alpha) = \mathcal{X}^{-1}(\zeta(t-\alpha), \eta(t-\alpha))$ (note that this is possible because the coordinate change of (9.27) is assumed to be valid for $s \in [t-\alpha, t]$). The coordinate transformation of (9.27) is not restrictive from an application point of view (one can easily verify that Assumption 9.1 holds for the two chemical reactors with recycle of Subsect. 9.2.2 and the two applications studied in Sects. 9.10–9.11. The assumption that $L_g L_{\bar{f}}^{r-1} h(x) \neq 0$ for all $x(s) \in \mathbb{R}^n$ and $s \in [t-\alpha, t]$ is necessary in order to guarantee that the controller which will be synthesized is well-posed in the sense that it does not generate infinite control action for any values of the states of the process (compare with the structure of the controller given in Theorem 9.9).

To proceed with the controller design, we need to impose the following stability requirement on the η−subsystem of the system of (9.28) which will allow addressing the controller synthesis task on the basis of the low-order ζ-subsystem.

Assumption 9.2 *The dynamical system:*

$$\begin{aligned} \dot{\eta}_1 &= \Psi_1(\zeta(t), \eta(t), \zeta(t-\alpha), \eta(t-\alpha)) \\ &\vdots \\ \dot{\eta}_{n-r} &= \Psi_{n-r}(\zeta(t), \eta(t), \zeta(t-\alpha), \eta(t-\alpha)) \end{aligned} \tag{9.29}$$

is input-to-state stable (see Definition 9.4 for a precise statement of this concept) with respect to the input $\zeta_t(\xi)$.

Loosely speaking, the above assumption states that if the state of the ζ−subsystem is bounded, then the state of the η−subsystem will also remain bounded (see Remark 9.12 for an interpretation of the η-subsystem). In practice, Assumption 9.2 can be verified by linearizing the system of (9.29) with $\zeta_t(\xi) = 0$ around the operating steady-state and computing the eigenvalues of the resulting linear system. If all of these eigenvalues are in the left-half of the complex plane, then [269] Assumption 9.2 is satisfied locally (i.e., for sufficiently small initial conditions and $\zeta_t(\xi)$). An application of this approach for checking Assumption 9.2 is discussed in Subsect. 9.11.2.

Using Assumption 9.2, the controller synthesis problem can be now addressed on the basis of the ζ−subsystem. Specifically, applying the following preliminary feedback law:

$$u = \frac{1}{L_g L_{\tilde{f}}^{r-1} h(\mathcal{X}^{-1}(\zeta,\eta))} \times (\tilde{u} - L_{\tilde{f}}^{r} h(\mathcal{X}^{-1}(\zeta,\eta)) - p_r(\zeta(t),\eta(t),\zeta(t-\alpha),\eta(t-\alpha))) \tag{9.30}$$

where $\tilde{u}$ is an auxiliary input, to the system of (9.28) in order to cancel all the nonlinear terms that can be cancelled by using a feedback which utilizes measurements of $x(s)$ for $s \in [t-\alpha, t]$, we obtain the following modified system:

$$\begin{aligned}
\dot{\zeta}_1 &= \zeta_2 + p_1(\zeta(t),\eta(t),\zeta(t-\alpha),\eta(t-\alpha)) \\
&\vdots \\
\dot{\zeta}_{r-1} &= \zeta_r + p_{r-1}(\zeta(t),\eta(t),\zeta(t-\alpha),\eta(t-\alpha))) \\
\dot{\zeta}_r &= \tilde{u} \\
\dot{\eta}_1 &= \Psi_1(\zeta(t),\eta(t),\zeta(t-\alpha),\eta(t-\alpha)) \\
&\vdots \\
\dot{\eta}_{n-r} &= \Psi_{n-r}(\zeta(t),\eta(t),\zeta(t-\alpha),\eta(t-\alpha))) \\
y &= \zeta_1
\end{aligned} \tag{9.31}$$

Introducing the notation:

$$\tilde{A} = \begin{bmatrix} 0 & 1 & 0 & \cdots & 0 & 0 \\ 0 & 0 & 1 & \cdots & 0 & 0 \\ 0 & 0 & 0 & \cdots & 0 & 0 \\ \vdots & \vdots & \vdots & \ddots & \vdots & \vdots \\ 0 & 0 & 0 & \cdots & 0 & 0 \end{bmatrix}, \quad b = \begin{bmatrix} 0 \\ 0 \\ 0 \\ \vdots \\ 1 \end{bmatrix},$$

$$p(\zeta(t),\eta(t),\zeta(t-\alpha),\eta(t-\alpha)) = \begin{bmatrix} p_1(\zeta(t),\eta(t),\zeta(t-\alpha),\eta(t-\alpha)) \\ p_2(\zeta(t),\eta(t),\zeta(t-\alpha),\eta(t-\alpha)) \\ \vdots \\ p_{r-1}(\zeta(t),\eta(t),\zeta(t-\alpha),\eta(t-\alpha)) \\ 0 \end{bmatrix} \tag{9.32}$$

the ζ-subsystem of the system of (**??**) can be written in the following compact form:

$$\begin{aligned}
\dot{\zeta} &= \tilde{A}\zeta + b\tilde{u} + p(\zeta(t),\eta(t),\zeta(t-\alpha),\eta(t-\alpha)) \\
y &= \zeta_1
\end{aligned} \tag{9.33}$$

The controller synthesis task has been now reduced to the one of synthesized $\tilde{u}$ to stabilize the ζ-subsystem and force the output y to asymptotically follow the reference input, v. To develop a solution to this problem, we will need to make the following assumption on the growth of the vector $p(\zeta(t),\eta(t),\zeta(t-\alpha),\eta(t-\alpha))$.

Assumption 9.3 *Let*
$\bar{e}(s) = [(h(x(s)) - v(s))\ (L_{\tilde{f}}h(x(s)) - v^{(1)}(s))\ \cdots\ (L_{\tilde{f}}^{r-1}h(x(s)) - v^{(r-1)}(s))]^T$,

$s \in [t-\alpha, t]$ *where* $v^{(k)}$ *denotes the* $k-th$ *time-derivative of the reference input* v. *There exist positive real numbers* a_1, a_2 *such that the following bound can be written:*

$$\|p(\bar{e}(t) + \bar{v}(t), \eta(t), \bar{e}(t-\alpha) + \bar{v}(t-\alpha), \eta(t-\alpha))\|^2 \leq a_1\bar{e}^2(t) + a_2\bar{e}^2(t-\alpha) \tag{9.34}$$

The above assumption on the growth of the vector $p(\bar{e}(t) + \bar{v}(t), \eta(t), \bar{e}(t-\alpha) + \bar{v}(t-\alpha), \eta(t-\alpha))$ does not need to hold globally (i.e., for any $\bar{e}(t), \eta(t)$), and thus, it is satisfied by most practical problems (see, for example, the applications studied in Sects. 9.10–9.11). Furthermore, Assumption 9.3 will allow us to synthesize a linear auxiliary feedback law of the form $\tilde{u} = K\bar{e}$ to stabilize the ζ-subsystem and enforce output tracking. The synthesis of such a $\tilde{u}$ will be performed by using the method of Lyapunov functionals. Specifically, $\tilde{u}$ will be designed so that the time-derivative of the following Lyapunov functional:

$$V(\bar{e}_t(\xi)) = \bar{e}^T P \bar{e} + a^2 \int_{t-\alpha}^{t} \bar{e}^T(s)\bar{e}(s)ds \tag{9.35}$$

where a is a positive real number, calculated along the state of the closed-loop $\zeta-$subsystem is negative definite. The incorporation of the integral term $\int_{t-\alpha}^{t} \bar{e}^T(s)\bar{e}(s)ds$ in the functional of (9.35) allows accounting for the distributed parameter (delayed) nature of the system of (9.2) in the controller design stage and synthesizing a controller that enforces the requested properties in the closed-loop system independently of the size of the state delay.

We are now in a position to state the main controller synthesis result of this section. Theorem 9.9 that follows provides the formula of the controller and conditions under which stability and output tracking is guaranteed in the closed-loop system (the proof of the theorem can be found in Appendix F).

Theorem 9.9. *Consider the system of nonlinear differential difference equations of (9.2) with* $\tilde{\alpha} = 0$, *for which Assumptions 9.1–9.3 hold. Then, if the matrix equation:*

$$\tilde{A}^T P + P\tilde{A} - 2P^T b R_2^{-1} b^T P + (a^2 + a_1)I + P^2 = -R_1 \tag{9.36}$$

where $a^2 > a_2$, *and* R_1, R_2 *are positive definite matrices, has a unique positive definite solution for* P, *the nonlinear state feedback controller:*

$$\begin{aligned} u &= \mathcal{A}(x(t), \bar{v}(t), x(t-\alpha), \bar{v}(t-\alpha)) \\ &:= \frac{1}{L_g L_{\tilde{f}}^{r-1} h(x)} \Big(-R_2^{-1} b^T P \bar{e}(t) + v^{(r)}(t) - L_{\tilde{f}}^r h(x) \\ &\quad - p_r(x(t), \bar{v}(t), x(t-\alpha), \bar{v}(t-\alpha)) \Big) \end{aligned} \tag{9.37}$$

enforces: (a) local exponential stability, and (b) asymptotic output tracking in the closed-loop system, independently of the size of the state delay.

Remark 9.10. Regarding the structure, implementation and closed-loop properties of the nonlinear state feedback controller of (9.37), several remarks are in order: a) it uses measurements of the states of the process at t and $t-\alpha$ (i.e., $x(t)$ and $x(t-\alpha)$), and thus, it belongs to the class of the requested control laws of (9.25), b) its practical implementation requires the use of memory lines to store the values of x in the time interval $[t-\alpha, t]$, and c) it enforces stability and asymptotic output tracking in the closed-loop system *independently of the size of the state delay.*

Remark 9.11. In order to apply the result of Theorem 9.9 to a chemical process application, one has to initially verify Assumptions 9.1–9.3 of the theorem on the basis of the process model and compute the parameters a_1 and a_2. Then, a, R_1, R_2 should be chosen so that $a^2 > a_2$ and the matrices R_1, R_2 are positive definite to ensure that (9.36) has a unique positive definite solution for P. Regarding the role of R_1, R_2 on closed-loop properties, we note that R_1 determines the speed of the closed-loop output response (namely, "larger" (in terms of the smallest eigenvalue) R_1 means faster response), while R_2 determines the penalty that should be imposed on the manipulated input in achieving stabilization and output tracking ("larger" R_2 means larger penalty on the control action). If these assumptions are satisfied, the synthesis formula of (9.37) can be directly used to derive the explicit form of the controller (see Sects. 9.10–9.11 for the application of this procedure to two chemical process examples).

Remark 9.12. In analogy to the case of nonlinear ODE systems (see, for example, [126,157]), one can show that the η-subsystem of (9.29) represents the inverse dynamics of the DDE system of (9.2). Moreover, the η-subsystem of (9.29) with $\zeta_t(\xi) = 0$, i.e.:

$$
\begin{aligned}
\dot{\eta}_1 &= \Psi_1(0, \eta(t), 0, \eta(t-\alpha)) \\
&\vdots \\
\dot{\eta}_{n-r} &= \Psi_{n-r}(0, \eta(t), 0, \eta(t-\alpha))
\end{aligned}
\tag{9.38}
$$

represents the zero dynamics of the DDE system of (9.2) (i.e., the dynamics of (9.2) when the output is set identically equal to zero). Linearizing the zero dynamics of (9.38) around the zero solution, one can show that the eigenvalues of the resulting linear system are identical to the zeros (which are infinite) of the linear DDE system of (9.3).

Remark 9.13. Applying the proposed method for the synthesis of state feedback controllers for DDE systems to linear systems of the form of (9.3) with $\tilde{\alpha} = 0$, we end up with the following controller synthesis formula:

$$u = [wA^{r-1}c]^{-1}\left(-R_2^{-1}b^T P\bar{e}(t) + v^{(r)}(t) - wA^r x(t) - wA^{r-1}Bx(t-\alpha)\right) \tag{9.39}$$

where P is the solution of (9.36). The linear controller of (9.39) can be thought of as an extension of linear control laws of the form:

$$u = F_1 x(t) + F_2 x(t-\alpha) \tag{9.40}$$

where F_1, F_2 are constant vectors of appropriate dimensions, which were considered in the context of stabilization of linear DDE systems (e.g., [206,233]), to the problems of stabilization with output tracking. We note that the usual approach followed in the literature for the design of the vectors F_1, F_2 is based on the method of Lyapunov functionals.

Remark 9.14. A control problem which has attracted significant attention in the area of nonlinear process control is the one of specifying the explicit formula of a controller that enforces a linear response between the controlled output and the reference input in the closed-loop system. This problem was solved in [158] for systems of nonlinear ODEs, and more recently, in [53] for systems of nonlinear hyperbolic PDEs. In this remark, we formally pose this problem for DDE systems of the form of (9.2) with $\tilde{\alpha} = 0$ and show that it leads to the synthesis of a nonlinear feedback controller which is non-realizable, and thus, it cannot be implemented in practice. Specifically, we seek to design a nonlinear feedback controller that enforces the following linear input/output response:

$$\gamma_r \frac{d^r y}{dt^r} + \cdots + \gamma_1 \frac{dy}{dt} + y = v \tag{9.41}$$

where r is the 'relative order' of the system of (9.2) (i.e., the smallest derivative of y which depends explicitly on the manipulated input u) and $\gamma_1, \gamma_2, \dots, \gamma_r$ are adjustable parameters, in the closed-loop system. Calculating the time-derivatives of the output y, up to order r, in the system of (9.2), we obtain the following expressions:

$$\begin{aligned}
y &= h(x) \\
\frac{dy}{dt} &= \psi_1(x(t), x(t-\alpha)) \\
\frac{d^2y}{dt^2} &= \psi_2(x(t), x(t-\alpha), x(t-2\alpha)) \\
&\vdots \\
\frac{d^r y}{dt^r} &= \psi_r(x(t), x(t-\alpha), \dots, x(t-r\alpha)) + \phi(x(t), x(t-\alpha), \dots, x(t-r\alpha))u
\end{aligned} \tag{9.42}$$

where $\psi_1(x(t), x(t-\alpha)), \dots, \psi_r(x(t), x(t-\alpha), \dots, x(t-r\alpha)), \phi(x(t), x(t-\alpha), \dots, x(t-r\alpha))$ are smooth nonlinear functions whose specific form is omitted for brevity. Substituting the expressions for the time-derivatives of y in

(9.41), assuming that the term $\phi(x(t), x(t-\alpha), \ldots, x(t-r\alpha)) \neq 0$ and solving for u, we obtain the following expression for the controller that enforces the linear response of (9.41) in the closed-loop system:

$$\begin{aligned} u(t) = {} & \frac{1}{\phi(x(t), x(t-\alpha), \ldots, x(t-r\alpha))}\Big((v - h(x(t)) \\ & - \sum_{\nu=1}^{r} \gamma_\nu \psi_\nu(x(t), x(t-\alpha), \ldots, x(t-\nu\alpha)\Big) \end{aligned} \tag{9.43}$$

The above controller clearly uses measurements of $x(t-\nu\alpha)$ with $2 \leq \nu \leq r$ which may not be available, and thus, it cannot be implemented in practice. To illustrate this point, we consider the representation of the controller of (9.43) for $t = 0$ i.e.:

$$\begin{aligned} u(0) = {} & \frac{1}{\phi(x(0), x(-\alpha), \ldots, x(-r\alpha))}\Big(v - h(x(0)) \\ & - \sum_{\nu=1}^{r} \gamma_\nu \psi_\nu(x(0), x(-\alpha), \ldots, x(-\nu\alpha)\Big) \end{aligned} \tag{9.44}$$

Clearly, the calculation of the initial control action, $u(0)$, requires values of the state, such as $x(-2\alpha), \ldots, x(-r\alpha)$, which are not included in the initial data of the system of (9.2), and thus, $u(0)$, cannot be realized. We finally note that an extension of the initial conditions in the past as a remedy to this problem is not meaningful, since, from a mathematical point of view, it leads to an ill-defined DDE system, while, from a practical point of view, it may require past knowledge of the state at an *arbitrarily large* time interval which, in general, there is no guarantee that exists.

9.7 Nonlinear State Observer Design for DDE Systems with State Delay

In this section, we consider nonlinear DDE systems of the form of (9.2) with $\tilde{\alpha} = 0$ and focus on the design of nonlinear state observers that use measurements of process output, $y(s)$, for $s \in [t-\alpha, t]$ to produce estimates of the state variables $x(t)$, with guaranteed exponential convergence of the error between the actual and the estimated values of $x(t)$ to zero. Specifically, we consider the design of state observers with the following general state-space description:

$$\begin{aligned} \dot{\omega} = {} & A\omega(t) + B\omega(t-\alpha) + f(\omega(t), \omega(t-\alpha)) + g(\omega(t), \omega(t-\alpha))u \\ & + \mathcal{I}(y(t) - h(\omega(t))), \\ & \omega(\xi) = \bar{\omega}(\xi),\ \xi \in [-\alpha, 0),\ \ \omega(0) = \bar{\omega}_0 \end{aligned} \tag{9.45}$$

where $\omega \in \mathbb{R}^n$ is the observer state, $\bar{\omega}(\xi)$ is a smooth vector function defined in $\xi \in [-\alpha, 0)$, $\bar{\omega}_0$ is a constant vector and $\mathcal{I}(\cdot)$ is a bounded nonlinear integral operator, mapping $\mathbb{R}$ into $\mathcal{C}$. The system of (9.45) consists of a replica of the system of (9.2) and the term $\mathcal{I}(y(t) - h(\omega(t)))$ which will be designed so that the system of (9.45) is locally exponentially stable and the discrepancy between $x(t)$ and $\omega(t)$ tends exponentially to zero. The derivation of the explicit form of the integral operator $\mathcal{I}(\cdot)$ will be performed by working with the infinite dimensional formulation, (9.10), of the nonlinear DDE system of (9.2) presented in Sect. 9.10, and using a spectral decomposition of the DDE system and nonlinear observer design results for finite-dimensional systems.

We initially transform the system of (9.10) into an interconnection of a finite-dimensional system that describes the dynamics of the eigenmodes of $\mathcal{A}$ corresponding to the unstable eigenvalues of $\sigma(\mathcal{A})$, and an infinite-dimensional system that describes the dynamics of the remaining eigenmodes of $\mathcal{A}$. Let $H = \{\lambda : \lambda \in \sigma(\mathcal{A}) \text{ and } Re\lambda \geq 0\}$ and assume, in order to simplify the development, that $\kappa(\lambda) = 1$ for each $\lambda \in \sigma(\mathcal{A})$ (i.e., the multiplicity of all the eigenvalues is assumed to be one; the usual case in most practical applications). Let m be the number of eigenvalues included in H (note that m is always finite). Also, let the elements of H be ordered as $(\lambda_1, \lambda_2, \ldots, \lambda_m)$, where $Re\lambda_1 \geq Re\lambda_2 \geq \cdots Re\lambda_m$. Clearly, the eigenfunctions (column vectors) in the $n \times m$ matrix $\Phi_H = [\phi_1, \phi_2, \ldots, \phi_m]$ form a basis in $\mathcal{C}_{\lambda \geq 0}$. Also, let $\Psi_H = [\psi_1, \psi_2, \ldots, \psi_m]^T$ be the basis in $\mathcal{C}^*_{\lambda \geq 0}$ chosen so that $(\Psi_H, \Phi_H) = I$, where $(\Psi_H, \Phi_H) = [(\psi_i, \phi_j)]$, ψ_i and ϕ_j being the i-th element and j-th element of Ψ_H and Φ_H, respectively. Defining the orthogonal projection operators P_p and P_n such that $x_t^p = P_p x_t$, $x_t^n = P_n x_t$ (note that $x_t^p = \Phi_H(\Psi_H, x_t)$), the state x_t of the system of (9.10) can be decomposed as:

$$x_t = x_t^p + x_t^n = P_p x_t + P_n x_t \tag{9.46}$$

Applying P_p and P_n to the system of (9.10) and using the above decomposition for x_t, the system of (9.10) can be equivalently written in the following form:

$$\begin{aligned}
\frac{dx_t^p}{dt} &= \mathcal{A}_p x_t^p + f_p(P(x_t^p + x_t^n), Q(x_t^p + x_t^n)) + g_p(P(x_t^p + x_t^n), Q(x_t^p + x_t^n))u \\
\frac{\partial x_t^n}{\partial t} &= \mathcal{A}_n x_t^n + f_n(P(x_t^p + x_t^n), Q(x_t^p + x_t^n)) + g_n(P(x_t^p + x_t^n), Q(x_t^p + x_t^n))u \\
y &= h(P(x_t^p + x_t^n)) \\
x_t^p(0) &= P_p x(0) = P_p \bar{\eta}, \quad x_t^n(0) = P_n x(0) = P_n \bar{\eta}
\end{aligned} \tag{9.47}$$

where $\mathcal{A}_p = P_p \mathcal{A} P_p$, $g_p = P_p g$, $f_p = P_p f$, $\mathcal{A}_n = P_n \mathcal{A} P_n$, $g_n = P_n g$ and $f_n = P_n f$ and the notation $\frac{\partial x_t^n}{\partial t}$ is used to denote that the state x_t^n belongs in an infinite-dimensional space. In the above system, $\mathcal{A}_p = \Phi_H A_p$, where A_p is a matrix of dimension $m \times m$ whose eigenvalues are the ones included in H, f_p and f_n are Lipschitz vector functions, and $\mathcal{A}_n$ is an infinite range matrix which is stable (this follows from the fact that H includes all the eigenvalues of, $\sigma(\mathcal{A})$, which are in the closed right half of the complex plane). Neglecting the x_n-subsystem, the following finite-dimensional system is obtained:

$$\begin{aligned} \frac{dx_t^p}{dt} &= \mathcal{A}_p x_t^p + f_p(Px_t^p, Qx_t^p) + g_p(Px_t^p, Qx_t^p)u \\ y_p &= h(Px_t^p) \end{aligned} \tag{9.48}$$

where the subscript p in y_p denotes that the output is associated with an approximate finite dimensional system.

Assumption 9.4 that follows states that the system of (9.48) is observable and is needed in order to design a nonlinear state observer for the system of (9.2) (the reader may refer to [254] for a precise definition of the concept of observability for nonlinear finite-dimensional systems; see also [36] for definitions and characterizations of observability for linear DDE systems).

Assumption 9.4 *The pair $[h(x_t^p)\ \mathcal{A}_p x_t^p + f_p(x_t^p, 0)]$ is locally observable in the sense that there exists a nonlinear gain column vector $L(x_t^p)$ of dimension m (where m is the number of unstable eigenvalues of $\mathcal{A}$) so that the finite-dimensional dynamical system:*

$$\frac{d\omega_t^p}{dt} = \mathcal{A}_p \omega_t^p + f_p(P\omega_t^p, Q\omega_t^p) + L(\omega_t^p)(y_p - h(P\omega_t^p)) \tag{9.49}$$

is locally exponentially stable.

Theorem 9.15 that follows provides a nonlinear distributed state observer (the proof of the theorem can be found in Appendix F).

Theorem 9.15. *Referring to the system of (9.10) with $u(t) \equiv 0$ and suppose that Assumption 9.4 holds. Then, if there exists a positive real number a_1 such that $\|\bar{\omega} - \bar{\eta}\|_2 \leq a_1$, the nonlinear infinite-dimensional dynamical system:*

$$\begin{aligned} \frac{d\omega_t}{dt} &= \mathcal{A}\omega_t + f(P\omega_t, Q\omega_t) + \Phi_H L(\Psi_H, \omega_t)(y(t) - h(P\omega_t)) , \\ &\omega_0(\xi) = \bar{\omega}, \qquad \omega_0(0) = \bar{\omega}_0 \end{aligned} \tag{9.50}$$

is a local exponential observer for the system of (9.10) in the sense that the estimation error, $e_t = \omega_t - x_t$, tends exponentially to zero.

The abstract dynamical system of (9.50) can be simplified by utilizing a procedure based on the method of characteristics for first-order hyperbolic PDE systems (this procedure is detailed in Appendix F of this dissertation) to obtain the following nonlinear integro-differential equation system representation for the state observer:

$$\begin{aligned}\dot{\omega} &= A\omega(t) + B\omega(t-\alpha) + f(\omega(t), \omega(t-\alpha)) + g(\omega(t), \omega(t-\alpha))u \\ &+\Phi_H(0)L(\Psi_H, \tilde{\omega}(\xi, t))(y(t) - h(\omega(t))) \\ &+B\int_0^\alpha \Phi_H(\xi-\alpha)L(\Psi_H, \tilde{\omega}(\xi, t))[y(t-\xi) - h(\omega(t-\xi))]d\xi\,, \\ &\omega(\xi) = \bar{\omega}(\xi),\ \xi \in [-\alpha, 0),\ \ \omega(0) = \bar{\omega}_0\end{aligned} \tag{9.51}$$

Remark 9.16. Referring to the state observer of (9.51), it is worth noting that: a) it includes an integral of the observer error, which is expected because the proposed observer design method accounts explicitly for the distributed parameter nature of the time-delay system of (9.2), and b) it consists of a replica of the process model and a nonlinear integral gain acting on the discrepancy between the actual and the estimated output, and thus, it can be thought of as the analogue of nonlinear Luenberger-type observers (e.g., [144, 254]) developed for nonlinear ODE systems in the case of nonlinear DDE systems.

Remark 9.17. For open-loop *stable* systems, the nonlinear gain $L(\Psi_H, \tilde{\omega}(\xi, t))$ can be set identically equal to zero and the distributed state observer of (9.51) reduces to an open-loop DDE observer of the form:

$$\dot{\omega} = A\omega(t) + B\omega(t-\alpha) + f(\omega(t), \omega(t-\alpha)) + g(\omega(t), \omega(t-\alpha))u \tag{9.52}$$

Remark 9.18. For open-loop *unstable* systems, the practical implementation of the state observer of (9.51) involves the design of a nonlinear gain, $L(\Psi_H, \tilde{\omega}(\xi, t))$, on the basis of a nonlinear finite dimensional system (Assumption 9.4). However, in most practical applications, the construction of a nonlinear gain requires performing extensive computations (for example, series solutions of nonlinear partial differential equations [144]), and thus, it cannot be easily performed. A computationally efficient way to address this problem is to design a constant gain, L, on the basis of a linear DDE system resulting from the linearization of the nonlinear DDE system around an operating steady-state and evaluating its validity through computer simulations (see the implementation of the observer of (9.51) in the reactor-separator system studied in Sect. 9.9). We note that the only computations needed to design a constant observer gain are: (a) the computation of the eigenvalues of the characteristic equation $det(\lambda I - A - Be^{-\lambda\alpha}) = 0$ which are in the right-half of the complex plane (this can be done by using standard algorithms, for example, [184,185]) and (b) the computation of the eigenfunctions from the formula $\phi_\lambda = e^{\lambda\xi}\phi_\lambda(0)$, where $\phi_\lambda(0)$ satisfies the equation $(\lambda I - A - Be^{-\lambda\alpha})\phi_\lambda(0) = 0$.

Remark 9.19. To illustrate the application of the result of Theorem 9.15, consider the numerical DDE example of Remark 9.3 (9.13) with $y = x_2$ as the output. The eigenvectors $\phi_j(\xi)$ corresponding to the two unstable eigenvalues of the delay operator of (9.14) are:

$$\phi_1(\xi) = \begin{bmatrix} -0.78 \\ 0.62 \end{bmatrix} e^{0.58\xi},\ \ \phi_2(\xi) = \begin{bmatrix} -0.68 \\ -0.73 \end{bmatrix} e^{0.21\xi} \tag{9.53}$$

The following constant observer gain:

$$L = \begin{bmatrix} -1.0 \\ -16.0 \end{bmatrix} \tag{9.54}$$

was found to satisfy Assumption 9.4 and yields the following nonlinear state observer:

$$\begin{aligned} \dot{\omega} = & \begin{bmatrix} -2.0 & 3.5 \\ 3.0 & -3.0 \end{bmatrix} \omega(t) + \begin{bmatrix} -2.0 & 0.0 \\ 0.0 & 0.0 \end{bmatrix} \omega(t-3) + \begin{bmatrix} 4\omega_1\omega_2 - 3\omega_1^2 - \omega_2^2 \\ 0 \end{bmatrix} \\ & + \begin{bmatrix} -0.78 & -0.68 \\ 0.62 & -0.73 \end{bmatrix} \begin{bmatrix} -1.0 \\ -16.0 \end{bmatrix} (x_2 - \omega_2) \\ & + \begin{bmatrix} -2.0 & 0.0 \\ 0.0 & 0.0 \end{bmatrix} \int_0^3 \begin{bmatrix} -0.78e^{0.58(\xi-3.0)} & -0.68e^{0.21(\xi-3.0)} \\ 0.62e^{0.58(\xi-3.0)} & -0.73e^{0.21(\xi-3.0)} \end{bmatrix} \\ & \times \begin{bmatrix} -1.0 \\ -16.0 \end{bmatrix} (x_2(t-\xi) - \omega_2(t-\xi))d\xi \end{aligned} \tag{9.55}$$

Remark 9.20. Applying the proposed observer design method to linear DDE systems of the form of (9.3) with $\tilde{\alpha} = 0$, we obtain the following linear state observer:

$$\begin{aligned} \dot{\omega} &= A\omega(t) + B\omega(t-\alpha) + cu + \Phi_H(0)L(wx(t) - w\omega(t)) \\ &\quad + B\int_0^{\alpha} \Phi_H(\xi - \alpha)L[wx(t-\xi) - w\omega(t-\xi)]d\xi, \\ &\omega(\xi) = \bar{\omega}(\xi),\ \xi \in [-\alpha, 0),\ \ \omega(0) = \bar{\omega}_0 \end{aligned} \tag{9.56}$$

9.8 Nonlinear Output Feedback Control of DDE Systems with State Delay

In this section, we consider DDE systems of the form of (9.2) with $\tilde{\alpha} = 0$ and address the problem of synthesizing distributed output feedback controllers that enforce local exponential stability and asymptotic output tracking in the closed-loop system, independently of the size of the state delay. The requisite output feedback controllers will be synthesized employing combination of the developed distributed state feedback controllers with distributed state observers.

Theorem 9.21 that follows provides a state-space realization of the distributed output feedback controller and the properties that it enforces in the closed-loop system (the proof of the theorem can be found in Appendix F).

Theorem 9.21. *Consider the system of nonlinear differential difference equations of (9.2) with $\tilde{\alpha} = 0$, for which the Assumptions 9.1–9.4 hold. Then, if* *there exists a positive real number a_0 such that $\|\bar{\omega}-\bar{\eta}\|_2 \leq a_0$ and the matrix equation of (9.36) has a unique positive definite solution for P, the distributed output feedback controller:*

$$
\begin{aligned}
\dot{\omega} &= A\omega(t) + B\omega(t-\alpha) + f(\omega(t),\omega(t-\alpha)) \\
&\quad + \Phi_H(0)L(\Psi_H,\tilde{\omega}(\xi,t))(y(t)-h(\omega(t))) \\
&\quad + B\int_0^{\alpha}\Phi_H(\xi-\alpha)L(\Psi_H,\tilde{\omega}(\xi,t))[y(t-\xi)-h(\omega(t-\xi))]d\xi \\
&\quad + g(\omega(t),\omega(t-\alpha)) \times \frac{1}{L_g L_{\bar{f}}^{r-1}h(\omega)}\Big(-R_2^{-1}b^T P\bar{e}(t) + v^{(r)}(t) - L_{\bar{f}}^r h(\omega) \\
&\quad - p_r(\omega(t),\bar{v}(t),\omega(t-\alpha),\bar{v}(t-\alpha))\Big) \\
&\quad \omega(\xi) = \bar{\omega}(\xi),\ \xi \in [-\alpha,0),\ \ \omega(0) = \bar{\omega}_0 \\
u &= \frac{1}{L_g L_{\bar{f}}^{r-1}h(\omega)}\Big(-R_2^{-1}b^T P\bar{e}(t) + v^{(r)}(t) - L_{\bar{f}}^r h(\omega) \\
&\quad - p_r(\omega(t),\bar{v}(t),\omega(t-\alpha),\bar{v}(t-\alpha))\Big)
\end{aligned}
\tag{9.57}
$$

where $\bar{e} = [(h(\omega)-v)\ (L_{\bar{f}}h(\omega)-v^{(1)})\ \cdots\ (L_{\bar{f}}^{r-1}h(\omega)-v^{(r-1)})]^T$, a) guarantees local exponential stability of the closed-loop system, and b) enforces asymptotic output tracking, independently of the size of the state delay.

Remark 9.22. For open-loop stable systems, $L(\Psi_H,\tilde{\omega}(\xi,t))$ can be set equal to zero and the distributed output feedback controller of (9.57) can be simplified to:

$$
\begin{aligned}
\dot{\omega} &= A\omega(t) + B\omega(t-\alpha) + f(\omega,\omega(t-\alpha)) + g(\omega(t),\omega(t-\alpha)) \\
&\quad \times \frac{1}{L_g L_{\bar{f}}^{r-1}h(\omega)}\Big(-R_2^{-1}b^T P\bar{e}(t) + v^{(r)}(t) - L_{\bar{f}}^r h(\omega) \\
&\quad - p_r(\omega(t),\bar{v}(t),\omega(t-\alpha),\bar{v}(t-\alpha))\Big) \qquad (9.58) \\
\bar{u} &= \frac{1}{L_g L_{\bar{f}}^{r-1}h(\omega)}\Big(-R_2^{-1}b^T P\bar{e}(t) + v^{(r)}(t) - L_{\bar{f}}^r h(\omega) \\
&\quad - p_r(\omega(t),\bar{v}(t),\omega(t-\alpha),\bar{v}(t-\alpha))\Big)
\end{aligned}
$$

Remark 9.23. Referring to the result of Theorem 9.21, we note that no consistent initialization requirement has been imposed on the observer states in order to prove local exponential stability and asymptotic output tracking in the closed-loop system (i.e., it is not necessary that $\bar{\omega}(\xi) = \bar{\eta}(\xi)$, $\xi \in [-\alpha,0)$ and $\bar{\omega}_0 = \bar{\eta}_0$).

Remark 9.24. The exponential stability of the closed-loop system guarantees that in the presence of small modeling errors (i.e., unknown model parameters and external disturbances) and initialization errors of the observer states, the states of the closed-loop system will remain bounded. Furthermore, it is possible to implement a linear error feedback controller around the $(y-v)$ loop to ensure asymptotic offsetless output tracking in the closed-loop system, in the presence of *constant* parametric uncertainty, external disturbances and initialization errors. In this case, one can use calculations similar to the ones in [64] to derive a mixed-error and output feedback controller, which possesses integral action, (i.e., a controller of the form of (9.57) with $\bar{e} = [(y(t)-v)\ (L_{\bar{f}}h(\omega)-v^{(1)})\ \cdots\ (L_{\bar{f}}^{r-1}h(\omega)-v^{(r-1)})]^T)$, that enforces exponential stability and asymptotic offsetless output tracking in the closed-loop system in the presence of constant modeling errors and disturbances.

9.9 Nonlinear Output Feedback Control for DDE Systems with State and Measurement Delay

In this section, we consider DDE systems of the form of (9.2) with $\tilde{\alpha} > 0$ and address the problem of synthesizing distributed output feedback controllers that enforce local exponential stability and asymptotic output tracking in the closed-loop system, independently of the size of the state delay. In order to account for the presence of the measurement delay in the controller design, the requisite output feedback controller will be obtained by working within a Smith Predictor framework [117, 162]. Within this framework, the state feedback controller is synthesized on the basis of an auxiliary output $\bar{y}$, which represents the prediction of the output if there were no dead-time on the output, and can be obtained by adding a corrective signal δy to the on-line measurement of the actual output y:

$$\bar{y} = y + \delta y$$

Assuming that the open-loop DDE system is observable, the corrective signal, δy, is obtained through a *closed-loop* Smith-type predictor; which for the problem in question is a nonlinear DDE system driven by the manipulated input, that simulates the difference in the responses between the process model without output dead-time and the process model with output dead-time:

$$\begin{aligned}
\dot{\omega} &= A\omega(t) + B\omega(t-\alpha) + f(\omega(t),\omega(t-\alpha)) + g(\omega(t),\omega(t-\alpha))u \\
&\quad + \Phi_H(0)L(\Psi_H,\tilde{\omega}(\xi,t))(y(t-\tilde{\alpha}) - h(\omega(t-\tilde{\alpha}))) \\
&\quad + B\int_0^{\alpha-\tilde{\alpha}} \Phi_H(\xi-\alpha)L(\Psi_H,\tilde{\omega}(\xi,t))[y(t-\xi-\tilde{\alpha}) - h(\omega(t-\xi-\tilde{\alpha}))]d\xi, \\
\omega(\xi) &= \bar{\omega}(\xi),\ \xi \in [-\alpha,0),\ \ \omega(0) = \bar{\omega}_0
\end{aligned} \tag{9.59}$$

The resulting output feedback controller and the properties that it enforces in the closed-loop system are given in Theorem 9.25 below (the proof of the theorem is similar to the one of theorem of Theorem 9.21 and will be omitted for brevity).

Theorem 9.25. *Consider the nonlinear DDE system of (9.2) with $\tilde{\alpha} > 0$, for which Assumptions 9.1–9.3 hold, and an observability property similar to the one of Assumption 9.4 holds for $0 \leq \tilde{\alpha} \leq \alpha$. Then, if there exists a positive real number a_0 such that $\|\bar{\omega} - \bar{\eta}\|_2 \leq a_0$ and the matrix equation of (9.36) has a unique positive definite solution for P, the distributed output feedback controller:*

$$\begin{aligned}
\dot{\omega} = {} & A\omega(t) + B\omega(t-\alpha) + f(\omega(t),\omega(t-\alpha)) \\
& + \Phi_H(0)L(\Psi_H,\tilde{\omega}(\xi,t))(y(t-\tilde{\alpha}) - h(\omega(t-\tilde{\alpha}))) \\
& + B\int_0^{\alpha-\tilde{\alpha}} \Phi_H(\xi-\alpha)L(\Psi_H,\tilde{\omega}(\xi,t))[y(t-\xi-\tilde{\alpha}) - h(\omega(t-\xi-\tilde{\alpha}))]d\xi \\
& + g(\omega(t),\omega(t-\alpha)) \times \frac{1}{L_g L_{\tilde{f}}^{r-1}h(\omega)}\Big(-R_2^{-1}b^T P\bar{e} + v^{(r)}(t) - L_{\tilde{f}}^r h(\omega) - \\
& - p_r(\omega(t),\bar{v}(t),\omega(t-\alpha),\bar{v}(t-\alpha))\Big)
\end{aligned}$$

$$\begin{aligned}
\omega(\xi) &= \bar{\omega}(\xi),\ \xi \in [-\alpha,0),\ \ \omega(0) = \bar{\omega}_0 \\
\bar{u} &= \frac{1}{L_g L_{\tilde{f}}^{r-1}h(\omega)}\Big(-R_2^{-1}b^T P\bar{e} + v^{(r)}(t) - L_{\tilde{f}}^r h(\omega) \\
&\quad - p_r(\omega(t),\bar{v}(t),\omega(t-\alpha),\bar{v}(t-\alpha))\Big)
\end{aligned} \tag{9.60}$$

where $\bar{e} = [(y(t-\tilde{\alpha}) - v + h(\omega(t)) - h(\omega(t-\tilde{\alpha})))\ (L_{\tilde{f}}h(\omega) - v^{(1)})\ \cdots\ (L_{\tilde{f}}^{r-1}h(\omega) - v^{(r-1)})]^T$, *enforces: a) local exponential stability, and b) asymptotic output tracking in the closed-loop system, independently of the size of the state delay.*

Remark 9.26. We note that no restrictions have been imposed on the stability properties of the open-loop DDE system (e.g., open-loop stable system) in order to derive the result of Theorem 9.25. This is because the predictor of (9.59) which is used to produce values of the corrective signal δy is a closed-loop one (we remark that the use of an open-loop predictor would require to assume that the open-loop DDE system is stable).

Remark 9.27. For open-loop stable DDE systems with measurement delay but no state delay (i.e., $\tilde{\alpha} > 0$ and $\alpha = 0$), the output feedback controller of (9.60) simplifies to:

$$\dot{\omega} = A\omega(t) + B\omega(t) + f(\omega(t),\omega(t)) + g(\omega(t),\omega(t))$$
$$\times \frac{1}{L_g L_{\tilde{f}}^{r-1} h(\omega)} \left(-R_2^{-1} b^T P\bar{e} + v^{(r)}(t) - L_{\tilde{f}}^r h(\omega) - p_r(\omega(t),\bar{v}(t),\omega(t),\bar{v}(t))\right)$$
$$\bar{u} = \frac{1}{L_g L_{\tilde{f}}^{r-1} h(\omega)} \left(-R_2^{-1} b^T P\bar{e} + v^{(r)}(t) - L_{\tilde{f}}^r h(\omega) - p_r(\omega(t),\bar{v}(t),\omega(t),\bar{v}(t))\right) \tag{9.61}$$

where $\bar{e} = [(y(t-\tilde{\alpha}) - v + h(\omega(t)) - h(\omega(t-\tilde{\alpha})))\ (L_{\tilde{f}} h(\omega) - v^{(1)})\ \cdots\ (L_{\tilde{f}}^{r-1} h(\omega) - v^{(r-1)})]^T$. The above controller is a Smith-predictor based open-loop output feedback controller similar to the one developed in [162].

Remark 9.28. The nonlinear distributed output feedback of (9.57–9.60) are infinite dimensional ones, due to the infinite dimensional nature of the observers of (9.51–9.59), respectively. Therefore, finite-dimensional approximation of these controllers have to be derived for on-line implementation. This task can be performed utilizing standard discretization techniques such as finite differences. We note that it is well-established (e.g., [251]) that as the number of discretization points increases, the closed-loop system resulting from the DDE model plus an approximate finite-dimensional controller converges to the closed-loop system resulting from the DDE model plus the infinite-dimensional controller, guaranteeing the well-posedness of the approximate finite-dimensional controller.

9.10 Application to a Reactor-Separator System with Recycle

In this section, the nonlinear control method is applied to an exothermic reactor-separator process with recycle and a fluidized catalytic cracker and is shown to outperform nonlinear controller designs that do not account for the presence of dead time associated with the recycle loop and the pipes transferring material from the reactor to the regenerator and vice versa, respectively.

9.10.1 Process Description – Control Problem Formulation

Consider the process, shown in Fig. 9.2, which consists of a reactor and a separator [171]. An irreversible reaction of the form $A \rightarrow B$, where A is the reactant species and B is the product species, takes place in the reactor. The reaction is exothermic and a cooling jacket is used to remove heat from the reactor. The reaction rate is assumed to be of first-order and is given by:

$$r = k_0 \exp\left(-\frac{E}{RT}\right) C_A$$

where k_0 and E denote the pre-exponential constant and activation energy of the reaction, and T and C_A denote the temperature and concentration

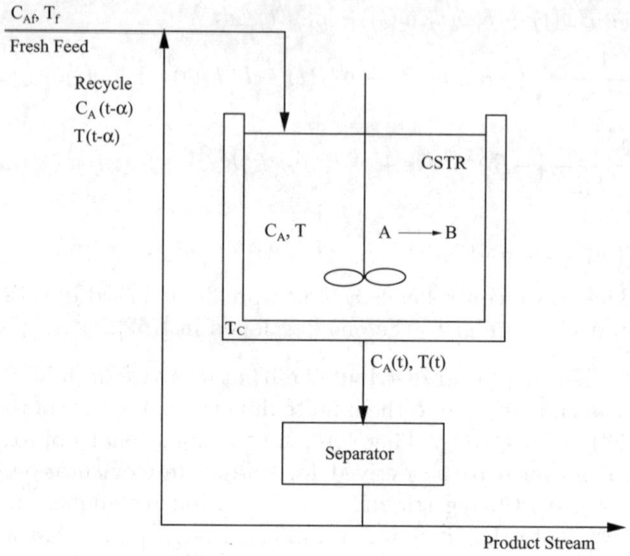

Fig. 9.2. A reactor-separator process with recycle

of species A in the reactor. The outlet of the reactor is fed to a separator where the unreacted species A is separated from the product B. The unreacted amount of species A is fed back to the reactor through a recycle loop; this allows increasing the overall conversion of the reaction and minimizing reactant wastes. The inlet stream to the reactor consists of a fresh feed of pure A, at flow-rate λF, concentration C_{Af} and temperature T_f, and of the recycle stream at flow-rate $(1-\lambda)F$, concentration $C_A(t-\alpha)$ and temperature $T(t-\alpha)$, where F is the total reactor flow-rate, λ is the recirculation coefficient (it varies from zero to one, with zero corresponding to total recycle and zero fresh feed and one corresponding to no recycle) and α is the recycle loop dead time. Under the assumptions of constant volume of the reacting liquid, V, negligible heat loses, constant density, ρ, and heat capacity, c_p, of the reacting liquid, and constant jacket temperature, T_c, a process dynamic model can be derived from mass and energy balances and consists of the following two nonlinear differential difference equations:

$$
\begin{aligned}
\frac{dC_A}{dt} &= \frac{\lambda F}{V} C_{Af} - \frac{F}{V} C_A + \frac{(1-\lambda)F}{V} C_A(t-\alpha) - k_0 \exp\left(-\frac{E}{RT}\right) C_A \\
\frac{dT}{dt} &= \frac{\lambda F}{V} T_f - \frac{F}{V} T + \frac{(1-\lambda)F}{V} T(t-\alpha) + \frac{(-\Delta H)}{\rho c_p} k_0 \exp\left(-\frac{E}{RT}\right) C_A \\
&\quad - \frac{UA}{V\rho c_p}(T - T_c)
\end{aligned}
\tag{9.62}
$$

Table 9.1. Process parameters of the reactor-separator system

V	$= 0.10$	m^3
F	$= 13.3330 \times 10^{-3}$	$m^3\ sec^{-1}$
C_{Af}	$= 0.7090$	$kmol\ m^{-3}$
λ	$= 0.25$	
E	$= 1.20 \times 10^4$	$kcal\ kmol^{-1}$
k_0	$= 1.0 \times 10^7$	sec^{-1}
R	$= 1.987$	$kcal\ kmol^{-1}\ K^{-1}$
ΔH_0	$= -5.0 \times 10^4$	$kcal\ kmol^{-1}$
c_p	$= 0.001$	$kcal\ kg^{-1}\ K^{-1}$
ρ	$= 1000.0$	$kg\ m^{-3}$
T_f	$= 300.0$	K
T_c	$= 295.0$	K
U	$= 30.0$	$kcal\ m^{-2}\ sec^{-1}\ K^{-1}$
A	$= 1.0$	m^2

where ΔH denotes the enthalpy of the reaction, U denotes the heat transfer coefficient, and A denotes the heat transfer area. The values of the process parameters are given in Table 9.1. For these values the corresponding steady-state is:

$$C_{As} = 0.5000\ mol/L, \quad T_s = 296.16^o K, \quad C_{Afs} = 0.7090\ mol/L \tag{9.63}$$

where the subscript s denotes the steady-state value. It was verified, through computation of the eigenvalues of the open-loop system, that this steady-state is a stable one.

The control objective for the process is formulated as the one of regulating the temperature of the reactor, T, by manipulating the inlet concentration of the fresh feed C_{Af}. Setting $x_1 = C_A$, $x_2 = T$, $y = x_2$ and $u = C_{Af} - C_{Afs}$, the process model of (9.62) can be written in the form of (9.26) with:

$$\tilde{f}(x(t)) = \begin{bmatrix} \dfrac{\lambda F}{V} C_{Afs} - \dfrac{F}{V} x_1 - k_0 \exp\left(-\dfrac{E}{Rx_2}\right) x_1 \\ \dfrac{\lambda F}{V} T_f - \dfrac{F}{V} x_2 + \dfrac{(-\Delta H)}{\rho c_p} k_0 \exp\left(-\dfrac{E}{Rx_2}\right) x_1 - \dfrac{UA}{V \rho c_p}(x_2 - T_c) \end{bmatrix}, \tag{9.64}$$

$$g(x(t), x(t-\alpha)) = \begin{bmatrix} \dfrac{\lambda F}{V} \\ 0 \end{bmatrix}, \quad \bar{p}(x(t), x(t-\alpha)) = \begin{bmatrix} \dfrac{(1-\lambda)F}{V} x_1(t-\alpha) \\ \dfrac{(1-\lambda)F}{V} x_2(t-\alpha) \end{bmatrix} \tag{9.65}$$

9.10.2 State Feedback Controller Design

For the system of (9.62), Assumption 9.1 is satisfied with $r = 2$ and the coordinate transformation of (9.27) takes the form:

$$\begin{bmatrix} \zeta_1 \\ \zeta_2 \end{bmatrix} = \mathcal{X}(x) = \begin{bmatrix} h(x) \\ L_{\tilde{f}}h(x) \end{bmatrix}$$

$$= \begin{bmatrix} x_2 \\ \frac{\lambda F}{V}T_f - \frac{F}{V}x_2 + \frac{(-\Delta H)}{\rho c_p}k_0 \exp\left(-\frac{E}{Rx_2}\right)x_1 - \frac{UA}{V\rho c_p}(x_2 - T_c) \end{bmatrix} \tag{9.66}$$

Using the above coordinate change, the process dynamic model can be equivalently written as:

$$\begin{aligned} \dot{\zeta}_1 &= \zeta_2 + \frac{(1-\lambda)F}{V}\zeta_1(t-\alpha) \\ \dot{\zeta}_2 &= L_{\tilde{f}}^2 h(x) + L_g L_{\tilde{f}} h(x) u + p_2(\zeta_1(t), \zeta_2(t), \zeta_1(t-\alpha), \zeta_2(t-\alpha)) \\ y &= \zeta_1 \end{aligned} \tag{9.67}$$

where the explicit form of the term $p_2(\zeta_1(t), \zeta_2(t), \zeta_1(t-\alpha), \zeta_2(t-\alpha))$ is omitted for brevity. For the above system Assumption 9.2 is trivially satisfied, while Assumption 9.3 is satisfied with $p(\bar{e}(t)+\bar{v}(t), \bar{e}(t-\alpha)+\bar{v}(t-\alpha)) = [0.1e_1(t-\alpha) \quad 0]^T$, $a_1 = 0$ and $a_2 = 0.01$. Utilizing the result of Theorem 9.9, the following matrix equation can be formed:

$$\tilde{A}^T P + P\tilde{A} - 2P^T b R_2^{-1} b^T P + a^2 I + P^2 = -R_1 \tag{9.68}$$

with $R_2 = 1.0$, $a = 0.101$ ($a^2 > a_2$), and

$$\tilde{A} = \begin{bmatrix} 0 & 1 \\ 0 & 0 \end{bmatrix}, \quad b = \begin{bmatrix} 0 \\ 1 \end{bmatrix}, \quad R_1 = \begin{bmatrix} 0.001 & 0 \\ 0 & 0.001 \end{bmatrix} \tag{9.69}$$

Equation (9.68) has a unique positive definite solution for P of the form:

$$P = \begin{bmatrix} 0.055 & -0.119 \\ -0.119 & 0.513 \end{bmatrix} \tag{9.70}$$

which leads to the following nonlinear state feedback controller:

$$\begin{aligned} u = \frac{1}{L_g L_{\tilde{f}} h(x)} \Big(& -0.119(x_2 - v) - 0.513(L_{\tilde{f}} h(x) + 0.1v(t-\alpha)) \\ & -L_{\tilde{f}}^2 h(x) - p_2(x(t), \bar{v}(t), x(t-\alpha), \bar{v}(t-\alpha)) \Big) \end{aligned} \tag{9.71}$$

9.10.3 State Observer and Output Feedback Controller Design

In order to avoid the lengthy computations involved in the design of a nonlinear observer gain, we will design a nonlinear state observer of the form of (9.51) with a constant observer gain, L, for the system of (9.62) (see also discussion in Remark 9.18). Computing the linearization of the system of (9.62) around the unstable steady state, $C_{As} = 2.36\ mol/L$, $T_s = 320.0^oK$ ($v = 320.0^oK$ will be the new value of the reference input in the simulations discussed in the next subsection), we obtain the following linear DDE system:

$$\dot{x} = Ax(t) + Bx(t-\alpha) \tag{9.72}$$

with:

$$A = \begin{bmatrix} -0.1970 & -0.00885 \\ 3181.8 & 142.23 \end{bmatrix}, \quad B = \begin{bmatrix} 0.1 & 0 \\ 0 & 0.1 \end{bmatrix} \tag{9.73}$$

The system of (9.72) was found to possess two real unstable eigenvalues, $\lambda_1 = 142.036$ and $\lambda_2 = 0.031$ and infinitely many stable eigenvalues. The eigenfunctions corresponding to the unstable eigenvalues were found to be:

$$\phi_1(\xi) = \begin{bmatrix} 0.00006 \\ -1.0 \end{bmatrix} e^{\lambda_1 \xi}, \quad \phi_2(\xi) = \begin{bmatrix} -0.0447 \\ 1.0 \end{bmatrix} e^{\lambda_2 \xi} \tag{9.74}$$

The following constant observer gain:

$$L = \begin{bmatrix} -361.0 \\ -50.0 \end{bmatrix} \tag{9.75}$$

was found to satisfy Assumption 9.4 and yields the following nonlinear output feedback controller:

$$\begin{aligned} \dot{\omega} &= \tilde{f}(\omega(t), \omega(t-\alpha)) + g(\omega, \omega(t-\alpha))u + \bar{p}(\omega(t), \omega(t-\alpha)) \\ &\quad + \Phi_H(0)L(y(t) - h(\omega(t))) + B\int_0^{\alpha} \Phi_H(\xi - \alpha)L[y(t-\xi) - h(\omega(t-\xi))]d\xi \\ u &= \frac{1}{L_g L_{\tilde{f}} h(\omega)} \Big(-0.119(y(t) - v) - 0.513(L_{\tilde{f}} h(\omega) + 0.1v(t-\alpha)) \\ &\quad - L_{\tilde{f}}^2 h(\omega) - p_2(\omega(t), \bar{v}(t), \omega(t-\alpha), \bar{v}(t-\alpha))\Big) \end{aligned} \tag{9.76}$$

where $\Phi_H(\xi) = [\phi_1(\xi)\ \phi_2(\xi)]$. When both state and measurement delays are included in the system of (9.62), the following nonlinear output feedback controller was employed in the simulations described in the next subsection:

$$\begin{aligned} \dot{\omega} &= \tilde{f}(\omega(t), \omega(t-\alpha)) + g(\omega, \omega(t-\alpha))u + \bar{p}(\omega(t), \omega(t-\alpha)) \\ &\quad + \Phi_H(0)L(y(t) - h(\omega(t))) \\ &\quad + B\int_0^{\alpha - \tilde{\alpha}} \Phi_H(\xi - \alpha)L[y(t-\xi-\tilde{\alpha}) - h(\omega(t-\xi-\tilde{\alpha}))]d\xi \\ u &= \frac{1}{L_g L_{\tilde{f}} h(\omega)} \Big(-0.119(y(t-\tilde{\alpha}) - v + \omega_2(t) - \omega_2(t-\tilde{\alpha}))) - 0.513(L_{\tilde{f}} h(\omega) \\ &\quad + 0.1v(t-\alpha)) - L_{\tilde{f}}^2 h(\omega) - p_2(\omega(t), \bar{v}(t), \omega(t-\alpha), \bar{v}(t-\alpha))\Big) \end{aligned} \tag{9.77}$$

Note that according to the discussion of Remark 9.24, the controllers of (9.76) and (9.77) possess integral action.

9.10.4 Closed-Loop System Simulations

We performed several sets of simulation runs to evaluate the stabilization and output tracking capabilities of the output feedback controllers of (9.76–9.77) and compare their performance with nonlinear controllers that do not account for the presence of recycle loop dead time in the model of (9.62). In all the simulation runs, the process was initially assumed to be at the steady state of (9.63) and the user-friendly software package SIMULINK was used to simulate the closed-loop DDE system (SIMULINK is a toolbox of the mathematical software MATLAB that includes a delay function which can be readily used to simulate differential equations with time delays). The computation of the integrals in the controllers of (9.76–9.77) was performed by discretizing the interval $[-\alpha, 0)$ into ten equispaced intervals using finite differences (further increase on the number of discretization intervals was found to lead to negligible differences on the results). In the first two sets of simulation runs, a $8.84^o K$ increase in the reference input value (i.e., $v = 305.0^o K$) was imposed at time $t = 0$ *sec*. The new reference input value corresponds to a stable steady state.

In the first set of simulation runs, we initially considered the process of (9.62) with $\alpha = 40$ *sec* and $\tilde{\alpha} = 0$ *sec* under the output feedback controller of (9.76) with $L = 0$ (due to operation at a stable region). Figure 9.3 shows the closed-loop output and manipulated input profiles (solid lines).

It is clear that the proposed controller drives quickly the output to the new reference input value, achieving an excellent transient response. For the sake

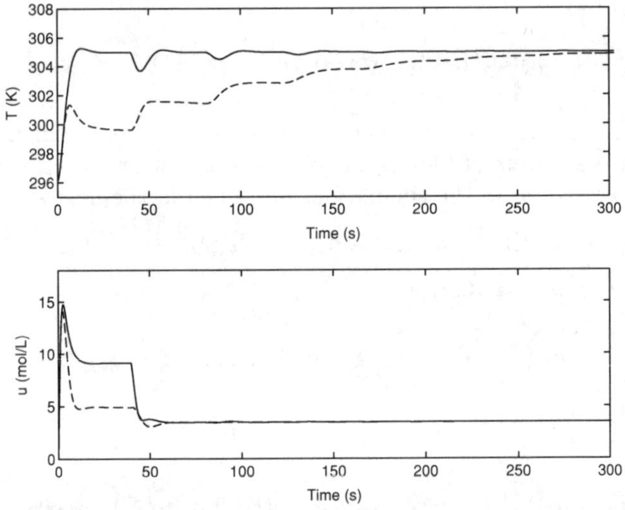

Fig. 9.3. Closed-loop output and manipulated input profiles with $\alpha = 40$ *sec* and $\tilde{\alpha} = 0$ *sec* under the controller of (9.76) with $L = 0$(*solid lines*) and the controller of (9.77) with $L = 0$ and $\alpha = 0$ *sec* (*dashed lines*) – operation in stable region

of comparison, we also implemented on the process the same output feedback controller with $\alpha = 0$ *sec*. The closed-loop output and manipulated input profiles under this controller are also displayed in Fig. 9.3 (dashed lines). This controller yields a very poor transient response driving the output (dashed line) to the new reference input value very slowly.

In the second set of simulation runs, we initially considered the process of (9.62) with $\alpha = 40$ *sec* and $\tilde{\alpha} = 12$ *sec* under the output feedback controller of (9.77) with $L = 0$. The resulting closed-loop output and manipulated input profiles are presented in Fig. 9.4 (solid lines). The proposed controller, after the initial delay in the output of $t = 12$ *sec*, which is caused by the presence of measurement delay, regulates successfully the output to the new reference input value. We also implemented on the process the same controller with $\alpha = 0$ *sec*. The closed-loop output and manipulated input profiles are also shown in Fig. 9.4 (dashed lines). The transient performance of the closed-loop system is clearly inferior to the one obtained by the proposed controller.

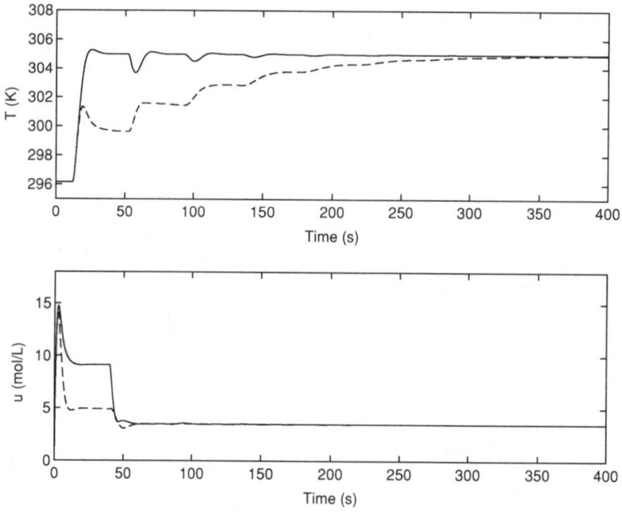

Fig. 9.4. Closed-loop output and manipulated input profiles with $\alpha = 40$ *sec* and $\tilde{\alpha} = 12$ *sec* under the controller of (9.77) with $L = 0$ (*solid lines*) and the controller of (9.77) with $L = 0$, $\alpha = 0$ *sec* and $\tilde{\alpha} = 12$ *sec* (*dashed lines*) – operation in stable region

In the next three sets of simulation runs, a $23.84^o K$ increase in the reference input value (i.e., $v = 320.0^o K$) was imposed at time $t = 0$ *sec*. The new reference input value corresponds to an unstable steady state. We initially considered the process of (9.62) with $\alpha = 40$ *sec* and $\tilde{\alpha} = 0$ *sec* under the controller of (9.76). Figure 9.5 shows the closed-loop output and manipulated input profiles. It is clear that the proposed controller drives quickly the output to the new reference input value. For the sake of comparison, we

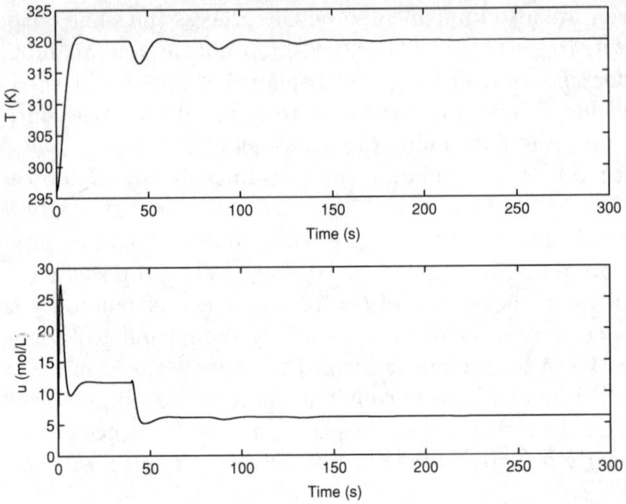

Fig. 9.5. Closed-loop output and manipulated input profiles with $\alpha = 40$ *sec* and $\tilde{\alpha} = 0$ *sec* under the controller of (9.76) – operation in unstable region

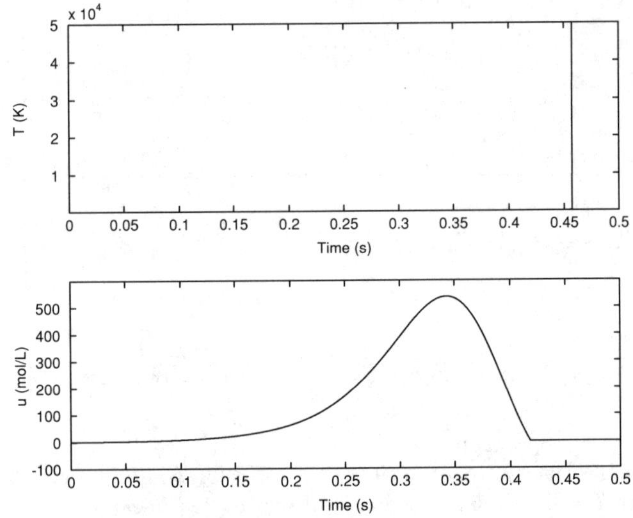

Fig. 9.6. Closed-loop output and manipulated input profiles with $\alpha = 40$ *sec* and $\tilde{\alpha} = 0$ *sec* under the controller of (9.76) with $\alpha = 0$ *sec* – operation in unstable region

also implemented on the process the same output feedback controller with $\alpha = 0$ *sec*. This controller led to an unstable closed-loop system (see the closed-loop output and manipulated input profiles in Fig. 9.6).

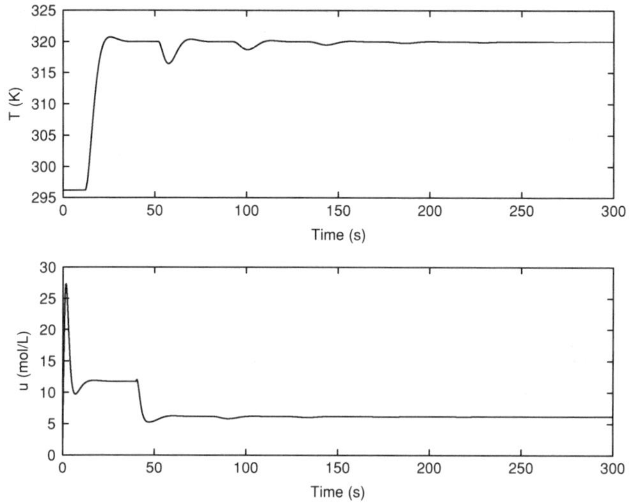

Fig. 9.7. Closed-loop output and manipulated input profiles with $\alpha = 40$ *sec* and $\tilde{\alpha} = 12$ *sec* under the the controller of (9.77) – operation in unstable region

Then, the process of (9.62) with $\alpha = 40$ *sec* and $\tilde{\alpha} = 12$ *sec* was considered under the output feedback controller of (9.77) and Fig. 9.7 shows the resulting closed-loop output and manipulated input profiles. The proposed controller, after the initial delay in the output of $t = 12$ *sec*, which is caused by the measurement delay, regulates successfully the output to the new reference input value. The same controller with $\alpha = 0$ *sec* was also implemented on the process. Again, this controller led to an unstable closed-loop system. Finally, we considered the process of (9.62) with $\alpha = 40$ *sec* and $\tilde{\alpha} = 12$ *sec* and studied the robustness properties of the controller of (9.76) in the presence of a $5^o K$ increase in the value of the temperature of the fresh feed. Figure 9.8 shows the closed-loop output and manipulated input profiles. Despite the presence of a significant disturbance and operation in unstable region, the controller drives the output of the closed-loop system close to the reference input value, exhibiting very good robustness properties.

9.11 Application to a Fluidized Catalytic Cracker

9.11.1 Process Modeling – Control Problem Formulation

In this section, we illustrate the implementation of the developed control methodology on another important chemical engineering process, the fluidized catalytic cracking (FCC) unit shown in Fig. 9.9. The FCC unit consists of a cracking reactor, where the cracking of high boiling gas oil fractions into lighter hydrocarbons (e.g., gasoline) and the carbon formation reactions (undesired

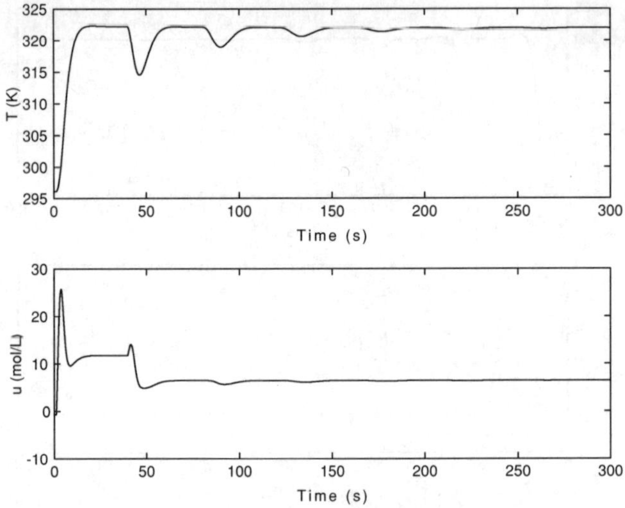

Fig. 9.8. Closed-loop output and manipulated input profiles with $\alpha = 40$ *sec* and $\tilde{\alpha} = 0$ *sec* under the controller of (9.76) in the presence of modeling error – operation in unstable region

reactions) take place and a regenerator, where the carbon removal reactions take place. The reader may refer to: a) [13, 71, 194] for a detailed discussion of the features of the FCC unit, b) [14] for an analysis of the issue of the control structure and c) [125, 202] and [56] for application of linear and nonlinear control methods to the FCC unit, respectively. Unfortunately, in all of these studies, the dead-time associated with the pipes transferring material from the reactor to the regenerator and vice versa were not accounted for both in modeling and controller design.

Under the standard modeling assumptions, of well-mixed reactive catalyst in the reactor, small-size catalyst particles, constant solid holdup in reactor and regenerator, uniform and constant pressure in reactor and regenerator, the process dynamic model takes the form [71]:

$$
\begin{aligned}
V_{ra}\frac{dC_{cat}}{dt} &= -60F_{rc}C_{cat}(t) + 50R_{cf}(C_{cat}(t), C_{rc}(t-\alpha_1), T_{ra}) \\
V_{ra}\frac{dC_{sc}}{dt} &= 60F_{rc}[C_{rc}(t-\alpha_1) - C_{sc}(t)] + 50R_{cf}(C_{cat}(t), C_{rc}(t-\alpha_1), T_{ra}) \\
V_{ra}\frac{dT_{ra}}{dt} &= 60F_{rc}[T_{rg}(t-\alpha_1) - T_{ra}(t)] + 0.875\frac{S_f}{S_c}D_{tf}R_{tf}[T_{fp} - T_{ra}(t)] \\
&\quad + 0.5\frac{(-\Delta H_{cr})}{S_c}R_{oc}(C_{cat}(t), C_{rc}(t-\alpha_1), T_{ra}) \\
&\quad + 0.875\frac{(-\Delta H_{fv})}{S_c}D_{tf}R_{tf}
\end{aligned}
$$

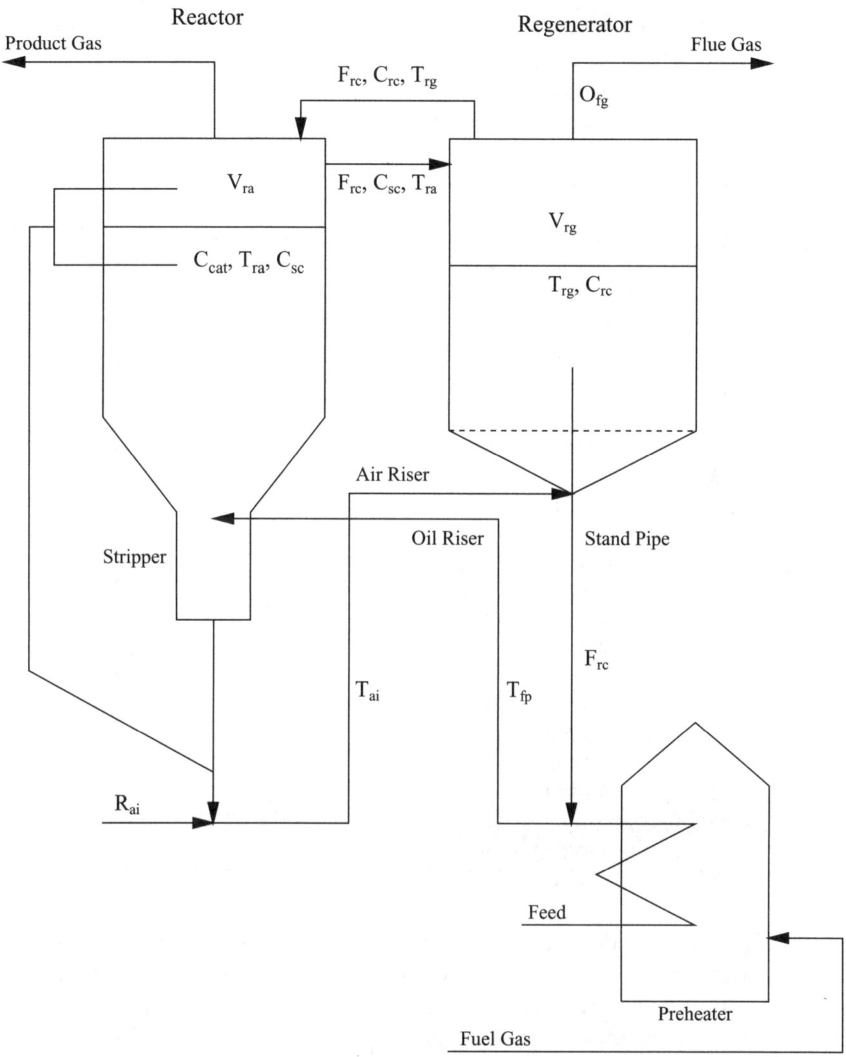

Fig. 9.9. A fluidized catalytic cracking unit

$$V_{rg}\frac{dC_{rc}}{dt} = 60F_{rc}[C_{sc}(t-\alpha_2) - C_{rc}(t)] - 50R_{cb}(C_{rc}(t), T_{rg}(t))$$

$$\begin{aligned} V_{rg}\frac{dT_{rg}}{dt} &= 60F_{rc}[T_{ra}(t-\alpha_2) - T_{rg}(t)] + 0.5\frac{S_a}{S_c}R_{ai}[T_{ai} - T_{rg}(t)] \\ &\quad -0.5\frac{(-\Delta H_{rg})}{S_c}R_{cb}(C_{rc}(t), T_{rg}(t)) \end{aligned} \tag{9.78}$$

Table 9.2. Process parameters of the fluidized catalytic cracking unit

E_{cc}	$= 18000.0$	$Btu\ lb^{-1}\ mole^{-1}$
E_{cr}	$= 27000.0$	$Btu\ lb^{-1}\ mole^{-1}$
E_{or}	$= 63000.0$	$Btu\ lb^{-1}\ mole^{-1}$
k_{cc}	$= 8.59$	$Mlb\ hr^{-1}\ psia^{-1} ton^{-1} (wt\%)^{-1.06}$
k_{cr}	$= 11600$	$Mbbl\ day^{-1}\ psia^{-1} ton^{-1} (wt\%)^{-1.15}$
k_{or}	$= 3,5 \times 10^{10}$	$Mlb\ hr^{-1}\ psia^{-1} ton^{-1}$
V_{rg}	$= 200.0$	ton
V_{ra}	$= 60.0$	ton
F_{rc}	$= 40.0$	$ton\ hr^{-1}$
T_{fps}	$= 744.0$	F
T_{ai}	$= 175.0$	F
P_{rg}	$= 25.0$	$psia$
P_{ra}	$= 40.0$	$psia$
ΔH_{fv}	$= 60.0$	$Btu\ lb^{-1}$
ΔH_{cr}	$= 77.3$	$Btu\ lb^{-1}$
ΔH_{rg}	$= 10561.0$	$Btu\ lb^{-1}$
S_a	$= 0.3$	$Btu\ lb^{-1} F^{-1}$
S_c	$= 0.3$	$Btu\ lb^{-1} F^{-1}$
S_f	$= 0.7$	$Btu\ lb^{-1} F^{-1}$
R_{tf}	$= 100.0$	$Mbbl/day$
D_{tf}	$= 7.0$	$lb\ gal^{-1}$
α	$= 0.3$	hr
R_{ai}	$= 400.0$	$Mlb\ min^{-1}$
$(C_{cat})_{s1}$	$= 0.8723$	$wt\%$
$(C_{sc})_{s1}$	$= 1.5696$	$wt\%$
$(C_{rc})_{s1}$	$= 0.6973$	$wt\%$
$(T_{ra})_{s1}$	$= 930.62$	F
$(T_{rg})_{s1}$	$= 1155.96$	F

where C_{cat}, C_{sc}, C_{rc} denote the concentrations of catalytic carbon on spent catalyst, the total carbon on spent catalyst, and carbon on regenerated catalyst, T_{ra}, T_{rg} denote the temperatures in the reactor and the regenerator, R_{ai} is the air flow rate in the regenerator, R_{tf} is the total feed flow rate, D_{tf} is the density of total feed, V_{ra}, V_{rg} denote the catalyst holdup of the reactor and the regenerator, $\Delta H_{rg}, \Delta H_{cr}$ are the heat of regeneration and cracking, ΔH_{fv} is the heat of feed vaporization, F_{rc} denotes the circulation flow rate of catalyst from reactor to regenerator and vice-versa, S_a, S_c, S_f denote specific heats of the air, the catalyst, and the feed, T_{fp}, T_{ai} denote the inlet temperatures of the feed in the reactor and of the air in the regenerator, and R_{cf}, R_{oc}, R_{cb} denote the reaction rates of total carbon forming, of gas-oil cracking, and of coke burning. The analytic expressions for the reaction rates R_{cf}, R_{oc}, R_{cb} can be found in [71]. The presence of time delay in the terms $C_{rc}(t - \alpha_1), T_{rg}(t - \alpha_1)$ is due to dead-time in the pipes transferring regenerated catalyst from the regenerator to the reactor, and the time delay in the terms $C_{sc}(t-\alpha_2), T_{ra}(t-\alpha_2)$ is due to dead time in pipes transferring spent catalyst from the reactor to the regenerator. Even though the proposed method can be readily applied to the case where $\alpha_1 \neq \alpha_2$, we pick in order to simplify our development, $\alpha_1 = \alpha_2 = \alpha = 0.3\ hr$. The values of the remaining process parameters and the corresponding steady-state values are given in Table 9.2.

The control objective is formulated as the one of regulating the temperature in the regenerator, T_{rg}, by manipulating the temperature of the inlet air in the regenerator, T_{ai}. By setting $x = [x_1\ x_2\ x_3\ x_4\ x_5]^T = [T_{rg}\ C_{rc}\ T_{ra}\ C_{sc}\ C_{cat}]^T$, $u = T_{ai} - T_{ais}$, $y = T_{rg}$, the process model of (9.78) can be written in the form of (9.26) with:

$$\tilde{f}(x(t)) = \begin{bmatrix} \tilde{f}_1 \\ \tilde{f}_2 \\ \tilde{f}_3 \\ \tilde{f}_4 \\ \tilde{f}_5 \end{bmatrix} = \begin{bmatrix} \dfrac{0.5S_aR_{ai}}{S_cV_{rg}}T_{ais} - \left(\dfrac{60F_{rc}}{V_{rg}} + 0.5\dfrac{S_aR_{ai}}{S_cV_{rg}}\right)T_{rg}(t) \\ -0.5\dfrac{(-\Delta H_{rg,n})}{S_cV_{rg}}R_{cb}(C_{rc}(t), T_{rg}(t)) \\ -\dfrac{60F_{rc}}{V_{rg}}C_{rc}(t) - \dfrac{50}{V_{rg}}R_{cb}(C_{rc}(t), T_{rg}(t)) \\ -\dfrac{60F_{rc}}{V_{ra}}T_{ra}(t) + 0.875\dfrac{S_f}{S_cV_{ra}}D_{tf}R_{tf}[T_{fp} - T_{ra}(t)] \\ +0.875\dfrac{(-\Delta H_{fv})}{S_cV_{ra}}D_{tf}R_{tf} \\ -\dfrac{60F_{rc}}{V_{ra}}C_{sc}(t) \\ -\dfrac{60F_{rc}}{V_{ra}}C_{cat}(t) \end{bmatrix} \tag{9.79}$$

$$\bar{p}(x(t), x(t-\alpha)) = \begin{bmatrix} \bar{p}_1 \\ \bar{p}_2 \\ \bar{p}_3 \\ \bar{p}_4 \\ \bar{p}_5 \end{bmatrix} = \begin{bmatrix} \frac{60F_{rc}}{V_{rg}} T_{ra}(t-\alpha) \\ \frac{60F_{rc}}{V_{rg}} C_{sc}(t-\alpha) \\ \frac{60F_{rc}}{V_{ra}} T_{rg}(t-\alpha) \\ +0.5\frac{(-\Delta H_{cr})}{S_c V_{ra}} R_{oc}(C_{cat}(t), C_{rc}(t-\alpha), T_{ra}(t)) \\ \frac{60F_{rc}}{V_{ra}} C_{rc}(t-\alpha) \\ +\frac{50}{V_{ra}} R_{cf}(C_{cat}(t), C_{rc}(t-\alpha), T_{ra}(t)) \\ \frac{50}{V_{ra}} R_{cf}(C_{cat}(t), C_{rc}(t-\alpha), T_{ra}(t)) \end{bmatrix} \tag{9.80}$$

$$g(x(t), x(t-\alpha)) = \begin{bmatrix} g_1 \\ g_2 \\ g_3 \\ g_4 \\ g_5 \end{bmatrix} = \begin{bmatrix} \frac{0.5 S_a R_{ai}}{S_c V_{rg}} \\ 0 \\ 0 \\ 0 \\ 0 \end{bmatrix} \tag{9.81}$$

9.11.2 State Feedback Controller Design

For the system of (9.78), Assumption 9.1 is satisfied with $r = 1$, and the coordinate transformation of (9.27) takes the form $[\zeta \; \eta_1 \; \eta_2 \; \eta_3 \; \eta_4]^T = [T_{rg} \; C_{rc} \; T_{ra} \; C_{sc} \; C_{cat}]^T$ and yields the following system:

$$\begin{aligned} \dot{\zeta} &= L_{\tilde{f}} h(\mathcal{X}^{-1}(\zeta, \eta)) + L_g h(\mathcal{X}^{-1}(\zeta, \eta))u(t) + p_1(\zeta(t), \eta(t), \zeta(t-\alpha), \eta(t-\alpha)) \\ \dot{\eta} &= \Psi(\zeta(t), \eta(t), \zeta(t-\alpha), \eta(t-\alpha)) \end{aligned} \tag{9.82}$$

where the explicit form of $\Psi(\zeta(t), \eta(t), \zeta(t-\alpha), \eta(t-\alpha))$ is omitted for brevity. To verify Assumption 9.2, we consider the η-subsystem of (9.82) with $\zeta(t) = \zeta(t-\alpha) = \zeta_s = 1155.96^o F$ i.e., the system:

$$\begin{bmatrix} \dot{\eta}_1 \\ \\ \dot{\eta}_2 \\ \\ \dot{\eta}_3 \\ \\ \dot{\eta}_4 \end{bmatrix} = \begin{bmatrix} -\dfrac{60F_{rc}}{V_{rg}}\eta_1 - \dfrac{50}{V_{rg}}R_{cb}(\eta_1,\zeta_s) + \dfrac{60F_{rc}}{V_{rg}}\eta_3(t-\alpha) \\ \\ -\dfrac{60F_{rc}}{V_{ra}}\eta_2 + 0.875\dfrac{S_f}{S_cV_{ra}}D_{tf}R_{tf}[T_{fp}-\eta_2] \\ +0.875\dfrac{(-\Delta H_{fv})}{S_cV_{ra}}D_{tf}R_{tf} + 0.5\dfrac{(-\Delta H_{cr})}{S_cV_{ra}}R_{oc}(\eta_4,\eta_1(t-\alpha),\eta_2) \\ \\ -\dfrac{60F_{rc}}{V_{ra}}\eta_3 + \dfrac{60F_{rc}}{V_{ra}}\eta_1(t-\alpha) + \dfrac{50}{V_{ra}}R_{cf}(\eta_4,\eta_1(t-\alpha),\eta_2) \\ \\ -\dfrac{60F_{rc}}{V_{ra}}\eta_4 + \dfrac{50}{V_{ra}}R_{cf}(\eta_4,\eta_1(t-\alpha),\eta_2) \end{bmatrix} \tag{9.83}$$

The linearization of the above system around the steady-state, $C_{rc} = 0.6973$ $wt\%$, $T_{ra} = 930.62^oF$, $C_{sc} = 1.5696$ $wt\%$, $C_{cat} = 0.8723$ $wt\%$, was found to be exponentially stable, which implies that the η-subsystem of (9.82) possesses a local input-to-state stability property with respect to $\zeta_t(\xi)$. Therefore, Assumption 9.2 holds and the controller synthesis problem can be addressed on the basis of the ζ-subsystem which is given below:

$$\begin{aligned} \dot{\zeta} = & -\left(\frac{60F_{rc}}{V_{rg}} + \frac{0.5S_aR_{ai}}{S_cV_{rg}}\right)\zeta(t) - 0.5\frac{(-\Delta H_{rg,n})}{S_cV_{rg}}R_{cb}(\eta_1(t),\zeta(t)) \\ & + \left(\frac{0.5S_aR_{ai}}{S_cV_{rg}}\right)u(t) + \left(\frac{60F_{rc}}{V_{rg}}\right)\eta_2(t-\alpha) \end{aligned} \tag{9.84}$$

Setting $e = \zeta - v$ where v is the desired set point and using a preliminary control law of the form of (9.30), the above system becomes:

$$\dot{e}(t) = -R_2^{-1}b^TPe(t) \tag{9.85}$$

For the above system, Assumption 9.4 is trivially satisfied since $p(x(t), x(t-\alpha)) = 0$. Utilizing the results of Theorem 9.9, the following equation can be formed:

$$\tilde{A}^TP + P\tilde{A} - 2P^TbR_2^{-1}b^TP + a^2 + P^2 = -R_1 \tag{9.86}$$

with $R_2 = 1.0$, $a^2 = 0.5$ ($a^2 > a_2 = 0$), and

$$\tilde{A} = 0\,, \qquad b = 1\,, \qquad R_1 = 0.5 \tag{9.87}$$

Equation (9.86) has a unique positive definite solution for P of the form:

$$P = 1.0 \tag{9.88}$$

which leads to the following nonlinear state feedback controller:

$$u = \frac{1}{L_g h(x(s))}\Big(-(x_1 - v) - L_{\tilde{f}}h(x(s)) - p_1(x(t), x(t-\alpha))\Big) \tag{9.89}$$

9.11.3 State Observer and Output Feedback Controller Design

Since the open-loop process is stable, we set the observer gain L equal to zero and derive the following nonlinear output feedback controller using the result of Theorem 9.21:

$$\begin{aligned} \dot{\omega} &= \tilde{f}(\omega(t)) + g(\omega, \omega(t-\alpha))u + \bar{p}(\omega(t), \omega(t-\alpha)) \\ u &= \frac{1}{L_g h(\omega)}\Big(-(y(t) - v) - L_{\tilde{f}}h(\omega) - p_1(\omega(t), \bar{v}(t), \omega(t-\alpha), \bar{v}(t-\alpha))\Big) \end{aligned} \tag{9.90}$$

When both state and measurement delays are included in the system of (9.78), the following nonlinear output feedback controller was derived by using the result of Theorem 9.25 and employed in the simulations described in the next subsection:

$$\begin{aligned} \dot{\omega} &= \tilde{f}(\omega(t)) + g(\omega, \omega(t-\alpha))u + \bar{p}(\omega(t), \omega(t-\alpha)) \\ u &= \frac{1}{L_g h(\omega)}\Big(-(y(t-\tilde{\alpha}) - v + \omega_1(t) - \omega_1(t-\tilde{\alpha})) - L_{\tilde{f}}h(\omega) \\ &\quad -p_1(\omega(t), \bar{v}(t), \omega(t-\alpha), \bar{v}(t-\alpha))\Big) \end{aligned} \tag{9.91}$$

Note that the controllers of (9.90) and (9.91) possess integral action.

9.11.4 Closed-Loop System Simulations

We performed several sets of simulation runs to evaluate the performance of the output feedback controllers of (9.90–9.91) and compare their performance with nonlinear controllers that do not account for the presence of time delays in the model of (9.78). In all the simulation runs, the process was initially ($t = 0.0\ hr$) assumed to be at the steady-state shown in Table 9.2 and the MATLAB toolbox SIMULINK was used to simulate the closed-loop DDE system.

In the first simulation run, we considered the process of (9.78) with $\alpha = 0.3\ hr$ and $\tilde{\alpha} = 0\ hr$ (i.e., no measurement delay is present) under the output feedback controller of (9.90). Figure 9.10 shows the output and manipulated input profiles, for a 44^oF increase in the reference input. It is clear that the proposed controller drives quickly the output to the new reference input value, compensating for the effect of the dead times associated with the pipes

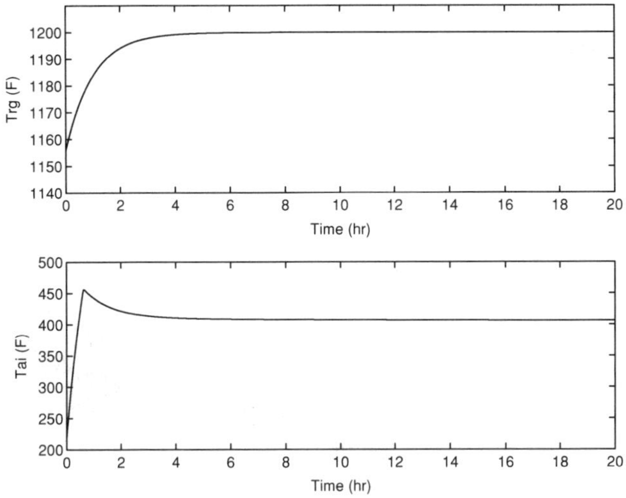

Fig. 9.10. Closed-loop output and manipulated input profiles with $\alpha = 0.3hr$ and $\tilde{\alpha} = 0hr$ under the controller of (9.90)

transferring material from the reactor to the regenerator and vice versa. For the sake of comparison, we also implemented on the process the same output feedback controller with $\alpha = 0\ hr$ and $\tilde{\alpha} = 0\ hr$ (i.e., we did not account for the presence of time delays in the design of the controller).

Figure 9.11 shows the output and manipulated input profiles; this controller leads to an unstable closed-loop system because it does not compensate for the detrimental effect of the time delays.

In the next simulation run, we considered the process of (9.78) with $\alpha = 0.3\ hr$ and $\tilde{\alpha} = 0.05\ hr$ (i.e., significant measurement delay is present) under the output feedback controller of (9.91).

Figure 9.12 shows the output and manipulated input profiles, for a 44^oF increase in the reference input. Clearly, the controller of (9.91) drives the output of the closed-loop system, after an initial delay caused by the measurement dead time, to the new reference input value.

Finally, we considered the process of (9.78) with $\alpha = 0.3\ hr$ and $\tilde{\alpha} = 0.05\ hr$ and studied the robustness properties of the controller of (9.89) in the presence of modeling errors. In particular we simultaneously considered 5% error in the values of: (a) the catalyst circulation rate, F_{rc}, (b) the temperature of the feed in the reactor, T_{fp}, and (c) the inflow air rate in the regenerator, R_{ai}.

Figure 9.13 shows the output and manipulated input profiles, for a 44^oF increase in the reference input. Despite the presence of a significant modeling errors the proposed controller drives the output of the closed-loop system to the new reference input value, exhibiting very good robustness properties.

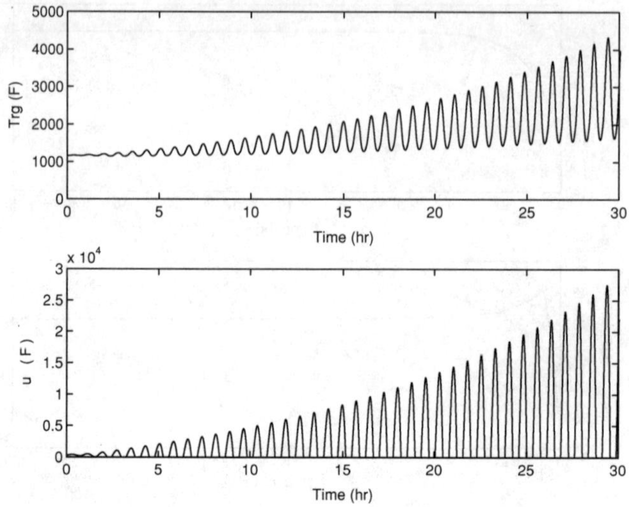

Fig. 9.11. Closed-loop output and manipulated input profiles with $\alpha = 0.3hr$ and $\tilde{\alpha} = 0hr$ under the controller of (9.90) with $\tilde{\alpha} = 0.05hr$

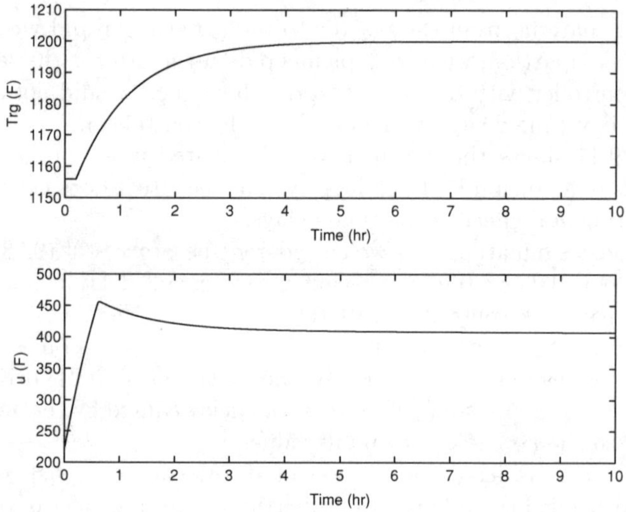

Fig. 9.12. Closed-loop output and manipulated input profiles with $\alpha = 0.3hr$ and $\tilde{\alpha} = 0.05hr$ under the controller of (9.91)

Summarizing, the results of both simulation studies clearly show that it is necessary to compensate for the effect dead-time associated with pipes transferring material from one unit to an other, as well as that the proposed control methodology is a very efficient tool for this purpose.

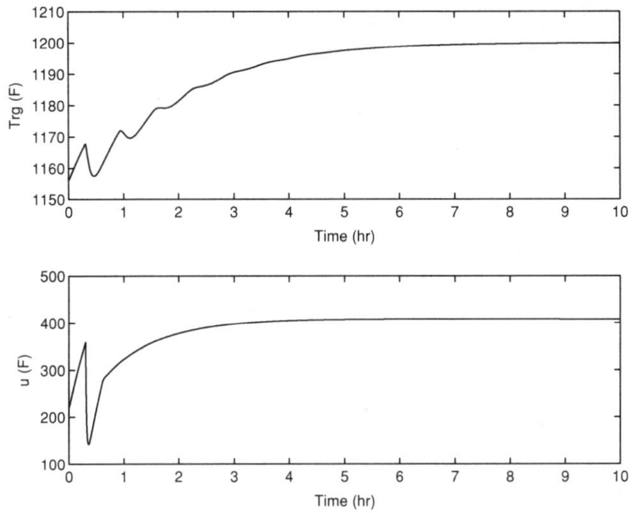

Fig. 9.13. Closed-loop output and manipulated input profiles with $\alpha = 0.3hr$ and $\tilde{\alpha} = 0.05hr$ under the controller of (9.91) in the presence of modeling errors

Remark 9.29. We finally note that even though the FCC unit exhibits two-time-scale behavior [56, 71] owing to the significantly different hold ups of the regenerator and reactor (i.e., the reactor dynamics are significantly faster than the regenerator dynamics), which, in general, may cause controller ill-conditioning (see, for example, [56] for a discussion on this issue), in the present study, the presence of time-scale multiplicity in the FCC process model was not taken into account in the controller design problem addressed in Subsects. 9.11.2–9.11.3 because the control problem that we formulated does not lead to the synthesis of an ill-conditioned controller.

9.12 Conclusions

In this chapter, we presented a methodology for the synthesis of nonlinear output feedback controllers for nonlinear DDE systems which include time delays in the states, the control actuator and the measurement sensor. Initially, DDE systems with state delays were considered and a novel combination of geometric and Lyapunov-based techniques was employed for the synthesis of nonlinear state feedback controllers that guarantee stability and enforce output tracking in the closed-loop system, independently of the size of the state delay. Then, the problem of designing nonlinear distributed state observers, which reconstruct the state of the DDE system while guaranteeing that the discrepancy between the actual and the estimated state tends exponentially to zero, was addressed and solved by using spectral decomposition techniques for DDE systems. The state feedback controllers and the distributed state observers were combined to yield distributed output feedback controllers that enforce stability and asymptotic output tracking in the closed-loop system, independently of the size of the time-delay. For DDE systems with state, control actuator and measurement delays, the output feedback controller was synthesized on the basis of an auxiliary output constructed within a Smith-predictor framework. The nonlinear control method was successfully applied to an exothermic reactor-separator process with recycle and a fluidized catalytic cracker and was shown to outperform nonlinear controller designs that do not account for the presence of dead time associated with the recycle loop and the pipes transferring material from the reactor to the regenerator and vice versa, respectively.

A

Proofs of Chapter 3

Proof of Theorem 3.3:
The proof is divided into two parts. In Part 1 we establish that the controller of (3.18) globally asymptotically stabilizes the origin of the closed-loop system and show that the closed-loop output satisfies the relation of (3.23). In Part 2, we prove that the controller of (3.18) is optimal with respect to the cost functional of (3.24). To simplify the development of the proof, we will consider only the case of a single uncertain variable, i.e., $q = 1$ in (3.1). The extension to the case when $q > 1$ is conceptually straightforward.

Part 1: Consider the representation of the closed-loop system in the transformed coordinates of (3.14). We now follow a three-step procedure to establish closed-loop asymptotic stability and output tracking. Initially, we use a Lyapunov argument to show that, starting from any initial condition, the states of the closed-loop e-subsystem converge asymptotically to the origin and derive bounds that capture the evolution of the states of the e and η subsystems. We then invoke a small-gain argument to show that the trajectories of the e-η interconnected system remain bounded for all times starting from any initial condition. Finally, we show that the states of the closed-loop system of (3.14) converge to the origin and that the output satisfies the relation of (3.23).

Step 1: Substituting the control law of (3.18) into (3.14) and computing the time-derivative of V along the trajectories of the closed-loop e-subsystem, we get:

$$\begin{aligned}\dot{V} &= L_{\bar{f}}V + L_{\bar{g}}Vu + L_{\bar{w}}V\theta \\ &= -c_0(L_{\bar{g}}V)^2 - \sqrt{(L_{\bar{f}}V)^2 + (L_{\bar{g}}V)^4} + L_{\bar{w}}V\theta \\ &\quad - \left(\frac{\rho + \chi\theta_b \|L_w L_f^{r-1}h(x)\|}{(L_g L_f^{r-1}h(x))^2 \left(\dfrac{\|L_{\bar{g}}V\|}{\|L_g L_f^{r-1}h(x)\|} + \phi \right)} \right) (L_{\bar{g}}V)^2\end{aligned} \tag{A.1}$$

for $L_{\bar{g}}V \neq 0$. Substituting the expressions $L_{\bar{g}}V = 2\left[(L_g L_f^{r-1}h(x)\right] b^T Pe$ and $L_{\bar{w}}V = 2\left[L_w L_f^{r-1}h(x)\right] b^T Pe$ into the above equation, and using the fact that $c_0 > 0$, we obtain:

$$\dot{V} \leq -c_0(L_{\bar{g}}V)^2 - \sqrt{(L_{\bar{f}}V)^2 + (L_{\bar{g}}V)^4} - \left(\frac{\rho + \chi\theta_b\|L_w L_f^{r-1}h(x)\|}{(\|2b^T Pe\| + \phi)}\right)\|2b^T Pe\|^2$$
$$+\|2b^T Pe\|\|L_w L_f^{r-1}h(x)\|\theta_b$$
$$= -c_0(L_{\bar{g}}V)^2 - \sqrt{(L_{\bar{f}}V)^2 + (L_{\bar{g}}V)^4} - \left(\frac{\rho\|2b^T Pe\|^2}{\|2b^T Pe\| + \phi}\right)$$
$$-\left(\frac{\theta_b\|L_w L_f^{r-1}h(x)\|\|2b^T Pe\|((\chi-1)\|2b^T Pe\| - \phi)}{\|2b^T Pe\| + \phi}\right) \tag{A.2}$$

From the last inequality and the fact that $\rho > 0$ and $\chi > 2$, it is clear that whenever $\|2b^T Pe\| > \phi(\chi-1)^{-1}$, the time-derivative of V satisfies:

$$\dot{V} \leq -c_0(L_{\bar{g}}V)^2 - \sqrt{(L_{\bar{f}}V)^2 + (L_{\bar{g}}V)^4} \tag{A.3}$$

which is strictly negative when $L_{\bar{g}}V \neq 0$. For the case when $\|2b^T Pe\| \leq \phi(\chi - 1)^{-1}$, we have $\|2b^T Pe\| \leq \phi(\frac{1}{2}\chi - 1)^{-1}$; and, therefore, from the assumption of vanishing uncertainty, we have that the function $L_w L_f^{r-1}h(x)$ satisfies the bound $\|L_w L_f^{r-1}h(x)\| \leq \delta\|2b^T Pe\|$ for all $\|2b^T Pe\| \leq \phi(\frac{1}{2}\chi - 1)^{-1}$, where δ is a positive constant. Substituting this bound, together with the fact that $\phi - (\chi - 1)\|2b^T Pe\| < \phi$, into the last inequality of (A.2) yields:

$$\dot{V} \leq -\sqrt{(L_{\bar{f}}V)^2 + (L_{\bar{g}}V)^4} + \left(\frac{(\theta_b\delta\phi - \rho)\|2b^T Pe\|^2}{\|2b^T Pe\| + \phi}\right) \tag{A.4}$$

$\forall\ \|2b^T Pe\| < \phi(\chi-1)^{-1}$. If ϕ is small enough to satisfy the bound $\phi \leq \dfrac{\rho}{\delta\theta_b} := 2\phi^*$, then it's clear from the above equation that $\dot{V}$ satisfies (A.3), irrespective of the value of $\|2b^T Pe\|$. Consider now the case when $L_{\bar{g}}V = 0$. In this case, we have from (3.18) that $u = 0$ and, hence, computing the time-derivative of $V = e^T Pe$, we get:

$$\dot{V} \leq e^T\left(A^T P + PA\right)e + 2b^T Pe\left(\bar{f}_r(e,\eta,\bar{v}) + \bar{w}_r(e,\eta,\bar{v})\theta\right) \tag{A.5}$$

where the functions, $\bar{f}_r$ and $\bar{w}_r$ are given by $\bar{f}_r(e,\eta,\bar{v}) = L_f^r h(T^{-1}(e,\eta,\bar{v})) - v^{(r)}$ and $\bar{w}_r(e,\eta,\bar{v}) = L_w L_f^{r-1}(T^{-1}(e,\eta,\bar{v}))$. From the fact that P satisfies $A^T P + PA - Pbb^T P < 0$, we have that there exists a positive-definite matrix, Q, such that $A^T P + PA - Pbb^T P = -Q$. Substituting this relation into (A.5) yields:

$$\dot{V} \leq -e^T Qe + (b^T Pe)^2 + 2b^T Pe\left(\bar{f}_r(e,\eta,\bar{v}) + \bar{w}_r(e,\eta,\bar{v})\theta\right) \tag{A.6}$$

From the assumption that $L_g L_f^{r-1} h(T^{-1}(e, \eta, \bar{v})) \neq 0$ for all $x \in \mathbb{R}^n$ (Assumption 3.1), we have that $L_{\bar{g}} V = 0 \implies b^T P e = 0$; and, consequently, $\dot{V} = -e^T Q e < 0 \; \forall \, e \neq 0$.

From the above analysis, it is clear that $\dot{V} < 0 \quad \forall \, e \neq 0$. Consequently, there exists a function β_e of class $\mathcal{KL}$ (see chapter 5 in [148] for details) such that the following ISS inequality holds for the e states of the system of (3.14):

$$\|e(t)\| \leq \beta_e \left(\|e(0)\|, t\right) \qquad\qquad \forall \, t \geq 0 \tag{A.7}$$

and the origin of the e-subsystem is asymptotically stable. From Assumption 3.2, we have that the η-subsystem of (3.14) possesses an ISS property with respect to e which implies that there exists a function β_η of class $\mathcal{KL}$ and a function γ_η of class $\mathcal{K}$ such that the following ISS inequality holds:

$$\|\eta(t)\| \leq \beta_\eta(\|\eta(0)\|, t) + \gamma_\eta(\|e\|^s) \qquad\qquad \forall \, t \geq 0 \tag{A.8}$$

uniformly in θ and $\bar{v}$.

Step 2: We now analyze the behavior of the interconnected dynamical system comprised of the e and η states of the system of (3.14) for which the inequalities of (A.7) and (A.8) hold. In order to proceed with our analysis, we define the following positive real numbers: $\delta_e = D^e + \phi^*$, $\delta_\eta = D^\eta + \gamma_\eta(\delta_e) + \phi^*$, $D^e = \beta_e(\bar{\delta}_e, 0)$, $D^\eta = \beta_\eta(\bar{\delta}_\eta, 0)$, $\phi^* = \dfrac{\rho}{2\delta\theta_b}$ and $\bar{\delta}_e, \bar{\delta}_\eta$ are any positive real numbers. Then, using a contradiction argument similar to the one used in [57,58], one can prove that if $\phi \in (0, \phi^*]$, then, starting from any initial state that satisfies $\|e(0)\| \leq \bar{\delta}_e$, $\|\eta(0)\| \leq \bar{\delta}_\eta$, the evolution of the e and η states satisfies $\|e(t)\| \leq \delta_e$ and $\|\eta(t)\| \leq \delta_\eta$, respectively, for all times; and that the states are therefore bounded. This argument is omitted for brevity.

Step 3: Having established in the previous two steps that the origin of the closed-loop e-subsystem is globally asymptotically stable and that the η subsystem, with e as input, is ISS, we can apply the results of Lemma 5.6 in [148] directly to conclude that the origin of the interconnected e-η system of (3.14) is globally asymptotically stable. It follows then from Assumption 3.1 that the origin of the closed-loop system of (3.1) is globally asymptotically stable. The asymptotic output tracking result can be obtained by taking the limsup of both sides of (A.7), which yields:

$$\limsup_{t \to \infty} \|e(t)\| = 0 \implies \limsup_{t \to \infty} \|y(t) - v(t)\| = 0 \tag{A.9}$$

Part 2: In this part, we prove that the control law of (3.18) is optimal with respect to a meaningful cost functional of the form of (3.24). We proceed in two steps. In the first step, we show that the cost functional defined in (3.24) is a meaningful one by proving that the weights, $l(e)$ and $R(x)$, are positive-definite and strictly positive functions, respectively. In the second step, we show that the stabilizing control law of (3.18) minimizes this cost functional.

Step 1: From its definition in Theorem 3.3, $l(e)$ is given by:

$$l(e) = -L_{\bar{f}}V + \frac{1}{4}L_{\bar{g}}VR^{-1}(x)L_{\bar{g}}V - \|L_{\bar{w}}V\|\,\theta_b \tag{A.10}$$

Consider first the case when $L_{\bar{g}}V \neq 0$. Direct substitution of the expression for $R^{-1}(x)$ given in Theorem 3.3 into the above equation yields:

$$\begin{aligned}
l(e) &= \frac{1}{2}c_0(L_{\bar{g}}V)^2 + \frac{1}{2}\left[-L_{\bar{f}}V + \sqrt{(L_{\bar{f}}V)^2 + (L_{\bar{g}}V)^4}\right] - \|L_{\bar{w}}V\|\theta_b \\
&\quad + \frac{1}{2}\frac{\left(\rho + \chi\theta_b\|L_w L_f^{r-1}h(x)\|\right)\|2b^TPe\|^2}{\|2b^TPe\| + \phi} \\
&= \frac{1}{2}c_0(L_{\bar{g}}V)^2 + \frac{1}{2}\left[-L_{\bar{f}}V + \sqrt{(L_{\bar{f}}V)^2 + (L_{\bar{g}}V)^4}\right] + \frac{1}{2}\left(\frac{\rho\|2b^TPe\|^2}{\|2b^TPe\| + \phi}\right) \\
&\quad + \frac{\theta_b\|2b^TPe\|\|L_w L_f^{r-1}h(x)\|\left[(\frac{1}{2}\chi - 1)\|2b^TPe\| - \phi\right]}{\|2b^TPe\| + \phi}
\end{aligned} \tag{A.11}$$

where we used the expression for $L_{\bar{w}}V$ and the fact that $c_0 > 0$ to derive the above inequality. Note that the second term on the right-hand side of the above equation is strictly positive when $L_{\bar{g}}V \neq 0$. From this and the fact that $\rho > 0$ and $\chi > 2$, it's clear that whenever $\|2b^TPe\| > \phi(\frac{1}{2}\chi - 1)^{-1}$, we have:

$$l(e) \geq \frac{1}{2}c_0(L_{\bar{g}}V)^2 \;\geq\; k_1\|b^TPe\|^2 \tag{A.12}$$

where $k_1 = \frac{1}{2}c_0k_2^2 > 0$, $k_2 = \min_{x\in\mathbb{R}^n}\|L_gL_f^{r-1}h(x)\|$. To analyze the sign of $l(e)$ when $\|2b^TPe\| \leq \phi(\frac{1}{2}\chi - 1)^{-1}$, we use the growth bound $\|L_wL_f^{r-1}h(x)\| \leq \delta\|2b^TPe\|$ together with the fact that $(\frac{1}{2}\chi - 1)\|2b^TPe\| - \phi > -\phi$ to write:

$$l(e) \geq \frac{1}{2}c_0(L_{\bar{g}}V)^2 + \frac{\left(\frac{\rho}{2} - \theta_b\delta\phi\right)\|2b^TPe\|^2}{\|2b^TPe\| + \phi} \quad \forall\,\|2b^TPe\| \leq \phi(\frac{1}{2}\chi - 1)^{-1} \tag{A.13}$$

Clearly, if ϕ is small enough to satisfy $\phi \leq \dfrac{\rho}{2\theta_b\delta} := \phi^*$, then we have that $l(e)$ satisfies (A.12), irrespective of the value of $\|2b^TPe\|$. Consider now the case when $L_{\bar{g}}V = 0$. In this case, we have from (A.10), and the fact that $V = e^TPe$ with P satisfying $A^TP + PA - Pbb^TP = -Q$ for some positive-definite matrix, Q, that:

$$\begin{aligned}
l(e) &= -L_{\bar{f}}V - \|L_{\bar{w}}V\|\theta_b \\
&= -e^T\left(A^TP + PA\right)e - 2b^TPe\left(\bar{f}_r(e,\eta,\bar{v})\right) - \|2b^TPe\|\|\bar{w}_r(e,\eta,\bar{v})\|\theta_b \\
&= e^TQe - (b^TPe)^2 - 2b^TPe\left(\bar{f}_r(e,\eta,\bar{v})\right) - \|2b^TPe\|\|\bar{w}_r(e,\eta,\bar{v})\|\theta_b
\end{aligned} \tag{A.14}$$

where $\bar{f}_r$ and $\bar{w}_r$ were defined following (A.5). Recall that $L_{\bar{g}}V = 0 \implies b^T Pe = 0$; and, consequently in this case, $l(e) = e^T Qe > 0 \ \forall \ e \neq 0$. Using the identities: (a) $e^T Qe \geq \lambda_{min}(Q)\|e\|^2$, where $\lambda_{min}(Q) > 0$ is the minimum eigenvalue of the matrix Q, (b) $\|Pe\|^2 \leq \lambda_{max}(P^2)\|e\|^2$, where $\lambda_{max}(P^2) > 0$ is the maximum eigenvalue of the matrix P^2, and (c) $\|b^T Pe\|^2 \leq \|Pe\|^2$ (recall from (3.13) that $\|b\| = 1$), we finally have $l(e) \geq k_3\|b^T Pe\|^2$, where $k_3 = \lambda_{min}(Q)\lambda_{max}^{-1}(P^2) > 0$. Combining this result with that obtained in (A.12) for the case when $L_{\bar{g}}V \neq 0$, we conclude that $l(e) > 0 \ \ \forall \ e \neq 0$ and that $l(e) \geq k\|b^T Pe\|^2$, where $k = \min\{k_1, k_3\}$. Note also that $R(x) > 0$ for all $x \in \mathbb{R}^n$. Therefore, the cost functional of (3.24) is a meaningful one.

Step 2: In this step, we prove that control law of (3.18) minimizes the cost functional of (3.24). Substituting:

$$\alpha = u + \frac{1}{2}R^{-1}(x)L_{\bar{g}}V \tag{A.15}$$

into (3.24), we get the following chain of equalities:

$$\begin{aligned}
J_v &= \int_0^\infty (l(e) + uR(x)u)dt \\
&= \int_0^\infty \left(-L_{\bar{f}}V + \frac{1}{4}L_{\bar{g}}VR^{-1}(x)L_{\bar{g}}V - \|L_{\bar{w}}V\|\theta_b + \alpha R\alpha - \alpha(L_{\bar{g}}V)\right) dt \\
&\quad + \int_0^\infty \left(\frac{1}{4}L_{\bar{g}}VR^{-1}\left(L_{\bar{g}}V\right)\right) dt \\
&= \int_0^\infty \left(-L_{\bar{f}}V - L_{\bar{g}}Vu - \|L_{\bar{w}}V\|\theta_b\right) dt + \int_0^\infty \alpha R(x)\alpha \ dt \\
&= -\int_0^\infty \sup_{\theta\in\mathcal{W}} \left(\dot{V}\right) dt + \int_0^\infty \alpha R(x)\alpha \ dt
\end{aligned}$$

Since $R(x) > 0$ for all $x \in \mathbb{R}^n$, it is clear that the minimum of J_v is achieved when $\alpha(t) \equiv 0$ (which proves that the controller of (3.18) minimizes the cost of (3.24)) and is equal to:

$$J_v^* = -\int_0^\infty \sup_{\theta\in\mathcal{W}} \left(\dot{V}\right) dt \ \leq \ V(e(0)) - \lim_{t\to\infty} V(e(t)) \ = \ V(e(0)) \tag{A.16}$$

where we have used the fact that $\lim_{t\to\infty} V(e(t)) = 0$, which follows from asymptotic stability of the origin, to derive the last equality. To complete the proof of optimality, we need to show that $J_v^* = V(e(0))$. To this end, consider the uncertain variable $\theta_\Delta \in \mathcal{W}$ where for every $e_\Delta(0) \in \mathbb{R}^r$, every $u \in \mathbb{R}$, and every $\Delta > 0$, we have:

$$\int_0^\infty \dot{V}\left(e_\Delta, u, \theta_\Delta\right) dt \geq \int_0^\infty \sup_{\theta\in\mathcal{W}} \dot{V}\left(e_\Delta, u, \theta\right) dt - \Delta \tag{A.17}$$

(The existence of θ_Δ follows from the properties of $\theta(t)$). From the inequality of (A.17), we obtain:

$$V(e(0)) = -\int_0^\infty \dot{V} dt \le -\int_0^\infty \sup_{\theta \in \mathcal{W}} \dot{V}\left(e_\Delta, u, \theta\right) dt + \Delta \tag{A.18}$$

Combining the inequalities of (A.18) and (A.16), we get:

$$V(e(0)) - \Delta \le J_v^* \le V(e(0)) \tag{A.19}$$

for arbitrarily small Δ. Clearly, if $J_v^* = V(e(0))$, the proof of optimality is complete. Otherwise, there exists $\mu > 0$ such that $J_v^* + \mu = V(e(0))$. Since Δ is arbitrary, we can choose it to be sufficiently small such that $\Delta < \mu$ and the inequality of (A.19) is violated. Thus, the only way to satisfy the inequality of (A.19) for arbitrary Δ is to set $J_v^* = V(e(0))$. This completes the proof of the theorem. △

Proof of Theorem 3.13:

The proof of this theorem follows that of Theorem 3.3. We will only highlight the differences.

Part 1: In this part, we establish that the controller of (3.28) enforces global boundedness of the closed-loop trajectories and that the closed-loop output satisfies (3.29).

Step 1: Substituting the control law of (3.28) into (3.14) and computing the time-derivative of V along the trajectories of the closed-loop e-subsystem, it is straightforward to show that $\dot{V}$ satisfies (A.3) whenever $\|2b^T Pe\| > \phi(\chi - 1)^{-1}$. To analyze the sign of $\dot{V}$ when $\|2b^T Pe\| \le \phi(\chi - 1)^{-1}$, we use the bound $\|L_w L_f^{r-1} h(x)\| \le \delta \|2b^T Pe\| + \mu$ and the fact that $\phi - (\chi - 1)\|2b^T Pe\| < \phi$ to obtain the following estimates:

$$\begin{aligned} &\theta_b \|L_w L_f^{r-1} h(x)\| \|2b^T Pe\| \left(\phi - (\chi - 1)\|2b^T Pe\|\right) \\ &\le \theta_b \|L_w L_f^{r-1} h(x)\| \|2b^T Pe\| \phi \\ &\le \theta_b \delta \phi \|2b^T Pe\|^2 + \theta_b \mu \phi \|2b^T Pe\| \end{aligned} \tag{A.20}$$

$\forall \; \|2b^T Pe\| \le \phi(\chi - 1)^{-1}$. Substituting the estimates of (A.20) directly into (A.2) yields:

$$\begin{aligned} \dot{V} \le & -c_0 \left(L_{\bar{g}} V\right)^2 - \sqrt{(L_{\bar{f}} V)^2 + (L_{\bar{g}} V)^4} \\ & + \frac{(-\rho + \delta \theta_b \phi)\|2b^T Pe\|^2}{\|2b^T Pe\| + \phi} + \frac{\theta_b \mu \phi \|2b^T Pe\|}{\|2b^T Pe\| + \phi} \end{aligned} \tag{A.21}$$

It is clear from the above equation, together with the fact that $\theta_b \mu \phi > 0$ and $\dfrac{\|2b^T Pe\|}{\|2b^T Pe\| + \phi} \le 1$, that if $\phi \le \dfrac{\rho}{\delta \theta_b} := \phi_1^*$, then $\dot{V}$ satisfies:

$$\dot{V} \leq -c_0 \left(L_{\bar{g}}V\right)^2 - \sqrt{(L_{\bar{f}}V)^2 + (L_{\bar{g}}V)^4} + \bar{\phi} \tag{A.22}$$

irrespective of the value of $\|2b^T Pe\|$, where $\bar{\phi} = \theta_b \mu \phi$. Note that the function described by the first two terms on the right-hand side of the above equation is negative-definite, since for all $e \neq 0$, $L_{\bar{g}}V = 0 \implies L_{\bar{f}}V < 0$. Consequently, there exists a function $\alpha_1(\cdot)$ of class $\mathcal{K}$ such that:

$$\dot{V} \leq -\alpha_1(\|e\|) + \bar{\phi} \tag{A.23}$$

Choosing ϕ small enough such that $\bar{\phi} \leq \frac{1}{2}\alpha_1(\|e\|)$, we have:

$$\dot{V} \leq -\tfrac{1}{2}\alpha_1(\|e\|) \; < \; 0 \quad \forall \, \|e\| \geq \alpha_1^{-1}(2\bar{\phi}) \tag{A.24}$$

This inequality shows that $\dot{V}$ is negative outside the set $B_1 := \{e \in \mathrm{I\!R}^r : \|e\| \leq \alpha_1^{-1}(2\bar{\phi})\}$. A direct application then of the result of Theorem 5.1 and its corollaries in [148] allows us to conclude that starting from any initial condition, the following ISS inequality holds for the e states of the closed-loop system of (3.14) and (3.28):

$$\|e(t)\| \leq \bar{\beta}_e \left(\|e(0)\|, t\right) + \bar{\gamma}_e(\phi) \qquad \forall \, t \geq 0 \tag{A.25}$$

where $\bar{\beta}_e$ is a class $\mathcal{KL}$ function and $\bar{\gamma}_e$ is a class $\mathcal{K}_\infty$ function. From Assumption 3.4, we have that the η states of the system of (3.14) possess an ISS property with respect to e and θ:

$$\begin{aligned} \|\eta(t)\| &\leq \bar{\beta}_\eta \left(\|\eta(0)\|, t\right) + \bar{\gamma}_\eta \left(\|[e^T \; \theta]^T\|^s\right) \\ &\leq \bar{\beta}_\eta \left(\|\eta(0)\|, t\right) + \bar{\gamma}_{\eta_1} \left(\|e\|^s\right) + \bar{\gamma}_{\eta_2} \left(\|\theta\|^s\right) \end{aligned} \tag{A.26}$$

uniformly in $\bar{v}$, where $\bar{\gamma}_{\eta_1}$, $\bar{\gamma}_{\eta_2}$ are class $\mathcal{K}$ functions defined as $\bar{\gamma}_{\eta_1}(s) = \bar{\gamma}_{\eta_2}(s) = \bar{\gamma}_\eta(2s)$.

Step 2: We now analyze the behavior of the interconnected closed-loop dynamical system of (3.14) and (3.28), comprised of the e and η states for which the inequalities of (A.25) and (A.26) hold. We first define the following positive real numbers: $\delta_e = D^{\bar{e}} + \phi^*$, $\delta_\eta = D^{\bar{\eta}} + \bar{\gamma}_{\eta 1}(\delta_e) + \phi^*$, $D^{\bar{e}} = \bar{\beta}_e(\bar{\delta}_e, 0) + \bar{\gamma}_e(\phi^*)$, $D^{\bar{\eta}} = \bar{\beta}_\eta(\bar{\delta}_\eta, 0) + \bar{\gamma}_{\eta_2}(\theta_b)$, $\phi^* = \min\{\frac{1}{2}\phi_1^*, \bar{\gamma}_e^{-1}(d)\}$ where $d > 0$ is arbitrary and $\bar{\delta}_e, \bar{\delta}_\eta$ are any positive real numbers. Then, using a contradiction argument similar to the one used in [57,58], one can show that if $\phi \in (0, \phi^*]$, the evolution of the states e and η, starting from any initial states that satisfy $\|e(0)\| \leq \bar{\delta}_e$, $\|\eta(0)\| \leq \bar{\delta}_\eta$, satisfies the inequalities $\|e(t)\| \leq \delta_e$, $\|\eta(t)\| \leq \delta_\eta$ for all times. Finally, for $\phi \in (0, \phi^*]$ and for any initial states, taking the limsup of both sides of (A.25) as $t \to \infty$, we have:

$$\begin{aligned} \limsup_{t \to \infty} \|y(t) - v(t)\| &\leq \limsup_{t \to \infty} \|e(t)\| \\ &\leq \limsup_{t \to \infty} \left(\bar{\beta}_e(\|e(0)\|, t) + \bar{\gamma}_e(\phi)\right) \; \leq \; \bar{\gamma}_e(\phi^*) \; \leq \; d \end{aligned} \tag{A.27}$$

Part 2: In this part, we prove that the control law of (3.28) is optimal with respect to a meaningful cost functional of the form of (3.30). We proceed in two steps. In the first step, we show that the cost functional defined in (3.30) is a meaningful one. In the second step, we show that the stabilizing control law of (3.28) minimizes this cost functional.

Step 1: From its definition in Theorem 3.13, $\bar{l}(e)$ is given by:

$$\bar{l}(e) = -L_{\bar{f}}V + \frac{1}{4}L_{\bar{g}}V\bar{R}^{-1}(x)L_{\bar{g}}V - \|L_{\bar{w}}V\|\theta_b \tag{A.28}$$

Substitution the expression for $\bar{R}^{-1}(x)$, given in Theorem 3.13, into the above equation, and performing some algebraic manipulations (similar to those in (A.11)), it can be shown that, whenever $\|2b^T Pe\| > \phi(\frac{1}{2}\chi - 1)^{-1}$, $\bar{l}(e)$ satisfies:

$$\bar{l}(e) \geq \frac{1}{2}\left[-L_{\bar{f}}V + \sqrt{(L_{\bar{f}}V)^2 + (L_{\bar{g}}V)^4}\right] \tag{A.29}$$

which is positive-definite since, away from the origin, $L_{\bar{f}}V$ and $L_{\bar{g}}V$ do not vanish together. To analyze the sign of $\bar{l}(e)$ when $\|2b^T Pe\| < \phi(\frac{1}{2}\chi - 1)^{-1}$, we use the bound $\|L_w L_f^{r-1}h(x)\| \leq \delta\|2b^T Pe\| + \mu$ and the fact that $(\frac{1}{2}\chi - 1)\|2b^T Pe\| - \phi > -\phi$ to obtain the following estimates:

$$\begin{aligned}
&\theta_b\|L_w L_f^{r-1}h(x)\|\|2b^T Pe\| \left(\phi - (\tfrac{1}{2}\chi - 1)\|2b^T Pe\|\right) \\
&\geq -\theta_b\|L_w L_f^{r-1}h(x)\|\|2b^T Pe\|\phi \\
&\geq -\theta_b\delta\phi\|2b^T Pe\|^2 - \theta_b\mu\phi\|2b^T Pe\|
\end{aligned} \tag{A.30}$$

$\forall\ \|2b^T Pe\| \leq \phi(\frac{1}{2}\chi - 1)^{-1}$. Substituting the estimates of (A.30) directly into (A.11) yields:

$$\begin{aligned}
\bar{l}(e) \geq &\frac{1}{2}\left[-L_{\bar{f}}V + \sqrt{(L_{\bar{f}}V)^2 + (L_{\bar{g}}V)^4}\right] \\
&+ \frac{(\frac{1}{2}\rho - \delta\theta_b\phi)\|2b^T Pe\|^2}{\|2b^T Pe\| + \phi} - \frac{\theta_b\mu\phi\|2b^T Pe\|}{\|2b^T Pe\| + \phi}
\end{aligned} \tag{A.31}$$

From the above equation, it is clear that if $\phi \leq \dfrac{\rho}{2\delta\theta_b} := \phi_2^*$, $\bar{l}(e)$ satisfies:

$$\bar{l}(e) \geq \frac{1}{2}\left[-L_{\bar{f}}V + \sqrt{(L_{\bar{f}}V)^2 + (L_{\bar{g}}V)^4}\right] - \bar{\phi} \tag{A.32}$$

irrespective of the value of $\|2b^T Pe\|$. Therefore, there exists a function $\alpha_2(\cdot)$ of class $\mathcal{K}$ such that:

$$\begin{aligned}
\bar{l}(e) &\geq \alpha_2(\|e\|) - \bar{\phi} \\
&\geq \tfrac{1}{2}\alpha_2(\|e\|) > 0 \quad \forall\ \|e\| \geq \alpha_2^{-1}(2\bar{\phi})
\end{aligned} \tag{A.33}$$

The last inequality implies that $\bar{l}(e)$ is positive outside the set $B_2 := \{e \in \mathbb{R}^r : \|e\| \leq \alpha_2^{-1}(2\bar{\phi})\}$. Note that this set is completely contained within the set Γ since $\|e\| \leq \alpha_2^{-1}(2\bar{\phi}) \leq \alpha_2^{-1}(2\bar{\phi}^*) \leq \epsilon \Longrightarrow e^T P e \leq \lambda_{max}(P)\|e\|^2 \leq \lambda_{max}(P)\epsilon^2$ where $\epsilon := \max\{\alpha_1^{-1}(2\bar{\phi}^*), \alpha_2^{-1}(2\bar{\phi}^*)\}$ and $\lambda_{max}(P) > 0$ is the maximum eigenvalue of the matrix P. Similarly, the set B_1 – defined right after (A.24) – is also contained within Γ, and therefore $\dot{V} < 0$ on and outside the boundary of Γ. From its definition in Theorem 3.13, T_f is the minimum time for the trajectories of the closed-loop system to reach and enter Γ without ever leaving again (note that Γ is a level set of V). Therefore, we have that $\|e(t)\| \geq \alpha_2^{-1}(2\bar{\phi}) \;\; \forall\, t \in [0, T_f]$. Hence, $\bar{l}(e(t)) > 0 \;\; \forall\, t \in [0, T_f]$. Note also that $\bar{R}(x) > 0$ for all x. Therefore the cost functional of (3.30) is a meaningful one.

Step 2: In this step, we prove that control law of (3.28) minimizes the cost functional of (3.30). Substituting:

$$\alpha = u + \frac{1}{2}\bar{R}^{-1}(x) L_{\bar{g}} V \tag{A.34}$$

into (3.30), we get the following chain of equalities:

$$\begin{aligned}
J_n &= V(e(T_f)) + \int_0^{T_f} \big(l(e(t)) + u(t)\bar{R}(x(t))u(t)\big)\, dt \\
&= V(e(T_f)) + \int_0^{T_f} \left(-L_{\bar{f}}V + \frac{1}{2}L_{\bar{g}}V\bar{R}^{-1}(x)L_{\bar{g}}V - \|L_{\bar{w}}V\|\theta_b - L_{\bar{g}}V\alpha\right) dt \\
&\quad + \int_0^{T_f} \alpha\bar{R}(x)\alpha\; dt \\
&= V(e(T_f)) - \int_0^{T_f} \sup_{\theta\in\mathcal{W}} \left(\dot{V}\right) dt + \int_0^{T_f} \alpha R(x)\alpha \;\; dt
\end{aligned} \tag{A.35}$$

Note that:

$$J_n^* = V(e(T_f)) - \int_0^{T_f} \sup_{\theta\in\mathcal{W}} \left(\dot{V}\right) dt \;\leq\; V(e(T_f)) - \int_0^{T_f} \dot{V} dt \;=\; V(e(0)) \tag{A.36}$$

Since $\bar{R}(x) > 0$ for all $x \in \mathbb{R}^n$, it is clear that the minimum of J_n is J_n^*. This minimum is achieved when $\alpha(t) \equiv 0$ which proves that the controller of (3.28) minimizes the cost of (3.30). Finally, using an argument similar to that used in (A.17–A.19, we have that $J_n^* = V(e(0))$. This completes the proof of the theorem. △

Proof of Theorem 3.20:

The proof of this theorem consists of three parts. In the first part, we use a singular perturbation formulation to represent the closed-loop system and show that the origin of the resulting fast subsystem is globally exponentially stable. In the second part, we focus on the closed-loop reduced system and derive ISS bounds for its states. Then, using the result of Lemma 2.18, we establish that these ISS bounds continue to hold up to an arbitrarily small offset, for arbitrarily large initial conditions and uncertainty. The resulting ISS inequalities are then analyzed to establish semi-global boundedness and local exponential stability of the full closed-loop system, which is then used to establish (3.47), provided that ϕ and ϵ are sufficiently small. Finally, in the third part, we establish the near-optimality of the controller of (3.46) with respect to the cost of (3.24).

Part 1: Defining the auxiliary error variables, $\hat{e}_i = L^{r-i}(y^{(i-1)} - \tilde{y}_i)$, $i = 1, \ldots, r$, the vector $e_o = [\hat{e}_1 \; \hat{e}_2 \; \cdots \; \hat{e}_r]^T$, the vector of controller parameters $\phi_c = [c_0 \; \rho \; \chi \; \phi \; \theta_b]^T$, the parameter $\epsilon = \frac{1}{L}$, the matrix $\tilde{A}$ and the vector $\tilde{b}$:

$$\tilde{A} = \begin{bmatrix} -a_1 & 1 \, 0 \cdots 0 \\ -a_2 & 0 \, 1 \cdots 0 \\ \vdots & \vdots \; \vdots \; \ddots \; \vdots \\ -a_{r-1} & 0 \, 0 \cdots 1 \\ -a_r & 0 \, 0 \cdots 0 \end{bmatrix}, \quad \tilde{b} = \begin{bmatrix} 0 \\ 0 \\ \vdots \\ 0 \\ 1 \end{bmatrix} \tag{A.37}$$

the system of (3.1), under the controller of (3.46), takes the form:

$$\begin{aligned} \epsilon \dot{e}_o &= \tilde{A} e_o + \epsilon \tilde{b} \Omega(x, \hat{x}, \theta, \phi_c, \bar{v}) \\ \dot{\omega}_1 &= \Psi_1(sat(\tilde{y}), \omega) \\ &\;\;\vdots \\ \dot{\omega}_{n-r} &= \Psi_{n-r}(sat(\tilde{y}), \omega) \\ \dot{x} &= f(x) + g(x) p(\hat{x}, \phi_c, \bar{v}) + \sum_{k=1}^{q} w_k(x) \theta_k \end{aligned} \tag{A.38}$$

where $\hat{x} = \mathcal{X}^{-1}(sat(y_d - \Delta(\epsilon) e_o), \omega)$ and $y_d = [y^{(0)} \; y^{(1)} \; \cdots \; y^{(r-1)}]^T$, $\Delta(\epsilon)$ is a diagonal matrix whose i-th diagonal element is ϵ^{r-i}, and $\Omega(x, \hat{x}, \theta, \phi_c, \bar{v})$ is a Lipschitz function of its argument. Owing to the presence of the small parameter, ϵ, that multiplies the time-derivative, $\dot{e}_o$, the system of (A.38) is a two-time-scale one. Defining the fast time-scale, $\tau = \frac{t}{\epsilon}$, and setting $\epsilon = 0$, the closed-loop fast subsystem takes the form:

$$\frac{de_o}{d\tau} = \tilde{A} e_o \tag{A.39}$$

Since the constant matrix $\tilde{A}$ is Hurwitz, the origin of the system of (A.39) is globally exponentially stable.

Part 2: In this part of the proof, we initially derive ISS bounds for the states of the system of (A.38) when $\epsilon = 0$, in appropriately transformed coordinates, and then use the result of Lemma 2.18 to show that these bounds hold up to an arbitrarily small offset, for initial conditions and uncertainty in an arbitrarily large compact set, provided that ϵ is sufficiently small. To this end, we first write the closed-loop system in the following form:

$$\begin{aligned}
\epsilon \dot{e}_o &= \tilde{A} e_o + \epsilon \tilde{b} \Omega(e, \eta, \hat{x}, \theta, \phi_c, \bar{v}) \\
\dot{\omega}_1 &= \Psi_1(sat(\tilde{y}), \omega) \\
&\vdots \\
\dot{\omega}_{n-r} &= \Psi_{n-r}(sat(\tilde{y}), \omega) \\
\dot{e}_1 &= e_2 \\
&\vdots \\
\dot{e}_{r-1} &= e_r \\
\dot{e}_r &= L_f^r h(\mathcal{X}^{-1}(e, \eta, \bar{v})) - v^{(r)} + L_g L_f^{r-1} h(\mathcal{X}^{-1}(e, \eta, \bar{v})) p(\hat{x}, \phi_c, \bar{v}) \\
&\quad + \sum_{k=1}^{q} L_{wk} L_f^{r-1} h(\mathcal{X}^{-1}(e, \eta, \bar{v})) \theta_k \\
\dot{\eta}_1 &= \Psi_1(e, \eta, \bar{v}) \\
&\vdots \\
\dot{\eta}_{n-r} &= \Psi_{n-r}(e, \eta, \bar{v}) \\
y &= e_1 + v
\end{aligned} \tag{A.40}$$

where $e = [e_1 \ e_2 \ \cdots \ e_r]^T$, $e_i = \zeta_i - v^{(i-1)}$, $i = 1, \ldots, r$, and Ψ_i, $i = 1, \ldots, n-r$, are Lipschitz functions of their arguments.

Step 1: Consider the system of (A.40) with $\epsilon = 0$. In order to analyze the dynamic behavior of the resulting closed-loop slow (reduced) system, we initially need to show that for the system of (A.40), $\eta(0) = \omega(0) + O(\epsilon)$ implies $\eta(t) = \omega(t) + O(\epsilon)$, $\forall \ t \geq 0$. To this end, consider the following singularly perturbed system:

$$
\begin{aligned}
\epsilon \dot{e}_o &= \tilde{A} e_o + \epsilon \tilde{b} \Omega(e, \eta, \hat{x}, \theta, \phi_c, \bar{v}) \\
\dot{\omega}_1 &= \Psi_1(sat(\tilde{y}), \omega) \\
&\vdots \\
\dot{\omega}_{n-r} &= \Psi_{n-r}(sat(\tilde{y}), \omega) \\
\dot{\eta}_1 &= \Psi_1(e, \eta, \bar{v}) \\
&\vdots \\
\dot{\eta}_{n-r} &= \Psi_{n-r}(e, \eta, \bar{v})
\end{aligned}
\tag{A.41}
$$

It is straightforward to verify that the above system satisfies the assumptions of Theorem 1 reported in [149]. Applying the result of this theorem, we have that there exists a positive real number, ϵ_0, such that for any positive real number, δ_ω, satisfying:

$$
\delta_\omega \geq \max_{\|x\| \leq \delta_x} \left\{ \sum_{\nu=1}^{n-r} \|\chi_\nu(x)\| \right\} \tag{A.42}
$$

where $\chi_\nu(x)$, $\nu = 1, \ldots, n-r$, are the functions defined in Assumption 3.5, the states (η, ω) of this system, starting from any initial condition that satisfies $\eta(0) = \omega(0) + O(\epsilon)$ (with $\max\{\|\eta(0)\|, \|\omega(0)\|\} \leq \delta_\omega$), if $\epsilon \in (0, \epsilon_0]$, satisfy $\eta(t) = \omega(t) + O(\epsilon)$, $\forall\, t \geq 0$.

Since $\eta(t) = \omega(t)$ $\forall\, t \geq 0$ and $e_o = 0$ when $\epsilon = 0$, the reduced (slow) closed-loop system (i.e., the system of (A.40) with $\epsilon = 0$) reduces to the one studied in the Proof of Theorem 3.3, where it was shown that the evolution of the e-subsystem satisfies:

$$
\|e(t)\| \leq \beta_e \left(\|e(0)\|, t\right) \tag{A.43}
$$

$\forall\, t \geq 0$. From Assumption 3.6, we have that the η states of the reduced (slow) closed-loop system of (A.40) satisfy the following ISS inequality:

$$
\|\eta(t)\| \leq K_\eta \|\eta(0)\| e^{-at} + \gamma_\eta(\|e\|^s) \tag{A.44}
$$

$\forall\, t \geq 0$, uniformly in $\bar{v}$, where $a > 0$, $K_\eta \geq 1$ and $\gamma_\eta(\cdot)$ is a class $\mathcal{K}$ function of its argument. Realizing that the reduced closed-loop slow system is an interconnection of the e and η subsystems, we show now that the origin of the reduced closed-loop system is asymptotically stable. From (A.43–A.44), we have that the solutions of the reduced closed-loop system satisfy:

$$
\|\eta(t)\| \leq K_\eta \|\eta(s)\| e^{-a(t-s)} + \gamma_\eta(\sup_{s \leq \tau \leq t} \|e(\tau)\|) \tag{A.45}
$$

$$\|e(t)\| \leq \beta_e \left(\|e(s)\|, t-s\right) \tag{A.46}$$

where $t \geq s \geq 0$. Applying (A.45) with $s = \frac{t}{2}$, we obtain:

$$\|\eta(t)\| \leq K_\eta \left\|\eta\left(\tfrac{t}{2}\right)\right\| e^{-a\frac{t}{2}} + \gamma_\eta \left(\sup_{\frac{t}{2} \leq \tau \leq t} \|e(\tau)\| \right) \tag{A.47}$$

To estimate $\eta\left(\frac{t}{2}\right)$, we apply (A.45) with $s = 0$ and t replaced by $\frac{t}{2}$ to obtain:

$$\left\|\eta\left(\tfrac{t}{2}\right)\right\| \leq K_\eta \|\eta(0)\| e^{-a\frac{t}{2}} + \gamma_\eta \left(\sup_{0 \leq \tau \leq \frac{t}{2}} \|e(\tau)\| \right) \tag{A.48}$$

From (A.46), we have:

$$\sup_{0 \leq \tau \leq \frac{t}{2}} \|e(\tau)\| \leq \beta_e \left(\|e(0)\|, 0\right) \tag{A.49}$$

$$\sup_{\frac{t}{2} \leq \tau \leq t} \|e(\tau)\| \leq \beta_e \left(\|e(0)\|, \tfrac{t}{2}\right) \tag{A.50}$$

Substituting (A.48–A.50) into (A.47) and using the inequalities $\|\eta(0)\| \leq \|\pi(0)\|$, $\|e(0)\| \leq \|\pi(0)\|$ and $\|\pi(t)\| \leq \|\eta(t)\| + \|e(t)\|$, where $\pi(0) = [e^T(0) \;\; \eta^T(0)]^T$, we obtain:

$$\|\pi(t)\| \leq \beta \left(\|\pi(0)\|, t\right) \tag{A.51}$$

where

$$\begin{aligned} \beta\left(\|\pi(0)\|, t\right) = & \left[K_\eta \|\pi(0)\| e^{-a\frac{t}{2}} + \gamma_\eta \left(\beta_e\left(\|\pi(0)\|, 0\right)\right)\right] e^{-a\frac{t}{2}} \\ & + \gamma_\eta \left(\beta_e\left(\|\pi(0)\|, \tfrac{t}{2}\right)\right) + \beta_e\left(\|\pi(0)\|, t\right) \end{aligned} \tag{A.52}$$

It can be easily verified that β is a class $\mathcal{KL}$ function. Hence, the origin of the reduced closed-loop system is asymptotically stable (i.e., $\pi(t) \to 0$ as $t \to \infty$). Therefore, there exists a class $\mathcal{KL}$ function, $\bar{\beta}_\eta(\cdot, \cdot)$, such that:

$$\|\eta(t)\| \leq \bar{\beta}_\eta \left(\|\eta(0)\|, t\right), \quad \forall\, t \geq 0 \tag{A.53}$$

To show exponential stability, we proceed as follows. Repeating the same calculations of Step 1 in Part 1 of the Proof of Theorem 3.3, it is straightforward to show that, when $L_{\bar{g}}V \neq 0$, there exists a positive real number ϕ^* such that if $\phi \leq \phi^*$, $\dot{V}$ – computed along the trajectories of the reduced closed-loop e-subsystem – satisfies:

$$\begin{aligned} \dot{V} & \leq -\frac{1}{2} c_0 (L_{\bar{g}}V)^2 \\ & \leq -2c_0 \|Pe\|^2 \|L_g L_f^{r-1} h(x)\|^2 \\ & \leq -2c_0 \lambda_{max}(P^2) \|e\|^2 \|L_g L_f^{r-1} h(x)\|^2 \end{aligned} \tag{A.54}$$

where we have used the fact that $\|L_{\bar{g}}V\| \leq 2\|Pe\|\|L_gL_f^{r-1}h(x)\|$ in deriving the second inequality in (A.54). Since $\|L_gL_f^{r-1}h(x)\|$ is a continuous function of x and $L_gL_f^{r-1}h(x) \neq 0 \; \forall \, x \in \mathbb{R}^n$, there exists a positive real number, κ_1, such that $\|L_gL_f^{r-1}h(x)\|^2 \geq \kappa_1$ and, therefore, $\dot{V} \leq -\kappa_2\|e\|^2$, where $\kappa_2 = 2c_0\kappa_1\lambda_{max}(P^2) > 0$. Recall also from the analysis of (A.6) in the Proof of Theorem 3.3 that, when $L_{\bar{g}}V = 0$, $\dot{V} \leq -e^TQe \leq -\lambda_{min}(Q)\|e\|^2$, where Q is a positive-definite matrix. Therefore, we have $\dot{V} \leq -\kappa_3\|e\|^2$, where $\kappa_3 = \max\{\kappa_2, \lambda_{min}(Q)\}$, and there exist real numbers, $k_1 \geq 1$ and $a_1 > 0$, such that the following inequality holds for the reduced closed-loop e-subsystem:

$$\|e(t)\| \leq k_1\|e(0)\|e^{-a_1 t} \quad \forall \, t \geq 0 \tag{A.55}$$

From the above inequality and the fact that the origin of the η-subsystem, with $e = 0$, is exponentially stable (from Assumption 3.6), it follows that the e-η interconnected, reduced closed-loop system of (A.40) is locally exponentially stable. Therefore, there exists a positive real number, r, such that:

$$\|\pi(t)\| \leq k_3\|\pi(t_0)\|e^{-a_2(t-t_0)} \quad \forall \; \|\pi(t_0)\| \leq r \tag{A.56}$$

for some $k_3 \geq 1$ and $a_2 > 0$.

Finally, we note that since the static component of the controller of (3.46) with $\hat{x} = x$ enforces global asymptotic stability in the reduced (slow) closed-loop system, the ζ states of this system satisfy a bound of the following form, $\forall \, t \geq 0$:

$$\|\zeta(t)\| \leq \beta_\zeta(\delta_\zeta, t) \tag{A.57}$$

where β_ζ is a class $\mathcal{KL}$ function and δ_ζ is the maximum value of the norm of the vector $[h(x) \; L_fh(x) \; \cdots \; L_f^{r-1}h(x)]$, for $\|x\| \leq \delta_x$. Based on the above bound and following the results of [149, 268], we disregard estimates of $\tilde{y}$, obtained from the high-gain observer, with norm $\|\tilde{y}\| > \beta_\zeta(\delta_\zeta, 0)$. Hence, we set $sat(\cdot) = \min\{1, \frac{\zeta_{max}}{\|\cdot\|}\}(\cdot)$ where ζ_{max} is the maximum value of the vector $[\zeta_1 \; \zeta_2 \; \cdots \; \zeta_r]$, for $\|\zeta\| \leq \beta_\zeta(\delta_\zeta, 0)$.

Step 2: In order to apply the result of Lemma 2.18, we need to define a set of positive real numbers $\{\bar{\delta}_e, \bar{\delta}_\eta, \delta_\theta, \delta_{\bar{v}}, \delta_\eta, \delta_e, d, d_e, d_\eta\}$, where δ_θ and $\delta_{\bar{v}}$ were specified in the statement of the theorem, d, d_e and d_η are arbitrary positive real numbers,

$$\begin{aligned} \bar{\delta}_e &\geq \max_{\|x\| \leq \delta_x, \|\bar{v}\|^s \leq \delta_{\bar{v}}} \left\{ \left\| \sum_{k=1}^{r} (v^{(k-1)} - L_{\tilde{f}}^{k-1}h(x)) \right\| \right\} \\ \bar{\delta}_\eta &\geq \max_{\|x\| \leq \delta_x, \|\bar{v}\|^s \leq \delta_{\bar{v}}} \left\{ \sum_{\nu=1}^{n-r} \|\chi_\nu(x)\| \right\} \end{aligned} \tag{A.58}$$

$\delta_e > \beta_e(\bar{\delta}_e, 0) + d_e$, $\delta_\eta > \bar{\beta}_\eta(\bar{\delta}_\eta, 0) + d_\eta$. First, consider the singularly perturbed system comprised of the states (ζ, e_o) of the closed-loop system of (A.40).

This system is in standard form, possesses a globally exponentially stable fast subsystem, and its corresponding reduced system is globally asymptotically stable.

Therefore, using the result of Lemma 2.18, given any pair of positive real numbers (d, δ), with $\delta = \max\{\delta_\zeta, \delta_\theta, \delta_{\bar{v}}, \delta_\eta\}$, there exists a positive real number, $\epsilon^\zeta(\phi)$, such that if $\epsilon \in (0, \epsilon^\zeta(\phi)]$, and $\|\zeta(0)\| \leq \delta_\zeta$, $\|\tilde{y}(0)\| \leq \delta_\zeta$, $\|\theta\|^s \leq \delta_\theta$, $\|\bar{v}\|^s \leq \delta_{\bar{v}}$, $\|\eta\| \leq \delta_\eta$, then:

$$\begin{aligned} \|\zeta(t)\| &\leq \beta_\zeta(\delta_\zeta, t) + d \\ \|e_o(t)\| &\leq k_4 \|e_o(0)\| e^{-a_3 \frac{t}{\epsilon}} + d \end{aligned} \tag{A.59}$$

for some $k_4 \geq 1$ and $a_3 > 0$.

Consider now the singularly perturbed system comprised of the states (e, e_o) of the system of (A.40). This system is in standard form, possesses a globally exponentially stable fast subsystem, and its corresponding reduced system is globally asymptotically stable. These properties allow the application of the result of Lemma 2.18, with $\delta = \max\{\bar{\delta}_e, \delta_\zeta, \delta_\theta, \delta_{\bar{v}}, \delta_\eta\}$ and $d = d_e$, to obtain the existence of a positive real number, $\epsilon^e(\phi)$, such that if $\epsilon \in (0, \epsilon^e(\phi)]$, and $\|e(0)\| \leq \bar{\delta}_e$, $\|\tilde{y}(0)\| \leq \delta_\zeta$, $\|\theta\|^s \leq \delta_\theta$, $\|\bar{v}\|^s \leq \delta_{\bar{v}}$, $\|\eta\|^s \leq \delta_\eta$, then:

$$\|e(t)\| \leq \beta_e\left(\|e(0)\|, t\right) + d_e \tag{A.60}$$

Similarly, it can be shown that Lemma 2.18, with $\delta = \max\{\bar{\delta}_\eta, \delta_\zeta, \delta_\theta, \delta_{\bar{v}}, \delta_e\}$ and $d = d_\eta$, can be applied to the system comprised of the states (η, e_o) of the system of (A.40). Thus, we have that there exist positive real numbers $\epsilon^\eta(\phi)$ such that if $\epsilon \in (0, \epsilon^\eta(\phi)]$ and $\|\eta(0)\| \leq \bar{\delta}_\eta$, $\|\tilde{y}(0)\| \leq \delta_\zeta$, $\|\theta\|^s \leq \delta_\theta$, $\|\bar{v}\|^s \leq \delta_{\bar{v}}$, $\|e\|^s \leq \delta_e$, then:

$$\|\eta(t)\| \leq \bar{\beta}_\eta\left(\|\eta(0)\|, t\right) + d_\eta \tag{A.61}$$

Finally, the inequalities of (A.43–A.44) and (A.59–A.61) can be manipulated using a small-gain theorem type argument, similar to the one used in the Proof of Theorem 1 in [58], to show that given the following set of positive real numbers $\{\bar{\delta}_e, \bar{\delta}_\eta, \delta_\zeta, \delta_\theta, \delta_{\bar{v}}, d, d_e, d_\eta\}$ and with $\phi \in (0, \phi^*]$, there exists a positive real number, $\tilde{\epsilon}(\phi) := \min\{\epsilon^\zeta(\phi), \epsilon^e(\phi), \epsilon^\eta(\phi)\}$, such that if $\epsilon \in (0, \min\{\epsilon_0, \tilde{\epsilon}(\phi)\}]$, and $\|e(0)\| \leq \bar{\delta}_e$, $\|\eta(0)\| \leq \bar{\delta}_\eta$, $\|\tilde{y}(0)\| \leq \delta_\zeta$, $\|\theta\|^s \leq \delta_\theta$, $\|\bar{v}\|^s \leq \delta_{\bar{v}}$, then the states of the closed-loop system are bounded. For brevity, this argument will not be repeated here.

Step 3: So far in the proof, we have established that the states of the singularly perturbed closed-loop system of (A.40) are semi-globally bounded. In this section we show that the origin of the closed-loop system is (semi-globally) asymptotically stable. To this end, we note – from the inequalities of (A.59–A.61) – that, as t increases, the trajectories will be ultimately bounded with ultimate bounds that depend on d, d_e, and d_η. Since d, d_e, and d_η are arbitrary, we can choose them small enough such that after a sufficiently large time, say $\tilde{t}$, the trajectories of the closed-loop system are confined within a small compact neighborhood of the origin. Obviously, $\tilde{t}$ depends on both the initial condition and the desired size of the neighborhood, but is independent of ϵ. Let $\tilde{d} := \max\{d, d_e, d_\eta\} \leq r/4$, where r was defined in (A.56), and $\tilde{t}$ be the earliest time such that $\max\{\beta_e\left(\|e(0)\|, \tilde{t}\right), \bar{\beta}_\eta\left(\|\eta(0)\|, \tilde{t}\right), k_4\|e_o(0)\|e^{-a_3\frac{\tilde{t}}{\epsilon}}\} \leq \tilde{d}$. Then it can be easily verified that:

$$\|\pi(t)\| \leq r, \quad \|e_o(t)\| \leq r, \quad \forall\, t \geq \tilde{t} \tag{A.62}$$

Recall from (A.56) and (A.39) that both the reduced (slow) and fast closed-loop subsystems are exponentially stable within the balls of (A.62). Therefore, a direct application of the result of Theorem 9.3 in [148] yields that there exists $\bar{\epsilon} > 0$ such that if $\epsilon \in (0, \bar{\epsilon}]$, then:

$$\|\Pi(t)\| \leq \alpha_1\|\Pi(\tilde{t})\|e^{-\beta_1(t-\tilde{t})} \tag{A.63}$$

for some $\alpha_1 \geq 1$ and $\beta_1 > 0$, where $\Pi(t) = [e^T(t)\ \eta^T(t)\ \omega^T(t)\ e_o^T(t)]^T$ and $\epsilon \in (0, \min\{\epsilon_0, \bar{\epsilon}, \tilde{\epsilon}\}]$. Consequently, the states of the closed-loop system of (A.40) converge to the origin as $t \to \infty$, which together with the boundedness of the close-loop states established in Step 2, implies that the origin of the closed-loop system is asymptotically (and locally exponentially) stable. The asymptotic output tracking result can now be obtained by simply noting that from (A.63) that we have:

$$\limsup_{t\to\infty} \|e(t)\| = 0 \tag{A.64}$$

and therefore

$$\limsup_{t\to\infty} \|e_1(t)\| = \limsup_{t\to\infty} \|y(t) - v(t)\| = 0 \tag{A.65}$$

<u>*Part 3*</u>: In this part, we show that the controller of (3.46) is near-optimal with respect to the cost of (3.24) in the sense that the cost achieved by this controller tends to the minimum cost for the state feedback problem as $\epsilon \to 0$. To this end, we adopt the following three-step procedure. In Step 1, we show that, for the reduced (slow) closed-loop system, the controller of (3.46) with $\epsilon = 0$ (i.e., $\hat{x} = x$) is optimal with respect to a cost functional of the form of (3.24) with a minimum cost of $V(e(0))$. In Step 2, we establish the closeness of solutions of the reduced and full closed-loop systems on the infinite time-interval. Finally, in Step 3, we exploit the closeness of solutions result to show that the minimum cost achieved by the controller of (3.46) in the full closed-loop system approaches the optimal cost for the reduced system under state feedback, when the observer gain is sufficiently large.
<u>Step 1</u>: In this step we show that the static component of the controller of (3.46) with $\epsilon = 0$ and $\hat{x} = x$ minimizes the cost functional associated with the reduced (slow) closed-loop system which has the form of (3.24). To this

end, consider the closed-loop system of (A.40) with $\epsilon = 0$, where u of (3.46) is applied with $\hat{x} = x$. We note that the resulting reduced closed-loop system is identical to the one studied in the Proof of Theorem 3.3 (state feedback problem) and satisfies the assumptions stated therein. Applying the result of this theorem, we have that the static component of the controller of (3.46) minimizes a cost functional of the form:

$$J_r = \int_0^{\infty} (l(\bar{e}) + uR(\bar{x})u)\, dt \tag{A.66}$$

where $\bar{x} = \mathcal{X}^{-1}(\bar{e}, \bar{\eta}, \bar{v})$ refers to the solution of the reduced closed-loop system under state feedback control. Following the same treatment presented in Part 3 of the Proof of Theorem 3.3, it can be shown that upon substitution of the static component, with $u = p(\bar{x})$, of (3.46) in the above expression, the minimum cost obtained is:

$$J_r^* = \int_0^{\infty} (l(\bar{e}) + p(\bar{x})R(\bar{x})p(\bar{x}))\, dt = V(\bar{e}(0)) \tag{A.67}$$

Step 2: In this step, we show that starting from an arbitrarily large compact set of initial conditions and for arbitrarily large uncertainties, the following estimates hold:

$$\begin{aligned} z(t,\epsilon) - \bar{z}(t) &= O(\epsilon) \quad \forall\ t \geq 0 \\ e_o(t,\epsilon) &= O(\epsilon) \quad \forall\ t \geq t_b \end{aligned} \tag{A.68}$$

for some $t_b > 0$ where $\bar{z}(t) = [\bar{e}^T(t)\ \bar{\eta}^T(t)\ \bar{\omega}^T(t)]^T$ is the solution of the reduced (slow) problem obtained by setting $\epsilon = 0$ in (A.40). The strategy we shall adopt to prove the above estimates involves, first, the application of Tikhonov's theorem to establish the closeness of solutions on a finite time-interval whose size can be made arbitrarily large by selecting ϵ sufficiently small. Then, once the trajectories of the closed-loop system become confined within an appropriately small ball around the origin, we apply the result of Theorem 9.4 reported in [148] to obtain the estimates of (A.68).

Referring to the system of (A.40), it is straightforward to show that this system satisfies the conditions of Theorem 9.1 in [148]. Applying the result of this theorem to the system of (A.40), we conclude that there exists a finite time, $t_1 > 0$, and positive constants. K, L , and $\bar{\bar{\epsilon}}$, such that if $\epsilon \in (0, \bar{\bar{\epsilon}}]$ and $\|e(0)\| \leq \bar{\delta}_e$, $\|\eta(0)\| \leq \bar{\delta}_\eta$, $\|\tilde{y}(0)\| \leq \delta_\zeta$, $\|\theta\|^s \leq \delta_\theta$, $\|\bar{v}\|^s \leq \delta_{\bar{v}}$, then:

$$\|z(t,\epsilon) - \bar{z}(t)\| \leq \epsilon K[1 + t_1] \exp{(Lt_1)} \tag{A.69}$$

and the estimate:

$$z(t,\epsilon) - \bar{z}(t) = O(\epsilon) \tag{A.70}$$

holds uniformly for $t \in [0, t_1]$. Furthermore, given any $t_b > 0$, there is $\epsilon_b \leq \bar{\bar{\epsilon}}$ such that the estimate:

$$e_o(t,\epsilon) = O(\epsilon) \tag{A.71}$$

holds uniformly for $t \in [t_b, t_1]$ whenever $\epsilon < \bar{\bar{\epsilon}}$. In order for the above estimates to hold on the infinite time-interval, we require that $t_1 \geq \tilde{t}$ where $\tilde{t}$ was defined in Step 3 of Part 2 of the proof. From (A.69), it is clear that t_1 can be made arbitrarily large – while maintaining the estimate of (A.70) – by selecting ϵ sufficiently small. Therefore, there exists a positive constant, ϵ^{**}, such that if $\epsilon \in (0, \epsilon^{**}]$, we have $t_1 \geq \tilde{t}$ and the trajectories of the closed-loop system lie within the neighborhood of the origin defined in (A.56) for all $t \geq t_1$. It can be easily verified that, within this neighborhood, the closed-loop system of (A.40) satisfies the assumptions of Theorem 9.4 in [148]. Applying the result of this theorem, we obtain the existence of a positive constant, ϵ', such that if $\epsilon \in (0, \epsilon']$, the estimates of (A.68) hold $\forall$ $\|z(t)\| \leq r$ and $\|e_o(t)\| \leq r$. Finally, we now have that if $\epsilon \in (0, \epsilon^*]$ where $\epsilon^* := \min\{\epsilon', \epsilon^{**}, \bar{\bar{\epsilon}}, \bar{\epsilon}, \tilde{\epsilon}, \epsilon_0\}$, then (A.68) is satisfied for $\|e(0)\| \leq \bar{\delta}_e$, $\|\eta(0)\| \leq \bar{\delta}_\eta$, $\|\tilde{y}(0)\| \leq \delta_\zeta$, $\|\theta\|^s \leq \delta_\theta$, $\|\bar{v}\|^s \leq \delta_{\bar{v}}$.

Step 3: In this step, we exploit the closeness of solutions result obtained in Step 2 and combine it with the optimality result established in Step 1 to prove that the output feedback controller of (3.46) is near-optimal in the sense that it achieves a cost for the full closed-loop system that approaches the minimum cost for the reduced (state feedback) problem provided that the gain of the observer is sufficiently large. To this end, consider re-writing the cost functional of (3.24) as follows:

$$J_v(u) = \int_0^{t_b} (l(e) + uR(x)u)\, dt + \int_{t_b}^{\infty} (l(e) + uR(x)u)\, dt \tag{A.72}$$

From Step 2, we have that the following estimates hold for $t \geq t_b$:

$$\begin{aligned} e(t) &= \bar{e}(t) + O(\epsilon) \\ x(t) &= \bar{x}(t) + O(\epsilon) \\ \hat{x}(t) &= \bar{x}(t) + O(\epsilon) \end{aligned} \tag{A.73}$$

It follows then from the continuity properties of the functions $l(\cdot)$, $u(\cdot)$, $R(\cdot)$ that, for $t \geq t_b$ and as $\epsilon \longrightarrow 0$, $l(e(t)) \longrightarrow l(\bar{e}(t))$, $p(\hat{x}(t)) \longrightarrow p(\bar{x}(t))$, $R(x(t)) \longrightarrow R(\bar{x}(t))$, and therefore:

$$\int_{t_b}^{\infty} (l(e) + p(\hat{x})R(x)p(\hat{x}))\, dt \longrightarrow \int_{t_b}^{\infty} (l(\bar{e}) + p(\bar{x})R(\bar{x})p(\bar{x}))\, dt \tag{A.74}$$

From the stability of the full closed-loop system established in Part 2 of the proof, together with the continuity properties of the functions $l(\cdot)$, $u(\cdot)$, $R(\cdot)$, we have that there exists a positive real number, M, such that $\|l(e) + p(\hat{x})R(x)p(\hat{x})\| \leq M$, for $\|e(0)\| \leq \bar{\delta}_e$, $\|\eta(0)\| \leq \bar{\delta}_\eta$, $\|\tilde{y}(0)\| \leq \delta_\zeta$, $\|\theta\|^s \leq \delta_\theta$, $\|\bar{v}\|^s \leq \delta_{\bar{v}}$, and with $\phi \in (0, \phi^*]$ and $\epsilon \in (0, \epsilon^*]$. Using the fact that $t_b = O(\epsilon)$, we get:

$$\begin{aligned}\int_0^{t_b} (l(e) + p(\hat{x})R(x)p(\hat{x}))\, dt &\leq \int_0^{t_b} M dt \\ &\leq \quad M\epsilon \\ &= \quad O(\epsilon)\end{aligned} \tag{A.75}$$

Similarly, from the stability of the reduced closed-loop system (state feedback problem) established in Part 1 and the fact that $t_b = O(\epsilon)$, there exists a positive real number M' such that:

$$\begin{aligned}\int_0^{t_b} (l(\bar{e}) + p(\bar{x})R(\bar{x})p(\bar{x}))\, dt &\leq \int_0^{t_b} M' dt \\ &\leq \quad M'\epsilon \\ &= \quad O(\epsilon)\end{aligned} \tag{A.76}$$

Combining (A.74–A.76), we have that as $\epsilon \longrightarrow 0$:

$$J_{v,o}^* := \int_0^{\infty} (l(e) + p(\hat{x})R(x)p(\hat{x}))\, dt \longrightarrow \int_0^{\infty} (l(\bar{e}) + p(\bar{x})R(\bar{x})p(\bar{x}))\, dt \tag{A.77}$$

and, therefore, from (A.67), we finally have:

$$J_{v,o}^* \longrightarrow V(\bar{e}(0)) \quad as\ \epsilon \longrightarrow 0 \tag{A.78}$$

This completes the proof of the theorem. △

Proof of Theorem 3.27:
The proof of this theorem follows closely the proof of Theorem 3.20. We will only point out the differences. In the first part, we establish the global exponential stability of the fast closed-loop subsystem and derive ISS bounds for the states of the reduced closed-loop system. Then using Lemma 2.18, we show that the derived ISS bounds continue to hold, up to an arbitrarily small offset, for arbitrarily large initial conditions and uncertainty. In the second part of the proof, the resulting ISS inequalities are studied, using techniques similar to those used in [58, 132] to show boundedness of the closed-loop trajectories and establish the inequality of (3.50). In third part of the proof, we apply Tikhonov's theorem to obtain closeness of solutions (between the full and reduced closed-loop systems) on the finite time-interval and then use this result to prove the finite-horizon near-optimality of the output feedback controller of (3.49).
Part 1: The global exponential stability of the closed-loop fast subsystem was established in Step 1 of Part 1 of the Proof of Theorem 3.20 and will not be repeated here. Instead, we focus on the reduced system and derive ISS bounds that capture the evolution of its states. To this end, we consider again the system of (A.40) with $\epsilon = 0$. Using the same argument presented in Step

2 of Part 1 in the Proof of Theorem 3.20, one can again show that for the system of (A.40), $\omega(0) = \eta(0) + O(\epsilon)$ implies $\eta(t) = \omega(t) + O(\epsilon)$ $\forall$ $t \geq 0$, which yields that $\omega(t) = \eta(t)$ when $\epsilon = 0$. Therefore, the reduced closed-loop system (i.e., the system of (A.40) with $\epsilon = 0$) is identical to the one studied in the Proof of Theorem 3.13, where it was shown that the evolution of the e-subsystem satisfies:

$$\|e(t)\| \leq \bar{\beta}_e \left(\|e(0)\|, t\right) + \bar{\gamma}_e(\phi) \tag{A.79}$$

$\forall\, t \geq 0$, where $\bar{\beta}_e$ is a class $\mathcal{KL}$ function and $\bar{\gamma}_e$ is a class $\mathcal{K}_\infty$ function. From Assumption 3.7, we have that the η states of the reduced closed-loop system possess an ISS property with respect to e:

$$\|\eta(t)\| \leq K_\eta \|\eta(0)\| e^{-at} + \bar{\gamma}_\eta \left(\|e\|^s\right) \tag{A.80}$$

uniformly in $\bar{v}$,where $\bar{\gamma}_\eta$ is a class $\mathcal{K}$ function. Realizing that the reduced closed-loop system, for which the inequalities of (A.79–A.80) hold, is identical to the system studied in the Proof of Theorem 3.13, we note that the static component of the controller of (3.49) with $\hat{x} = x$ enforces global ultimate boundedness in the reduced closed-loop system, and thus, the ζ states of the reduced closed-loop system satisfy the following bound:

$$\|\zeta(t)\| \leq \beta_\zeta(\delta_\zeta, t) + \tilde{\gamma}(\phi) \quad \forall\, t \geq 0 \tag{A.81}$$

where β_ζ is a class $\mathcal{KL}$ function, $\tilde{\gamma}$ is a class $\mathcal{K}_\infty$ function and δ_ζ is the maximum value of the vector $[h(x)\; L_f h(x)\; \cdots\; L_f^{r-1} h(x)]$ for $\|x\| \leq \delta_x$.

Based on the above bound and following the results of [149,268], we disregard estimates of $\tilde{y}$, obtained from the high-gain observer, with norm $\|\tilde{y}\| > \beta_\zeta(\delta_\zeta, 0) + d$ where $d > \tilde{\gamma}(\phi)$. Hence, we set $sat(\cdot) = \min\{1, \frac{\zeta_{max}}{\|\cdot\|}\}(\cdot)$ where ζ_{max} is the maximum value of the vector $[\zeta_1\; \zeta_2\; \cdots\; \zeta_r]$ for $\|\zeta\| \leq \beta_\zeta(\delta_\zeta, 0) + d$.

We are now in a position to apply the result of Lemma 2.18. To this end, we define the following positive real numbers, $\{\bar{\delta}_e,\, \bar{\delta}_\eta,\, \delta_\theta,\, \delta_{\bar{v}},\, \delta_\eta,\, \delta_e,\, d_e,\, d_\eta\}$, where δ_θ and $\delta_{\bar{v}}$ were specified in the statement of the theorem, $\bar{\delta}_e$ and $\bar{\delta}_\eta$ were defined in (A.58), d_e and d_η are arbitrary, $\delta_e > \bar{\beta}_e(\bar{\delta}_e, 0) + \bar{\gamma}_e(\phi) + d_e$, $\delta_\eta > K_\eta \bar{\delta}_\eta + \bar{\gamma}_\eta(\delta_e) + d_\eta$.

First, consider the singularly perturbed system comprised of the states (ζ, e_o) of the closed-loop system. This system is in standard form, possesses a globally exponentially stable fast subsystem, and its corresponding reduced system satisfies (A.81). These properties allow a direct application of the result of Lemma 2.18 with $\delta = \max\{\delta_\zeta, \delta_\theta, \delta_{\bar{v}}, \delta_\eta\}$ and $d = \phi$, to obtain the existence of a positive real number $\epsilon^\zeta(\phi)$ such that if $\epsilon \in (0, \epsilon^\zeta(\phi)]$, and $\|\zeta(0)\| \leq \delta_\zeta$, $\|\tilde{y}(0)\| \leq \delta_\zeta$, $\|\theta\|^s \leq \delta_\theta$, $\|\bar{v}\|^s \leq \delta_{\bar{v}}$, $\|\eta\| \leq \delta_\eta$, then:

$$\|\zeta(t)\| \leq \beta_\zeta(\delta_\zeta, t) + \tilde{\gamma}(\phi) + \phi \tag{A.82}$$

Let $\bar{\phi}$ be such that $\tilde{\gamma}(\bar{\phi}) + \bar{\phi} \leq d$. Consider now the singularly perturbed system comprised of the states (e, e_o) of the system of (A.40). This system is in standard form, possesses a globally exponentially stable fast subsystem, and its corresponding reduced system satisfies (A.79). Applying the result of Lemma 2.18 with $\delta = \max\{\bar{\delta}_e, \delta_\zeta, \delta_\theta, \delta_{\bar{v}}, \delta_\eta\}$ and $d = d_e = \phi$, we obtain the existence of a positive real number $\epsilon^e(\phi)$ such that if $\epsilon \in (0, \epsilon^e(\phi)]$, and $\|e(0)\| \leq \bar{\delta}_e$, $\|\tilde{y}(0)\| \leq \delta_\zeta$, $\|\theta\|^s \leq \delta_\theta$, $\|\bar{v}\|^s \leq \delta_{\bar{v}}$, $\|\eta\|^s \leq \delta_\eta$, then:

$$\|e(t)\| \leq \bar{\beta}_e\left(\|e(0)\|, t\right) + \bar{\gamma}_e(\phi) + \phi \tag{A.83}$$

Let $\tilde{\phi}$ be such that $\bar{\gamma}_e(\tilde{\phi}) + \tilde{\phi} \leq d$. Similarly, it can be shown that Lemma 2.18, with $\delta = \max\{\bar{\delta}_\eta, \delta_\zeta, \delta_\theta, \delta_{\bar{v}}, \delta_e\}$ and $d = d_{\bar{\eta}}$, can be applied to the system comprised of the states (η, e_o) of the system of (A.40). Thus, we have that there exist positive real numbers $\epsilon^\eta(\phi)$ such that if $\epsilon \in (0, \epsilon^\eta(\phi)]$ and $\|\eta(0)\| \leq \bar{\delta}_\eta$, $\|\tilde{y}(0)\| \leq \delta_\zeta$, $\|\theta\|^s \leq \delta_\theta$, $\|\bar{v}\|^s \leq \delta_{\bar{v}}$, $\|e\|^s \leq \delta_e$, then:

$$\|\eta(t)\| \leq K_\eta \|\eta(0)\| e^{-at} + \bar{\gamma}_\eta(\|e\|^s) + d_\eta \tag{A.84}$$

Finally, the inequalities of (A.79–A.84) can be manipulated using a small-gain theorem type argument, similar to the one used in the Proof of Theorem 1 in [58], to show that given the set of positive real numbers $\{\bar{\delta}_e, \bar{\delta}_\eta, \delta_\zeta, \delta_\theta, \delta_{\bar{v}}, d\}$ and with $\phi^* := \min\{\bar{\phi}, \tilde{\phi}\}$ and with $\phi \in (0, \phi^*]$, there exists $\epsilon^*(\phi) \in (0, \min\{\epsilon^\zeta(\phi), \epsilon^e(\phi), \epsilon^\eta(\phi)\}]$ such that if $\epsilon \in (0, \epsilon^*(\phi)]$, and $\|e(0)\| \leq \bar{\delta}_e$, $\|\eta(0)\| \leq \bar{\delta}_\eta$, $\|e_o(0)\| \leq \delta_{e_o}$, $\|\theta\|^s \leq \delta_\theta$, $\|\bar{v}\|^s \leq \delta_{\bar{v}}$, then $\|\zeta(t)\| \leq \beta_\zeta(\delta_\zeta, 0) + d$ and the remaining states of the closed-loop system are bounded and its output satisfies the relation of (3.50). For brevity, this argument will not be repeated here.

Part 3: In this part of the proof, we show that the output feedback controller of (3.49) is near-optimal over a finite time-interval for the class of systems considered in (3.1) and produces a finite cost which tends to the optimal cost for the reduced closed-loop system (state feedback problem) as $\epsilon \longrightarrow 0$. To this end, we adopt a similar three-step procedure to the one employed in Part 3 of the Proof of Theorem 3.20. In the first step, we show that, for the reduced system, the controller of (3.49) with $\epsilon = 0$ is optimal with respect to a cost functional of the form of (3.30). In Step 2, we use Tikhonov's theorem to establish closeness of the solutions between the reduced and full closed-loop systems on a finite time-interval. Finally, in Step 3, we combine the results of Steps 1 and 2 to show that the cost associated with the full closed-loop system tends to the optimal cost for the reduced system as $\epsilon \to 0$.

Step 1: In this step we show that the controller of (3.49) with $\epsilon = 0$ and $\hat{x} = x$ minimizes the cost functional associated with the reduced system. To this end, consider the closed-loop system of (A.40) with $\epsilon = 0$. We note that the resulting slow system is identical to the one studied in Theorem 3.13 under state feedback and satisfies the assumptions stated therein. Applying the result of this theorem, we obtain that the static component of the controller of (3.49) minimizes the cost functional:

$$\bar{J}_r = \lim_{t \to T_f} V(\bar{e}(t)) + \int_0^{T_f} \left(l(\bar{e}) + u(\bar{x})R(\bar{x})u(\bar{x})\right) dt \tag{A.85}$$

where, as before, $\bar{x} = \mathcal{X}^{-1}(\bar{e}, \bar{\eta}, \bar{v})$ refers to the solution of the reduced closed-loop system under state feedback and T_f is defined in Theorem 3.13. Following the same treatment presented in Part 3 of the Proof of Theorem 3.13, it can be readily shown that upon substitution of the static component of the controller of (3.49) in the above expression (i.e., $u = p(\bar{x})$), the minimum cost obtained is:

$$\bar{J}_r^* := \lim_{t \to T_f} V(\bar{e}(t)) + \int_0^{T_f} (l(\bar{e}) + p(\bar{x})R(\bar{x})p(\bar{x}))\, dt = V(\bar{e}(0)) \tag{A.86}$$

Step 2: Referring to the system of (A.40), it is straightforward to show that this system satisfies the conditions of Theorem 9.1 in [148]. Applying the result of this theorem to the system of (A.40), we conclude that there exists a finite time, $t_1 > 0$, and positive constants, K, L, and $\bar{\bar{\epsilon}}$, such that if $\epsilon \in (0, \bar{\bar{\epsilon}}]$ and $\|e(0)\| \leq \bar{\delta}_e$, $\|\eta(0)\| \leq \bar{\delta}_\eta$, $\|\tilde{y}(0)\| \leq \delta_\zeta$, $\|\theta\|^s \leq \delta_\theta$, $\|\bar{v}\|^s \leq \delta_{\bar{v}}$, then:

$$\|z(t, \epsilon) - \bar{z}(t)\| \leq \epsilon K[1 + t_1] exp(L t_1) \tag{A.87}$$

and the estimate:

$$z(t, \epsilon) - \bar{z}(t) = O(\epsilon) \tag{A.88}$$

holds uniformly for $t \in [0, t_1]$ where $\bar{z}(t) = [\bar{e}^T(t)\ \bar{\eta}^T(t)\ \bar{\omega}^T(t)]^T$ is the solution of the reduced (slow) problem. Furthermore, given any $t_b > 0$, there is $\epsilon_b \leq \bar{\bar{\epsilon}}$ such that the estimate:

$$e_o(t, \epsilon) = O(\epsilon) \tag{A.89}$$

holds uniformly for $t \in [t_b, t_1]$ whenever $\epsilon < \bar{\bar{\epsilon}}$.

Step 3: In this step, we exploit the closeness of solutions result obtained in Step 1 and combine it with the optimality result established in Step 2 to prove that the output feedback controller of (3.49) is near-optimal in the sense that the cost associated with the full closed-loop system approaches the optimal cost for the reduced system under state feedback, provided that the gain of the observer is sufficiently large. Note first that in order to guarantee near-optimality, we must require $t_1 \geq T_f$. This requirement guarantees that the solutions of the reduced and full closed-loop systems remain close during the time interval over which the state feedback controller (i.e. u of (3.49) with $\epsilon = 0$ and $\hat{x} = x$) is optimal. From (A.87), it is clear that t_1 can be made arbitrarily large – while maintaining the estimate of (A.88) – by selecting ϵ sufficiently small (or equivalently by selecting the gain of the observer sufficiently large). Therefore, there exists a positive constant, ϵ^{**}, such that if $\epsilon \in (0, \epsilon^{**}]$, we have $t_1 \geq T_f$. Therefore, we set ϵ^* in Theorem 3.27 as $\epsilon^* := min\{\epsilon_0, \tilde{\epsilon}, \bar{\bar{\epsilon}}, \epsilon^{**}\}$. To proceed with the proof of near-optimality, consider rewriting the cost functional of (3.30) as follows:

$$J_n(u) = \lim_{t \to T_f} V(e(t)) + \int_0^{t_b} (l(e) + uR(x)u)\, dt + \int_{t_b}^{T_f} (l(e) + uR(x)u)\, dt \tag{A.90}$$

From Step 2, we have that the following estimates hold for $t \in [t_b, t_1]$:

$$\begin{aligned} e(t) &= \bar{e}(t) + O(\epsilon) \\ x(t) &= \bar{x}(t) + O(\epsilon) \\ \hat{x}(t) &= \bar{x}(t) + O(\epsilon) \end{aligned} \tag{A.91}$$

It follows then from the continuity properties of the functions, $V(\cdot)$, $l(\cdot)$, $u(\cdot)$, and $R(\cdot)$ that for $t \in [t_b, t_1]$, and as $\epsilon \longrightarrow 0$, $V(e(t)) \longrightarrow V(\bar{e}(t))$, $l(e(t)) \longrightarrow l(\bar{e}(t))$, $p(\hat{x}(t)) \longrightarrow p(\bar{x}(t))$, $R(x(t)) \longrightarrow R(\bar{x}(t))$, and therefore:

$$\begin{aligned} &\lim_{t \to T_f} V(e(t)) + \int_{t_b}^{T_f} (l(e) + p(\hat{x})R(x)p(\hat{x}))\, dt \longrightarrow \\ &\lim_{t \to T_f} V(\bar{e}(t)) + \int_{t_b}^{T_f} (l(\bar{e}) + p(\bar{x})R(\bar{x})p(\bar{x}))\, dt \end{aligned} \tag{A.92}$$

From the boundedness of the trajectories of the closed-loop system established in Part 2, there exists a positive real number, M, such that $\|l(e) + p(\hat{x})R(x)p(\hat{x})\| \leq M$, for $\|e(0)\| \leq \bar{\delta}_e$, $\|\eta(0)\| \leq \bar{\delta}_\eta$, $\|\tilde{y}(0)\| \leq \delta_\zeta$, $\|\theta\|^s \leq \delta_\theta$, $\|\bar{v}\|^s \leq \delta_{\bar{v}}$, and with $\phi \in (0, \phi^*]$ and $\epsilon \in (0, \epsilon^*]$. Using the fact that $t_b = O(\epsilon)$, we get:

$$\int_0^{t_b} (l(e) + p(\hat{x})R(x)p(\hat{x}))\, dt \leq \int_0^{t_b} M dt \leq M\epsilon = O(\epsilon) \tag{A.93}$$

Similarly, from the stability of the reduced closed-loop system under state feedback established in Part 2 and the fact that $t_b = O(\epsilon)$, there exists a positive real number M' such that:

$$\int_0^{t_b} (l(\bar{e}) + p(\bar{x})R(\bar{x})p(\bar{x}))\, dt \leq \int_0^{t_b} M' dt \leq M'\epsilon = O(\epsilon) \tag{A.94}$$

Combining (A.92–A.94), we obtain that as $\epsilon \longrightarrow 0$:

$$\begin{aligned} J_{n,o}^* &:= \lim_{t \to T_f} V(e(t)) + \int_0^{T_f} (l(e) + p(\hat{x})R(x)p(\hat{x}))\, dt \\ &\longrightarrow \lim_{t \to T_f} V(\bar{e}(t)) + \int_0^{T_f} (l(\bar{e}) + p(\bar{x})R(\bar{x})p(\bar{x}))\, dt \end{aligned} \tag{A.95}$$

and, from (A.86), we finally have:

$$J_{n,o}^* \longrightarrow V(\bar{e}(0)) \quad as \ \epsilon \longrightarrow 0 \tag{A.96}$$

This completes the proof of the theorem. △

Proof of Theorem 3.31:
The proof of this theorem is somewhat analogous to the Proof of Theorem 3.27. We will only highlight the differences. In the first part of the proof, the global exponential stability of the fast closed-loop subsystem is established. In the second part, we focus on the reduced closed-loop system and derive ISS bounds for its states. Then, we use the result of Lemma 2.18 to establish that these ISS bounds continue to hold up to an arbitrarily small offset, for arbitrarily large initial conditions, uncertainty and rate of change of uncertainty. In the third part, the resulting ISS inequalities are studied, using a small-gain type argument to show boundedness of the closed-loop trajectories and establish the inequality of (3.66), provided that ϕ and $\bar{\epsilon}$ are sufficiently small.

Part 1: Defining the auxiliary error variables, $\hat{e}_i = L^{r-i}(y^{(i-1)} - \tilde{y}_i)$, $i = 1, \ldots, r$, the vector $e_o = [\hat{e}_1 \ \hat{e}_2 \ \cdots \ \hat{e}_r]^T$, the vector of controller parameters $\phi_c = [c_0 \ \rho \ \chi \ \phi \ \theta_b]^T$, and the parameter $\mu = \frac{1}{L}$, the system of (3.60), under the controller of (3.65), takes the form:

$$
\begin{aligned}
\mu \dot{e}_o &= \tilde{A} e_o + \mu \tilde{b} \Omega(x, \hat{x}, \theta, \phi_c, \bar{v}) \\
\dot{\omega}_1 &= \Psi_1(sat(\tilde{y}), \omega) \\
&\vdots \\
\dot{\omega}_{n-r} &= \Psi_{n-r}(sat(\tilde{y}), \omega) \\
\dot{x} &= f(x) + g(x)p(\hat{x}, \phi_c, \bar{v}) + w_1(x)\theta + Q_1(x)[z - C(x, \hat{x}, \theta, \phi_c, \bar{v})] \\
\epsilon \dot{z} &= Q_2(x)[z - C(x, \hat{x}, \theta, \phi_c, \bar{v})]
\end{aligned}
\tag{A.97}
$$

where the matrix, $\tilde{A}$, and the vector, $\tilde{b}$ were defined in (A.37), $\hat{x} = \mathcal{X}^{-1}(sat(y_d - \Delta(\mu)e_0), \omega)$ and $y_d = [y^{(0)} \ y^{(1)} \ \cdots \ y^{(r-1)}]^T$, $\Delta(\mu)$ is a diagonal matrix whose i-th diagonal element is μ^{r-i}, $\Omega(x, \hat{x}, \theta, \phi_c, \bar{v})$ is a Lipschitz function of its argument and $C(x, \hat{x}, \theta, \phi_c, \bar{v}) = -[Q_2(x)]^{-1}[f_2(x) + g_2(x)p(\hat{x}, \phi_c, \bar{v}) + w_2(x)\theta]$. Owing to the presence of the small parameters, μ and ϵ, that multiply the time-derivatives, $\dot{e}_o$ and $\dot{z}$, respectively, the system of (A.97) can be, in general, a three-time-scale one. Therefore, the results proved in [147] will be used to establish asymptotic stability of the fast dynamics. Defining $\bar{\epsilon} := \max\{\mu, \epsilon\}$, multiplying the z-subsystem of (A.97) with $\frac{\bar{\epsilon}}{\epsilon}$ and the e_o-subsystem with $\frac{\bar{\epsilon}}{\mu}$, introducing the fast time-scale, $\bar{\tau} = \frac{t}{\bar{\epsilon}}$, and setting $\bar{\epsilon} = 0$, the closed-loop fast subsystem takes the form:

$$
\begin{aligned}
\frac{de_o}{d\bar{\tau}} &= \frac{\bar{\epsilon}}{\mu} \tilde{A} e_o \\
\frac{dz}{d\bar{\tau}} &= \frac{\bar{\epsilon}}{\epsilon} Q_2(x)[z - C(x, \hat{x}, \theta, \phi_c, \bar{v})]
\end{aligned}
\tag{A.98}
$$

The above system possesses a triangular (cascaded) structure, the matrix $\tilde{A}$ is Hurwitz and the matrix $Q_2(x)$ is Hurwitz uniformly in $x \in \mathbb{R}^n$. Therefore, the system of (A.98) satisfies Assumption 3 in [147] and, since $\frac{\bar{\epsilon}}{\mu} \geq 1$, $\frac{\bar{\epsilon}}{\epsilon} \geq 1$, the same approach as in [147] can be used to show that it possesses a globally exponentially stable equilibrium manifold of the form $e_o = 0$, $z_s = C(x, \hat{x}, \theta, \phi_c, \bar{v})$, for all values of μ and ϵ.

Part 2: In this part of the proof, we initially derive ISS bounds, when $\bar{\epsilon} = 0$, for the states of the system of (A.97), in appropriately transformed coordinates, and then use the result of Lemma 2.18 to show that these bounds hold up to an arbitrarily small offset, for initial conditions, uncertainty and rate of change of uncertainty in an arbitrarily large compact set, provided that $\bar{\epsilon}$ is sufficiently small. Defining the variables, $e_z = z - C(x, \hat{x}, \theta, \phi_c, \bar{v})$, $x = \mathcal{X}^{-1}(\zeta, \eta)$, $e_i = \zeta_i - v^{(i-1)}$, $i = 1, \ldots, r$, $e = [e_1 \; e_2 \; \cdots \; e_r]^T$ and $\bar{v} = [v \; v^{(1)} \; \cdots \; v^{(r-1)}]^T$, the closed-loop system can be written as:

$$
\begin{aligned}
\mu \dot{e}_o &= \tilde{A} e_o + \mu \tilde{b} \Omega(e, \eta, \hat{x}, \theta, \phi_c, \bar{v}) \\
\dot{\omega}_1 &= \Psi_1(sat(\tilde{y}), \omega) \\
&\vdots \\
\dot{\omega}_{n-r} &= \Psi_{n-r}(sat(\tilde{y}), \omega) \\
\dot{e}_1 &= e_2 + e_z \bar{\Psi}_1(e, \eta, \bar{v}, \theta) \\
&\vdots \\
\dot{e}_r &= L_f^r h(x) - v^{(r)} + L_g L_f^{r-1} h(x) p(\hat{x}, \phi_c, \bar{v}) + L_w L_f^{r-1} h(x) \theta \\
&\quad + e_z \bar{\Psi}_r(e, \eta, \bar{v}, \theta) \\
\dot{\eta}_1 &= \Psi_1(e, \eta, \bar{v}) + e_z \bar{\Psi}_{r+1}(e, \eta, \bar{v}, \theta) \\
&\vdots \\
\dot{\eta}_{n-r} &= \Psi_{n-r}(e, \eta, \bar{v}) + e_z \bar{\Psi}_n(e, \eta, \bar{v}, \theta) \\
\epsilon \dot{z} &= Q_2(e, \eta, \bar{v}) e_z
\end{aligned}
\tag{A.99}
$$

where $\bar{\Psi}_i$, $i = 1, \ldots, n$ are Lipschitz functions of their arguments.

Step 1: Consider the system of (A.99) with $\bar{\epsilon} = \max\{\epsilon, \mu\} = 0$. In order to analyze the dynamic behavior of the resulting reduced closed-loop system, we initially need to show that, for the system of (A.99), $\omega(0) = \eta(0) + O(\bar{\epsilon})$ implies $\eta(t) = \omega(t) + O(\bar{\epsilon})$, $\forall \; t \geq 0$. To this end, consider the singularly perturbed system comprised of the states, e_o, ω, η, z, of the system of (A.99). For this system, it is straightforward to verify that it satisfies the assumptions of Theorem 1 reported in [149]. Applying the result of this theorem, we have that there exists a positive real number, ϵ_0, such that for any positive real number δ_ω satisfying $\delta_\omega \geq \max_{\|x\| \leq \delta_x} \left\{ \sum_{\nu=1}^{n-r} \|\chi_\nu(x)\| \right\}$, where $\chi_\nu(x)$, $\nu = 1, \ldots, n-$

r are the functions defined in Assumption 3.5, the states (η, ω) of this system, starting from any initial conditions that satisfy $\eta(0) = \omega(0) + O(\bar{\epsilon})$ (with $\max\{\|\eta(0)\|, \|\omega(0)\|\} \leq \delta_\omega$), if $\bar{\epsilon} \in (0, \epsilon_0]$, satisfy $\eta(t) = \omega(t) + O(\bar{\epsilon})$, $\forall\, t \geq 0$. Since $\eta(t) = \omega(t)\ \forall\, t \geq 0$, $e_o = 0$ and $e_z = 0$, when $\bar{\epsilon} = 0$, the closed-loop slow system reduces to the one studied in the Proof of Theorem 3.13 (see also the Proof of Theorem 3.27), where it was shown that the e states satisfy:

$$\|e(t)\| \leq \bar{\beta}_e\left(\|e(0)\|, t\right) + \bar{\gamma}_e(\phi) \qquad \forall\, t \geq 0 \tag{A.100}$$

where $\bar{\beta}_e$ is a class $\mathcal{KL}$ function and $\bar{\gamma}_e$ is a class $\mathcal{K}_\infty$ function. From Assumption 3.7, we have that the η states of the reduced closed-loop system possess an ISS property with respect to e:

$$\|\eta(t)\| \leq K_\eta \|\eta(0)\| e^{-at} + \bar{\gamma}_\eta\left(\|e\|^s\right) \tag{A.101}$$

uniformly in $\bar{v}$, where $\bar{\gamma}_\eta$ is a class $\mathcal{K}$ function. Realizing that the reduced closed-loop system, for which the inequalities of (A.100–A.101) hold, is identical to the system studied in the Proof of Theorem 3.13, we note that the static component of the controller of (3.65) with $\hat{x} = x$ (i.e., $p(x, \phi_c, \bar{v})$) enforces global ultimate boundedness in the reduced closed-loop system, and thus the ζ states of the reduced closed-loop system satisfy the following bound:

$$\|\zeta(t)\| \leq \beta_\zeta(\delta_\zeta, t) + \tilde{\gamma}(\phi) \quad \forall\, t \geq 0 \tag{A.102}$$

where β_ζ is a class $\mathcal{KL}$ function, $\tilde{\gamma}$ is a class $\mathcal{K}_\infty$ function and δ_ζ is the maximum value of the vector $[h(x)\ L_f h(x)\ \cdots\ L_f^{r-1} h(x)]$ for $\|x\| \leq \delta_x$. Based on the above bound and following the results of [149,268], we disregard estimates of $\tilde{y}$, obtained from the high-gain observer, with norm $\|\tilde{y}\| > \beta_\zeta(\delta_\zeta, 0) + d$ where $d > \tilde{\gamma}(\phi)$. Hence, we set $sat(\cdot) = \min\{1, \frac{\zeta_{max}}{\|\cdot\|}\}(\cdot)$ where ζ_{max} is the maximum value of the vector $[\zeta_1\ \zeta_2\ \cdots\ \zeta_r]$ for $\|\zeta\| \leq \beta_\zeta(\delta_\zeta, 0) + d$.

Step 2: We are now in a position to apply the result of Lemma 2.18. To this end, we define the following set of positive real numbers, $\{\bar{\delta}_e, \bar{\delta}_\eta, \delta_z, \delta_\theta, \delta_{\dot{\theta}}, \delta_{\bar{v}}, \delta_\eta, \delta_e, d_e, d_\eta\}$ where δ_z, δ_θ, $\delta_{\dot{\theta}}$, and $\delta_{\bar{v}}$ were specified in the statement of the theorem, $\bar{\delta}_e$ and $\bar{\delta}_\eta$ were defined in (A.58), d_e and d_η are arbitrary, $\delta_e > \bar{\beta}_e(\bar{\delta}_e, 0) + \bar{\gamma}_e(\phi) + d_e$, $\delta_\eta > K_\eta \bar{\delta}_\eta + \bar{\gamma}_\eta(\delta_e) + d_\eta$. First, consider the singularly perturbed system comprised of the states (ζ, e_o, z) of the closed-loop system. This system is in standard form, possesses a globally exponentially stable fast subsystem, and its corresponding reduced system satisfies (A.102). These properties allow the application of the result of Lemma 2.18 with $\delta = \max\{\delta_\zeta, \delta_z, \delta_\theta, \delta_{\bar{v}}, \delta_\eta\}$ and $d = \phi$, to obtain the existence of a positive real number $\epsilon^\zeta(\phi)$ such that if $\epsilon \in (0, \epsilon^\zeta(\phi)]$, and $\|\zeta(0)\| \leq \delta_\zeta$, $\|\tilde{y}(0)\| \leq \delta_\zeta$, $\|z(0)\| \leq \delta_z$, $\|\theta\|^s \leq \delta_\theta$, $\|\bar{v}\|^s \leq \delta_{\bar{v}}$, $\|\eta\|^s \leq \delta_\eta$, then:

$$\|\zeta(t)\| \leq \beta_\zeta(\delta_\zeta, t) + \tilde{\gamma}(\phi) + \phi \tag{A.103}$$

Let $\bar{\phi}$ be such that $\tilde{\gamma}(\bar{\phi}) + \bar{\phi} \leq d$. Consider now the singularly perturbed system comprised of the states (e, e_o, z) of the system of (A.99). This system is in standard form, possesses a globally exponentially stable fast subsystem, and its corresponding reduced system satisfies (A.100). These properties allow the application of the result of Lemma 2.18 with $\delta = \max\{\bar{\delta}_e, \delta_\zeta, \delta_z, \delta_\theta, \delta_{\bar{v}}, \delta_\eta\}$ and $d = d_e = \phi$, to obtain the existence of a positive real number $\epsilon^e(\phi)$ such that if $\epsilon \in (0, \epsilon^e(\phi)]$, and $\|e(0)\| \leq \bar{\delta}_e$, $\|\tilde{y}(0)\| \leq \delta_\zeta$, $\|z(0)\| \leq \delta_z$, $\|\theta\|^s \leq \delta_\theta$, $\|\bar{v}\|^s \leq \delta_{\bar{v}}$, $\|\eta\|^s \leq \delta_\eta$, then:

$$\|e(t)\| \leq \bar{\beta}_e\left(\|e(0)\|, t\right) + \bar{\gamma}_e(\phi) + \phi \tag{A.104}$$

Let $\tilde{\phi}$ be such that $\bar{\gamma}_e(\tilde{\phi}) + \tilde{\phi} \leq d$. Similarly, it can be shown that Lemma 2.18, with $\delta = \max\{\bar{\delta}_\eta, \delta_\zeta, \delta_z, \delta_\theta, \delta_{\dot{\theta}}, \delta_{\bar{v}}, \delta_e\}$ and $d = d_{\bar{\eta}}$, can be applied to the system comprised of the states (η, e_o, z) of the system of (A.99). Thus, we have that there exist positive real numbers $\epsilon^\eta(\phi)$ such that if $\epsilon \in (0, \epsilon^\eta(\phi)]$ and $\|\eta(0)\| \leq \bar{\delta}_\eta$, $\|\tilde{y}(0)\| \leq \delta_\zeta$, $\|z(0)\| \leq \delta_z$, $\|\theta\|^s \leq \delta_\theta$, $\|\dot{\theta}\| \leq \delta_{\dot{\theta}}$, $\|\bar{v}\|^s \leq \delta_{\bar{v}}$, $\|e\|^s \leq \delta_e$, then:

$$\|\eta(t)\| \leq K_\eta\|\eta(0)\|e^{-at} + \bar{\gamma}_\eta(\|e\|^s) + d_\eta \tag{A.105}$$

<u>*Part 3*</u>: Finally, the inequalities of (A.100–A.105) can be manipulated using a small-gain theorem type argument similar to the one used in the Proof of Theorem 1 in [58] to show that the states of the closed-loop systems are bounded and its output satisfies the relation of (3.66). For brevity, this argument will not be repeated here. This completes the proof of the theorem. △

B

Proofs of Chapter 4

Proof of Theorem 4.1:
Consider the representation of the closed-loop system in terms of the transformed coordinates (e, η) introduced in (4.5–4.6):

$$\begin{aligned}
\dot{e} &= \bar{f}(e, \eta, \bar{v}) - \sum_{i=1}^{m} \bar{g}_i(e, \eta, \bar{v}) k(x, u_{max}, \theta_b, \rho, \chi, \phi) L_{\bar{g}_i} V + \sum_{k=1}^{q} \bar{w}_k(e, \eta, \bar{v}) \theta_k \\
\dot{\eta} &= \Psi(e, \eta, \bar{v}, \theta) \\
y_i &= e_1^{(i)} + v_i, \quad i = 1, \ldots, m
\end{aligned} \tag{B.1}$$

where $\Psi(\cdot) = [\Psi_1(\cdot) \ \cdots \ \Psi_{n-\sum r_i}(\cdot)]^T$. We now follow a four-step procedure to establish both asymptotic stability and reference-input tracking in the closed-loop system of (B.1). Initially, we show that the controller of (4.12–4.14) satisfies the constraints within the region described by the set $\Pi(\theta_b, u_{max})$ in (4.15). Then, using a Lyapunov argument we show that, starting from any initial condition that belongs to any invariant subset of Π, Ω, the state feedback controller of (4.12–4.14) asymptotically stabilizes the closed-loop e-subsystem and derive bounds that capture the evolution of the states of the e and η subsystems. Next, a small gain argument is invoked to show that the trajectories of the e-η interconnected closed-loop system remain bounded for all times. Finally, we show that the states of the full closed-loop system of (B.1) converge to the origin and that the outputs satisfy the relation of (4.16).

Step 1: To prove that the control law of (4.12–4.14) satisfies the constraints within the region described by the set $\Pi(\theta_b, u_{max})$, we have from (4.12–4.14) that:

$$\|u(x)\| \leq \|k(x, u_{max}, \theta_b, \rho, \chi, \phi)\| \, \|(L_G V)^T\|$$

$$\leq \frac{\left\| L_{\bar{f}}^* V + \sqrt{\left(L_{\bar{f}}^{**} V\right)^2 + (u_{max} \|(L_G V)^T\|)^4} \right\|}{\|(L_G V)^T\| \left[1 + \sqrt{1 + (u_{max} \|(L_G V)^T\|)^2}\right]} \tag{B.2}$$

From the definitions of $L_{\bar{f}}^* V$ and $L_{\bar{f}}^{**} V$ in (4.14) and the fact that $\rho > 0$, it is clear that if $L_{\bar{f}}^{**} V \leq u_{max} \|(L_G V)^T\|$, then we also have $L_{\bar{f}}^* V \leq u_{max} \|(L_G V)^T\|$. Therefore, for any $x \in \Pi$, the following estimates hold:

$$\begin{aligned} \left(L_{\bar{f}}^{**} V\right)^2 &\leq \left(u_{max} \|(L_G V)^T\|\right)^2 \\ L_{\bar{f}}^* V &\leq u_{max} \|(L_G V)^T\| \end{aligned} \tag{B.3}$$

Substituting the above estimates into (B.2) yields:

$$\|u(x)\| \leq \frac{u_{max} \|(L_G V)^T\| \left[1 + \sqrt{1 + (u_{max} \|(L_G V)^T\|)^2}\right]}{\|(L_G V)^T\| \left[1 + \sqrt{1 + (u_{max} \|(L_G V)^T\|)^2}\right]} = u_{max} \tag{B.4}$$

Step 2: Consider the smooth, positive-definite function, $V : \mathbb{R}^{\sum_i r_i} \rightarrow R_{\geq 0}$, $V = e^T P e$ as a Lyapunov function candidate for the e-subsystem of (B.1). Computing the time-derivative of V along the trajectories of the closed-loop e-subsystem, we get:

$$\begin{aligned} \dot{V} &= L_{\bar{f}} V + L_G V u + \sum_{k=1}^{q} L_{\bar{w}_k} V \theta_k \\ &= L_{\bar{f}} V - \left(\frac{L_{\bar{f}}^* V + \sqrt{\left(L_{\bar{f}}^{**} V\right)^2 + (u_{max} \|(L_G V)^T\|)^4}}{\left[1 + \sqrt{1 + (u_{max} \|(L_G V)^T\|)^2}\right]} \right) + \sum_{k=1}^{q} L_{\bar{w}_k} V \theta_k \\ &\leq L_{\bar{f}} V + \chi \sum_{k=1}^{q} \|L_{\bar{w}_k} V\| \theta_{bk} - \frac{L_{\bar{f}}^* V + \sqrt{\left(L_{\bar{f}}^{**} V\right)^2 + (u_{max} \|(L_G V)^T\|)^4}}{\left[1 + \sqrt{1 + (u_{max} \|(L_G V)^T\|)^2}\right]} \end{aligned} \tag{B.5}$$

After performing some algebraic manipulations, the above inequality can be re-written as:

$$\dot{V} \leq \alpha(e,\eta,\bar{v}) + \left(\frac{\sum_{k=1}^{q} \theta_{bk} \left\| L_{\bar{w}_k} V \right\| \left(\frac{\phi - (\chi - 1) \left\| 2Pe \right\|}{\left\| 2Pe \right\| + \phi} \right) - \rho \left(\frac{\left\| 2Pe \right\|^2}{\left\| 2Pe \right\| + \phi} \right)}{\left[1 + \sqrt{1 + (u_{max} \|(L_G V)^T\|)^2} \right]} \right) \tag{B.6}$$

where

$$\alpha(e,\eta,\bar{v}) = \frac{\left(L_{\bar{f}} V + \chi \sum_{k=1}^{q} \theta_{kb} \left\| L_{\bar{w}_k} V \right\| \right) \sqrt{1 + (u_{max} \|(L_G V)^T\|)^2}}{\left[1 + \sqrt{1 + (u_{max} \|(L_G V)^T\|)^2} \right]} - \frac{\sqrt{\left(L_{\bar{f}}^{**} V \right)^2 + (u_{max} \|(L_G V)^T\|)^4}}{\left[1 + \sqrt{1 + (u_{max} \|(L_G V)^T\|)^2} \right]} \tag{B.7}$$

To analyze the sign of $\dot{V}$ in (B.6), we will initially study the sign of the term, $\alpha(e,\eta,\bar{v})$, on the right-hand side. It is clear that the sign of this term depends on the sign of the term $L_{\bar{f}} V + \chi \sum_{k=1}^{q} \theta_{kb} \left\| L_{\bar{w}_k} V \right\|$. To this end, we consider the following two cases:

Case 1: $L_{\bar{f}}^{**} V \leq 0$

Since $L_{\bar{f}}^{**} V = L_{\bar{f}} V + \rho \left\| 2Pe \right\| + \chi \sum_{k=1}^{q} \|L_{\bar{w}_k} V\| \theta_{bk}$ and ρ is a positive real number, the fact that $L_{\bar{f}}^{**} V \leq 0$ implies that $L_{\bar{f}} V + \chi \sum_{k=1}^{q} \left\| L_{\bar{w}_k} V \right\| \theta_{bk} \leq 0$. As a result, we have that $\alpha(e,\eta,\bar{v}) \leq 0$ and the time-derivative of V in this case satisfies the following bound:

$$\dot{V} \leq \frac{\sum_{k=1}^{q} \theta_{bk} \left\| L_{\bar{w}_k} V \right\| \left(\frac{\phi - (\chi - 1) \left\| 2Pe \right\|}{\left\| 2Pe \right\| + \phi} \right) - \rho \left(\frac{\left\| 2Pe \right\|^2}{\left\| 2Pe \right\| + \phi} \right)}{\left[1 + \sqrt{1 + (u_{max} \|(L_G V)^T\|)^2} \right]} := \beta(e,\eta,\bar{v}) \tag{B.8}$$

Case 2: $0 < L_{\bar{f}}^{**} V \leq u_{max} \left\| (L_G V)^T \right\|$

In this case, we have:

$$\left(L_{\bar{f}}^{**} V \right)^2 \leq \left(u_{max} \|(L_G V)^T\| \right)^2 \tag{B.9}$$

and therefore:

$$
\begin{aligned}
& -\sqrt{\left(L_{\bar{f}}^{**}V\right)^2 + (u_{max}\|(L_GV)^T\|)^4} \\
&= -\sqrt{\left(L_{\bar{f}}^{**}V\right)^2 + (u_{max}\|(L_GV)^T\|)^2(u_{max}\|(L_GV)^T\|)^2} \qquad \text{(B.10)} \\
&\leq -\left(L_{\bar{f}}^{**}V\right)\sqrt{1+(u_{max}\|(L_GV)^T\|)^2}
\end{aligned}
$$

Substituting the estimate of (B.10) into the expression for $\dot{V}$ in (B.6–B.7) yields:

$$
\begin{aligned}
\dot{V} &= \frac{-\rho\|2Pe\|\sqrt{1+(u_{max}\|(L_GV)^T\|)^2}}{\left[1+\sqrt{1+(u_{max}\|(L_GV)^T\|)^2}\right]} + \beta(e,\eta,\bar{v}) \qquad \text{(B.11)} \\
&\leq \beta(e,\eta,\bar{v})
\end{aligned}
$$

From the above analysis, it is clear that whenever $L_{\bar{f}}^{**}V \leq u_{max}\|(L_GV)^T\|$, the inequality of (B.8) holds. To guarantee that the state of the closed-loop system satisfies $L_{\bar{f}}^{**}V \leq u_{max}\|(L_GV)^T\|$ for all time, we confine the initial conditions within an invariant subset, Ω, embedded within the region described by (4.15). Therefore, starting from any $x(0) \in \Omega$, the inequality of (B.8) holds. Referring to this inequality, note that since $\chi > 1$ and $\rho > 0$, it is clear that, whenever $\|2Pe\| > \dfrac{\phi}{\chi - 1}$, the first term on the right-hand side is strictly negative, and therefore $\dot{V}$ satisfies:

$$
\begin{aligned}
\dot{V} &\leq -\frac{\rho\|2Pe\|^2}{(\|2Pe\|+\phi)\left[1+\sqrt{1+(u_{max}\|(L_GV)^T\|)^2}\right]} \\
&\leq -\frac{k_1\|e\|^2}{(\|2Pe\|+\phi)\left[1+\sqrt{1+(u_{max}\|(L_GV)^T\|)^2}\right]} \qquad \text{(B.12)}
\end{aligned}
$$

where $k_1 = 2\rho\lambda_{min}(P^2) > 0$. To study the behavior of $\dot{V}$ when $\|2Pe\| \leq \phi/(\chi-1)$, we first note that since the functions $\bar{w}_k(e,\eta,\bar{v})$ are smooth and vanish when $e = 0$, then there exists positive real constants ϕ_1^*, δ_k, $k = 1,\ldots,q$, such that if $\phi \leq \phi_1^*$, the bound $\|\bar{w}_k(e,\eta,\bar{v})\| \leq \delta_k\|e\|$, holds for $\|2Pe\| \leq \phi/(\chi-1)$. Using this bound, we obtain the following estimates:

$$\begin{aligned}\sum_{k=1}^{q}\theta_{bk}\left\|L_{\bar{w}_k}V\right\|(\phi-(\chi-1)\left\|2Pe\right\|) &\leq \sum_{k=1}^{q}\theta_{bk}\left\|L_{\bar{w}_k}V\right\|\phi \\ &= \sum_{k=1}^{q}\theta_{bk}\left\|\bar{w}_k\right\|\left\|2Pe\right\|\phi \\ &\leq k_2\phi\sum_{k=1}^{q}\theta_{bk}\delta_k\|e\|^2 \quad \forall\ \|2Pe\| \leq \frac{\phi}{\chi-1}\end{aligned} \tag{B.13}$$

where $k_2 = 2\sqrt{\lambda_{max}(P^2)} > 0$. Substituting the estimate of (B.13) directly into (B.8), we get:

$$\dot{V} \leq \left(\frac{\left(k_2\phi\sum_{k=1}^{q}\theta_{bk}\delta_k - 2k_1\right)\|e\|^2}{(\|2Pe\|+\phi)\left[1+\sqrt{1+(u_{max}\|(L_GV)^T\|)^2}\right]}\right) \quad \forall\ \|2Pe\| \leq \frac{\phi}{\chi-1} \tag{B.14}$$

If ϕ is sufficiently small to satisfy the bound $\phi \leq \dfrac{k_1}{k_2\sum_{k=1}^{q}\delta_k\theta_{bk}} := \phi_2^*$, then it is clear from (B.12–B.14) that the last inequality in (B.12) is satisfied, irrespective of the value of $\|2Pe\|$. In summary, we have that for any initial condition in the invariant set Ω (where (4.15) holds $\forall\ t \geq 0$), there exists $\phi^* := \min\{\phi_1^*, \phi_2^*\}$ such that if $\phi \leq \phi^*$, $\dot{V}$ satisfies:

$$\dot{V} \leq \frac{-k_1\|e\|^2}{(\|2Pe\|+\phi)\left[1+\sqrt{1+(u_{max}\|(L_GV)^T\|)^2}\right]} < 0 \quad \forall\, e \neq 0 \tag{B.15}$$

Consequently, there exists a function β_e of class $\mathcal{KL}$ (see [148] for details) such that the following ISS inequality holds for the e states of the system of (B.1):

$$\|e(t)\| \leq \beta_e\left(\|e(0)\|, t\right) \qquad \forall\, t \geq 0 \tag{B.16}$$

and the origin of the e-subsystem is asymptotically stable. From Assumption 4.2, we have that the η-subsystem of (B.1) possesses an ISS property with respect to e which implies that there exists a function, β_η, of class $\mathcal{KL}$ and a function, γ_η, of class $\mathcal{K}$ such that the following ISS inequality holds:

$$\|\eta(t)\| \leq \beta_\eta(\|\eta(0)\|, t) + \gamma_\eta(\|e\|^s) \qquad \forall\, t \geq 0 \tag{B.17}$$

uniformly in θ, $\bar{v}$. Using the inequalities of (B.16–B.17), it can be shown by means of a small gain argument, similar to that used in [57,58], that the origin of the full closed-loop system is asymptotically stable for all initial conditions inside the invariant set Ω. The asymptotic output tracking result can finally be obtained by simply taking the limsup of both sides of (B.16) which yields:

$$\limsup_{t\to\infty} \|e(t)\| = 0 \tag{B.18}$$

and, hence:

$$\limsup_{t\to\infty} \|e_1^{(i)}(t)\| = \limsup_{t\to\infty} \|y_i(t) - v_i(t)\| = 0, \quad i = 1, \ldots, m \tag{B.19}$$

This completes the proof of the theorem. △

Proof of Theorem 4.17:

The proof of this theorem consists of three parts. In the first part, we use a singular perturbation formulation to represent the closed-loop system and show that the resulting fast subsystem is globally exponentially stable. In the second part, we focus on the closed-loop reduced (slow) system and derive bounds for its states. Then, in the third part, we use a technical lemma proved in [50], to establish that these bounds continue to hold up to an arbitrarily small offset, for arbitrarily large compact subsets of the stability region obtained under state feedback. The resulting bounds are then analyzed to establish asymptotic and local exponential stability of the full closed-loop system, which is then used to establish (4.25), provided that ϕ and $\bar{\epsilon}$ are sufficiently small.

Part 1: Defining the auxiliary error variables $\hat{e}_j^{(i)} = L_i^{r_i - j}(y_i^{(j-1)} - \tilde{y}_j^{(i)})$, $j = 1, \ldots, r_i$, the vectors $e_o^{(i)} = [\hat{e}_1^{(i)} \; \hat{e}_2^{(i)} \; \cdots \; \hat{e}_{r_i}^{(i)}]^T$, $e_o = [e_o^{(1)^T} \; e_o^{(2)^T} \; \cdots \; e_o^{(m)^T}]^T$, the parameters $\epsilon_i = \dfrac{1}{L_i}$, the matrices $\tilde{A}_i$ and the vector $\tilde{b}$:

$$\tilde{A}_i = \begin{bmatrix} -a_1^{(i)} & 1\,0\cdots 0 \\ -a_2^{(i)} & 0\,1\cdots 0 \\ \vdots & \vdots\;\vdots\;\ddots\;\vdots \\ -a_{r_i-1}^{(i)} & 0\,0\cdots 1 \\ -a_{r_i}^{(i)} & 0\,0\cdots 0 \end{bmatrix}, \quad \tilde{b} = \begin{bmatrix} 0 \\ 0 \\ \vdots \\ 0 \\ 1 \end{bmatrix} \tag{B.20}$$

where $i = 1, \ldots, m$, the system of (4.1) under the controller of (4.24) takes the following form:

$$\begin{aligned} \epsilon_i \dot{e}_o^{(i)} &= \tilde{A}_i e_o^{(i)} + \epsilon_i \tilde{b} \psi_i(x, \hat{x}, \theta, \rho, \chi, \phi, \bar{v}), \quad i = 1, \ldots, m \\ \dot{\omega} &= \Psi(sat(\tilde{y}), \omega) \\ \dot{x} &= f(x) - \sum_{i=1}^{m} g_i(x) k(\hat{x}, u_{max}, \theta_b, \rho, \chi, \phi) \widehat{L_{\bar{g}_i} V} + \sum_{k=1}^{q} w_k(x)\theta_k(t)) \end{aligned} \tag{B.21}$$

where $\psi_i(\cdot)$ is a Lipschitz function of its argument. Owing to the presence of the small parameters ϵ_i that multiply the time-derivatives, $\dot{e}_o^{(i)}$, the system of (B.21) can be, in general, a multiple-time-scale system. Therefore, the results proved in [147] will be used to establish exponential stability of the fast dynamics. Defining $\bar{\epsilon} = \max\{\epsilon_i\}$, $i = 1, \ldots, m$, multiplying each $e_o^{(i)}$-subsystem by $\bar{\epsilon}/\epsilon$, introducing the fast time-scale $\bar{\tau} = \frac{t}{\bar{\epsilon}}$ and setting $\bar{\epsilon} = 0$, the closed-loop fast subsystem takes the form:

$$\frac{de_o}{d\bar{\tau}} = \frac{\bar{\epsilon}}{\epsilon}\tilde{A}_i e_o, \ i = 1, \ldots, m \tag{B.22}$$

The above system possesses a triangular (cascaded) structure, with the constant matrices $\tilde{A}_i$ being Hurwitz. Therefore, it satisfies Assumption 3 in [147] and, since $\bar{\epsilon}/\epsilon \geq 1$, the same approach, as in [147], can be used to deduce that the origin of this system is globally exponentially stable, i.e., there exists real numbers, $k_3 \geq 1$, $a_3 > 0$ such that:

$$\|e_o(\bar{\tau})\| \leq k_3 \|e_o(0)\| e^{-a_3\bar{\tau}} \ \ \forall\, \bar{\tau} \geq 0 \tag{B.23}$$

Part 2: Using the coordinate change of (4.5) and (4.23), together with the notation introduced thereafter, we can re-write the closed-loop system of (B.21) in the following form:

$$\begin{aligned}
\epsilon_i \dot{e}_o &= \tilde{A}_i e_o + \epsilon_i \tilde{b}\psi_i(e, \eta, \hat{x}, \theta, \chi, \rho, \phi, \bar{v}) \\
\dot{\omega} &= \Psi(sat(\tilde{y}), \omega) \\
\dot{e} &= \bar{f}(e, \eta, \bar{v}) - \sum_{i=1}^{m} \bar{g}_i(e, \eta, \bar{v}) k(\hat{x}, u_{max}, \theta_b, \rho, \chi, \phi) \widehat{L_{\bar{g}_i} V} + \sum_{k=1}^{q} \bar{w}_k(e, \eta, \bar{v})\theta_k \\
\dot{\eta} &= \Psi(e, \eta, \bar{v}) \\
y_i &= e_1^{(i)} + v_i, \ \ i = 1, \ldots, m
\end{aligned} \tag{B.24}$$

Consider the above system with $\bar{\epsilon} = \max\{\epsilon_i\} = 0$. Using the result of Theorem 1 in [149], it can be shown that $\eta(0) = \omega(0) + O(\bar{\epsilon})$ implies $\eta(t) = \omega(t) + O(\bar{\epsilon})$, $\forall\, t \geq 0$ and therefore $\eta(t) = \omega(t)\ \forall\, t \geq 0$, when $\bar{\epsilon} = 0$. The closed-loop reduced (slow) system of (B.24) therefore reduces to the one studied in the Proof of Theorem 4.1 (see (B.1)) under state feedback, where we have already shown that given any initial condition such that $\|x(0)\| \leq \delta_s$, there exists $\phi^* > 0$ such that if $\phi \leq \phi^*$ and $\|x(0)\| \leq \delta_s$, the origin of the closed-loop system is asymptotically stable. Consequently, the denominator expression in (B.15) is bounded and there exist real numbers, $k_e > 0$, $k_1 \geq 1$, $a_1 > 0$ such that if $\|e(0)\| \leq \delta_e$, $\|\eta(0)\| \leq \delta_\eta$, $\|\bar{v}\|^s \leq \delta_{\bar{v}}$, where $\delta_s = T^{-1}(\delta_e, \delta_\eta, \delta_{\bar{v}})$, we have that $\dot{V}$ satisfies $\dot{V} \ \leq \ -k_e\|e\|^2$ and the e states of the closed-loop system satisfy:

$$\|e(t)\| \leq k_1 \|e(0)\| e^{-a_1 t} \ \forall\, t \geq 0 \tag{B.25}$$

which shows that the origin of the e-subsystem is exponentially stable. From this result and the fact that the η-subsystem (with $e = 0$) is locally exponentially stable, we have that the e-η interconnected closed-loop reduced (slow) system is also locally exponentially stable (see [148] for details). Therefore, given the set of positive real numbers, $\{\delta_e, \delta_\eta, \delta_{\bar{v}}\}$, there exists $b > 0$ such that if $\|e(0)\| \leq \delta_e$, $\|\eta(0)\| \leq \delta_\eta$, $\|\bar{v}\|^s \leq \delta_{\bar{v}}$, the following bound holds:

$$\|\pi(t)\| \leq k_2 \|\pi(t_0)\| e^{-a_2(t-t_0)} \ \forall \ t \geq t_0 \tag{B.26}$$

for all $\|\pi(t_0)\| \leq b$, where $\pi(t) = [e^T(t) \ \ \eta^T(t)]^T$, for some $k_2 \geq 1$, $a_2 > 0$. Note that since the static component of the controller of (4.24) with $\hat{x} = x$ enforces asymptotic stability in the closed-loop slow system, the state, ζ, of the closed-loop slow system satisfies a bound of the following form $\forall\, t \geq 0$:

$$\|\zeta(t)\| \leq \beta_\zeta(\delta_\zeta, t) \tag{B.27}$$

where β_ζ is a class $\mathcal{KL}$ function and δ_ζ is the maximum value of the norm of the vector $[l_1^T(x) \ l_2^T(x) \ \cdots \ l_m^T(x)]^T$ for $\|x\| \leq \delta_b$, where $l_i(x) = [h_i(x) \ L_f h_i(x) \ \cdots L_f^{r_i-1} h_i(x)]^T$. Based on the above bound and following the results of [149, 268], we disregard estimates of $\tilde{y}$, obtained from the high-gain observer, with norm $\|\tilde{y}\| > \beta_\zeta(\delta_\zeta, 0)$. Hence, we set $sat(\cdot) = \min\{1, \frac{\zeta_{max}}{\|\cdot\|}\}(\cdot)$ where ζ_{max} is the maximum value of the vector $[\zeta_1 \ \zeta_2 \ \ \cdots \ \zeta_r]$ for $\|\zeta\| \leq \beta_\zeta(\delta_\zeta, 0)$.

Part 3: Having analyzed the stability properties of both the fast and slow closed-loop systems in Parts 1 and 2, respectively, it can be shown, with the aid of calculations similar to those performed in [57, 79], that the inequalities of (B.23) (derived for the fast system), and (B.17–B.25) (derived for the slow system) continue to hold, for the states of the full closed-loop system, up to an arbitrarily small offset, d, for initial conditions in large compact subsets ($\Omega_b \subset \Omega$) where $\Omega_b := \{x \in \mathbb{R}^m : \ \|x\| \leq \delta_b\}$ and $\beta(\delta_b, 0) + d \leq \delta_s$, provided that the singular perturbation parameter, $\bar{\epsilon}$, is sufficiently small. The requirement that $\beta(\delta_b, 0) + d \leq \delta_s$ guarantees that during the initial boundary layer (when the fast states have not decayed yet), the slow states of the closed-loop system remain within the invariant region Ω. Therefore, given the pair (δ_b, d), the set $(\delta_\theta, \delta_{\bar{v}}, \delta_\zeta)$, and with $\phi \in (0, \phi^*]$, there exists $\bar{\epsilon}^{(1)} > 0$ such that if $\bar{\epsilon} \in (0, \bar{\epsilon}^{(1)}]$, $\|x(0)\| \leq \delta_b$, $\|\tilde{y}\| \leq \delta_\zeta$, $\|\theta_k\|^s \leq \delta_\theta$, $\|\bar{v}\|^s \leq \delta_{\bar{v}}$, then, for all $t \geq 0$, the states of the closed-loop singularly perturbed system satisfy:

$$\begin{aligned} \|\pi(t)\| &\leq k_2 \|\pi(0)\| e^{-a_2 t} + d \\ \|e_o(t)\| &\leq k_3 \|e_o(0)\| e^{-a_3 \frac{t}{\epsilon}} + d \end{aligned} \tag{B.28}$$

The above inequalities imply that the trajectories of the closed-loop singularly perturbed system will be bounded. Furthermore, as t increases, they will be ultimately bounded with an ultimate bound that depends on d. Since d is arbitrary, we can choose it small enough such that after a sufficiently large time, say $\tilde{t}$, the trajectories of the closed-loop system are confined within a small compact neighborhood of the origin of the closed-loop system. Obviously, $\tilde{t}$ depends on both the initial conditions and the desired size of the neighborhood, but is independent of $\bar{\epsilon}$. For reasons that will become obvious shortly, we choose $d = b/2$ and let $\tilde{t}$ be the smallest time such that $\max\{k_2\|\pi(0)\|e^{-a_2\tilde{t}}, k_3\|e_o(0)\|e^{-a_3\tilde{t}/\epsilon}\} \leq d$. Then it can be easily verified that:

$$\|\pi(t)\| \leq b, \;\; \|e_o(t)\| \leq b \;\; \forall\, t \geq \tilde{t} \tag{B.29}$$

Recall from (B.23) and (B.26) that both the fast and slow subsystems are exponentially stable within the ball of (B.29). Then, a direct application of the result of Theorem 9.3 in [148] can be performed to show that there exists $\bar{\epsilon}^{(2)}$ such that if $\bar{\epsilon} \leq \bar{\epsilon}^{(2)}$, the singularly perturbed closed-loop system is locally exponentially stable and, therefore, once inside the ball of (B.29), the closed-loop trajectories converge to the origin as $t \to \infty$.

To summarize, we have that given the pair of positive real numbers (δ_b, d) such that $\beta(\delta_b, 0) + d \leq \delta_s$, given the set of positive real numbers $(\delta_\theta,\, \delta_{\bar{v}},\, \delta_\zeta)$, and with $\phi \in (0.\phi^*]$, there exists $\bar{\epsilon}^* := \min\{\bar{\epsilon}^{(1)}, \bar{\epsilon}^{(2)}\}$ such that if $|x(0)| \leq \delta_b$, $|\tilde{y}(0)| \leq \delta_\zeta$, $\|\theta\| \leq \delta_\theta$, $\|\bar{v}\| \leq \delta_{\bar{v}}$, and $\bar{\epsilon} \in (0, \bar{\epsilon}^*]$, the closed-loop trajectories are bounded and converge to the origin as time tends to infinity, i.e., the closed-loop system is asymptotically stable. The asymptotic output tracking result can then be established by noting that from (B.26) we have:

$$\limsup_{t\to\infty} \|e(t)\| = 0 \tag{B.30}$$

and therefore:

$$\limsup_{t\to\infty} \|e_1^{(i)}(t)\| = \limsup_{t\to\infty} \|y_i(t) - v_i(t)\| = 0, \quad i = 1, \ldots, m \tag{B.31}$$

This completes the proof of the theorem. △

C

Proofs of Chapter 5

Proof of Theorem 5.7:
To prove this theorem, we proceed in two steps. In the first step we show that, starting from any initial condition within the set $\Omega(u_{max})$, the bounded control law of (5.7–5.8) asymptotically stabilizes the constrained closed-loop. Then, in the second step, we show that switching between the model predictive controller of (5.3–5.5) and the bounded controller, according to the rule of (5.13–5.14), guarantees asymptotic stability in the switched closed-loop system, starting from any initial condition that belongs to $\Omega(u_{max})$.
Step 1: Consider first the system of (5.1), under the control law of (5.7–5.8). Evaluating the time-derivative of the Lyapunov function along the closed-loop trajectories, we have:

$$\begin{aligned}
\dot{V} &= L_fV + L_gVu \\
&= L_fV - L_gV\frac{L_f^*V + \sqrt{(L_f^*V)^2 + (u_{max}\|(L_gV)^T\|)^4}}{\|(L_gV)^T\|^2\left[1 + \sqrt{1 + (u_{max}\|(L_gV)^T\|)^2}\right]}(L_gV)^T \\
&= \frac{-\rho x'Px + L_fV\sqrt{1+(u_{max}\|(L_gV)^T\|)^2} - \sqrt{(L_f^*V)^2 + (u_{max}\|(L_gV)^T\|)^4}}{\left[1+\sqrt{1+ (u_{max}\|(L_gV)^T\|)^2}\right]}
\end{aligned} \tag{C.1}$$

From the above equation, and the fact that $L_f^*V < 0 \Longrightarrow L_fV < 0$ (since $\rho > 0$), we have that V satisfies:

$$\dot{V} \leq \frac{-\rho x'Px}{\left[1 + \sqrt{1 + (u_{max}\|(L_gV)^T\|)^2}\right]} < 0 \quad \forall\, x \neq 0 \tag{C.2}$$

Furthermore, whenever $0 < L_f^*V \leq u_{max}\|(L_gV)^T\|$, we have:

$$\sqrt{(L_f^*V)^2 + (u_{max}\|(L_gV)^T\|)^4} \geq L_fV\sqrt{1 + (u_{max}\|(L_gV)^T\|)^2}$$

Substituting the above estimate into (C.1), we have that $\dot{V}$ satisfies (C.2). To summarize, we see that whenever $L_f^*V \leq u_{max}\|(L_gV)^T\|$, we have $\dot{V} < 0$. Since $\Omega(u_{max})$ is taken to be the largest invariant set where this inequality holds for all $x \neq 0$, then starting from any initial state $x(0) \in \Omega(u_{max})$, the inequality of (5.9) holds for all times and consequently:

$$\dot{V} < 0 \;\; \forall \; x \; \in \; \Omega(u_{max})\backslash\{0\} \tag{C.3}$$

which implies that the origin of the closed-loop system, under the control law of (5.7–5.8) is asymptotically stable.
Step 2: Consider the switched system of (5.11), subject to the switching rule of (5.13–5.14), with any initial state $x(0) \in \Omega(u_{max})$. From the definition of T_s given in Theorem 5.7, it is clear that if T_s is a finite number, then $\dot{V}(x^M(t)) < 0 \;\; \forall\, 0 \leq t < T$, where the notation $x^M(t)$ denotes the closed-loop state under MPC at time t, which implies that $x(t) \in \Omega(u_{max}) \;\; \forall\, 0 \leq t < T_s$ (or that $x(T^-) \in \Omega(u_{max})$). This fact, together with the continuity of the solution of the switched system, $x(t)$, (which follows from the fact that the right hand side of (5.11) is continuous in x and piecewise continuous in time (only finite number of switches is allowed over any finite time interval)) implies that, upon switching (instantaneously) to the bounded controller at $t = T_s$, we have $x(T_s) \in \Omega(u_{max})$ and $u(t) = b(x(t))$ for all $t \geq T_s$. Therefore, from our analysis in Step 1 we conclude that $\dot{V}(x^b(t)) < 0 \;\; \forall\, t \geq T_s$. In summary, the switching rule of (5.13–5.14) guarantees that, starting from any $x(0) \in \Omega(u_{max})$, $\dot{V}(x(t)) < 0 \;\; \forall \; 0 \neq x \in \Omega(u_{max}), \;\; \forall \; t \geq 0$. which implies that the origin of the switched closed-loop system is asymptotically stable. Note that if no such T_s exists, then we simply have $\dot{V}(x^M(t)) < 0 \;\;\; \forall \;\; t \geq 0$ and the origin of the closed-loop system is, again, asymptotically stable. This completes the proof of the theorem. △

Proof of Theorem 5.14:
To prove this theorem, we first note that if neither T_b nor T_N exists, then no switching takes place and we simply have $\dot{V}(x^M(t)) < 0 \;\;\; \forall \;\; t \geq 0$ implying that the origin of the closed-loop system is asymptotically stable under the model predictive controller of (5.3–5.5). Therefore, in what follows we consider only the case when at least one of these numbers exists. From the definition of T_N given in Theorem 5.14, it is clear that (if this number exists) the Lyapunov function under the model predictive controller, $V(x^M(t))$, changes sign no more than N times over the time interval $[0, T_N)$. Since the set $\Omega(u_{max})$ is not necessarily invariant under the control law of (5.3–5.5), the trajectory of the closed-loop system can possibly leave $\Omega(u_{max})$ at some time during the interval $[0, T_N)$ which implies that in this case there exists (at least one) finite time $t_1 \in [0, T_N)$ such that $V(x^M(t_1)) = c_{max}$. From the definition of T_b given in Theorem 5.14, it is clear that the smallest such t_1 is equal to T_b, and that

$T_b \in [0, T_N)$, i.e. $\min\{T_b, T_N\} = T_b$. If, on the other hand, the closed-loop trajectory under the model predictive controller remains within $\Omega(u_{max})$ for all times in $[0, T_N)$, then we have $V(x^M(t)) \leq c_{max} \quad \forall\ t \in [0, T_N)$ and we set $\min\{T_b, T_N\} = T_N$.
From the above analysis we can deduce that, starting from any initial state $x(0) \in \Omega(u_{max})$, the closed-loop trajectory satisfies $x^M(t) \in \Omega(u_{max}) \quad \forall\ 0 \leq t < \min\{T_b, T_N\}$. This fact, together with continuity of the solution of the switched closed-loop system implies that, upon switching (instantaneously) to the bounded controller at $t = \min\{T_b, T_N\} := T_p$, we have $x(T_p) \in \Omega(u_{max})$ and $u(t) = b(x(t))$ for all $t \geq T_p$. Then, by repeating the arguments of Step 1 in the Proof of Theorem 5.7, we can show that $\dot{V}(x^b(t)) < 0 \quad \forall\, t \geq T_p$ and that the origin of the switched closed-loop system is therefore asymptotically stable for any $x(0) \in \Omega(u_{max})$. This completes the proof of the theorem. $\triangle$

Proof of Proposition 5.25:
The proof consists of two parts. In the first part, we establish that the bounded state feedback control law of (5.7–5.8) enforces asymptotic stability for all initial conditions within Ω. In the second part, we compute an estimate of the tolerable measurement error, e_m, which guarantees that a state trajectory starting within Ω remains within it for all $\|e\| \leq e_m$, for all $t \geq 0$.
Part 1: We have already shown in Step 1 of Part 1 of the Proof of Theorem 5.7 that the time–derivative of the V, along the trajectories of the closed–loop system of (5.22) and (5.7–5.8), satisfies:

$$\begin{aligned}\dot{V}(x) &= L_fV(x) + L_gV(x)u(x)\\ &\leq \frac{-\rho x'Px}{\left[1+\sqrt{1+(2u_{max}\|B'Px\|)^2}\right]}\end{aligned} \tag{C.4}$$

for all $x \in \Phi$, and hence for all $x \in \Omega$, where Φ and Ω were defined in (5.9–5.10), respectively. Since the denominator term in (C.4) is bounded on Ω, there exists a positive real number, ρ^*, such that:

$$\dot{V} \leq -\rho^* x'Px \tag{C.5}$$

for all $x \in \Omega$, which implies that the origin of the closed–loop system, under the control law of (5.7–5.8), is asymptotically stable, with Ω as an estimate of the domain of attraction.
Part 2: In this part, we analyze the behavior of $\dot{V}$ on the boundary of Ω (i.e., the level surface described by $V(x) = c_{max}$) under bounded measurement errors, $\|e\| \leq e_m$. To this end, we have:

$$\begin{aligned}\dot{V}(x) &= L_fV(x) + L_gV(x)u(x-e) \\ &= L_fV(x) + L_gV(x)u(x) + L_gV(x)\left[u(x-e) - u(x)\right] \\ &\leq -\rho^* c_{max} + \|L_gV\|\|u(x-e) - u(x)\| \\ &\leq -\rho^* c_{max} + M\|u(x-e) - u(x)\|\end{aligned} \tag{C.6}$$

for all $x = V^{-1}(c_{max})$, where $M = \max\limits_{V(x)=c_{max}} (\|L_gV(x)\|)$ (note that M exists since $\|L_gV(\cdot)\|$ is continuous and the maximization is considered on a closed set). Since $u(\cdot)$ is continuous, then given any positive real number r such that $\mu = \dfrac{\rho^* c_{max} - r}{M} > 0$, there exists $e_m > 0$ such that if $\|(x-e)-x\| = \|e\| \leq e_m$, then $\|u(x-e) - u(x)\| \leq \mu$ and, consequently:

$$\dot{V}(x) \leq -\rho^* c_{max} + M\mu \ = \ -r < 0 \tag{C.7}$$

for all $x = V^{-1}(c_{max})$. This implies that for all measurement errors such that $\|e\| \leq e_m$, we have $\dot{V} < 0$ on the boundary of Ω. Therefore, under the bounded controller, any closed–loop state trajectory, starting within Ω, cannot escape this region, i.e., $x(t) \in \Omega \ \forall t \geq 0$. This completes the proof of the proposition. △

Proof of Proposition 5.27:

The proof is divided into two parts. In the first part, we show that given $\delta_b > 0$ (the size of the output feedback stability region) such that $\Omega_b \subset \Omega$, there exists a choice of β such that the closed–loop trajectories are bounded for all $x(0)$ and $\hat{x}(0)$ belonging in Ω_b. In the second part, we use a Lyapunov argument, together with boundedness of the states of the closed–loop system and of the estimation error, to show that the state is ultimately bounded with a bound that depends on the norm of the estimation error; and, therefore, converges to zero as the error tends to zero.

Part 1: From (5.26), we have that the error dynamics are given by $\dot{e} = (A - LC)e$, where $A - LC$ is Hurwitz, and all the eigenvalues of the matrix $A - LC$ satisfy $\lambda \leq -\beta$. It then follows that an estimate of the form $\|e(t)\| \leq \kappa(\beta)\|e(0)\|\exp(-\beta t)$ holds for some $\kappa(\beta) > 0$, for all $t \geq 0$. Given any positive real number, δ_b, such that $\Omega_b = \{x \in \mathbb{R}^n : \|x\|_P^2 \leq \delta_b\} \subset \Omega$, let $T_{min} = \min\{t \geq 0 : V(x(0)) = \delta_b, V(x(t)) = c_{max}, u(t) \in \mathcal{U}\}$ (i.e., T_{min} is the shortest time during which the closed–loop state trajectory can reach the boundary of Ω starting from the boundary of Ω_b using *any* admissible control action). Furthermore, let $e_{max}(0) = \max\limits_{x,\hat{x} \in \Omega_b} \|x(0) - \hat{x}(0)\|$ ($e_{max}(0)$ therefore is the largest possible initial error given that both the states and state estimates are initialized within Ω_b). Choose T_d such that $0 < T_d < T_{min}$ and let β^* be such that $e_m \leq \kappa(\beta^*)e_{max}(0)\exp(-\beta^* T_d)$ (the existence of such a β^* follows from the fact that $\kappa(\beta)$ is polynomial in β). For any choice of $\beta \geq \beta^*$, therefore, it follows that $\|e(T_{min})\| \leq e_m$ (since $\|e(T_d)\| \leq e_m$, $T_{min} \geq T_d$ and the bound on the norm of the estimation error decreases monotonically with time for all $t \geq 0$). This implies that the norm of the estimation error decays to a value less than e_m before the closed–loop state trajectory, starting within Ω_b, could reach the boundary of Ω. It then follows from Proposition 5.25 that the closed–loop state trajectory cannot escape Ω for all $t \geq 0$, i.e., the trajectories are bounded and $\|x(t)\|_p^2 \leq c_{max}\ \forall\, t \geq 0$.

Part 2: To prove asymptotic stability, we note that for all $x \in \Omega$:

$$\begin{aligned}\dot{V}(x) &= L_fV(x) + L_gV(x)u(x-e)\\ &= L_fV(x) + L_gV(x)u(x) + \|L_gV(x)\|\|u(x-e) - u(x)\| \\ &\leq -\rho^*\|x\|_p^2 + M\|u(x-e) - u(x)\|\end{aligned} \tag{C.8}$$

The term $\|u(x-e) - u(x)\|$ is continuous and vanishes when $e = 0$. Therefore, since both x and e are bounded, there exists a positive real number φ such that $\|u(x-e) - u(x)\| \leq \varphi\|e\|$ for all $\|x\|_p^2 \leq c_{max}$, $\|e\| \leq e_m$. Substituting this estimate into (C.8) yields:

$$\begin{aligned}\dot{V}(x) &\leq -\rho^*\|x\|_p^2 + M\varphi\|e\|\\ &\leq -\frac{\rho^*}{2}\|x\|_p^2\ \forall\ \|x\|_p \geq \sqrt{\frac{2M\varphi\|e\|}{\rho^*}} := \gamma_1(\|e\|)\end{aligned} \tag{C.9}$$

where $\gamma_1(\cdot)$ is a class $\mathcal{K}$ function (a continuous function $\alpha(\cdot)$ is said to belong to class $\mathcal{K}$ if it is strictly increasing and $\alpha(0) = 0$; see also [148]). The above inequality implies that $\dot{V}$ is negative outside some residual set whose size depends on $\|e\|$. Using the result of Theorem 5.1-Corollary 5.2 in [148], this implies that, for any $x(0) \in \Omega_b$, there exists a class $\mathcal{KL}$ function $\bar{\beta}(\cdot,\cdot)$ and a class $\mathcal{K}$ function $\gamma_2(\cdot)$, such that:

$$\|x(t)\| \leq \bar{\beta}(\|x(0)\|, t) + \gamma_2(\sup_{\tau\geq 0}\|e(\tau)\|)\ \forall\, t \geq 0 \tag{C.10}$$

implying that the x subsystem of (5.29), with e as input, is input-to-state stable (recall, from Proposition 5.25, that $x(t) \in \Omega_b\ \forall t \geq 0$). Noting also that the origin of the e-subsystem of (5.29) is asymptotically stable ($\lim\limits_{t\to\infty}\|e(t)\| = 0$), and using Lemma 5.6 in [148], we get that the origin of the interconnected system of (5.29) is asymptotically stable. This completes the proof of the proposition. △

Proof of Proposition 5.31:

Since the error dynamics obey a bound of the form $\|e(t)\| \leq \kappa(\beta)e_{max}(0)$ $\exp(-\beta t)$, substituting T_d^* into this expression yields $\|e(T_d^*)\| := e^* \leq \epsilon\sqrt{c_{max}/\lambda_{max}(P)}$. Then, for all $t \geq T_d^*$, if $\hat{x}'(t)P\hat{x}(t) \leq \delta_s$ for some $\delta_s > 0$, we can write:

$$
\begin{aligned}
x'(t)Px(t) &= (\hat{x}(t)+e(t))'P(\hat{x}(t)+e(t)) \\
&= \hat{x}'(t)P\hat{x}(t) + 2\hat{x}'(t)Pe(t) + e'(t)Pe(t) \\
&\leq \delta_s + 2\|P\hat{x}(t)\|\|e(t)\| + \|e(t)\|_p^2 \\
&\leq \delta_s + 2\sqrt{\frac{\lambda_{max}(P^2)\delta_s}{\lambda_{min}(P)}}\, e^* + \lambda_{max}(P)e^{*2} \\
&:= f(\delta_s)
\end{aligned}
\tag{C.11}
$$

Note that $f(\cdot)$ is a continuous, monotonically increasing function of $\delta_s \geq 0$, with $f(0) = \epsilon^2 c_{max} < c_{max}$ and $f(c_{max}) > c_{max}$. This implies that there exists $0 < \delta_s^* < c_{max}$ such that, for all $\delta_s \leq \delta_s^*$, the relation $f(\delta_s) \leq c_{max}$ holds, i.e., $x'(t)Px(t) \leq c_{max}$. This completes the proof of the proposition. $\triangle$

Proof of Theorem 5.33:
Given $x(0) \in \Omega_b$, $\hat{x}(0) \in \Omega_b$, $\beta \geq \beta^*$ (see Proposition 5.27), and $\delta_s \leq \delta_s^*$ (see Proposition 5.31), we consider the following cases:

Case 1: Consider first the case when $T_m = \infty$ (i.e., MPC is never feasible). Then, we have from (5.32) that $i(t) = 1$ and $u(\hat{x}(t)) = b(\hat{x}(t))$ for all $t \geq 0$. It then follows from Proposition 5.27 that the origin of the closed–loop system is asymptotically stable.

Case 2: Now, consider the case when $T_m < \infty$ and $T_f = \infty$. Since $T_m \geq T_d$ and $\|e(T_d)\| \leq e_m$ (see Part 1 of the Proof of Proposition 5.27), we have that $\|e(T_m)\| \leq e_m$. Since only the bounded controller is implemented, i.e. $i(t) = 1$, for $0 \leq t < T_m$, it follows from Proposition 5.25 that $x(t) \in \Omega$ for all $0 \leq t < T_m$ (or that $x(T_m^-) \in \Omega$). This fact, together with the continuity of the solution of the switched closed–loop system – which follows from the fact that the right hand side of (5.30) is continuous in x and piecewise continuous in time since only a finite number of switches is allowed over any finite time interval – implies that, upon switching (instantaneously) to the model predictive controller at $t = T_m$, we have $x(T_m) \in \Omega$. Since $T_f = \infty$, then from the definition of T_f in Theorem 5.33 it follows that $u(\hat{x}(t)) = M_s(\hat{x}(t))$ and $\dot{V}(\hat{x}(t)) < 0$ for all $t \geq T_m$. This implies that $\hat{x}(t) \in \Omega_s$ for all $t \geq T_m$ and, consequently from the definition of T_m (note that $T_m \geq T_d^*$ where T_d^* was defined in Proposition 5.31), that $x(t) \in \Omega$ for all $t \geq T_m$ (i.e., the closed–loop state trajectory is bounded). Furthermore, since $\hat{x}(t) = x(t) - e(t)$ and $\lim_{t\to\infty} \|e(t)\| = 0$, it follows that $\lim_{t\to\infty} \hat{x}'(t)P\hat{x}(t) = 0$; and, hence, $\lim_{t\to\infty} x'(t)Px(t) = 0$. The origin of the switched closed–loop system is therefore asymptotically stable.

Case 3: Finally, consider the case when $T_m < \infty$ and $T_f < \infty$. From the analysis in case 2 above, we have that $x(t) \in \Omega \quad \forall\, 0 \leq t < T_f$ (or that $x(T_f^-) \in \Omega$). This fact, together with the continuity of the solution of the switched closed–loop system, implies that, upon switching (instantaneously) to the bounded controller at $t = T_f$, we have $x(T_f) \in \Omega$ and $u(t) = b(\hat{x}(t))$ for all $t \geq T_f$. We also have $\|e(t)\| \leq e_m$ for all $t \geq T_f$ since $T_f > T_d$. Therefore, from Proposition 5.25 it is guaranteed that $x(t) \in \Omega$ for all $t \geq T_f$. Finally, following the same arguments presented in Part 2 of the Proof of Proposition 5.27, it can be shown that $\|x(t)\| \to 0$ as $t \to \infty$, which together with boundedness of the state trajectory, establishes asymptotic stability of the origin of the closed–loop system. This completes the proof of the theorem. $\triangle$

D

Proofs of Chapter 6

Proof of Theorem 6.2:
Step 1: Substituting the control law of (6.3–6.4) into the system of (6.1) and evaluating the time-derivative of the Lyapunov function along the trajectories of the closed–loop system, it can be shown that $\dot{V}_k < 0$ for all $x \in \Phi_k(u_{max})$ (where $\Phi_k(u_{max})$ was defined in (6.5)). Since $\Omega_k(u_{max})$ (defined in (6.6)) is an invariant subset of $\Phi_k(u_{max}) \bigcup \{0\}$, it follows that for any $x(0) \in \Omega_k(u_{max})$, the origin of the closed–loop system, under the control law of (6.3–6.4), is asymptotically stable.
Step 2: Consider the switched closed–loop system of (6.12), subject to the switching rule of (6.13–6.14), with any initial state $x(0) \in \Omega_k(u_{max})$. From the definition of $\bar{T}$ given in Theorem 6.2, it is clear that if $\bar{T}$ is a finite number, then $\dot{V}_k(x^M(t)) < 0 \;\; \forall \; 0 \leq t < \bar{T}$, where the notation $x^M(t)$ denotes the closed–loop state under MPC at time t, which implies that $x(t) \in \Omega_k(u_{max}) \;\; \forall \; 0 \leq t < \bar{T}$ (or that $x(\bar{T}^-) \in p\Omega_k(u_{max})$). This fact, together with the continuity of the solution of the switched system, $x(t)$, (following from the fact that the right hand side of (6.12) is continuous in x and piecewise continuous in time) implies that, upon switching (instantaneously) to the bounded controller at $t = \bar{T}$, we have $x(\bar{T}) \in \Omega_k(u_{max})$ and $u(t) = b_k(x(t))$ for all $t \geq \bar{T}$. Therefore, from our analysis in Step 1, we conclude that $\dot{V}_k(x^{b_k}(t)) < 0 \;\; \forall \; t \geq \bar{T}$. In summary, the switching rule of (6.13–6.14) guarantees that, starting from any $x(0) \in \Omega_k(u_{max})$, $\dot{V}_k(x(t)) < 0 \;\; \forall \; x \neq 0, \; x \in \Omega_k(u_{max}), \; \forall \; t \geq 0$, which implies that the origin of the switched closed–loop system is asymptotically stable. Note that if no such $\bar{T}$ exists, then we simply have from (6.13–6.14) that $\dot{V}_k(x^M(t)) < 0 \;\; \forall \, t \geq 0$, and the origin of the closed–loop system is also asymptotically stable. This completes the proof of the theorem. $\triangle$

Proof of Theorem 6.11:
The proof of this theorem, for the case when only one bounded controller is used as the fall–back controller (i.e., $p = 1$), is same as that of Theorem 6.2. To simplify the proof, we prove the result only for the case when the family of

fall–back controllers consists of 2 controllers (i.e., $p = 2$). Generalization of the proof to the case of a family of p controllers, where $2 < p < \infty$, is conceptually straightforward. Also, without loss of generality, we consider the case when the optimization problem in MPC is feasible for all times (i.e., $T^i = \infty$), since if it is not (i.e., $T^i < \infty$), then the switching time is simply taken to be the minimum of $\{T^i, T^f\}$, as stated in Theorem 6.11.

Without loss of generality, let the closed–loop system be initialized within the stability region of the first bounded controller, $\Omega_1(u_{max})$, under MPC. Then one of the following scenarios will take place:

Case 1: $\dot{V}_1(x^M(t)) < 0$ for all $t \geq 0$. The switching law of (6.15–6.17) dictates in this case that MPC be implemented for all times. Since $x(0) \in \Omega_1(u_{max})$, where Ω_1 is a level set of V_1, and $\dot{V}_1 < 0$, then the state of the closed–loop system is bounded and converges to the origin as $t \to \infty$.

Case 2: $\dot{V}_1(x^M(T_1)) \geq 0$ and $x^M(T_1) \in \Omega_1(u_{max})$, for some finite $T_1 > 0$. In this case, one of the following scenarios will occur:

(a) If $x(T_1)$ is outside of $\Omega_2(u_{max})$, then it follows from (6.15–6.17) that the supervisor will set $T^* = T_f = T_1$ and switch to the first bounded controller at $t = T_1$, using V_1 as the CLF, which enforces asymptotic stability as discussed in step 2 of the proof of Theorem 6.2.

(b) If $x(T_1) \in \Omega_2(u_{max})$ and $\dot{V}_2(x(T_1)) < 0$ (i.e., $0 < T_1 < T_2$), then the supervisor will keep MPC in the closed–loop system at T_1. If T_2 is a finite number, then the supervisor will set $T^* = T_f = T_2$ (since $T_1^* = 0$ for all $t > T_1$ from (6.16)) at which time it will switch to the second bounded controller, using V_2 as the CLF. Since $\dot{V}_2(x^M(t)) < 0$ for all $T_1 \leq t < T_2$, and $x(T_1) \in \Omega_2(u_{max})$, then $x(T^{*^-}) \in \Omega_2(u_{max})$. By continuity of the solution of the closed–loop system, it follows that $x(T^*) \in \Omega_2(u_{max})$; and since $\Omega_2(u_{max})$ is the stability region corresponding to V_2, then implies that upon implementation of the corresponding bounded controller for all future times, asymptotic stability is achieved. Note that if T_2 does not exist (or $T_2 = \infty$), then we simply have $x(T_1) \in \Omega_2(u_{max})$ and $\dot{V}_2(x(T_1)) < 0$ for all $t \geq T_1$, which implies that the origin of the closed–loop system is again asymptotically stable.

(c) If $0 < T_2 < T_1$ (i.e., $x(T_2) \in \Omega_2(u_{max})$, $\dot{V}_1(x(T_2)) < 0$ and $\dot{V}_2(x(T_2)) \geq 0$), then it follows from (6.15–6.17) that the supervisor will set $T^* = T_f = T_1$ (since $T_2^* = 0$ for all $t > T_2$ from (6.16)) and switch to the first bounded controller, using V_1 as the CLF. Since $\dot{V}_1(x^M(t)) < 0$ for all $t < T_1$, and $x(0) \in \Omega_1(u_{max})$, then $x(T^{*^-}) \in \Omega_1(u_{max})$. By continuity of the solution of the closed–loop system, we have $x(T^*) \in \Omega_1(u_{max})$, and since $\Omega_1(u_{max})$ is the stability region corresponding to V_1, this implies that upon implementation of the first bounded controller for the remaining time, asymptotic closed–loop stability is achieved. This completes the proof of the theorem. △

Proof of Theorem 6.20: The proof draws upon the fact that if the state of the closed–loop system resides in Ω at the time that the bounded controller is switched in, i.e., $x(T_{switch}) \in \Omega$, then boundedness of closed–loop state, and convergence to the set $\mathbb{D}$ under the bounded robust controller are guaranteed (for a proof, see [81,82]). Therefore, we need only show that, under the switching scheme of (6.34–6.35), the closed–loop state is always within Ω at the time the bounded controller is switched in. For this purpose, we enumerate the four possible values that T_{switch} could take, and show that $x(T_{switch}) \in \Omega$ in all cases.

- Case 1: If $T_{switch} = T_s$, then from the definition of T_s in (6.34), $x(T_s^-) \in \Omega$. By continuity of the solution of the system of (6.24) (which follows from the fact that the right hand side of (6.33) is continuous in x and piecewise continuous in time, since only a finite number of switches is allowed), we have $x(T_s) \in \Omega$, i.e., $x(T_{switch}) \in \Omega$.
- Case 2: If $T_{switch} = T_D$, then from the definition of T_D, $x(T_D) \in \mathbb{D} \subset \Omega$; hence $x(T_{switch}) \in \mathbb{D} \subset \Omega$.
- Case 3: If $T_{switch} = T_{design}$, then from the definition of T_{switch}, $T_{design} \leq T_s$. To prove that $x(T_{design}) \in \Omega$, we proceed by contradiction. Assume $x(T_{design}) \notin \Omega$. Then $V(x(T_{design})) > c^{max}$ (from the definition of Ω). By continuity of the solution, continuity of $V(\cdot)$ and the fact that $V(x_0) \leq c^{max}$, there exists a time $0 \leq T_s' \leq T_{design}$ for which $V(x(T_s^{'-})) = c^{max}$. Since T_s is the *earliest* time for which (6.34) holds, then it must be that $T_s \leq T_s' \leq T_{design}$, which leads to a contradiction ($T_{design} \leq T_s$). Therefore, once again, $x(T_{switch}) \in \Omega$.
- Case 4: If $T_{switch} = T_{inf}$, the same argument as in Case 3 can be used (replacing T_{design} by T_{inf}) to show that $x(T_{switch}) \in \Omega$.

This completes the proof of the theorem, $\triangle$

Proof of Theorem 6.27: Similar to the proof of Theorem 6.20, the proof of this theorem also draws upon the fact that if the state of the closed–loop system resides in Ω_L^0 at the time that the fth bounded controller, for which $x(T_{switch}) \in \Omega_f^0$, is switched in (and kept in the closed–loop for all future times), then under the bounded robust controller f, the region Ω_f^0 is invariant (for a proof, see [81,82]), and since $\Omega_f^0 \subset \Omega_L^0$, the closed–loop state evolves in Ω_L^0 for all future times.

We now show that for $t \leq T_{switch}$ (i.e., up until the time that the switching between the controllers is governed by (6.38–6.39)), the closed–loop state, initialized within Ω_L^0, stays within Ω_L^0. We only need to show that the switching logic ensures that the bounded controller is switched in before the states have escaped Ω_L^0 under MPC. To this end, we first note that $c_{k,j}^{max} < c_{k,0}^{max}$, for all $k = 1, \ldots, l$. Therefore $\Omega_L^j \subset \Omega_L^0$. This implies that if $x \in \Omega_L^j$ then $x \in \Omega_L^0$.

Consider any time, t, when $T_j \leq t < T_{j+1} < T_{switch}$, for some $j \geq 1$. Between T_j and T_{j+1}, we have from (6.39) that MPC is switched in (and the bounded controller is switched out) only when $x(t) \in \Omega_L^j$. Now consider the time T_{j+1} at which time the bounded controller is switched back in, either due to (6.38) or because MPC is infeasible. In the first case, $x(T_{j+1}) \in \Omega_L^j \subset \Omega_L^0$. In the second case, since the right hand side of the equation 6.36 is a piecewise continuous function of time and a continuous function of its states, and the closed–loop system is initialized within Ω_L^j under MPC, any time the MPC optimization problem is infeasible, $x(t) \in \Omega_L^j \subset \Omega_L^0$.

At the time $t = T_{switch}$, the state of the closed–loop system can be shown to be within Ω_L^0 following the same line of reasoning used in the proof of Theorem 6.20, with the additional possibility of $T_{switch} = T_N$. As shown above, $x(t) \in \Omega_L^0$ for all $T_{j-1} \leq t \leq T_j$, and therefore $x \in \Omega_L^0$ for $T_{N-1} \leq t \leq T_N$ and therefore $x(T_N) \in \Omega_L^0$ implying once again that $x(T_{switch}) \in \Omega_L^0$.

For all $t \geq T_{switch}$, the closed–loop state evolves under some bounded controller $u(x) = b_k(x)$, for which $x(T_{switch}) \in \Omega_k^0$. Therefore, $\limsup\limits_{t \to \infty} \|x(t)\| \leq d_k$ (see [81,82] for the proof). Note that since d^{max} is chosen to be the largest of $d_k, k = 1, \ldots, l$, $\|x(t)\| \leq d_k$ implies $\|x(t)\| \leq d^{max}$; hence $\limsup_{t \to \infty} \|x(t)\| \leq d^{max}$. This completes the proof of Theorem 6.27. △

E

Proofs of Chapter 7

Proof of Theorem 7.1:
To prove this theorem, we proceed in two steps. In the first step we show that, for each individual mode (without switching), the control law of (7.6–7.7) satisfies the constraints within the region described by (7.9) and that, starting from any initial condition within the set, Ω_i^*, the corresponding feedback control law robustly asymptotically stabilizes the origin of the i-th closed-loop subsystem. In the second step, we use this fact together with MLF stability analysis to show that the switching laws of (7.10–7.11) enforce asymptotic stability in the switched uncertain closed-loop system, starting from any initial condition that belongs to any of the sets, Ω_i^*, $i \in \mathcal{I}$.

Step 1: To prove that the control law of (7.6–7.7) satisfies the constraints within the region described by the inequality of (7.9), we need consider only the case when $\|(L_{G_i}V_i)^T\| \neq 0$ (since when $\|(L_{G_i}V_i)^T\| = 0$, $u_i = 0$ and the constraints are trivially satisfied). For this case, we have from (7.6–7.7):

$$
\begin{aligned}
\|u_i(x)\| &\leq \|k_i(V_i, u_i^{max}, \theta_{bi}, \chi_i, \phi_i)\| \left\|(L_{G_i}V_i)^T\right\| \\
&\leq \frac{\left\| L_{f_i}^* V_i + \sqrt{\left(L_{f_i}^{**}V_i\right)^2 + \left(u_i^{max} \|(L_{G_i}V)^T\|\right)^4} \right\|}{\|(L_{G_i}V_i)^T\| \left[1 + \sqrt{1 + (u_i^{max} \|(L_{G_i}V_i)^T\|)^2}\right]}
\end{aligned} \tag{E.1}
$$

From the definitions of $L_{f_i}^* V_i$ and $L_{f_i}^{**}V_i$ in (7.8) and the fact that $\rho > 0$, it is clear that if $L_{f_i}^{**}V_i \leq u_i^{max} \left\|(L_{G_i}V_i)^T\right\|$, then we also have $L_{f_i}^* V_i \leq u_i^{max} \left\|(L_{G_i}V_i)^T\right\|$. Therefore, for any x satisfying (7.9), the following estimates hold:

$$
\begin{aligned}
\left(L_{f_i}^{**}V_i\right)^2 &\leq \left(u_i^{max} \left\|(L_{G_i}V_i)^T\right\|\right)^2 \\
L_{f_i}^* V_i &\leq u_i^{max} \left\|(L_{G_i}V_i)^T\right\|
\end{aligned} \tag{E.2}
$$

Substituting the above estimates into (E.1) yields:

$$\|u_i(x)\| \leq \frac{u_i^{max}\left\|(L_{G_i}V_i)^T\right\|\left[1+\sqrt{1+(u_i^{max}\left\|(L_{G_i}V_i)^T\right\|)^2}\right]}{\left\|(L_{G_i}V_i)^T\right\|\left[1+\sqrt{1+(u_i^{max}\left\|(L_{G_i}V_i)^T\right\|)^2}\right]} = u_i^{max} \tag{E.3}$$

which shows that the constraints are satisfied. Consider now the i-th subsystem of the switched nonlinear system of (7.1). Substituting the control law of (7.6–7.8), evaluating the time-derivative of the Lyapunov function along the closed-loop trajectories, and using the fact that $\|(L_{G_i}V_i)^T\|^2 = (L_{G_i}V_i)(L_{G_i}V_i)^T$, we obtain:

$$\begin{aligned}\dot{V}_i &= L_{f_i}V_i + L_{G_i}V_i u_i + L_{W_i}V_i\theta_i \\ &\leq L_{f_i}V_i + \chi_i\|\left(L_{W_i}V_i\right)^T\|\theta_{bi} \\ &\quad - \left(\frac{L_{f_i}^{*}V_i + \sqrt{(L_{f_i}^{**}V_i)^2 + (u_i^{max}\|(L_{G_i}V_i)^T\|)^4}}{\|(L_{G_i}V_i)^T\|^2\left[1+\sqrt{1+(u_i^{max}\|(L_{G_i}V_i)^T\|)^2}\right]}\right)\end{aligned} \tag{E.4}$$

After performing some algebraic manipulations, the above inequality can be re-written as:

$$\dot{V}_i \leq \alpha_i(x) + \left(\frac{\theta_{bi}\left\|(L_{W_i}V_i)^T\right\|\left(\frac{\phi_i - (\chi - 1)\|x\|}{\|x\| + \phi_i}\right) - \rho_i\left(\frac{\|x\|^2}{\|x\| + \phi_i}\right)}{\left[1+\sqrt{1+(u_i^{max}\|(L_{G_i}V_i)^T\|)^2}\right]}\right) \tag{E.5}$$

where

$$\begin{aligned}\alpha_i(x) = &\left(\frac{\left(L_{f_i}V_i + \chi_i\theta_{bi}\left\|(L_{W_i}V_i)^T\right\|\right)\sqrt{1+(u_i^{max}\|(L_{G_i}V_i)^T\|)^2}}{\left[1+\sqrt{1+(u_i^{max}\|(L_{G_i}V_i)^T\|)^2}\right]}\right) \\ &-\left(\frac{-\sqrt{\left(L_{f_i}^{**}V_i\right)^2 + (u_i^{max}\|(L_{G_i}V_i)^T\|)^4}}{\left[1+\sqrt{1+(u_i^{max}\|(L_{G_i}V_i)^T\|)^2}\right]}\right)\end{aligned} \tag{E.6}$$

To analyze the sign of $\dot{V}_i$ in (E.5), we initially study the sign of the term, $\alpha_i(x)$, on the right-hand side. It is clear that the sign of this term depends on the sign of the term $L_{f_i}V_i + \chi_i\left\|(L_{W_i}V_i)^T\right\|\theta_{bi}$. To this end, we consider the following two cases:

Case 1: $L_{f_i}^{**}V_i \leq 0$

Since $L_{f_i}^{**}V_i = L_{f_i}V_i + \rho_i \|x\| + \chi_i \| (L_{W_i}V_i)^T \| \theta_{bi}$ and ρ_i is a positive real number, the fact that $L_{f_i}^{**}V_i \leq 0$ implies that $L_{f_i}V_i + \chi_i \|L_{W_i}V_i\| \theta_{bi} \leq 0$. As a result, we have that $\alpha_i(x) \leq 0$ and the time-derivative of V_i in this case satisfies the following bound:

$$\dot{V}_i \leq \left(\frac{\left\|(L_{W_i}V_i)^T\right\| \theta_{bi} \left(\dfrac{\phi_i - (\chi_i - 1)\|x\|}{\|x\| + \phi_i} \right) - \rho_i \left(\dfrac{\|x\|^2}{\|x\| + \phi_i} \right)}{\left[1 + \sqrt{1 + (u_i^{max}\|(L_{G_i}V_i)^T\|)^2}\right]} \right) := \beta_i(x) \tag{E.7}$$

Case 2: $0 < L_{f_i}^{**}V_i \leq u_i^{max} \left\|(L_{G_i}V_i)^T\right\|$

In this case, we have:

$$\left(L_{f_i}^{**}V_i\right)^2 \leq \left(u_i^{max}\|(L_{G_i}V_i)^T\|\right)^2 \tag{E.8}$$

and therefore:

$$\begin{aligned} & -\sqrt{\left(L_{f_i}^{**}V_i\right)^2 + (u_i^{max}\|(L_{G_i}V_i)^T\|)^4} \\ &= -\sqrt{\left(L_{f_i}^{**}V_i\right)^2 + (u_i^{max}\|(L_{G_i}V_i)^T\|)^2 (u_i^{max}\|(L_{G_i}V_i)^T\|)^2} \\ &\leq -\left(L_{f_i}^{**}V_i\right)\sqrt{1 + (u_i^{max}\|(L_{G_i}V_i)^T\|)^2} \end{aligned} \tag{E.9}$$

Substituting the estimate of (E.9) in the expression for $\dot{V}_i$ in (E.5–E.6) yields:

$$\begin{aligned} \dot{V}_i &= \frac{-\rho_i\|x\|\sqrt{1 + (u_i^{max}\|(L_{G_i}V_i)^T\|)^2}}{\left[1 + \sqrt{1 + (u_i^{max}\|(L_{G_i}V_i)^T\|)^2}\right]} + \beta_i(x) \\ &\leq \beta_i(x) \end{aligned} \tag{E.10}$$

From the above analysis, it is clear that whenever $L_{f_i}^{**}V_i \leq u_i^{max}\|(L_{G_i}V_i)^T\|$, the inequality of (E.7) holds. Since $\Omega_i^*(u_i^{max}, \theta_{bi})$ is taken to be the largest invariant set embedded within the region described by (7.9), we have that starting from any $x(0) \in \Omega_i^*$, the inequality of (E.7) holds. Referring to this inequality, note that since $\chi_i > 1$ and $\rho_i > 0$, it is clear that, whenever $\|x\| > \dfrac{\phi_i}{\chi_i - 1}$, the first term on the right-hand side is strictly negative, and therefore $\dot{V}_i$ satisfies:

$$\dot{V}_i \le -\frac{\rho_i \left\|x\right\|^2}{(\|x\| + \phi_i)\left[1 + \sqrt{1 + (u_i^{max}\|(L_{G_i}V_i)^T\|)^2}\right]} \tag{E.11}$$

To study the behavior of $\dot{V}_i$ when $\|x\| <\le \phi_i/(\chi_i - 1)$, we first note that since the entries of the matrix function, $W_i(x)$, and the entries of the row vector, $\nabla_x V_i$, are smooth and vanish when $x = 0$, then there exists positive real constants, ϕ_i^1, δ_i and δ_i', such that if $\phi_i \le \phi_i^1$, the bounds $\|W_i(x)\| \le \delta_i\|x\|$, $\|\nabla_x^T V_i\| \le \delta_i'\|x\|$ hold for $\|x\| \le \phi_i/(\chi_i - 1)$. Using this bound, we obtain the following estimates:

$$\begin{aligned}\left\|(L_{W_i}V_i)^T\right\| \theta_{bi}\left(\phi_i - (\chi_i - 1)\left\|x\right\|\right) &\le \left\|(L_{W_i}V_i)^T\right\| \theta_{bi}\phi_i \\ &= \|W_i(x)\| \left\|\nabla_x^T V_i\right\| \theta_{bi}\phi_i \\ &\le \phi_i\theta_{bi}\delta_i\delta_i'\|x\|^2 \qquad \forall \ \|x\| \le \frac{\phi_i}{\chi_i - 1}\end{aligned} \tag{E.12}$$

Substituting the estimate of (E.12) directly into (E.7), we get:

$$\dot{V}_i \le \left(\frac{(\phi_i\theta_{bi}\delta_i\delta_i' - \rho_i)\left\|x\right\|^2}{(\|x\| + \phi_i)\left[1 + \sqrt{1 + (u_i^{max}\|(L_{G_i}V_i)^T\|)^2}\right]}\right) \qquad \forall \ \|x\| \le \frac{\phi_i}{\chi_i - 1} \tag{E.13}$$

If ϕ_i is sufficiently small to satisfy the bound $\phi_i \le \dfrac{\rho_i}{\theta_{bi}\delta_i\delta_i'} := \phi_i^2$, then it is clear from (E.11–E.13) that the inequality in (E.11) is satisfied, irrespective of the value of $\|x\|$. In summary, we have that for any initial condition in the invariant set, Ω_i^*, there exists $\phi_i^* := \min\{\phi_i^1, \phi_i^2\}$ such that if $\phi_i \le \phi_i^*$, $\dot{V}_i$ satisfies:

$$\dot{V} \le \frac{-\rho_i \left\|x\right\|^2}{(\|x\| + \phi_i)\left[1 + \sqrt{1 + (u_i^{max}\|(L_{G_i}V_i)^T\|)^2}\right]} < 0 \ \forall\, x \ne 0,\ i = 1, \ldots, N \tag{E.14}$$

which implies that the origin of the individual closed-loop subsystems are asymptotically stable.

Step 2: Consider now the switched closed-loop system and, without loss of generality, suppose that $x(0) \in \Omega_i^*$ for some $i \in \mathcal{I}$. Then it follows from (E.14) above and the invariance of Ω_i^* that the Lyapunov function for this mode, V_i, decays monotonically, along the trajectories of the closed-loop system, for as long as mode i is to remain active, i.e., for all times such that $\sigma(t) = i$. If at any time T, such that $x(T) \in \Omega_j^*$ for some $j \in \mathcal{I}$, $j \ne i$, we set $\sigma(T^+) = j$ (i.e., activate mode j and its respective controller), then using the same argument, it is clear that the corresponding Lyapunov function for this mode, V_j, will also decay monotonically for as long as we keep $\sigma(t) = j$. Note that T, which is the time that mode i is switched out, is not known a priori but is rather determined by the evolution of the closed-loop continuous state. By tracking the closed-loop trajectory in this manner, we conclude that, starting from any $x(0) \in \Omega_i^*$ for any $i \in \mathcal{I}$ and as long as the i-th mode (and its controller) is activated only at a time when $x(t) \in \Omega_i^*$, we have that for all $i \in \mathcal{I}$, $k \in Z_+$:

$$\dot{V}_{\sigma(t_{i_k})} < 0 \quad \forall\ t \in [t_{i_k}, t_{i'_k}) \tag{E.15}$$

where t_{i_k} and $t_{i'_k}$ refer, respectively, to the times that the i-th mode is switched in and out for the k-th time, by the supervisor. Furthermore, from (7.11), we have that for any admissible switching time, t_{i_k}:

$$V_i(x(t_{i_k})) < V_i(x(t_{i'_{k-1}})) \tag{E.16}$$

which consequently implies that:

$$V_i(x(t_{i_k})) < V_i(x(t_{i_{k-1}})) \tag{E.17}$$

since $V_i(x(t_{i'_{k-1}})) < V_i(x(t_{i_{k-1}}))$ from (E.15). Using (E.15–E.17), a direct application of the MLF result of Theorem 2.3 in [39] can be performed to conclude that the switched closed-loop system, under the switching laws of Theorem 7.1, is Lyapunov stable. To prove asymptotic stability, we note that (E.16) also implies:

$$V_i(x(t_{i'_k})) < V_i(x(t_{i'_{k-1}})) \tag{E.18}$$

since $V_i(x(t_{i'_k})) < V_i(x(t_{i_k}))$ from (E.15). From the strict inequality in (E.18), it follows that for every (infinite) sequence of switching times, $t_{i'_1}, t_{i'_2}, \cdots$, such that $\sigma(t_{i_k}^+) = \sigma(t_{i'_k}^-) = i$, the sequence $V_{\sigma(t_{i'_1}^-)}, V_{\sigma(t_{i'_2}^-)}, \cdots$ is decreasing and positive, and therefore has a limit $L \geq 0$. We have:

$$\begin{aligned} 0 = L - L &= \lim_{k\to\infty} V_{\sigma(t_{i'_{k+1}}^-)}(x(t_{i'_{k+1}})) - \lim_{k\to\infty} V_{\sigma(t_{i'_k}^-)}(x(t_{i'_k})) \\ &= \lim_{k\to\infty} \left[V_i(x(t_{i'_{k+1}})) - V_i(x(t_{i'_k}))\right] \end{aligned} \tag{E.19}$$

Note that the term in brackets in the above equation is strictly negative for all nonzero x and zero only when $x = 0$ (from (E.18)) . Therefore, there exists a function α of class $\mathcal{K}$ (i.e., continuous, increasing, and zero at zero) such that:

$$\left[V_i(x(t_{i'_{k+1}})) - V_i(x(t_{i'_k}))\right] \leq -\alpha(\|x(t_{i'_k})\|) \tag{E.20}$$

Substituting the above estimate into (E.19), we have:

$$\begin{aligned} 0 &= \lim_{k\to\infty} \left[V_i(x(t_{i'_{k+1}})) - V_i(x(t_{i'_k}))\right] \\ &\leq \lim_{k\to\infty} \left[-\alpha(\|x(t_{i'_k})\|)\right] \leq 0 \end{aligned} \tag{E.21}$$

which implies that $x(t)$ converges to the origin. This fact, together with Lyapunov stability, implies that the origin of the switched closed-loop system is asymptotically stable. This concludes the proof of the theorem. △

Proof of Theorem 7.11:
The proof of this theorem shares several steps with the Proof of Theorem 7.1. We will highlight only the differences.

Step 1: We have already shown in Step 1 of the Proof of Theorem 7.1 that, for each mode considered separately, the evolution of the corresponding Lyapunov function, starting from any $x(0) \in \Omega_i^*$, obeys the growth bound of (E.11) whenever $\|x\| \geq \dfrac{\phi_i}{\chi_i - 1}$. Since the uncertain variables are non-vanishing, the bounds used in (E.12) to establish asymptotic convergence to the origin cannot be invoked here and no further conclusion can be made regarding the sign of $\dot{V}_i$ when $\|x\| < \dfrac{\phi_i}{\chi_i - 1}$. However, from (E.11), we conclude that $\dot{V}_i$ is negative-definite outside a ball of radius $\epsilon_i = \phi_i/(\chi_i - 1)$, which implies (see Theorem 5.1 and its corollaries in [148]) that, for any $x(0) \in \Omega_i^*$, there exists a finite time t_1 such that the solution to the i-th closed-loop system satisfies:

$$\begin{aligned} \|x(t)\| &\leq \beta_i(\|x_0\|, t), \quad \forall\, 0 \leq t < t_1 \\ \|x(t)\| &\leq b_i(\epsilon_i), \quad \forall\, t \geq t_1 \end{aligned} \tag{E.22}$$

where $\beta_i(\cdot,\cdot)$ is a class $\mathcal{KL}$ function and $b_i(\cdot)$ is of class $\mathcal{K}_\infty$. This implies that the state is ultimately bounded and that the ultimate bound can be made arbitrarily small by choosing ϵ_i to be sufficiently small.

Step 2: Having established boundedness of the trajectory of the individual closed-loop modes of the hybrid system, we proceed in this step to show boundedness of the overall switched closed-loop trajectory. To this end, given any positive real number, d, it follows from the properties of class $\mathcal{K}_\infty$ functions that there exist a set of positive real numbers, $\{\epsilon_1^*, \epsilon_2^*, \ldots, \epsilon_N^*\}$, such that:

$$b_1(\epsilon_1^*) \;=\; b_2(\epsilon_2^*) \;=\; \cdots \;=\; b_N(\epsilon_N^*) \;\leq\; d \tag{E.23}$$

where $\epsilon_i^* \geq \epsilon_i$, $\forall\, i \in \mathcal{I}$, which ensures that all modes share a common residual set. Since switching occurs only in regions where the various stability regions intersect (as required by (7.10)), we need consider only the case when the intersection, $\bigcap_i \Omega_i^*$, is nonempty and choose d such that the set $D = \{x \in \mathbb{R}^n : \|x\| \leq d\} \subset \bigcap_i \Omega_i^*$, $i = 1, \ldots, N$.

Without loss of generality, assume that $x(0) \in \Omega_i^*$ for some $i \in \mathcal{I}$. If this mode remains active for all times, then boundedness follows directly from the analysis in Step 1 above. If switching takes place, however, then it follows from the switching rules of (7.10–7.11) that for every admissible sequence of switching times, $t_{i_1}, t_{i_2}, \cdots$, such that $\sigma(t_{i_k}^+) = i$ and $\|x(t_{i_k})\| > d$, the positive sequence $V_{\sigma(t_{i_1})}, V_{\sigma(t_{i_2})}, \cdots$ is monotonically decreasing. This, together with the fact that the set D is non-empty and completely contained within Ω_i^*, implies the existence of a finite k^* such that $\|x(t_{i_k})\| \leq d$ for some $k \geq k^*$, $k \in Z_+$. Since N is finite, and only a finite number of switches are allowed over any finite time-interval, the preceding analysis implies that, if $\epsilon_i \leq \epsilon_i^*$, $\forall\, i \in \mathcal{I}$, there exists finite time, $t' > 0$, during which (at least) one of the modes will converge to D. Since D was chosen to be a common residual set for all the modes (E.23), it follows that $\|x(t)\| \leq d$ for all $t \geq t'$, regardless of which mode is switched in or out for $t \geq t'$. Therefore, the switched closed-loop trajectory is bounded for all times. This concludes the proof of the theorem. $\triangle$

Proof of Proposition 7.16:
The proof consists of two parts. In the first part, we establish that the bounded state feedback control law of (7.19–7.20) enforces asymptotic stability for all initial conditions in Ω_k with a certain robustness margin. In the second part, given the size of a ball around the origin that the system is required to converge to, d_k,we show the existence of a positive real number, Δ_k^*, such that if the discretization time Δ is chosen to be in the interval $(0, \Delta_k^*]$, then Ω_k remains invariant under discrete implementation of the bounded control law, and also that the state of the closed–loop system converges to the ball $\|x\| \leq d_k$.
Part 1: Substituting the control law of (7.19–7.20) into the system of (7.18) for a fixed $\sigma(t) = k$, it can be shown that:

$$\dot{V}_k(x) = L_{f_k}V_k(x) + L_{G_k}V_k(x)u(x) \leq \frac{-\rho_k V_k}{\left[1 + \sqrt{1 + (u_k^{max}\|(L_{G_k}V_k)^T(x)\|)^2}\right]} \tag{E.24}$$

for all $x \in \Omega_k$, where Ω_k was defined in (7.21). Since the denominator term in (E.24) is bounded in Ω_k, there exists a positive real number, ρ_k^*, such that $\dot{V}_k \leq -\rho_k^* V_k$ for all $x \in \Omega_k$, which implies that the origin of the closed–loop system, under the control law of (7.19–7.20), is asymptotically stable, with Ω_k as an estimate of the domain of attraction.
Part 2: Note that since $V_k(\cdot)$ is a continuous function of the state, one can find a finite, positive real number, $\delta_k^{'}$, such that $V_k(x) \leq \delta_k^{'}$ implies $\|x\| \leq d_k$. In the rest of the proof, we show the existence of a positive real number Δ_k^* such that all closed–loop state trajectories originating in Ω_k converge to the level set of V_k ($V_k(x) \leq \delta_k^{'}$) for any value of $\Delta \in (0, \Delta_k^*]$ and hence we have that $\limsup_{t\to\infty} \|x(t)\| \leq d_k$.

To this end, consider a "ring" close to the boundary of the stability region, described by $\mathcal{M}_k := \{x \in \mathbb{R}^n : (c_k^{max} - \delta_k) \leq V_k(x) \leq c_k^{max}\}$ (see also Figure 7.10), for a $0 \leq \delta_k < c_k^{max}$. Let the control action be computed for some $x(0) := x_0 \in \mathcal{M}_k$ and held constant until a time Δ_k^{**}, where Δ_k^{**} is a positive real number ($u_k(t) = u_k(x_0) := u_0 \ \forall\, t \in [0, \Delta_k^{**}]$). Then, $\forall\, t \in [0, \Delta_k^{**}]$,

$$\begin{aligned} \dot{V}_k(x(t)) &= L_{f_k}V_k(x(t)) + L_{G_k}V_k(x(t))u_0 \\ &= L_{f_k}V_k(x_0) + L_{G_k}V_k(x_0)u_0 + (L_{f_k}V_k(x(t)) - L_{f_k}V_k(x_0)) \\ &\quad +(L_{G_k}V_k(x(t))u_0 - L_{G_k}V_k(x_0)u_0) \end{aligned} \tag{E.25}$$

Since the control action is computed based on the states in $\mathcal{M}_k \subseteq \Omega_k$, $L_{f_k}V_k(x_0) + L_{G_k}V_k(x_0)u_0 \leq -\rho_k^* V_k(x_0)$. By definition, for all $x_0 \in \mathcal{M}_k$, $V_k(x_0) \geq c_k^{max} - \delta_k$, therefore $L_{f_k}V(x_0) + L_{G_k}V_k(x_0)u_0 \leq -\rho_k^*(c_k^{max} - \delta_k)$.

Since the function $f_k(\cdot)$ and the elements of the matrix $G_k(\cdot)$ are continuous, $\|u_k\| \leq u_k^{max}$, and $\mathcal{M}_k$ is bounded, one can find, for all $x_0 \in \mathcal{M}_k$ and a fixed Δ_k^{**}, a positive real number K_k^1, such that $\|x(t) - x_0\| \leq K_k^1 \Delta_k^{**}$ for all $t \leq \Delta_k^{**}$.

Since the functions $L_{f_k}V_k(\cdot)$, $L_{G_k}V_k(\cdot)$ are continuous, then given that $\|x(t) - x_0\| \leq K_k^1 \Delta_k^{**}$, one can find positive real numbers K_k^2 and K_k^3 such that $\|L_{f_k}V_k(x(t)) - L_{f_k}V_k(x_0)\| \leq K_k^3 K_k^1 \Delta_k^{**}$ and $\|L_{G_k}V_k(x(t))u_0 - L_{G_k}V_k(x_0)u_0\| \leq K_k^2 K_k^1 \Delta_k^{**}$. Using these inequalities in (E.25), we get:

$$\dot{V}_k(x(t)) \leq -\rho_k^*(c_k^{max} - \delta_k) + (K_k^1 K_k^2 + K_k^1 K_k^3)\Delta_k^{**} \tag{E.26}$$

For a choice of $\Delta_k^{**} < \dfrac{\rho_k^*(c_k^{max} - \delta_k) - \epsilon_k}{(K_k^1 K_k^2 + K_k^1 K_k^3)}$, where ϵ_k is a positive real number that satisfies

$$\epsilon_k < \rho_k^*(c_k^{max} - \delta_k) \tag{E.27}$$

we get that $\dot{V}_k(x(t)) \leq -\epsilon_k < 0$ for all $t \leq \Delta_k^{**}$. This implies that, given $\delta_k^{'}$, if we pick δ_k such that $c_k^{max} - \delta_k < \delta_k^{'}$ and find a corresponding value of Δ_k^{**}, then if the control action is computed for any $x \in \mathcal{M}_k$, and the 'hold' time is less than Δ_k^{**}, we get that $\dot{V}_k$ remains negative during this time, and therefore the state of the closed–loop system cannot escape Ω_k (since Ω_k is a level set of V_k). We now show the existence of $\Delta_k^{'}$ such that for all $x_0 \in \Omega_k^f := \{x \in \mathbb{R}^n : V_k(x_0) \leq c_k^{max} - \delta_k\}$, we have that $x(\Delta) \in \Omega_k^u := \{x_0 \in \mathbb{R}^n : V_k(x_0) \leq \delta_k^{'}\}$, where $\delta_k^{'} < c_k^{max}$, for any $\Delta \in (0, \Delta_k^{'}]$.

Consider $\Delta_k^{'}$ such that:

$$\delta'_k = \max_{V_k(x_0) \leq c_k^{max} - \delta_k,\ u_k \in \mathcal{U}_k,\ t \in [0, \Delta_k^{'}]} V_k(x(t)) \tag{E.28}$$

Since V_k is a continuous function of x, and x evolves continuously in time, then for any value of $\delta_k < c_k^{max}$, one can choose a sufficiently small $\Delta_k^{'}$ such that (E.28) holds. Let $\Delta_k^* = \min\{\Delta_k^{**},\ \Delta_k^{'}\}$. We now show that for all $x_0 \in \Omega_k^u$ and $\Delta_k \in (0, \Delta_k^*]$, $x(t) \in \Omega_k^u$ for all $t \geq 0$.

For all $x_0 \in \Omega_k^u \bigcap \Omega_k^f$, by definition $x(t) \in \Omega_k^u$ for $0 \leq t \leq \Delta_k$ (since $\Delta_k \leq \Delta_k^{'}$). For all $x_0 \in \Omega_k^u \backslash \Omega_k^f$ (and hence $x_0 \in \mathcal{M}_k$), $\dot{V}_k < 0$ for $0 \leq t \leq \Delta$ (since $\Delta_k \leq \Delta_k^{**}$). Since Ω_k^u is a level set of V_k, then $x(t) \in \Omega_k^u$ for $0 \leq t \leq \Delta_k$. Either way, for all initial conditions in Ω_k^u, $x(t) \in \Omega_k^u$ for all future times.

We note that for x such that $x \in \Omega_k \backslash \Omega_k^u$, negative definiteness of $\dot{V}_k$ is guaranteed for $\Delta \leq \Delta_k^* \leq \Delta_k^{**}$. Hence, all trajectories originating in Ω_k converge to Ω_k^u, which has been shown to be invariant under the bounded control law with a hold time Δ less than or equal to Δ_k^*, and therefore, for all $x_0 \in \Omega_k$, $\limsup_{t\to\infty} V_k(x(t)) \leq \delta_k^{'}$. Finally, since $V_k(x) \leq \delta_k^{'}$ implies $\|x\| \leq d_k$, we have that $\limsup_{t\to\infty} \|x(t)\| \leq d_k$. This completes the proof of Proposition 7.16. $\triangle$

Proof of Proposition 7.18:
From the proof of Proposition 7.16, we infer that given a positive real number, d_k, there exists an admissible manipulated input trajectory (provided by the bounded controller), and values of Δ_k^* and $\delta_k^{'}$, such that for any $\Delta \in (0, \Delta_k^*]$ and $x(0) \in \Omega_k$, $\limsup_{t\to\infty} V_k(x(t)) \leq \delta_k^{'}$ and $\limsup_{t\to\infty} \|x(t)\| \leq d_k$. The rest of the proof is divided in three parts. In the first part, we show that for all $x_0 \in \Omega_k$, the predictive controller of (7.23–7.28) is feasible. We then show that Ω_k is invariant under the predictive control algorithm of (7.23–7.28). Finally, we prove practical stability for the closed–loop system.

Part 1: Consider some $x_0 \in \Omega_k$ under the predictive controller of (7.23–7.28), with a prediction horizon $T = N\Delta$, where Δ is the hold time and $1 \leq N < \infty$ is the number of prediction steps. The initial condition can be such that either $V_k(x_0) \leq \delta^{'}$ or $\delta_k^{'} < V_k(x_0) \leq c_k^{max}$.
Case 1: If $\delta_k^{'} < V_k(x_0) \leq c_k^{max}$, the control input trajectory under the bounded controller of (7.19–7.20) provides a feasible solution to the constraint of (7.25) (see Proposition 7.16), given by $u(j\Delta) = b_k(x(j\Delta))$, $j = 1, \ldots, N$. Note that if $u = b_k$ for $t \in [0, \Delta]$, and $\Delta \in (0, \Delta_k^*]$, then $\dot{V}_k(x(t)) \leq -\epsilon_k < 0 \, \forall \, t \in [0, \Delta]$ and $b_k(x(t)) \in U_k$ (since b_k is computed using the bounded controller of (7.19–7.20)).
Case 2: If $V_k(x_0) \leq \delta_k^{'}$, once again we infer from Proposition 7.16 that the control input trajectory provided by the bounded controller of (7.19–7.20) provides a feasible initial guess, given by $u(j\Delta) = b_k(x(j\Delta))$, $j = 1, \ldots, N$ (recall from Proposition 7.16 that under the bounded controller of (7.19–7.20), if $V_k(x_0) \leq \delta_k^{'}$ then $V_k(x(t)) \leq \delta_k^{'} \, \forall \, t \geq 0$). This shows that for all $x_0 \in \Omega_k$, the Lyapunov–based predictive controller of (7.23–7.28) is feasible.
Part 2: As shown in Part 1, for any $x_0 \in \Omega_k \backslash \Omega_k^u$, the constraint of (7.25) in the optimization problem is feasible. Upon implementation, therefore, the value of the Lyapunov function decreases. Since Ω_k is a level set of V_k, the state trajectories cannot escape Ω_k. On the other hand, if $x_0 \in \Omega_k^u$, feasibility of the constraint of (7.26) guarantees that the closed–loop state trajectory stays in $\Omega_k^u \subset \Omega_k$. In both cases, Ω_k continues to be an invariant region under the Lyapunov–based predictive controller of (7.23–7.28).

Part 3: Finally, consider an initial condition $x_0 \in \Omega_k \backslash \Omega_k^u$. Since the optimization problem continues to be feasible, we have that $\dot{V}_k < 0$ for all $x(t) \notin \Omega_k^u$, i.e., $V_k(x(t)) > \delta_k^{'}$. All trajectories originating in Ω_k, therefore converge to Ω_k^u. For $x_0 \in \Omega_k^u$, the feasibility of the optimization problem implies $x(t) \in \Omega_k^u$, i.e., $V_k(x(t)) \leq \delta'_k$. Therefore, for all $x_0 \in \Omega_k$, $\limsup_{t\to\infty} V_k(x(t)) \leq \delta_k^{'}$. Also, since $V_k(x) \leq \delta_k^{'}$ implies $\|x\| \leq d_k$, we have that $\limsup_{t\to\infty} \|x(t)\| \leq d_k$. This completes the proof of Proposition 7.18. △

Proof of Theorem 7.24:

The proof of this theorem follows from the assumption of feasibility of the constraints of (7.32–7.34) at all times. Given the radius of the ball around the origin, d^{max}, the values of $\delta_k^{'}$ and Δ_k^*, for all $k \in \mathcal{K}$, are computed the same way as in the Proof of Proposition 7.16. Then, for the purpose of MPC implementation, a value of $\Delta_{k_r} \in (0, \Delta^*]$ is chosen where $\Delta^* = \min_{k=1,\ldots,p} \Delta_k^*$ and $t_{k_r^{out}} - t_{k_r^{in}} = l_{k_r} \Delta_{k_r}$ for some integer $l_{k_r} > 0$ (note that given any two positive real numbers $t_{k_r^{out}} - t_{k_r^{in}}$ and Δ^*, one can always find a positive real number $\Delta_{k_r} \leq \Delta^*$ such that $t_{k_r^{out}} - t_{k_r^{in}} = l_{k_r} \Delta_{k_r}$ for some integer $l_{k_r} > 0$).

Part 1: First, consider the case when the switching sequence is infinite. Let t be such that $t_{k_r^{in}} \leq t < t_{k_r^{out}}$ and $t_{m_j^{in}} = t_{k_r^{out}} < \infty$. Consider the active mode k. If $V_k(x) > \delta_k^{'}$, the continued feasibility of the constraint of (7.32) implies that $V_k(x(t_{k_r^{out}})) < V_k(x(t_{k_r^{in}}))$. The transition constraint of (7.34) ensures that if this mode is switched out and then switched back in, then $V_k(x(t_{k_{r+1}^{in}})) < V_k(x(t_{k_r^{in}}))$. In general, $V_k(x(t_{k_l^{in}})) < V_k(x(t_{k_{l-1}^{in}})) < \cdots < c_k^{max}$. Under the assumption of feasibility of the constraints of (7.32–7.34) for all future times, therefore, the value of $V_k(x)$ continues to decrease. If the mode corresponding to this Lyapunov function is not active, there exists at least some $j \in 1, \ldots, p$ such that mode j is active and Lyapunov function V_j continues to decrease until the time that $V_j \leq \delta_j^{'}$ (this happens because there are finite number of modes, even if the number of switches may be infinite). From this point onwards, the constraint of (7.33) ensures that V_j continues to be less than $\delta_j^{'}$; hence, $\limsup_{t\to\infty} \|x(t)\| \leq d^{max}$.

Part 2: For the case of a finite switching sequence, consider a t such that $t_{k_r^{in}} \leq t < t_{k_r^{out}} = \infty$. Under the assumption of continued feasibility of (7.32–7.34), $V_k(x(t_{k_r^{in}})) < V_k(x(t_{k_{r-1}^{in}})) < \cdots < c_k^{max}$. At the time of the switch to mode k, therefore, $x(t_{k_r^{in}}) \in \Omega_k$. From this point onwards, the Lyapunov-based predictive controller is implemented using the Lyapunov function V_k, and the constraint of (7.34) is removed, in which case the predictive controller of Theorem 7.24 reduces to the predictive controller of (7.23–7.27). Since the value of $\varDelta_{k_r}$ is chosen to be in $(0, \varDelta^*]$, where $\varDelta^* = \min_{k=1,\dots,p} \varDelta_k^*$, we have $\varDelta_{k_r} < \varDelta_k^*$, which guarantees feasibility and convergence to the ball $\|x\| \leq d^{max}$ for any value of the prediction horizon (hence, for a choice of $T = T_{design}$), and leads to $\limsup_{t\to\infty} \|x(t)\| \leq d^{max}$. This completes the proof of Theorem 7.24. $\triangle$

F

Proofs of Chapter 9

Proof of Theorem 9.7:
Under the controller of (9.21), the closed-loop system takes the form:

$$\begin{aligned}\dot{\omega} &= \mathcal{F}(\omega, h(x(t-\alpha)), v)\\ \dot{x} &= Ax(t) + Bx(t-\alpha) + f(x(t), x(t-\alpha))\\ &\quad +g(x(t), x(t-\alpha))\mathcal{P}(\omega, h(x(t-\alpha)), t),\\ x(\xi) &= \bar{\eta}(\xi), \qquad \xi \in [-\alpha, 0), \qquad x(0) = \bar{\eta}_0\end{aligned} \tag{F.1}$$

Introducing the extended state vector $\hat{\omega} = [\omega^T \; x^T]^T$, the above system can be written in the following compact form:

$$\dot{\hat{\omega}} = \bar{\mathcal{F}}(\hat{\omega}(t), \hat{\omega}(t-\alpha)) \tag{F.2}$$

where the specific form of the vector $\bar{\mathcal{F}}(\hat{\omega}(t), \hat{\omega}(t-\alpha))$ can be easily obtained by comparing (F.1) and (F.2), and it is omitted here for brevity. From the fact that the controller of (9.21) enforces local exponential stability and asymptotic output tracking in the closed-loop system when $\alpha = 0$, we have that the system:

$$\dot{\hat{\omega}} = \bar{\mathcal{F}}(\hat{\omega}(t), \hat{\omega}(t)) \tag{F.3}$$

is locally exponentially stable. For $t \geq 2\alpha$, the system of (F.2) can be rewritten as:

$$\begin{aligned}\dot{\hat{\omega}} &= \bar{\mathcal{F}}(\hat{\omega}(t), \hat{\omega}(t)) + [\bar{\mathcal{F}}(\hat{\omega}(t), \hat{\omega}(t-\alpha)) - \bar{\mathcal{F}}(\hat{\omega}(t), \hat{\omega}(t))]\\ &= \bar{\mathcal{F}}(\hat{\omega}(t), \hat{\omega}(t)) - \int_{t-\alpha}^{t} \frac{\partial \bar{\mathcal{F}}}{\partial \hat{\omega}}(\hat{\omega}(\theta), \hat{\omega}(\theta), t)\bar{\mathcal{F}}(\hat{\omega}(\theta-\alpha), \hat{\omega}(\theta))d\theta\end{aligned} \tag{F.4}$$

We consider the function $V = \hat{\omega}^T(t)\hat{\omega}(t)$ as a Lyapunov function candidate for the system of (F.4). Computing the time-derivative of V along the trajectory of the system of (F.4), we obtain:

$$\dot{V} = 2\hat{\omega}(t)\bar{\mathcal{F}}(\hat{\omega}(t),\hat{\omega}(t)) - 2\int_{t-\alpha}^{t} \hat{\omega}(t)\frac{\partial \bar{\mathcal{F}}}{\partial \hat{\omega}}(\hat{\omega}(\theta),\hat{\omega}(\theta),t)\bar{\mathcal{F}}(\hat{\omega}(\theta-\alpha),\hat{\omega}(\theta))d\theta \tag{F.5}$$

From the smoothness of the function $\bar{\mathcal{F}}$, we have that there exists an L such that $\left\|\frac{\partial \bar{\mathcal{F}}}{\partial \hat{\omega}}(\hat{\omega}(\theta),\hat{\omega}(\theta),t)\right\|_{\mathbb{R}^n} \leq L$, and the bound on $\dot{V}$ of (F.5), takes the form:

$$\begin{aligned}\dot{V} &\leq 2\hat{\omega}(t)\bar{\mathcal{F}}(\hat{\omega}(t),\hat{\omega}(t)) + 2L^2\int_{t-\alpha}^{t} \|\hat{\omega}(t)\hat{\omega}(\theta-\alpha)\|d\theta \\ &\leq 2\hat{\omega}(t)\bar{\mathcal{F}}(\hat{\omega}(t),\hat{\omega}(t)) + 2L^2\|\hat{\omega}(t)\|\int_{t-\alpha}^{t} \|\hat{\omega}(\theta-\alpha)\|d\theta\end{aligned} \tag{F.6}$$

Now, given a positive real number $\bar{q} > 1$, we consider the set of all $\hat{\omega}(t)$ that satisfy:

$$\hat{\omega}^2(t-\xi) \leq \bar{q}^2\hat{\omega}^2(t), \qquad 0 \leq \xi \leq 2\alpha \tag{F.7}$$

for which, we have that:

$$\dot{V} \leq 2\hat{\omega}(t)\bar{\mathcal{F}}(\hat{\omega}(t),\hat{\omega}(t)) + 2L^2\alpha\bar{q}\|\hat{\omega}(t)\|^2 \tag{F.8}$$

From the fact that the controller of (9.21) enforces local exponential stability and asymptotic output tracking in the closed-loop system when $\alpha = 0$, we have that the system:

$$\dot{\hat{\omega}} = \bar{\mathcal{F}}(\hat{\omega}(t),\hat{\omega}(t)) \tag{F.9}$$

is locally exponentially stable, which implies that:

$$2\hat{\omega}(t)\bar{\mathcal{F}}(\hat{\omega}(t),\hat{\omega}(t)) \leq -a\|\hat{\omega}(t)\|^2 \tag{F.10}$$

where a is a positive real number. Substituting the above bound into (F.8), we obtain:

$$\begin{aligned}\dot{V} &\leq -a\|\hat{\omega}(t)\|^2 + 2L^2\alpha\bar{q}\|\hat{\omega}(t)\|^2 \\ &\leq -(a - 2L^2\alpha\bar{q})\|\hat{\omega}(t)\|^2\end{aligned} \tag{F.11}$$

For $\alpha < \frac{a}{2L^2\bar{q}}$, we have that $\dot{V} \leq 0$, and using Theorem 4.2 from [108], we directly obtain that the system of (F.1) is exponentially stable. The proof that $\lim_{t\to\infty}\|y - v\| = 0$ is conceptually similar and will be omitted for brevity. △

Proof of Theorem 9.9:
Substitution of the controller of (9.37) into the system of (9.2) with $\tilde{\alpha} = 0$, yields the following system:

$$\begin{aligned}\dot{x} &= Ax + Bx(t-\alpha) + f(x,x(t-\alpha)) \\ &\quad + g(x,x(t-\alpha))\mathcal{A}(x(t),\bar{v}(t),x(t-\alpha),\bar{v}(t-\alpha)) \\ y &= h(x)\end{aligned} \tag{F.12}$$

Using that $f(x(t), x(t-\alpha)) = f_1(x(t)) + f_2(x(t), x(t-\alpha))$, $\tilde{f}(x(t)) = Ax(t) + f_1(x(t))$, and $\bar{p}(x(t), x(t-\alpha)) = Bx(t-\alpha) + f_2(x(t), x(t-\alpha))$, the above system can be written as:

$$\begin{aligned}\dot{x} &= \tilde{f}(x(t)) + g(x(t), x(t-\alpha))\mathcal{A}(x(t), \bar{v}(t), x(t-\alpha), \bar{v}(t-\alpha)) \\ &\quad + \bar{p}(x(t), x(t-\alpha)) \\ y &= h(x)\end{aligned} \tag{F.13}$$

Applying the coordinate transformation of (9.27) to the above system and setting, for ease of notation, $x(s) = \mathcal{X}^{-1}(\zeta(s), \eta(s))$, $s \in [t-\alpha, t]$, we obtain:

$$\begin{aligned}\dot{\zeta}_1 &= \zeta_2 + p_1(\zeta(t), \eta(t), \zeta(t-\alpha), \eta(t-\alpha)) \\ &\ \vdots \\ \dot{\zeta}_{r-1} &= \zeta_r + p_{r-1}(\zeta(t), \eta(t), \zeta(t-\alpha), \eta(t-\alpha)) \\ \dot{\zeta}_r &= L_{\tilde{f}}^r h(x) + L_g L_{\tilde{f}}^{r-1} h(x)\mathcal{A}(x(t), \bar{v}(t), x(t-\alpha), \bar{v}(t-\alpha)) \\ &\quad + p_r(\zeta(t), \eta(t), \zeta(t-\alpha), \eta(t-\alpha)) \\ \dot{\eta}_1 &= \Psi_1(\zeta(t), \eta(t), \zeta(t-\alpha), \eta(t-\alpha)) \\ &\ \vdots \\ \dot{\eta}_{n-r} &= \Psi_{n-r}(\zeta(t), \eta(t), \zeta(t-\alpha), \eta(t-\alpha)) \\ y &= \zeta_1\end{aligned} \tag{F.14}$$

Introducing, the variables $e_i = \zeta_i - v^{(i-1)}$, $i = 1, \ldots, r$, the system of (F.14) takes the form:

$$\begin{aligned}\dot{e}_1 &= e_2 + p_1(\bar{e}(t) + \bar{v}(t), \eta(t), \bar{e}(t-\alpha) + \bar{v}(t-\alpha), \eta(t-\alpha)) \\ &\ \vdots \\ \dot{e}_{r-1} &= e_r + p_{r-1}(\bar{e}(t) + \bar{v}(t), \eta(t), \bar{e}(t-\alpha) + \bar{v}(t-\alpha), \eta(t-\alpha)) \\ \dot{e}_r &= L_{\tilde{f}}^r h(x) - v^{(r)} + L_g L_{\tilde{f}}^{r-1} h(x)\mathcal{A}(x(t), \bar{v}(t), x(t-\alpha), \bar{v}(t-\alpha)) \\ &\quad + p_r(\bar{e}(t) + \bar{v}(t), \eta(t), \bar{e}(t-\alpha) + \bar{v}(t-\alpha), \eta(t-\alpha)) \\ \dot{\eta}_1 &= \Psi_1(\bar{e}(t) + \bar{v}(t), \eta(t), \bar{e}(t-\alpha) + \bar{v}(t-\alpha), \eta(t-\alpha)) \\ &\ \vdots \\ \dot{\eta}_{n-r} &= \Psi_{n-r}(\bar{e}(t) + \bar{v}(t), \eta(t), \bar{e}(t-\alpha) + \bar{v}(t-\alpha), \eta(t-\alpha))\end{aligned} \tag{F.15}$$

where $\bar{e} = [e_1\ e_2\ \cdots\ e_r]^T$. For the above system, we assume, without loss of generality, that when $\eta = 0$, $\bar{e} = 0$ is an equilibrium solution.

We now consider the $\bar{e}$-subsystem of the system of (F.15):

$$
\begin{aligned}
\dot{e}_1 &= e_2 + p_1(\bar{e}(t) + \bar{v}(t), \eta(t), \bar{e}(t-\alpha) + \bar{v}(t-\alpha), \eta(t-\alpha)) \\
&\vdots \\
\dot{e}_{r-1} &= e_r + p_{r-1}(\bar{e}(t) + \bar{v}(t), \eta(t), \bar{e}(t-\alpha) + \bar{v}(t-\alpha), \eta(t-\alpha)) \\
\dot{e}_r &= L_{\tilde{f}}^r h(x) - v^{(r)} + L_g L_{\tilde{f}}^{r-1} h(x) \mathcal{A}(x(t), \bar{v}(t), x(t-\alpha), \bar{v}(t-\alpha)) \\
&\quad + p_r(\bar{e}(t) + \bar{v}(t), \eta(t), \bar{e}(t-\alpha) + \bar{v}(t-\alpha), \eta(t-\alpha))
\end{aligned}
\tag{F.16}
$$

Using the explicit form of the controller formula of (9.37) and the definition for the matrix $\tilde{A}$ and the vectors b, p of (9.32), the above system can be written in the following compact form:

$$
\dot{\bar{e}} = \tilde{A}\bar{e} - bR_2^{-1}b^T P\bar{e} + p(\bar{e}(t) + \bar{v}(t), \eta(t), \bar{e}(t-\alpha) + \bar{v}(t-\alpha), \eta(t-\alpha))
\tag{F.17}
$$

We will now show that if (9.36) has a unique positive definite solution for P and the state of the η-subsystem of (F.15) is bounded, then the system of (F.17) is exponentially stable, which implies that $\lim_{t\to\infty} \|y - v\| = 0$. To establish this result, we consider the following smooth functional $V : \mathcal{C} \to \mathbb{R}_{\geq 0}$:

$$
V(\bar{e}_t(\xi)) = \bar{e}^T P \bar{e} + a^2 \int_{t-\alpha}^{t} \bar{e}^T(s)\bar{e}(s)ds
\tag{F.18}
$$

which clearly satisfies, $K_1\|\bar{e}(t)\|^2 \leq V(\bar{e}_t(\xi)) \leq K_2\|\bar{e}_t(\xi)\|^2$, for some positive constants K_1 and K_2. Computing the time-derivative of V along the trajectories of the system of (F.17), we obtain:

$$
\begin{aligned}
\dot{V} &= \dot{\bar{e}}^T P\bar{e} + \bar{e}^T P\dot{\bar{e}} + a^2(\bar{e}^T(t)\bar{e}(t) - \bar{e}^T(t-\alpha)\bar{e}(t-\alpha)) \\
&\leq (\bar{e}^T\tilde{A}^T - \bar{e}^T P^T bR_2^{-1}b^T \\
&\quad + p^T(\bar{e}(t) + \bar{v}(t), \eta(t), \bar{e}(t-\alpha) + \bar{v}(t-\alpha), \eta(t-\alpha)))P\bar{e} \\
&\quad + \bar{e}^T P(\tilde{A}\bar{e} - bR_2^{-1}b^T P\bar{e} \\
&\quad + p(\bar{e}(t) + \bar{v}(t), \eta(t), \bar{e}(t-\alpha) + \bar{v}(t-\alpha), \eta(t-\alpha))) \\
&\quad + a^2(\bar{e}^T(t)\bar{e}(t) - \bar{e}^T(t-\alpha)\bar{e}(t-\alpha)) \\
&\leq \bar{e}^T(\tilde{A}^T P + P\tilde{A} - 2P^T bR_2^{-1}b^T P + a^2 I_{n\times n})\bar{e} \\
&\quad + 2p^T(\bar{e}(t) + \bar{v}(t), \eta(t), \bar{e}(t-\alpha) + \bar{v}(t-\alpha), \eta(t-\alpha))P\bar{e} \\
&\quad - a^2\bar{e}^T(t-\alpha)\bar{e}(t-\alpha)
\end{aligned}
\tag{F.19}
$$

where $I_{n\times n}$ denotes the identity matrix of dimension $n \times n$. Using the inequality $2x^T y \leq x^T x + y^T y$ where x and y are column vectors, we obtain:

$$
\begin{aligned}
\dot{V} &\leq \bar{e}^T(\tilde{A}^T P + P\tilde{A} - 2P^T bR_2^{-1}b^T P + a^2 I_{n\times n})\bar{e} \\
&\quad + p^2(\bar{e}(t) + \bar{v}(t), \eta(t), \bar{e}(t-\alpha) + \bar{v}(t-\alpha), \eta(t-\alpha)) \\
&\quad + \bar{e}^T P^2\bar{e} - a^2\bar{e}^T(t-\alpha)\bar{e}(t-\alpha)
\end{aligned}
$$

$$\begin{aligned}&\leq \bar{e}^T(\tilde{A}^TP + P\tilde{A} - 2P^TbR_2^{-1}b^TP + a^2I_{n\times n} + P^2)\bar{e}\\ &\quad +p^2(\bar{e}(t)+\bar{v}(t),\eta(t),\bar{e}(t-\alpha)+\bar{v}(t-\alpha),\eta(t-\alpha))\\ &\quad -a^2\bar{e}^T(t-\alpha)\bar{e}(t-\alpha)\end{aligned} \tag{F.20}$$

Since the state of the η-subsystem of (F.15) is supposed to be bounded and assuming quadratic growth for the term $p^2(\bar{e}(t)+\bar{v}(t),\eta(t),\bar{e}(t-\alpha)+\bar{v}(t-\alpha),\eta(t-\alpha))$ (Assumption 9.3), we have that there exist positive real numbers a_1, a_2 such that the following bound can be written:

$$\|p(\bar{e}(t)+\bar{v}(t),\eta(t),\bar{e}(t-\alpha)+\bar{v}(t-\alpha),\eta(t-\alpha))\|^2 \leq a_1\bar{e}^2(t) + a_2\bar{e}^2(t-\alpha) \tag{F.21}$$

Substituting the above inequality on the bound for $\dot{V}$ in (F.20), we obtain:

$$\begin{aligned}\dot{V} &\leq \bar{e}^T(\tilde{A}^TP + P\tilde{A} - 2P^TbR_2^{-1}b^TP + a^2I_{n\times n} + P^2)\bar{e}\\ &\quad +a_1\bar{e}^2(t) + a_2\bar{e}^2(t-\alpha) - a^2\bar{e}^T(t-\alpha)\bar{e}(t-\alpha)\\ &\leq \bar{e}^T(\tilde{A}^TP + P\tilde{A} - 2P^TbR_2^{-1}b^TP + (a^2+a_1)I_{n\times n} + P^2)\bar{e}\\ &\quad -(a^2-a_2)\bar{e}^2(t-\alpha)\end{aligned} \tag{F.22}$$

Now, if $a^2 > a_2$ and there exists a positive definite symmetric matrix P which solves the following matrix equation:

$$\tilde{A}^TP + P\tilde{A} - 2P^TbR_2^{-1}b^TP + (a^2+a_1)I_{n\times n} + P^2 = -R_1 \tag{F.23}$$

where R_1 is a positive definite matrix, then:

$$\begin{aligned}\dot{V} &\leq -\bar{e}^TR_1\bar{e} - a_3\bar{e}^2(t-\alpha)\\ &\leq -\lambda_{min}(R_1)\bar{e}^2 - a_3\bar{e}^2(t-\alpha)\end{aligned} \tag{F.24}$$

where a_3 is a positive real number and $\lambda_{min}(R_1)$ denotes the smallest eigenvalue of the matrix R_1. Since V and $\dot{V}$ satisfy the assumptions of Theorem 9.5, we have that there exist positive real numbers K, a such that the state of the $\bar{e}-$subsystem of (F.16) is exponentially stable i.e., it satisfies:

$$\|\bar{e}(t)\| \leq Ke^{-at}\|\bar{e}(0)\| \tag{F.25}$$

for every value of the delay, α, and thus, $\lim_{t\to\infty}\|y-v\| = 0$.

To complete the proof of the theorem, we need to show that there exists a positive real number δ such that if $\max\{\|x_0(\xi)\|, \|\bar{v}\|^s\} \leq \delta$, then the state of the closed-loop system is exponentially stable and the output of the closed-loop system satisfies $\lim_{t\to\infty}\|y-v\| = 0$. To establish this result, we will analyze the behavior of the DDE system of (F.15) using a two-step small-gain theorem type argument [269]. In the first step, we will use a contradiction argument to show that the evolution of the states $\bar{e}, \eta$, starting from sufficiently small initial conditions i.e., $\|\bar{e}_0(\xi)\| \leq \delta_{\bar{e}}$ and $\|\eta_0(\xi)\| \leq \delta_\eta$ (where $(\delta_{\bar{e}}, \delta_\eta)$ are positive real numbers which can be explicitly computed as functions of δ from the coordinate change of (9.27)), satisfies the following inequalities:

$$\|\bar{e}(t)\| \leq \bar{\delta}_{\bar{e}}\,, \qquad \|\eta(t)\| \leq \bar{\delta}_{\eta}, \qquad \forall\, t \in [0, \infty) \tag{F.26}$$

where $\bar{\delta}_{\bar{e}} > K\delta_{\bar{e}}$ and $\bar{\delta}_{\eta} > \delta_{\eta}$ are positive real numbers which will be specified below. In the second step, we will use the boundedness result obtained from the first step and the exponentially decaying bound of (F.25) to prove that the state of the η-subsystem of (F.15) decays exponentially to zero.

From Assumption 9.3, we have that the $\eta-$subsystem of the system of (F.15) is input-to-state stable, which implies that there exist a function $\bar{\gamma}_{\bar{e}}$ of class Q, a function β_{η} of class KL and positive real numbers $\bar{\delta}_{\bar{e}}, \delta_{\eta}$ such that if $\|\eta_0(\xi)\| < \delta_{\eta}$ and $\|\bar{e}_t\|^s < \bar{\delta}_{\bar{e}}$, then:

$$\|\eta(t)\| \leq \|\eta_t(\xi)\| \leq \beta(\|\eta_0(\xi)\|, t) + \bar{\gamma}_{\bar{e}}(\|\bar{e}_t\|^s), \;\; \forall\, t \geq 0 \tag{F.27}$$

We will proceed by contradiction. Set $\bar{\delta}_{\eta} > \beta_{\eta}(\delta_{\eta}, 0) + \bar{\gamma}_{\bar{e}}(K\delta_{\bar{e}})$ and let $\bar{T}$ be the smallest time such that there is a $\hat{\delta}$ so that $t \in (\bar{T}, \bar{T} + \hat{\delta})$ implies either $\|\bar{e}(t)\| > \bar{\delta}_{\bar{e}}$ or $\|\bar{\eta}(t)\| > \bar{\delta}_{\eta}$. Then, for each $t \in [0, \bar{T}]$ the conditions of (F.26) hold.

Consider the functions $\bar{e}^{\bar{T}}(t)$, $\eta^{\bar{T}}(t)$ defined as follows:

$$\bar{e}^{\bar{T}}(t) = \begin{Bmatrix} \bar{e}(t) & t \in [0, \bar{T}] \\ 0 & t \in (\bar{T}, \infty) \end{Bmatrix}, \qquad \eta^{\bar{T}}(t) = \begin{Bmatrix} \eta(t) & t \in [0, \bar{T}] \\ 0 & t \in (\bar{T}, \infty) \end{Bmatrix} \tag{F.28}$$

Now, using (F.27) we have that:

$$\sup_{0 \leq t \leq \bar{T}} \left(\beta_{\eta}(\|\eta_0(\xi)\|, t) + \bar{\gamma}_{\bar{e}}(\|\bar{e}_t\|^s)\right) \leq \beta_{\eta}(\delta_{\eta}, 0) + \bar{\gamma}_{\bar{e}}(K\delta_{\bar{e}}) \tag{F.29}$$

$$\sup_{0 \leq t \leq \bar{T}} \left(K\|\bar{e}(0)\|e^{-at}\right) \leq K\|\bar{e}(0)\| \leq K\delta_{\bar{e}}$$

and that:

$$\begin{aligned} \|\bar{e}^{\bar{T}}\|^s &\leq K\delta_{\bar{e}} < \bar{\delta}_{\bar{e}} \\ \|\eta^{\bar{T}}\|^s &\leq \beta_{\eta}(\delta_{\eta}, 0) + \bar{\gamma}_{\bar{e}}(K\delta_{\bar{e}}) < \bar{\delta}_{\eta} \end{aligned} \tag{F.30}$$

By continuity, we have that there exist some positive real number $\bar{k}$ such that $\|\bar{e}^{\bar{T}+\bar{k}}(t)\|^s \leq \bar{\delta}_{\bar{e}}$ and $\|\eta^{\bar{T}+\bar{k}}(t)\|^s \leq \bar{\delta}_{\eta}$, $\forall\, t \in [0, \bar{T} + \bar{k}]$. This contradicts the definition of $\bar{T}$. Hence, (F.26) holds $\forall\, t \geq 0$.

Since the states, $(\bar{e}, \eta)$, of the closed-loop system of (F.15) are bounded and $\bar{e}(t)$ decays exponentially to zero, a direct application of the result of Theorem 2 in [269] implies that the η-subsystem of (F.15) is also exponentially stable, and thus, the system of (F.15) is exponentially stable and its output satisfies $\lim_{t \to \infty} \|y - v\| = 0$. △

Proof of Theorem 9.15:
We initially construct that dynamical system which describes the dynamics of the estimation error, $e_t = \omega_t - x_t$. Differentiating e_t and using the systems of (9.10) and (9.50), we obtain:

$$\begin{aligned}\frac{de_t}{dt} &= \mathcal{A}e_t + f(P(e_t + x_t), Q(e_t + x_t)) - f(Px_t, Qx_t) \\ &\quad + \Phi_H L(\Psi_H, e_t + x_t)(y(t) - h(P(e_t + x_t)))\end{aligned} \tag{F.31}$$

Computing the linearization of the above system and applying the spectral decomposition procedure, we obtain:

$$\begin{aligned}\frac{de_t^p}{dt} &= \mathcal{A}_p e_t^p - \Phi_H L(w(Pe_t^p + Pe_t^n)) \\ \frac{\partial e_t^n}{\partial t} &= \mathcal{A}_n e_t^n \\ e_t^p(0) &= P_p e(0) = P_p(\bar{\omega} - \bar{\eta}), \quad e_t^n(0) = P_n e(0) = P_n(\bar{\omega} - \bar{\eta})\end{aligned} \tag{F.32}$$

because $P_n \Phi_H L(\Psi_H, \omega_t)(y(t) - h(P\omega_t)) \equiv 0$ by construction, and $w = \dfrac{\partial h}{\partial x}(0)$. Since all the eigenvalues of the operator $\mathcal{A}_n$ lie in the left-half of the complex, this implies that the subsystem:

$$\frac{\partial e_t^n}{\partial t} = \mathcal{A}_n e_t^n \tag{F.33}$$

is exponentially stable. Therefore, the infinite-dimensional system of (F.32) possesses identical stability properties with the finite-dimensional system:

$$\frac{de_t^p}{dt} = \mathcal{A}_p e_t^p + \Phi_H L w P e_t^p \tag{F.34}$$

From the observability Assumption 9.4, however, we have that the above system is exponentially stable, which implies that the error system is exponentially to zero, and thus, the estimation error, $e_t = \omega_t - x_t$, tends locally exponentially to zero.

Nonlinear state observer simplification:
The abstract dynamical system of (9.50) can be simplified by utilizing the fact that:

$$\begin{aligned}&\mathcal{A}\omega_t(\xi) + \Phi_H L(\Psi_H, \phi)(y(t) - h(P\omega_t(\xi))) \\ &= \left\{\begin{array}{ll} \dfrac{d\phi(\xi)}{d\xi} + \Phi_H(\xi) L(\Psi_H, \phi)(y(t) - h(P\omega_t(\xi))), & \xi \in [-\alpha, 0) \\ & \\ A\omega_t(0) + B\phi(-\alpha) - \Phi_H(0) L(\Psi_H, \phi) h(P\omega_t(\xi)), & \xi = 0 \end{array}\right\}\end{aligned} \tag{F.35}$$

which implies that is equivalent to the following system of partial differential equations:

$$\frac{\partial \tilde{\omega}}{\partial t}(\xi,t) = \frac{\partial \tilde{\omega}}{\partial \xi}(\xi,t) + \Phi_H(\xi)L(\Psi_H,\tilde{\omega}(\xi,t))(y(t) - h(\tilde{\omega}(0,t))) \tag{F.36}$$

$$\frac{\partial \tilde{\omega}}{\partial t}(0,t) = A\tilde{\omega}(0,t) + B\tilde{\omega}(-\alpha,t) + \Phi_H(0)L(\Psi_H,\tilde{\omega}(\xi,t))(y(t) - h(\tilde{\omega}(0,t))) \tag{F.37}$$

where $\tilde{\omega}(\xi,t) = \omega_t(\xi)$. A further simplification can be performed by integrating the hyperbolic PDE of (F.36) along its characteristics to obtain a nonlinear integro-differential equation system for the observer. The characteristics of (F.36) are the family of lines with slope $\frac{dt}{d\xi} = -1$, i.e., the set of lines satisfying $t+\xi = c$, where c is a constant. Integrating the (F.36) along its characteristics, we obtain:

$$\frac{d\tilde{\omega}}{d\xi}(c-\xi,\xi) = -\Phi(\xi)L(\Psi_H,\tilde{\omega}(\xi,t))h(\tilde{\omega}(0,c-\xi)) + \Phi(\xi)L(\Psi_H,\tilde{\omega}(\xi,t))y(c-\xi) \tag{F.38}$$

so that:

$$\tilde{\omega}(c-\theta,\theta) = \tilde{\omega}(c,0) - \int_0^\theta \Phi(\xi)L(\Psi_H,\tilde{\omega}(\xi,t))(h(\tilde{\omega}(0,c-\xi)) - y(c-\xi))d\xi, \quad \theta \in [-\alpha,0], \tag{F.39}$$

or:

$$\begin{aligned}\tilde{\omega}(t,\theta) &= \tilde{\omega}(t+\theta,0) \\ &\quad - \int_0^\theta \Phi(\xi)L(\Psi_H,\tilde{\omega}(\xi,t)))(h(\tilde{\omega}(0,t+\theta-\xi)) - y(t+\theta-\xi))d\xi\end{aligned} \tag{F.40}$$

Finally,

$$\begin{aligned}\tilde{\omega}(t,-\alpha) &= \tilde{\omega}(t-\alpha,0) - \int_0^{-\alpha} \Phi(\xi)L(\Psi_H,\tilde{\omega}(\xi,t))(h(\tilde{\omega}(0,t-\alpha-\xi)) \\ &\quad -y(t-\alpha-\xi))d\xi \\ &= \tilde{\omega}(t-\alpha,0) + \int_0^{\alpha} \Phi(\xi-\alpha)L(\Psi_H,\tilde{\omega}(\xi,t))(h(\tilde{\omega}(0,t-\xi)) \\ &\quad -y(t-\xi))d\xi\end{aligned} \tag{F.41}$$

Substituting $\tilde{\omega}(t+\theta,0) = \omega(t+\theta)$ into (F.37), we obtain the following system:

$$\begin{aligned}\dot{\omega} &= A\omega(t) + B\omega(t-\alpha) + f(\omega(t),\omega(t-\alpha)) + g(\omega(t),\omega(t-\alpha))u \\ &\quad +\Phi_H(0)L(\Psi_H,\tilde{\omega}(\xi,t))(y(t) - h(\omega(t))) \\ &\quad +B\int_0^\alpha \Phi_H(\xi-\alpha)L(\Psi_H,\tilde{\omega}(\xi,t))[y(t-\xi) - h(\omega(t-\xi))]d\xi\end{aligned} \tag{F.42}$$

which is identical to the one of (9.51). △

Proof of Theorem 9.21:

Under the output feedback controller of (9.57), the closed-loop system takes the form:

$$\begin{aligned}\dot{\omega} &= A\omega(t) + B\omega(t-\alpha) + f(\omega(t),\omega(t-\alpha)) \\ &\quad + \Phi_H(0)L(\Psi_H,\tilde{\omega}(\xi,t))(y(t) - h(\omega(t))) \\ &\quad + B\int_0^\alpha \Phi_H(\xi-\alpha)L(\Phi_H,\tilde{\omega}(\xi,t))[y(t-\xi) - h(\omega(t-\xi))]d\xi \\ &\quad + g(\omega(t),\omega(t-\alpha))\frac{1}{L_gL_{\tilde{f}}^{r-1}h(\omega)}\left(-R_2^{-1}b^TP\bar{e}(t) + v^{(r)}(t) - L_{\tilde{f}}^r h(\omega)\right. \\ &\quad \left. - p_r(\omega(t),\bar{v}(t),\omega(t-\alpha),\bar{v}(t-\alpha))\right) \\ \dot{x} &= Ax(t) + Bx(t-\alpha) + f(x(t),x(t-\alpha)) \\ &\quad + g(x(t),x(t-\alpha))\frac{1}{L_gL_{\tilde{f}}^{r-1}h(\omega)}\left(-R_2^{-1}b^TP\bar{e}(t) + v^{(r)}(t) - L_{\tilde{f}}^r h(\omega)\right. \\ &\quad \left. - p_r(\omega(t),\bar{v}(t),\omega(t-\alpha),\bar{v}(t-\alpha))\right)\end{aligned} \tag{F.43}$$

Since we are interested in proving local exponential stability of the above system, we compute its linearization around the zero solution and use that $f(0,0)=0$ and $f(x(t),x(t-\alpha))$ includes higher-order terms, and $g(0,0)=c$, where c is a constant vector, to obtain the following linear system:

$$\begin{aligned}\dot{\omega} &= A\omega(t) + B\omega(t-\alpha) + \Phi_H(0)L(wx(t) - w\omega(t)) \\ &\quad + B\int_0^\alpha \Phi_H(\xi-\alpha)L[wx(t-\xi) - w\omega(t-\xi)]d\xi \\ &\quad + cu_{lin}(\omega(t),\bar{v}(t),\omega(t-\alpha),\bar{v}(t-\alpha)) \\ \dot{x} &= Ax(t) + Bx(t-\alpha) + cu_{lin}(\omega(t),\bar{v}(t),\omega(t-\alpha),\bar{v}(t-\alpha))\end{aligned} \tag{F.44}$$

where L is the linearization of the nonlinear vector $L(\Psi_H,\tilde{\omega}(\xi,t))$ around the zero solution, $w = \frac{\partial h}{\partial x}(0)$, and u_{lin} is the linearization of the term:

$$\begin{aligned}&\frac{1}{L_gL_{\tilde{f}}^{r-1}h(\omega)}(-R_2^{-1}b^TP\bar{e}(t) + v^{(r)}(t) - L_{\tilde{f}}^r h(\omega) \\ &- p_r(\omega(t),\bar{v}(t),\omega(t-\alpha),\bar{v}(t-\alpha)))\end{aligned} \tag{F.45}$$

around the zero solution. Introducing the error vector $e_r = x - \omega$, the closed-loop system of (F.44) can be written as:

$$\dot{e_r} = Ae_r(t) + Be_r(t-\alpha) + \Phi_H(0)Lwe_r(t) + B\int_0^{\alpha} \Phi_H(\xi-\alpha)Lwe_r(t-\xi)d\xi$$

$$\begin{aligned}\dot{x} &= Ax(t) + Bx(t-\alpha) \\ &\quad + cu_{lin}(x(t)+e_r(t), \bar{v}(t), x(t-\alpha)+e_r(t-\alpha), \bar{v}(t-\alpha))\end{aligned} \tag{F.46}$$

From the observability Assumption 9.3, we have that the error system:

$$\dot{e_r} = Ae_r(t) + Be_r(t-\alpha) + \Phi_H(0)Lwe_r(t) + B\int_0^{\alpha} \Phi_H(\xi-\alpha)Lwe_r(t-\xi)d\xi \tag{F.47}$$

is exponentially stable. Moreover, from the construction of the state feedback controller, we have that the system:

$$\begin{aligned}\dot{x} &= Ax(t) + Bx(t-\alpha) + cu_{lin}(x(t), \bar{v}(t), x(t-\alpha), \bar{v}(t-\alpha)) \\ y &= wx(t)\end{aligned} \tag{F.48}$$

is exponentially stable and the output asymptotically follows the reference input (Proof of Theorem 9.7). Therefore, we have that the linear system of (F.46) is an interconnection of two exponentially stable subsystems, which implies that it is also exponentially stable. Using the result of Theorem 1.1 in [108, Chapter 10] (which allows inferring the local stability properties of a nonlinear system based on its linearization) we obtain that the nonlinear closed-loop system of (F.43) is locally exponentially stable and the discrepancy between the output and the reference input asymptotically tends to zero. △

References

1. T. Alamo, D. Muoz de la Pea, D. Limon, and E. F. Camacho. Constrained min-max predictive control: a polynomial-time approach. In *Proceedings of 42nd IEEE Conference on Decision and Control*, pages 912–916, Maui, Hawaii, USA, 2003.
2. F. Allgöwer and H. Chen. Nonlinear model predictive control schemes with guaranteed stability. In R. Berber and C. Kravaris (Eds.), *NATO ASI on Nonlinear Model Based Process Control*, pages 465–494, Kluwer Academic Publishers, Dordrecht, The Netherlands, 1998.
3. F. Allgöwer and F. J. Doyle. Nonlinear process control - which way to the promised land? In *Proceedings of 5th International Conference on Chemical Process Control*, volume 93, pages 24–45, J. C. Kantor, C. E. Garcia and B. Carnahan Eds., AIChE Symposium Series No. 316, CACHE, AIChE, 1997.
4. F. Allgöwer, A. Rehm, and E. D. Gilles. An engineering perspective on nonlinear H^∞ control. In *Proceedings of 33rd IEEE Conference on Decision and Control*, pages 2537–2542, Orlando, FL, 1994.
5. R. Alur, C. Belta, F. Ivancic, V. Kumar, M. Mintz, G. Pappas, H. Rubin, and J. Schug. Hybrid modeling and simulation of biomolecular networks. In *Lecture Notes in Computer Science Series*, volume 2034, pages 19–32, Di Benedetto, M. D. and A. Sangiovanni-Vincentelli (Eds.), Berlin, Germany: Springer-Verlag, 2001.
6. R. Alur, C. Belta, V. Kumar, M. Mintz, G. J. Pappas, H. Rubin, and J. Schug. Modeling and analyzing biomolecular networks. *Computing in Science and Engineering*, 4:20–31, 2002.
7. J. Alvarez, J. J. Alvarez, M. A. Barron, and R. Suarez. The global stabilization of a two-input three-state polymerization reactor with saturated feedback. In *Proceedings of the American Control Conference*, pages 2917–2921, San Francisco, CA, 1993.
8. J. Alvarez, J. J. Alvarez, and R. Suarez. Nonlinear bounded control for a class of continuous agitated tank reactors. *Chem. Eng. Sci.*, 46:3235–3249, 1991.
9. C. Antoniades and P. D. Christofides. Feedback control of nonlinear differential difference equation systems. *Chem. Eng. Sci.*, 54:5677–5709, 1999.
10. C. Antoniades and P. D. Christofides. Robust control of nonlinear time-delay systems. *Appl. Math. & Comp. Sci.*, 9:811–838, 1999.

11. C. Antoniades and P. D. Christofides. Nonlinear feedback control of parabolic partial differential difference equation systems. *Int. J. Contr.*, 73:1572–1591, 2000.
12. H. B. Aradhye, J. F. Davis, and B. R. Bakshi. Art-2 and multiscale art-2 for on-line process fault detection - validation via industrial case studies and monte carlo simulation. *Annual Reviews in Control*, 26:113–127, 2002.
13. A. Arbel, I. H. Rinard, and R. Shinnar. Dynamics and control of fluidized catalytic crackers. 1. Modeling of the current generation of the FCC's. *Ind. & Eng. Chem. Res.*, 34:1228–1243, 1995.
14. A. Arbel, I. H. Rinard, and R. Shinnar. Dynamics and control of fluidized catalytic crackers. 3. Designing the control system: Choice of manipulated and measured variables for partial control. *Ind. & Eng. Chem. Res.*, 35:2215–2233, 1996.
15. Y. Arkun and J. P. Calvet. Robust stabilization of input/output linearizable systems under uncertainty and disturbances. *AIChE J.*, 38:1145–1154, 1992.
16. A. Armaou and P. D. Christofides. Wave suppression by nonlinear finite-dimensional control. *Chem. Eng. Sci.*, 55:2627–2640, 2000.
17. Z. Artstein. Stabilization with relaxed control. *Nonlinear Analysis*, 7:1163–1173, 1983.
18. K. J. Astrom and T. Hagglund. Automatic tuning of simple regulators with specification on the gain and phase margins. *Automatica*, 20:645–651, 1984.
19. K. J. Astrom and T. Hagglund. Revisiting the Ziegler-Nichols step response method for PID control. *J. Proc. Contr.*, 14:635–650, 2004.
20. K. J. Astrom, T. Hagglund, C. C. Hang, and W. K. Ho. Automatic tuning and adaption for PID controllers- A survey. *Contr. Eng. Pract.*, 1:699–714, 1993.
21. B. Aufderheide, V. Prasad, and B. W. Bequette. A comparison of fundamental model-based and multiple model predictive control. In *Proceedings of 40th IEEE Conference on Decision and Control*, pages 4863–4868, Orlando, FL, 2001.
22. T. A. Badgwell. Robust model predictive control of stable linear systems. *Int. J. Contr.*, 68:797–818, 1997.
23. T. A. Badgwell. Robust stability conditions for SISO model predictive control algorithms. *Automatica*, 33:1357–1361, 1997.
24. J. Baker and P. D. Christofides. Finite dimensional approximation and control of nonlinear parabolic PDE systems. *Int. J. Contr.*, 73:439–456, 2000.
25. A. Banerjee and Y. Arkun. Model predictive control of plant transitions using a new identification technique for interpolating nonlinear models. *J. Proc. Contr.*, 8:441–457, 1998.
26. J. Bao, W. Z. Zhang, and P. L. Lee. Passivity-based decentralized failure-tolerant control. *Ind. & Eng. Chem. Res.*, 41:5702–5715, 2002.
27. J. F. Bartee, K. F. Bloss, and C. Georgakis. Design of nonlinear reference system control structures. In *AIChE Annual Meeting, San Francisco, CA*, 1989.
28. P. I. Barton and C. C. Pantelides. Modeling of combined discrete/continuous processes. *AIChE Journal*, 40:966–979, 1994.
29. G. Bastin and D. Dochain. *On-line Estimation and Adaptive Control of Bioreactors.* Elsevier Science Pub. B. V., Amsterdam, 1990.
30. A. Batigun, K. R. Harris, and A. Palazoglu. Studies on the analysis of nonlinear processes via functional expansions. I. Solution of nonlinear ODEs. *Chem. Eng. Sci.*, 52:3183–3195, 1997.

31. A. Bemporad, F. Borreli, and M. Morari. Min-max control of constrained uncertain discrete-time linear systems. *IEEE Trans. Automat. Contr.*, 48:1600–1606, 2003.
32. A. Bemporad and M. Morari. Control of systems integrating logic, dynamics and constraints. *Automatica*, 35:407–427, 1999.
33. A. Bemporad and M. Morari. Control of systems integrating logic, dynamics and constraints. *Automatica*, 35:407–427, 1999.
34. A. R. Benaskeur and A. Desbiens. Backstepping-based adaptive PID control. *IEEE Proc. Contr. Theory Appl.*, 149:54–59, 2002.
35. W. B. Bequette. Nonlinear control of chemical processes: A review. *Ind. & Eng. Chem. Res.*, 30:1391–1413, 1991.
36. K. P. M. Bhat and H. N. Koivo. Modal characterizations of controllability and observability in time delay systems. *IEEE Trans. Automat. Contr.*, 21:292–293, 1976.
37. R. R. Bitmead, M. Gevers, and V. Wertz. *Adaptive Optimal Control-The Thinking Man's GPC.* Prentice-Hall, Englewood Cliffs, NJ, 1990.
38. M. Blanke, R. Izadi-Zamanabadi, S. A. Bogh, and C. P. Lunau. Fault-tolerant control systems – a holistic view. *Contr. Eng. Prac.*, 5:693–702, 1997.
39. M. S. Branicky. Multiple Lyapunov functions and other analysis tools for switched and hybrid systems. *IEEE Trans. Automat. Contr.*, 43:475–482, 1998.
40. R. Brockett. Minimum attention control. In *Proceedings of 36th Conference on Decision and Control*, pages 2628–2632, San Diego, CA, 1997.
41. C. Brosilow. Structure and design of smith predictors from the viewpoint of inferential control. In *Proceedings of Joint American Control Conference*, 1976.
42. J. P. Calvet and Y. Arkun. Feedforward and feedback linearization of nonlinear systems and its implementation using internal model control. *Ind. & Eng. Chem. Res.*, 27:1822, 1991.
43. P. J. Campo and M. Morari. Robust model predictive control. In *Proceedings of the American Control Conference*, pages 1021–1026, Green Valley, AZ, 1987.
44. W. Chang, R. Hwang, and J. Hsieh. A self tuning PID control for a class of nonlinear systems based on the Lyapunov approach. *J. Proc. Contr.*, 12:233–242, 2002.
45. H. Chen and F. Allgöwer. A quasi-infinite horizon nonlinear model predictive control scheme with guaranteed stability. *Automatica*, 34:1205–1217, 1998.
46. L. Chisci, P. Falugi, and G. Zappa. Gain-scheduling MPC of nonlinear systems. *Int. J. Robust & Nonlin. Contr.*, 13:295–308, 2003.
47. T. Chiu and P. D. Christofides. Nonlinear control of particulate processes. *AIChE J.*, 45:1279–1297, 1999.
48. D. Chmielewski and V. Manousiouthakis. On constrained infinite-time linear quadratic optimal control. *Syst. & Contr. Lett.*, 29:121–129, 1996.
49. P. D. Christofides. Output feedback control of nonlinear two-time-scale processes. *Ind. & Eng. Chem. Res.*, 37:1893–1909, 1998.
50. P. D. Christofides. Robust output feedback control of nonlinear singularly perturbed systems. *Automatica*, 36:45–52, 2000.
51. P. D. Christofides. *Model-Based Control of Particulate Processes.* Kluwer Academic Publishers, 2002.
52. P. D. Christofides and P. Daoutidis. Compensation of measurable disturbances in two-time-scale nonlinear systems. *Automatica*, 32:1553–1573, 1996.
53. P. D. Christofides and P. Daoutidis. Feedback control of hyperbolic PDE systems. *AIChE J.*, 42:3063–3086, 1996.

54. P. D. Christofides and P. Daoutidis. Feedback control of two-time-scale nonlinear systems. *Int. J. Contr.*, 63:965–994, 1996.
55. P. D. Christofides and P. Daoutidis. Finite-dimensional control of parabolic PDE systems using approximate inertial manifolds. *J. Math. Anal. Appl.*, 216:398–420, 1997.
56. P. D. Christofides and P. Daoutidis. Robust control of multivariable two-time-scale nonlinear systems. *J. Proc. Contr.*, 7:313–328, 1997.
57. P. D. Christofides and A. R. Teel. Singular perturbations and input-to-state stability. *IEEE Trans. Automat. Contr.*, 41:1645–1650, 1996.
58. P. D. Christofides, A. R. Teel, and P. Daoutidis. Robust semi-global output tracking for nonlinear singularly perturbed systems. *Int. J. Contr.*, 65:639–666, 1996.
59. P. Cominos and N. Munro. PID controllers: recent tuning methods and design to specifications. *IEEE Proc. Contr. Theory Appl.*, 149:17–25, 2002.
60. M. Corless. Control of uncertain nonlinear systems. *J. Dyn. Syst. Meas. & Contr.*, 115:362–372, 1993.
61. E. Coulibaly, S. Maiti, and C. Brosilow. Internal model predictive control. *Automatica*, 31:1471–1479, 1995.
62. F. A. Cuzzola, J. C. Geromel, and M. Morari. An improved approach for constrained robust model predictive control. *Automatica*, 38:1183–1189, 2002.
63. D. Da and M. Corless. Exponential stability of a class of nonlinear singularly perturbed systems with marginally stable boundary layer systems. In *Proceedings of American Control Conference*, pages 3101–3105, San Francisco, CA, 1993.
64. P. Daoutidis and P. D. Christofides. Dynamic feedforward/output feedback control of nonlinear processes. *Chem. Eng. Sci.*, 50:1889–2007, 1995.
65. P. Daoutidis and C. Kravaris. Dynamic output feedback control of multivariable nonlinear processes. *Chem. Eng. Sci.*, 49:433–447, 1994.
66. J. F. Davis, M. L. Piovoso, K. Kosanovich, and B. Bakshi. Process data analysis and interpretation. *Advances in Chemical Engineering*, 25:1–103, 1999.
67. G. De Nicolao, L. Magni, and R. Scattolini. On robustness of receding horizon control with terminal constraints. *IEEE Trans. Automat. Contr.*, 41:451–453, 1996.
68. N. M. C. De oliveira and L. T. Biegler. Constraint handling and stability properties of model-predictive control. *AIChE J.*, 40:1138–1155, 1994.
69. C. De Persis and A. Isidori. A geometric approach to nonlinear fault detection and isolation. *IEEE Trans. Automat. Contr.*, 46:853–865, 2001.
70. R. A. DeCarlo, M. S. Branicky, S. Pettersson, and B. Lennartson. Perspectives and results on the stability and stabilizability of hybrid systems. *Proceedings of the IEEE*, 88:1069–1082, 2000.
71. M. M. Denn. *Process Modeling.* Longam Inc., New York, 1986.
72. D. Dochain and M. Perrier. Control design for nonlinear wastewater treatment processes. *Wat. Sci. Tech.*, 18:283–294, 1993.
73. V. Dua, N. A. Bozinis, and E. N. Pistikopoulos. A multiparametric programming approach for mixed-integer quadratic engineering problems. *Comp. & Chem. Eng.*, 26:715–733, 2002.
74. S. Dubljević and N. Kazantzis. A new Lyapunov design approach for nonlinear systems based on Zubov's method. *Automatica*, 38:1999–2007, 2002.
75. N. H. El-Farra, A. Armaou, and P. D. Christofides. Analysis and control of palabolic PDE systems with input constraints. *Automatica*, 39:715–725, 2003.

76. N. H. El-Farra, T. Chiu, and P. D. Christofides. Analysis and control of particulate processes with input constraints. *AIChE J.*, 47:1849–1865, 2001.
77. N. H. El-Farra and P. D. Christofides. Feedback control of switched nonlinear systems using multiple Lyapunov functions. In *Proceedings of American Control Conference*, pages 3496–3502, Arlington, VA, 2001.
78. N. H. El-Farra and P. D. Christofides. Integrating robustness, optimality and constraints in control of nonlinear processes. *Chem. Eng. Sci.*, 56:1841–1868, 2001.
79. N. H. El-Farra and P. D. Christofides. Robust near-optimal output feedback control of nonlinear systems. *Int. J. Contr.*, 74:133–157, 2001.
80. N. H. El-Farra and P. D. Christofides. Switching and feedback laws for control of constrained switched nonlinear systems. In *Lecture Notes in Computer Science*, volume 2289, pages 164–178, Tomlin, C. J. and M. R. Greenstreet Eds., Berlin, Germany: Springer-Verlag, 2002.
81. N. H. El-Farra and P. D. Christofides. Bounded robust control of constrained multivariabe nonlinear processes. *Chem. Eng. Sci.*, 58:3025–3047, 2003.
82. N. H. El-Farra and P. D. Christofides. Coordinating feedback and switching for control of hybrid nonlinear processes. *AIChE J.*, 49:2079–2098, 2003.
83. N. H. El-Farra and P. D. Christofides. Robust inverse optimal control of nonlinear systems. *Int. J. Robust & Nonlin. Contr.*, 13:1371–1388, 2003.
84. N. H. El-Farra and P. D. Christofides. Coordinating feedback and switching for control of spatially-distributed processes. *Comp. & Chem. Eng.*, 28:111–128, 2004.
85. N. H. El-Farra, A. Gani, and P. D. Christofides. Analysis of mode transitions in biological networks. *AIChE J.*, 51, 2005.
86. N. H. El-Farra, A. Gani, and P. D. Christofides. Fault-tolerant control of process systems using communication networks. *AIChE J.*, 51:1665–1682, 2005.
87. N. H. El-Farra, P. Mhaskar, and P. D. Christofides. Hybrid control of uncertain process systems. In *Proceedings of 8th International Symposium on Process Systems and Engineering*, pages 814–819, Kumming, P.R. China, 2003.
88. N. H. El-Farra, P. Mhaskar, and P. D. Christofides. Hybrid predictive control of nonlinear systems: Method and applications to chemical processes. *Int. J. Robust & Nonlin. Contr.*, 14:199–225, 2004.
89. N. H. El-Farra, P. Mhaskar, and P. D. Christofides. Uniting bounded control and MPC for stabilization of constrained linear systems. *Automatica*, 40:101–110, 2004.
90. N. H. El-Farra, P. Mhaskar, and P. D. Christofides. Output feedback control of switched nonlinear systems using multiple Lyapunov functions. *Syst. & Contr. Lett.*, in press, 2005.
91. S. Engell, S. Kowalewski, C. Schulz, and O. Stursberg. Continuous-discrete interactions in chemical processing plants. *Proceedings of the IEEE*, 88:1050–1068, 2000.
92. C. A. Farschman, K. P. Viswanath, and E. B. Ydstie. Process systems and inventory control. *AIChE J.*, 44:1841–1857, 1998.
93. A. Fernando, C. Fontes, and L. Magni. Min-max model predictive control of nonlinear systems using discontinuous feedback. *IEEE Trans. Automat. Contr.*, 48:1750–1755, 2003.
94. B. Foss and S. J. Qin. Interpolating optimizing process control. *J. Proc. Contr.*, 7:129–138, 1997.

95. P. M. Frank and X. Ding. Survey of robust residual generation and evaluation methods in observer-based fault detection systems. *J. Proc. Contr.*, 7:403–424, 1997.
96. R. A. Freeman and P. V. Kokotovic. Robust control Lyapunov functions: The measurement feedback case. In *Proceedings of 33rd IEEE Conference on Decision and Control*, pages 3533–3538, Lake Buena Vista, FL, 1994.
97. R. A. Freeman and P. V. Kokotovic. *Robust Nonlinear Control Design: State-Space and Lyapunov Techniques.* Birkhauser, Boston, 1996.
98. C. E. Garcia and M. Morari. Internal model control. 1: A unifying review and some new results. *IEC Proc. Des. Dev.*, 21:308–322, 1982.
99. C. E. Garcia, D. M. Prett, and M. Morari. Model predictive control: Theory and practice–a survey. *Automatica*, 25:335–348, 1989.
100. V. Garcia-Onorio and B. E. Ydstie. Distributed, asynchronous and hybrid simulation of process networks using recording controllers. *Int. J. Robust & Nonlin. Contr.*, 14:227–248, 2004.
101. H. Genceli and M. Nikolaou. Robust stability analysis of constrained l(1)–norm model predictive control. *AIChE J.*, 39:1954–1965, 1993.
102. H. Genceli and M. Nikolaou. Robust stability analysis of constrained l1-norm model predictive control. *AIChE J.*, 39:1954–1965, 1993.
103. R. Ghosh and C. J. Tomlin. Lateral inhibition through delta-notch signaling: A piecewise affine hybrid model. In *Lecture Notes in Computer Science Series*, volume 2034, pages 232–246, Di Benedetto, M. D. and A. Sangiovanni-Vincentelli (Eds.), Berlin, Germany: Springer-Verlag, 2001.
104. G. Grimm, M. Messina, A. Teel, and S. Tuna. Nominally robust model predictive control with state constraints. In *Proceedings of 42nd IEEE Conference on Decision and Control*, pages 1413–1418, Maui, HI, 2003.
105. J. W. Grizzle and P. V. Kokotovic. Feedback linearization of sampled-data systems. *IEEE. Trans. Aut. Contr.*, 33:857–859, 1988.
106. I. E. Grossmann, S. A. van den Heever, and I. Harjukoski. Discrete optimization methods and their role in the integration of planning and scheduling. In *Proceedings of 6th International Conference on Chemical Process Control*, pages 124–152, Tucson, AZ, 2001.
107. G. Gu, P. P. Khargonekar, and E. B. Lee. Approximation of infinite dimensional systems. *IEEE Trans. Automat. Contr.*, 34:610–618, 1989.
108. J. K. Hale and S. M. Verduyn Lunel. *Introduction to Functional Differential Equations.* Springer-Verlag, New York, 1993.
109. K. M. Hangos, A. A. Alonso, J. D. Perkins, and E. B. Ydstie. Thermodynamic approach to the structural stability of process plants. *AIChE J.*, 45:802–816, 1999.
110. I. Harjunkoski, V. Jain, and I. E. Grossmann. Hybrid mixed-integer/constrained logic programming strategies for solving scheduling and combinatorial optimization problems. *Comp. & Chem. Eng.*, 24:337–343, 2000.
111. K. R. Harris and A. Palazoglu. Studies on the analysis of nonlinear processes via functional expansions. II. forced dynamic responses. *Chem. Eng. Sci.*, 52:3197–3207, 1997.
112. K. R. Harris and A. Palazoglu. Studies on the analysis of nonlinear processes via functional expansions. III. control design. *Chem. Eng. Sci.*, 53:4005–4022, 1998.
113. T. J. Harris, F. Boudreau, and J. F. MacGregor. Performance assessment of multivariable feedback controllers. *Automatica*, 32:1505–1518, 1996.

114. J. Hasty, F. Isaacs, M. Dolnik, D. McMillen, and J. J. Collins. Designer gene networks: Towards fundamental cellular control. *Chaos*, 11:207–220, 2001.
115. V. Hatzimanikatis, K. H. Lee, and J. E. Bailey. A mathematical description of regulation of the G1-S transition of the mammalian cell cycle. *Biotechnology and Bioengineering*, 65:631–637, 1999.
116. M. A. Henson. Nonlinear model predictive control: Current status and future directions. *Comp. & Chem. Eng.*, 23:187–202, 1998.
117. M. A. Henson and D. E. Seborg. Time delay compensation for nonlinear processes. *Ind. & Eng. Chem. Res.*, 33:1493–1500, 1994.
118. M. A. Henson and D. E. Seborg. *Nonlinear Process Control*. Prentice-Hall, Englewood Cliffs, NJ, 1997.
119. J. Hespanha, D. Liberzon, A. S. Morse, B. D. O. Anderson, T. S. Brinsmead, and F. De Bruyne. Multiple model adaptive control – Part 2: Switching. *Int. J. Robust & Nonlin. Contr.*, 11:479–496, 2001.
120. J. P. Hespanha and A. S. Morse. Stability of switched systems with average dwell time. In *Proceedings of 38th IEEE Conference on Decision and Control*, pages 2655–2660, Phoenix, AZ, 1999.
121. K. A. Hoo and J. C. Kantor. An exothermic continuous stirred tank reactor is feedback equivalent to a linear system. *Chem. Eng. Comm.*, 37:1–17, 1985.
122. K. Hrissagis and O. D. Crisalle. Mixed objective optimization for robust predictive controller synthesis. In *Proceedings of 5th International Conference on Chemical Process Control*, in press, Tahoe City, CA, 1997.
123. B. Hu, X. Xu, P. J. Antsaklis, and A. N. Michel. Robust stabilizing control law for a class of second-order switched systems. *Syst. & Contr. Lett.*, 38:197–207, 1999.
124. T. Hu and Z. Lin. Composite quadratic Lyapunov functions for constrained control systems. In *Proceedings of 41st IEEE Conference on Decision and Control*, pages 3500–3505, Las Vegas, NV, 2002.
125. I. Huq, M. Morari, and R. C. Sorensen. Modifications to model IV fluid catalytic cracking units to improve dynamic performance. *AIChE J.*, 41:1481–1499, 1995.
126. A. Isidori. *Nonlinear Control Systems: An Introduction*. Springer-Verlag, Berlin-Heidelberg, second edition, 1989.
127. A. Isidori. A necessary condition for nonlinear H_∞ control via measurement feedback. *Syst. & Contr. Lett.*, 23:169–177, 1994.
128. A. Isidori. Stabilization of nonlinear systems using output feedback. In *Dynamical Systems, Control, Coding, Computer Vision: New Trends, Interfaces, and Interplay*, pages 111–133, G. Picci and D. S. Gilliam Eds., Birkhäuser, 1999.
129. A. Isidori and A. Astolfi. Disturbance attenuation and H_∞ control via measurement feedback in a nonlinear system. *IEEE Trans. Automat. Control*, 37:1283–1293, 1992.
130. A. Isidori and W. Kang. H_∞ control via measurement feedback for general nonlinear systems. *IEEE Trans. Automat. Control*, 40:466–472, 1995.
131. N. F. Jerome and W. H. Ray. High performance multivariable control strategies for systems having time delays. *AIChE J.*, 32:914–931, 1986.
132. Z. P. Jiang, A. R. Teel, and L. Praly. Small-gain theorem for ISS systems and applications. *Math. Control Signals Systems*, 7:95–120, 1995.
133. R. E. Kalman. When is a control system optimal? *J. Basic Eng.*, 86:51–60, 1964.

134. I. Kanellakopoulos, P. V. Kokotovic, and R. Marino. An extended direct scheme for robust adaptive nonlinear control. *Automatica*, 27:247–255, 1991.
135. N. Kapoor and P. Daoutidis. Stabilization of systems with input constraints. *Int. J. Contr.*, 34:653–675, 1998.
136. N. Kapoor and P. Daoutidis. An observer-based anti-windup scheme for nonlinear systems with input constraints. *Int. J. Contr.*, 9:18–29, 1999.
137. N. Kapoor and P. Daoutidis. On the dynamics of nonlinear systems with input constraints. *Chaos*, 9:88–94, 1999.
138. N. Kapoor and P. Daoutidis. Stabilization of nonlinear processes with input constraints. *Comp. & Chem. Eng.*, 24:9–21, 2000.
139. N. Kapoor, A. R. Teel, and P. Daoutidis. An anti-windup design for linear systems with input saturation. *Automatica*, 34:559–574, 1998.
140. N. Kazantzis. A functional equations approach to nonlinear discrete-time feedback stabilization through pole-placement. *Syst. & Contr. Lett.*, 43:361–369, 2001.
141. N. Kazantzis and C. Kravaris. Energy-predictive control: a new synthesis approach for nonlinear process control. *Chem. Eng. Sci.*, 54:1697–1709, 1999.
142. N. Kazantzis and C. Kravaris. Nonlinear observer design using Lyapunov's auxiliary theorem. *Syst. & Contr. Lett.*, 34:241–247, 1999.
143. N. Kazantzis, C. Kravaris, C. Tseronis, and R. A. Wright. Optimal controller tuning for nonlinear processes. *Automatica*, 41:79–86, 2005.
144. N. Kazantzis, C. Kravaris, and R. A. Wright. Nonlinear observer design for process monitoring. *Ind. & Eng. Chem. Res.*, 39:408–419, 2000.
145. S. Keerthi and E. Gilbert. Optimal, infinite horizon feedback laws for a general class of constrained discrete time systems. *Journal of Optimization Theory and Applications*, 57:265–293, 1988.
146. H. Khalil. Robust servomechanism output feedback controller for feedback linearizable systems. *Automatica*, 30:1587–1599, 1994.
147. H. K. Khalil. Stability analysis of nonlinear multi-parameter singularly perturbed systems. *IEEE Trans. Automat. Contr.*, 32:260–263, 1987.
148. H. K. Khalil. *Nonlinear Systems.* Macmillan Publishing Company, New York, second edition, 1996.
149. H. K. Khalil and F. Esfandiari. Semiglobal stabilization of a class of nonlinear systems using output feedback. *IEEE Trans. Automat. Contr.*, 38:1412–1415, 1993.
150. H. N. Koivo and A. J. Koivo. Control and estimation of systems with time-delays. In *Distributed Parameter Systems,* W. H. Ray and D. G. Lainiotis (eds.), New York, NY, 1978.
151. P. Kokotovic and M. Arcak. Constructive nonlinear control: a historical perspective. *Automatica*, 37:637–662, 2001.
152. P. V. Kokotovic, O'Malley R. E., and P. Sannuti. Singular perturbations and order reduction in control theory- an overview. *Automatica*, 12:123–132, 1976.
153. P. V. Kokotovic, H. K. Khalil, and J. O'Reilly. *Singular Perturbations in Control: Analysis and Design.* Academic Press, London, 1986.
154. M. V. Kothare, P. J. Campo, M. Morari, and C. N. Nett. A unified framework for the study of anti-windup designs. *Automatica*, 30:1869–1883, 1994.
155. S. L. D. Kothare and M. Morari. Contractive model predictive control for constrained nonlinear systems. *IEEE Trans. Automat. Contr.*, 45:1053–1071, 2000.

156. X. D. Koutsoukos, P. J. Antsaklis, J. A. Stiver, and M. D. Lemmon. Supervisory control of hybrid systems. *Proceedings of the IEEE*, 88:1026–1049, 2000.
157. C. Kravaris and Y. Arkun. Geometric nonlinear control – an overview. In *Proceedings of 4th International Conference on Chemical Process Control*, pages 477–515, Y. Arkun and W. H. Ray Eds., Padre Island, TX, 1991.
158. C. Kravaris and C. B. Chung. Nonlinear state feedback synthesis by global input/output linearization. *AIChE J.*, 33:592–604, 1987.
159. C. Kravaris and J. C. Kantor. Geometric methods for nonlinear process control 1. Background. *Ind. & Eng. Chem. Res.*, 29:2295–2310, 1990.
160. C. Kravaris and J. C. Kantor. Geometric methods for nonlinear process control 2. Controller synthesis. *Ind. & Eng. Chem. Res.*, 29:2310–2323, 1990.
161. C. Kravaris and S. Palanki. Robust nonlinear state feedback under structured uncertainty. *AIChE J.*, 34:1119–1127, 1988.
162. C. Kravaris and R. A. Wright. Deadtime compensation for nonlinear processes. *AIChE J.*, 35:1535–1542, 1989.
163. M. Krstic and Z-H. Li. Inverse optimal design of input-to-state stabilizing nonlinear controllers. *IEEE Trans. Automat. Contr.*, 43:336–350, 1998.
164. N. Krstic, I. Kanellakopoulos, and P. Kokotovic. *Nonlinear and Adaptive Control Design*. Wiley, New York, first edition, 1995.
165. M. J. Kurtz, G. Y. Zhu, and M. A. Henson. Constrained output feedback control of a multivariable polymerization reactor. *IEEE Trans. Contr. Syst. Tech.*, 8:87–97, 2000.
166. W. Langson, I. Chryssochoos, S. V. Rakovic, and D. Q. Mayne. Robust model predictive control using tubes. *Automatica*, 40:125–133, 2004.
167. J. H. Lee and B. Cooley. Recent advances in model predictive control and other related areas. In *Proceedings of 5th International Conference on Chemical Process Control*, volume 93, pages 201–216, J. C. Kantor, C. E. Garcia and B. Carnahan Eds., AIChE Symposium Series No. 316, CACHE, AIChE, 1997.
168. J. H. Lee and Z. H. Yu. Worst-case formulations of model predictive control for systems with bounded parameters. *Automatica*, 33:763–781, 1997.
169. J. T. Lee, W. Cho, and T. F. Edgar. Control system design based on a first order plus time delay model. *J. Proc. Contr.*, 7:65–73, 1997.
170. Y. I. Lee and B. Kouvaritakis. Superposition in efficient robust constrained predictive control. *Automatica*, 38:875–878, 2002.
171. B. Lehman, I. Widjaya, and K. Shujaee. Vibrational control of chemical reactions in a CSTR with delayed recycle stream. *J. Math. Anal. Appl.*, 193:28–59, 1995.
172. G. Leitmann. On one approach to the control of uncertain systems. *J. Dyn. Syst. Meas. & Contr.*, 115:373–380, 1993.
173. A. Leonessa, W. M. Haddad, and V. Chellaboina. Nonlinear system stabilization via hierarchical switching control. *IEEE Trans. Automat. Contr.*, 46:17–28, 2001.
174. J. Levine and P. Rouchon. Quality control of binary distillation columns via nonlinear aggregated models. *Automatica*, 27:463–480, 1991.
175. B. Lewin. *Genes VII*. Oxford Univ. Press, Cambridge, UK, 2000.
176. D. Liberzon, E.D. Sontag, and Y Wang. Universal construction of feedback laws achieving ISS and integral-ISS disturbance attenuation. *Syst. & Contr. Lett.*, 46:111–127, 2002.
177. Y. Lin and E. D. Sontag. A universal formula for stabilization with bounded controls. *Systems & Control Letters*, 16:393–397, 1991.

178. W. Lu, E. B. Lee, and Q. T. Zhang. Balanced approximation of two-dimensional and delay-differential equations. *Int. J. Contr.*, 46:2199–2218, 1987.
179. L. Magni, G. De Nicolao, and R. Scattolini. A stabilizing model-based predictive control algorithm for nonlinear systems. *Automatica*, 37:1351–1362, 2001.
180. L. Magni, G. De Nicolao, R. Scattolini, and F. Allgöwer. Robust model predictive control for nonlinear discrete-time systems. *Int. J. Robust & Nonlin. Contr.*, 13:229–246, 2003.
181. L. Magni and R. Sepulchre. Stability margins of nonlinear receding-horizon control via inverse optimality. *Syst. & Contr. Lett.*, 32:241–245, 1997.
182. M. Mahmoud, J. Jiang, and Y. Zhang. *Active Fault Tolerant Control Systems: Stochastic Analysis and Synthesis*, volume 287 of *Lecture Notes in Control and Information Sciences*. Springer-Verlag, Heidelberg, 2003.
183. N. A. Mahmoud and H. K. Khalil. Asymptotic regulation of minimum phase nonlinear systems using output feedback. *IEEE Trans. Automat. Contr.*, 41:1402–1412, 1996.
184. A. Manitius and H. Tran. Computation of closed-loop eigenvalues associated with the optimal regulator problem for functional differential equations. *IEEE Trans. Automat. Contr.*, 30:1245–1248, 1985.
185. A. Manitius, H. Tran, G. Payre, and R. Roy. Computation of eigenvalues associated with functional differential equations. *SIAM J. Sci. Stat. Comput.*, 8:222–247, 1987.
186. V. Manousiouthakis and D. Chmielewski. On constrained infinite-time nonlinear optimal control. *Chem. Eng. Sci.*, 537:105–114, 2002.
187. I. M. Y. Mareels and D. J. Hill. Monotone stability of nonlinear feedback systems. *J. Math. Systems Estimation Control*, 2:275–291, 1992.
188. R. Marino and P. Tomei. Global adaptive output feedback control of nonlinear systems – Part 1: Linear parameterization. *IEEE Trans. Automat. Contr.*, 38:17–32, 1993.
189. R. Marino and P. Tomei. Global adaptive output feedback control of nonlinear systems – Part 2: nonlinear parameterization. *IEEE Trans. Automat. Contr.*, 38:33–48, 1993.
190. D. Q. Mayne and H. Michalska. Receeding horizon control of nonlinear systems. *IEEE Trans. Automat. Contr.*, 35:814–824, 1990.
191. D. Q. Mayne, J. B. Rawlings, C. V. Rao, and P. O. M. Scokaert. Constrained model predictive control: Stability and optimality. *Automatica*, 36:789–814, 2000.
192. D. Q. Mayne. Nonlinear model predictive control: An assessment. In *Proceedings of 5th International Conference on Chemical Process Control*, volume 93, pages 217–231, J. C. Kantor, C. E. Garcia and B. Carnahan Eds., AIChE Symposium Series No. 316, CACHE, AIChE, 1997.
193. M. W. McConley, B. D. Appleby, M. A. Dahleh, and E. Freon. A computationally efficient Lyapunov-based scheduling procedure for control of nonlinear systems with stability guarantees. *IEEE Trans. Automat. Contr.*, 45:33–49, 2000.
194. R. C. McFarlane, R. C. Reineman, J. F. Bartee, and C. Georgakis. Dynamic simulator for a model IV fluid catalytic cracking unit. *Comp. & Chem. Eng.*, 17:275–293, 1993.

195. P. Mhaskar, N. H. El-Farra, and P. D. Christofides. Hybrid predictive control of process systems. *AIChE J.*, 50:1242–1259, 2004.
196. P. Mhaskar, N. H. El-Farra, and P. D. Christofides. Predictive control of switched nonlinear systems with scheduled mode transitions. *IEEE Trans. Automat. Contr.,* in press, 2005.
197. P. Mhaskar, N. H. El-Farra, and P. D. Christofides. Robust hybrid predictive control of nonlinear systems. *Automatica*, 41:209–217, 2005.
198. P. Mhaskar, N. H. El-Farra, and P. D. Christofides. Stabilization of nonlinear systems with control and state constraints using Lyapunov-based predictive control. *Syst. & Contr. Lett.,* in press, 2005.
199. H. Michalska and D. Q. Mayne. Receeding horizon control of non-linear systems. *IEEE Trans. Automat. Contr.*, 35:814–824, 1990.
200. H. Michalska and D. Q. Mayne. Robust receding horizon control of constrained nonlinear systems. *IEEE Trans. Automat. Contr.*, 38:1623–1633, 1993.
201. H. Michalska and D. Q. Mayne. Moving horizon observers and observer-based control. *IEEE Trans. Automat. Contr.*, 40:995–1006, 1995.
202. J. J. Monge and C. Georgakis. The effect of operating variables on the dynamics of catalytic cracking processes. *Chem. Eng. Comm.*, 60:1–15, 1987.
203. L. A. Montestruque and P. J. Antsaklis. On the model-based control of networked systems. *Automatica*, 39:1837–1843, 2003.
204. C. F. Moore, C. L. Smith, and P. W. Murril. Improved algorithm for direct digital control. *Instr. & Contr. Syst.*, 43:70–79, 1970.
205. P. J. Moylan and B. D. O. Anderson. Nonlinear regulator and an inverse optimal control problem. *IEEE Trans. Automat. Contr.*, 18:460–465, 1973.
206. G. J. Nazaroff. Stability and stabilization of linear differential delay systems. *IEEE Trans. Automat. Contr.*, 18:317–318, 1973.
207. D. Nesic, A. R. Teel, and P. V. Kokotovic. Sufficient conditions for stabilization of sampled-data nonlinear systems via discrete-time approximations. *Syst. & Contr. Lett.*, 38:259–270, 1999.
208. H. Nijmeijer and A. J. van der Schaft. *Nonlinear Dynamical Control Systems.* Springer-Verlag, New York, 1990.
209. M. N. Nounou, B. R. Bakshi, P. K. Goel, and X. Shen. Bayesian principal component analysis. *J. Chemometrics*, 16:576–595, 2002.
210. B. Novak and J. J. Tyson. Modeling the cell division cycle: M-phase trigger, oscillations and size control. *Journal Theoretical Biology*, 165:101–134, 1993.
211. B. A. Ogunnaike and W. H. Ray. Multivariable controller design for linear systems having multiple time delays. *AIChE J.*, 25:1042–1057, 1979.
212. B. A. Ogunnaike and R. A. Wright. Industrial applications of nonlinear control. In *Proceedings of 5th International Conference on Chemical Process Control*, volume 93, pages 46–59, J. C. Kantor, C. E. Garcia and B. Carnahan Eds., AIChE Symposium Series No. 316, CACHE, AIChE, 1997.
213. N. M. C. Oliveira and L. T. Biegler. Constraint handling and stability properties of model-predictive control. *AIChE J.*, 40:1138–1150, 1994.
214. Z. Pan and T. Basar. H^∞-optimal control for singularly perturbed systems – Part I: Perfect state measurements. *Automatica*, 29:401–423, 1993.
215. H. Panagopoulos, K. J. Astrom, and T. Hagglund. Design of PID controllers based on constrained optimization. *IEEE Proc. Contr. Theory Appl.*, 149:32–40, 2002.

216. A. Papachristodoulou, S. Prajna, and J. C. Doyle. On the construction of lyapunov functions using the sum of squares decomposition. In *Proceedings of 41st IEEE Conference on Decision and Control*, pages 3482–3487, Las Vegas, NV, 2002.
217. J. R. Partington. Approximation of delay system by fourier-laguerre series. *Automatica*, 27:569–572, 1991.
218. R. J. Patton. Fault-tolerant control systems: The 1997 situation. In *Proceedings of the IFAC Symposium SAFEPROCESS 1997*, pages 1033–1054, Hull, United Kingdom, 1997.
219. E. N. Pistikopoulos, V. Dua, N. A. Bozinis, A. Bemporad, and M. Morari. On-line optimization via off-line parametric optimization tools. *Comp. & Chem. Eng.*, 26:175–185, 2002.
220. E. Polak and T. H. Yang. Moving horizon control of linear systems with input saturation and plant uncertainty – Part 1: Robustness. *Int. J. Contr.*, 58:613–638, 1993.
221. J. A. Primbs, V. Nevistic, and J. C. Doyle. A receding horizon generalization of pointwise min-norm controllers. *IEEE Trans. Automat. Contr.*, 45:898–909, 2000.
222. M. Ptashne. *A Genetic Switch: Gene Control and Phage λ.* Cell Press, Cambridge, MA, 1986.
223. S. J. Qin and T. J. Badgwell. A survey of industrial model predictive control technology. *Contr. Eng. Prac.*, 11:733–764, 2003.
224. Z. Qu. Robust control of nonlinear uncertain systems under generalized matching conditions. *Automatica*, 29:985–998, 1993.
225. C. V. Rao and J. B. Rawlings. Steady states and constraints in model predictive control. *AIChE J.*, 45:1266–1278, 1999.
226. C. V. Rao and J. B. Rawlings. Constrained process monitoring: Moving-horizon approach. *AIChE J.*, 48:97–109, 2002.
227. J. B. Rawlings. Tutorial overview of model predictive control. *IEEE Control Systems Magazine*, 20:38–52, 2000.
228. J. B. Rawlings and K. R. Muske. The stability of constrained receding horizon control. *IEEE Trans. Automat. Contr.*, 38:1512–1516, 1993.
229. W. H. Ray. *Advanced Process Control.* McGraw-Hill, New York, 1981.
230. P. Renard, D. Dochain, G. Bastin, H. Naveau, and E. J. Nyns. Adaptive control of anaerobic digestion processes: a pilot-scale application. *Biotech. Bioeng.*, 31:287–294, 1988.
231. J. Richalet, A. Rault, J. L. Testud, and J. Papon. Model predictive heuristic control: Applications to industrial processes. *Automatica*, 14:413–428, 1978.
232. D. E. Rivera, M. Morari, and S. Skogestad. Internal model control. 4. PID controller design. *Ind. Eng. Chem. Proc. Des. & Dev.*, 25:252–265, 1986.
233. D. W. Ross and I. Flugge-Lotz. An optimal control problem for systems described by differential-difference equations. *SIAM J. Opt. Contr.*, 7:609–623, 1969.
234. W. J. Rugh and J. S. Shamma. Research on gain scheduling. *Automatica*, 36:1401–1425, 2000.
235. A. Saberi and H. Khalil. Quadratic-type Lyapunov functions for singularly perturbed systems. *IEEE Trans. Automat. Contr.*, 29:542–500, 1984.
236. V. Sakizlis, N. M. P. Kakalis, V. Dua, J. D. Perkins, and E. N. Pistikopoulos. Design of robust model–based controllers via parametric programming. *Automatica*, 40:189–201, 2004.

237. M. Santillán and M. C. Mackey. Why the lysogenic state of phage λ is so stable: A mathematical modeling approach. *Biophysical Journal*, 86:75–84, 2004.
238. S. S. Sastry and A. Isidori. Adaptive control of linearizable systems. *IEEE Trans. Automat. Contr.*, 34:1123–1131, 1989.
239. A. T. Schwarm and M. Nikolaou. Chance-constrained model predictive control. *AIChE J.*, 45:1743–1752, 1999.
240. P. O. M. Scokaert and D. Q. Mayne. Min-max feedback model predictive control for constrained linear systems. *IEEE Trans. Automat. Contr.*, 43:1136–1142, 1998.
241. P. O. M. Scokaert, D. Q. Mayne, and J. B. Rawlings. Constrained linear quadratic regulation. *IEEE Trans. Automat. Contr.*, 43:1163–1169, 1998.
242. P. O. M. Scokaert, D. Q. Mayne, and J. B. Rawlings. Suboptimal model predictive control (feasibility implies stability). *IEEE Trans. Automat. Contr.*, 44:648–654, 1999.
243. P. O. M. Scokaert and J. B. Rawlings. Constrained linear quadratic regulation. *IEEE Trans. Automat. Contr.*, 43:1163–1169, 1998.
244. P. O. M. Scokaret and J. B. Rawlings. Feasibility issues in linear model predictive control. *AIChE J.*, 45:1649–1259, 1999.
245. R. Sepulchre, M. Jankovic, and P. Kokotovic. *Constructive Nonlinear Control.* Springer-Verlag, Berlin-Heidelberg, 1997.
246. D. Shi, N. H. El-Farra, P. Mhaskar, M. Li, and P. D. Christofides. Predictive control of particle size distribution in particulate processes. *Chem. Eng. Sci.*, in press, 2005.
247. P. Singstad, H. Nordhus, K. Strand, M. Lien, L. B. Lyngmo, and O. Moen. Multivariable nonlinear control of industrial LDPE reactors. In *Proceedings of American Control Conference*, pages 615–620, Chicago, Il, 1992.
248. M. Skliar and W. F. Ramirez. Implicit kalman filtering. *Int. J. Contr.*, 66:393–412, 1997.
249. S. Skogestad. Simple analytic rules for model reduction and PID controller tuning. *J. Proc. Contr.*, 13:291–309, 2003.
250. Smith. Closer control of loops with dead time. *Chem. Eng. Prog.*, 53:217–225, 1957.
251. M. A. Soliman and W. H. Ray. Optimal feedback control for linear quadratic systems having time delays. *Int. J. Contr.*, 15:609–627, 1972.
252. E. D. Sontag. Smooth stabilization implies coprime factorization. *IEEE Trans. Automat. Contr.*, 34:435–443, 1989.
253. E. D. Sontag. A 'universal' construction of Artstein's theorem on nonlinear stabilization. *Syst. & Contr. Lett.*, 13:117–123, 1989.
254. M. Soroush. Nonlinear state observer design with application to reactors. *Chem. Eng. Sci.*, 52:387–404, 1997.
255. M. Soroush and C. Kravaris. Nonlinear control of a batch polymerization reactor: an experimental study. *AIChE J.*, 38:1429–1443, 1992.
256. M. Soroush and C. Kravaris. Nonlinear control of a polymerization CSTR with singular characteristic matrix. *AIChE J.*, 40:980–990, 1994.
257. M. Soroush and S. Valluri. Optimal directionality compensation in processes with input saturation non-linearities. *Int. J. Contr.*, 72:1555–1564, 1999.
258. G. Stephanopoulos. *Chemical Process Control: An Introduction to Theory and Practice.* Prentice Hall, Englewood Cliffs, New Jersey, 1984.
259. D. Sun and K. A. Hoo. Dynamic transition control structure for a class of SISO nonlinear systems. *IEEE Trans. Contr. Syst. Tech.*, 7:622–629, 1999.

260. D. Sun and K. A. Hoo. A robust transition control structure for time-delay systems. *Int. J. Contr.*, 72:150–163, 1999.
261. H. J. Sussmann and P. V. Kokotovic. The peaking phenomenon and the global stabilization of nonlinear systems. *IEEE Trans. Automat. Contr.*, 36:424–440, 1991.
262. M. Sznaier, J. Cloutier, R. Hull, D. Jacques, and C. Mracek. Receding horizon control Lyapunov function approach to suboptimal regulation of nonlinear systems. *Journal of Guidance Control & Dynamics*, 23:399–405, 2000.
263. M. Sznaier and M. J. Damborg. Suboptimal control of linear systems with state and control inequality constraints. In *Proceedings of 26th IEEE Conference on Decision and Control*, pages 761–762, Los Angeles, CA, 1987.
264. M. Sznaier, R. Suarez, and J. Cloutier. Suboptimal control of constrained nonlinear systems via receding horizon constrained control Lyapunov functions. *Int. J. Robust & Nonlin. Contr.*, 13:247–259, 2003.
265. A. Teel. Global stabilization and restricted tracking for multiple integrators with bounded controls. *Syst. & Contr. Lett.*, 18:165–171, 1992.
266. A. Teel, R. Kadiyala, P. V. Kokotovic, and S. Sastry. Indirect techniques for adaptive input-output linearization of nonlinear systems. *Int. J. Contr.*, 53:193–222, 1991.
267. A. Teel and L. Praly. Global stabilizability and observability imply semi-global stabilizability by output feedback. *Syst. & Contr. Lett.*, 22:313–325, 1994.
268. A. Teel and L. Praly. Tools for semi-global stabilization by partial state and output feedback. *SIAM J. Contr. & Optim.*, 33:1443–1488, 1995.
269. A. R. Teel. Connections between razumikhin-type theorems and the ISS nonlinear small-gain theorem. *IEEE Trans. Automat. Contr.* submitted, 1996.
270. F. E. Thau. On the inverse optimal control problem for a class of nonlinear autonomous systems. *IEEE Trans. Automat. Contr.*, 12:674–681, 1967.
271. Y. Tipsuwan and M.-Y. Chow. Control methodologies in networked control systems. *Contr. Eng. Prac.*, 11:1099–1111, 2003.
272. B. D. Tyreus and W. L. Luyben. Tuning PI controllers for integrator dead time processes. *Ind. & Eng. Chem. Res.*, 31:2625–2628, 1992.
273. J. J. Tyson, A. Csikasz-Nagy, and B. Novak. The dynamics of cell cycle regulation. *BioEssays*, 24:1095–1109, 2002.
274. S. Valluri and M. Soroush. Analytical control of SISO nonlinear processes with input constraints. *AIChE J.*, 44:116–130, 1998.
275. S. Valluri and M. Soroush. A non-linear controller design method for processes with saturating actuators. *Int. J. Contr.*, 76:698–716, 2003.
276. S. Valluri, M. Soroush, and M. Nikravesh. Shortest-prediction-horizon nonlinear model-predictive control. *Chem. Eng. Sci.*, 53:273–292, 1998.
277. A. J. van der Schaft. L_2-gain analysis of nonlinear systems and nonlinear state feedback H_∞ control. *IEEE Trans. Automat. Contr.*, 37:770–783, 1992.
278. V. Venkatasubramanian, R. Rengaswamy, K. Yin, and S. N. Kavuri. Review of process fault diagnosis – Parts I, II, III. *Comp. & Chem. Eng.*, 27:293–346, 2003.
279. K. Vit. Smithlike predictor for control of parameter-distributed processes. *Int. J. Contr.*, 30:179–191, 1979.
280. P. Vuthandam, H. Genceli, and M. Nikolaou. Performance bounds for robust quadratic dynamic control with end condition. *AIChE J.*, 41:2083–2097, 1995.
281. B. Wahlberg. System identification using laguerre models. *IEEE Trans. Automat. Contr.*, 36:551–562, 1991.

282. G. C. Walsh, H. Ye, and L. G. Bushnell. Stability analysis of networked control systems. *IEEE Trans. Contr. Syst. Tech.*, 10:438–446, 2002.
283. Z. Wan and M. V. Kothare. Efficient scheduled stabilizing model predictive control for constrained nonlinear systems. *Int. J. Robust & Nonlin. Contr.*, 13:331–346, 2003.
284. Y. J. Wang and J. B. Rawlings. A new robust model predictive control method I: theory and computation. *J. Proc. Contr.*, 14:231–247, 2004.
285. M. C. Wellons and T. F. Edgar. The generalized analytical predictor. *Ind. & Eng. Chem. Res.*, 26:1523–1532, 1987.
286. J. R. Whiteley and J. F. Davis. Qualitative interpretation of sensor patterns. *IEEE Expert*, 8:54–63, 1992.
287. A. S. Willsky. A survey of design methods for failure detection in dynamic systems. *Automatica*, 12:601–611, 1998.
288. S. K. P. Wong and D. E. Seborg. A theoretical analysis of smith and analytical predictors. *AIChE J.*, 32:1597–1605, 1986.
289. R. A. Wright and C. Kravaris. Nonlinear control of pH processes using the strong acid equivalent. *Ind. & Eng. Chem. Res.*, 30:1561–1572, 1991.
290. R. A. Wright, C. Kravaris, and N. Kazantzis. Model based synthesis of nonlinear PI and PID controllers. *AIChE J.*, 47:1805–1818, 2001.
291. R. A. Wright, M. Soroush, and C. Kravaris. Strong acid equivalent control of pH processes: An experimental study. *Ind. & Eng. Chem. Res.*, 30:2439–2448, 1991.
292. B. Xu and Y. Liu. An improved razumikhin-type theorem and its applications. *IEEE Trans. Automat. Contr.*, 39:839–841, 1994.
293. Y. Xu and J. Hespanha. Communication logics for networked control systems. In *Proceedings of American Control Conference*, pages 572–577, Boston, MA, 2004.
294. E. C. Yamalidou and J. Kantor. Modeling and optimal control of discrete-event chemical processes using petri nets. *Comp. & Chem. Eng.*, 15:503–519, 1990.
295. G. H. Yang, S. Y. Zhang, J. Lam, and J. Wang. Reliable control using redundant controllers. *IEEE Trans. Automat. Contr.*, 43:1588–1593, 1998.
296. E. B. Ydstie. Certainty equivalence adaptive control: Paradigms puzzles and switching. In *Proceedings of 5th International Conference on Chemical Process Control*, volume 93, pages 9–23, J. C. Kantor, C. E. Garcia and B. Carnahan Eds., AIChE Symposium Series No. 316, CACHE, AIChE, 1997.
297. E. B. Ydstie. New vistas for process control: Integrating physics and communication networks. *AIChE J.*, 48:422–426, 2002.
298. E. B. Ydstie and A. A. Alonso. Process systems and passivity via the Clausius-Planck inequality. *Syst. & Contr. Lett.*, 30:253–264, 1997.
299. L. Zaccarian, A. R. Teel, and D. Nesic. On finite gain L_P stability of nonlinear sampled-data systems. *Syst. & Contr. Lett.*, 49:201–212, 2003.
300. E. Zafiriou. Robust model predictive control of processes with hard constraints. *Comp. & Chem. Eng.*, 14:359–371, 1990.
301. G Zames. On the input-output stability of time-varying nonlinear feedback systems – Part I. *IEEE Trans. Automat. Contr.*, 11:228–238, 1966.
302. A. Zheng. Robust stability analysis of constrained model predictive control. *J. Proc. Contr.*, 9:271–278, 1999.
303. J. G. Ziegler and N. B. Nichols. Optimum settings for automatic controllers. *Trans. ASME*, 64:759–768, 1942.

304. J. W. Zwolak, J. J. Tyson, and L. T. Watson. Finding all steady state solutions of chemical kinetic models. *Nonlinear Analysis-Real World Applications*, 5:801–814, 2004.

Lecture Notes in Control and Information Sciences

Edited by M. Thoma and M. Morari
Further volumes of this series can be found on our homepage:
springeronline.com

Vol. 257: Moallem, M.; Patel, R.V.; Khorasani, K.
Flexible-link Robot Manipulators
176 p. 2001 [1-85233-333-2]

Vol. 258: Isidori, A.; Lamnabhi-Lagarrigue, F.; Respondek, W. (Eds.)
Nonlinear Control in the Year 2000 Volume 1
616 p. 2001 [1-85233-363-4]

Vol. 259: Isidori, A.; Lamnabhi-Lagarrigue, F.; Respondek, W. (Eds.)
Nonlinear Control in the Year 2000 Volume 2
640 p. 2001 [1-85233-364-2]

Vol. 260: Kugi, A.
Non-linear Control Based on Physical Models
192 p. 2001 [1-85233-329-4]

Vol. 261: Talebi, H.A.; Patel, R.V.; Khorasani, K.
Control of Flexible-link Manipulators Using Neural Networks
168 p. 2001 [1-85233-409-6]

Vol. 262: Dixon, W.; Dawson, D.M.; Zergeroglu, E.; Behal, A.
Nonlinear Control of Wheeled Mobile Robots
216 p. 2001 [1-85233-414-2]

Vol. 263: Galkowski, K.
State-space Realization of Linear 2-D Systems with Extensions to the General nD (n>2) Case
248 p. 2001 [1-85233-410-X]

Vol. 264: Baños, A.; Lamnabhi-Lagarrigue, F.; Montoya, F.J
Advances in the Control of Nonlinear Systems
344 p. 2001 [1-85233-378-2]

Vol. 265: Ichikawa, A.; Katayama, H.
Linear Time Varying Systems and Sampled-data Systems
376 p. 2001 [1-85233-439-8]

Vol. 266: Stramigioli, S.
Modeling and IPC Control of Interactive Mechanical Systems – A Coordinate-free Approach
296 p. 2001 [1-85233-395-2]

Vol. 267: Bacciotti, A.; Rosier, L.
Liapunov Functions and Stability in Control Theory
224 p. 2001 [1-85233-419-3]

Vol. 268: Moheimani, S.O.R. (Ed)
Perspectives in Robust Control
390 p. 2001 [1-85233-452-5]

Vol. 269: Niculescu, S.-I.
Delay Effects on Stability
400 p. 2001 [1-85233-291-316]

Vol. 270: Nicosia, S. et al.
RAMSETE
294 p. 2001 [3-540-42090-8]

Vol. 271: Rus, D.; Singh, S.
Experimental Robotics VII
585 p. 2001 [3-540-42104-1]

Vol. 272: Yang, T.
Impulsive Control Theory
363 p. 2001 [3-540-42296-X]

Vol. 273: Colonius, F.; Grüne, L. (Eds.)
Dynamics, Bifurcations, and Control
312 p. 2002 [3-540-42560-9]

Vol. 274: Yu, X.; Xu, J.-X. (Eds.)
Variable Structure Systems: Towards the 21st Century
420 p. 2002 [3-540-42965-4]

Vol. 275: Ishii, H.; Francis, B.A.
Limited Data Rate in Control Systems with Networks
171 p. 2002 [3-540-43237-X]

Vol. 276: Bubnicki, Z.
Uncertain Logics, Variables and Systems
142 p. 2002 [3-540-43235-3]

Vol. 277: Sasane, A.
Hankel Norm Approximation for Infinte-Dimensional Systems
150 p. 2002 [3-540-43327-9]

Vol. 278: Chunling D. and Lihua X. (Eds.)
H_∞ Control and Filtering of Two-dimensional Systems
161 p. 2002 [3-540-43329-5]

Vol. 279: Engell, S.; Frehse, G.; Schnieder, E. (Eds.)
Modelling, Analysis, and Design of Hybrid Systems
516 p. 2002 [3-540-43812-2]

Vol. 280: Pasik-Duncan, B. (Ed)
Stochastic Theory and Control
564 p. 2002 [3-540-43777-0]

Vol. 281: Zinober A.; Owens D. (Eds.)
Nonlinear and Adaptive Control
416 p. 2002 [3-540-43240-X]

Vol. 282: Schröder, J.
Modelling, State Observation and Diagnosis of Quantised Systems
368 p. 2003 [3-540-44075-5]

Vol. 283: Fielding, Ch. et al. (Eds.)
Advanced Techniques for Clearance of Flight Control Laws
480 p. 2003 [3-540-44054-2]

Vol. 284: Johansson, M.
Piecewise Linear Control Systems
216 p. 2003 [3-540-44124-7]

Vol. 285: Wang, Q.-G.
Decoupling Control
373 p. 2003 [3-540-44128-X]

Vol. 286: Rantzer, A. and Byrnes C.I. (Eds.)
Directions in Mathematical Systems Theory and Optimization
399 p. 2003 [3-540-00065-8]

Vol. 287: Mahmoud, M.M.; Jiang, J.; Zhang, Y.
Active Fault Tolerant Control Systems
239 p. 2003 [3-540-00318-5]

Vol. 288: Taware, A. and Tao, G.
control of Sandwich Nonlinear Systems
393 p. 2003 [3-540-44115-8]

Vol. 289: Giarré, L. and Bamieh, B.
Multidisciplinary Research in Control
237 p. 2003 [3-540-00917-5]

Vol. 290: Borrelli, F.
Constrained Optimal Control
of Linear and Hybrid Systems
237 p. 2003 [3-540-00257-X]

Vol. 291: Xu, J.-X. and Tan, Y.
Linear and Nonlinear Iterative Learning Control
189 p. 2003 [3-540-40173-3]

Vol. 292: Chen, G. and Yu, X.
Chaos Control
380 p. 2003 [3-540-40405-8]

Vol. 293: Chen, G. and Hill, D.J.
Bifurcation Control
320 p. 2003 [3-540-40341-8]

Vol. 294: Benvenuti, L.; De Santis, A.; Farina, L. (Eds.)
Positive Systems: Theory
and Applications (POSTA 2003)
414 p. 2003 [3-540-40342-6]

Vol. 295: Kang, W.; Xiao, M.; Borges, C. (Eds.)
New Trends in Nonlinear Dynamics and Control,
and their Applications
365 p. 2003 [3-540-10474-0]

Vol. 296: Matsuo, T.; Hasegawa, Y.
Realization Theory of Discrete-Time
Dynamical Systems
235 p. 2003 [3-540-40675-1]

Vol. 297: Damm, T.
Rational Matrix Equations in Stochastic Control
219 p. 2004 [3-540-20516-0]

Vol. 298: Choi, Y.; Chung, W.K.
PID Trajectory Tracking Control
for Mechanical Systems
127 p. 2004 [3-540-20567-5]

Vol. 299: Tarn, T.-J.; Chen, S.-B.; Zhou, C. (Eds.)
Robotic Welding, Intelligence and Automation
214 p. 2004 [3-540-20804-6]

Vol. 300: Nakamura, M.; Goto, S.; Kyura, N.; Zhang, T.
Mechatronic Servo System Control
Problems in Industries and their Theoretical Solutions
212 p. 2004 [3-540-21096-2]

Vol. 301: de Queiroz, M.; Malisoff, M.; Wolenski, P. (Eds.)
Optimal Control, Stabilization
and Nonsmooth Analysis
373 p. 2004 [3-540-21330-9]

Vol. 302: Filatov, N.M.; Unbehauen, H.
Adaptive Dual Control: Theory and Applications
237 p. 2004 [3-540-21373-2]

Vol. 303: Mahmoud, M.S.
Resilient Control of Uncertain Dynamical Systems
278 p. 2004 [3-540-21351-1]

Vol. 304: Margaris, N.I.
Theory of the Non-linear Analog Phase Locked Loop
303 p. 2004 [3-540-21339-2]

Vol. 305: Nebylov, A.
Ensuring Control Accuracy
256 p. 2004 [3-540-21876-9]

Vol. 306: Bien, Z.Z.; Stefanov, D. (Eds.)
Advances in Rehabilitation
472 p. 2004 [3-540-21986-2]

Vol. 307: Kwon, S.J.; Chung, W.K.
Perturbation Compensator based Robust Tracking
Control and State Estimation of Mechanical Systems
158 p. 2004 [3-540-22077-1]

Vol. 308: Tarbouriech, S.; Abdallah, C.T.; Chiasson, J. (Eds.)
Advances in Communication Control Networks
358 p. 2005 [3-540-22819-5]

Vol. 309: Kumar, V.; Leonard, N.; Morse, A.S. (Eds.)
Cooperative Control
301 p. 2005 [3-540-22861-6]

Vol. 310: Janczak, A.
Identification of Nonlinear Systems Using Neural
Networks and Polynomial Models
197 p. 2005 [3-540-23185-4]

Vol. 311: Lamnabhi-Lagarrigue, F.; Loría, A.; Panteley, V. (Eds.)
Advanced Topics in Control Systems Theory
294 p. 2005 [1-85233-923-3]

Vol. 312: Henrion, D.; Garulli, A. (Eds.)
Positive Polynomials in Control
313 p. 2005 [3-540-23948-0]

Vol. 313: Li, Z.; Soh, Y.; Wen, C.
Switched and Impulsive Systems
277 p. 2005 [3-540-23952-9]

Vol. 314: Gil', M.I.
Explicit Stability Conditions for Continuous Systems
193 p. 2005 [3-540-23984-7]

Vol. 315: Herbordt, W.
Sound Capture for Human/Machine Interfaces:
Practical Aspects of Microphone
Array Signal Processing
286 p. 2005 [3-540-23954-5]

Vol. 316: R.V. Patel; F. Shapdey
Control of Redundant Robot Manipulators
224 p. 2005 [3-540-25071-9]

Vol. 317: Chuan Ma; W. Murray Wonham
Nonblocking Supervisory Control
of State Tree Structures
208 p. 2005 [3-540-25069-7]

Vol. 318: Eli Gershon; Uri Shaked; IsaacYaesh
H_∞ Control and Estimation
of State-multiplicative
Linear Systems 272 p. [1-85233-997-7]

Vol. 319: Hofbaur; W. Michael
Hybrid Estimation of Complex Systems
176 p. [3-540-25727-6]

Vol. 320: Thomas Steffen
Control Reconfiguration of Dynamical Systems
286 p. [3-540-25730-6]

Vol. 321: Wijesuriya P. Dayawansa; Anders Lindquist; Yishao Zhou
New Directions and Applications in Control Theory
416 p. [3-540-23953-7]

Printing and Binding: Strauss GmbH, Mörlenbach